AF892718

Lecture Notes in Electrical Engineering

Volume 1009

Series Editors

Leopoldo Angrisani, Department of Electrical and Information Technologies Engineering, University of Napoli Federico II, Naples, Italy
Marco Arteaga, Departament de Control y Robótica, Universidad Nacional Autónoma de México, Coyoacán, Mexico
Bijaya Ketan Panigrahi, Department of Electrical Engineering, Indian Institute of Technology Delhi, New Delhi, Delhi, India
Samarjit Chakraborty, Fakultät für Elektrotechnik und Informationstechnik, TU München, Munich, Germany
Jiming Chen, Zhejiang University, Hangzhou, Zhejiang, China
Shanben Chen, School of Materials Science and Engineering, Shanghai Jiao Tong University, Shanghai, China
Tan Kay Chen, Department of Electrical and Computer Engineering, National University of Singapore, Singapore, Singapore
Rüdiger Dillmann, Humanoids and Intelligent Systems Laboratory, Karlsruhe Institute for Technology, Karlsruhe, Germany
Haibin Duan, Beijing University of Aeronautics and Astronautics, Beijing, China
Gianluigi Ferrari, Università di Parma, Parma, Italy
Manuel Ferre, Centre for Automation and Robotics CAR (UPM-CSIC), Universidad Politécnica de Madrid, Madrid, Spain
Sandra Hirche, Department of Electrical Engineering and Information Science, Technische Universität München, Munich, Germany
Faryar Jabbari, Department of Mechanical and Aerospace Engineering, University of California, Irvine, CA, USA
Limin Jia, State Key Laboratory of Rail Traffic Control and Safety, Beijing Jiaotong University, Beijing, China
Janusz Kacprzyk, Systems Research Institute, Polish Academy of Sciences, Warsaw, Poland
Alaa Khamis, German University in Egypt El Tagamoa El Khames, New Cairo City, Egypt
Torsten Kroeger, Stanford University, Stanford, CA, USA
Yong Li, Hunan University, Changsha, Hunan, China
Qilian Liang, Department of Electrical Engineering, University of Texas at Arlington, Arlington, TX, USA
Ferran Martín, Departament d'Enginyeria Electrònica, Universitat Autònoma de Barcelona, Bellaterra, Barcelona, Spain
Tan Cher Ming, College of Engineering, Nanyang Technological University, Singapore, Singapore
Wolfgang Minker, Institute of Information Technology, University of Ulm, Ulm, Germany
Pradeep Misra, Department of Electrical Engineering, Wright State University, Dayton, OH, USA
Sebastian Möller, Quality and Usability Laboratory, TU Berlin, Berlin, Germany
Subhas Mukhopadhyay, School of Engineering and Advanced Technology, Massey University, Palmerston North, Manawatu-Wanganui, New Zealand
Cun-Zheng Ning, Department of Electrical Engineering, Arizona State University, Tempe, AZ, USA
Toyoaki Nishida, Graduate School of Informatics, Kyoto University, Kyoto, Japan
Luca Oneto, Department of Informatics, Bioengineering, Robotics and Systems Engineering, University of Genova, Genova, Genova, Italy
Federica Pascucci, Dipartimento di Ingegneria, Università degli Studi Roma Tre, Roma, Italy
Yong Qin, State Key Laboratory of Rail Traffic Control and Safety, Beijing Jiaotong University, Beijing, China
Gan Woon Seng, School of Electrical and Electronic Engineering, Nanyang Technological University, Singapore, Singapore
Joachim Speidel, Institute of Telecommunications, Universität Stuttgart, Stuttgart, Germany
Germano Veiga, Campus da FEUP, INESC Porto, Porto, Portugal
Haitao Wu, Academy of Opto-electronics, Chinese Academy of Sciences, Beijing, China
Walter Zamboni, DIEM—Università degli studi di Salerno, Fisciano, Salerno, Italy
Junjie James Zhang, Charlotte, NC, USA

The book series *Lecture Notes in Electrical Engineering* (LNEE) publishes the latest developments in Electrical Engineering—quickly, informally and in high quality. While original research reported in proceedings and monographs has traditionally formed the core of LNEE, we also encourage authors to submit books devoted to supporting student education and professional training in the various fields and applications areas of electrical engineering. The series cover classical and emerging topics concerning:

- Communication Engineering, Information Theory and Networks
- Electronics Engineering and Microelectronics
- Signal, Image and Speech Processing
- Wireless and Mobile Communication
- Circuits and Systems
- Energy Systems, Power Electronics and Electrical Machines
- Electro-optical Engineering
- Instrumentation Engineering
- Avionics Engineering
- Control Systems
- Internet-of-Things and Cybersecurity
- Biomedical Devices, MEMS and NEMS

For general information about this book series, comments or suggestions, please contact leontina.dicecco@springer.com.

To submit a proposal or request further information, please contact the Publishing Editor in your country:

China

Jasmine Dou, Editor (jasmine.dou@springer.com)

India, Japan, Rest of Asia

Swati Meherishi, Editorial Director (Swati.Meherishi@springer.com)

Southeast Asia, Australia, New Zealand

Ramesh Nath Premnath, Editor (ramesh.premnath@springernature.com)

USA, Canada

Michael Luby, Senior Editor (michael.luby@springer.com)

All other Countries

Leontina Di Cecco, Senior Editor (leontina.dicecco@springer.com)

**** This series is indexed by EI Compendex and Scopus databases. ****

Hariharan Muthusamy · János Botzheim ·
Richi Nayak
Editors

Robotics, Control and Computer Vision

Select Proceedings of ICRCCV 2022

Springer

Editors
Hariharan Muthusamy
National Institute of Technology
Uttarakhand
Srinagar, India

János Botzheim
Eötvös Loránd University
Budapest, Hungary

Richi Nayak
Queensland University of Technology
Brisbane, QLD, Australia

ISSN 1876-1100 ISSN 1876-1119 (electronic)
Lecture Notes in Electrical Engineering
ISBN 978-981-99-0235-4 ISBN 978-981-99-0236-1 (eBook)
https://doi.org/10.1007/978-981-99-0236-1

© The Editor(s) (if applicable) and The Author(s), under exclusive license to Springer Nature Singapore Pte Ltd. 2023
This work is subject to copyright. All rights are solely and exclusively licensed by the Publisher, whether the whole or part of the material is concerned, specifically the rights of translation, reprinting, reuse of illustrations, recitation, broadcasting, reproduction on microfilms or in any other physical way, and transmission or information storage and retrieval, electronic adaptation, computer software, or by similar or dissimilar methodology now known or hereafter developed.
The use of general descriptive names, registered names, trademarks, service marks, etc. in this publication does not imply, even in the absence of a specific statement, that such names are exempt from the relevant protective laws and regulations and therefore free for general use.
The publisher, the authors, and the editors are safe to assume that the advice and information in this book are believed to be true and accurate at the date of publication. Neither the publisher nor the authors or the editors give a warranty, expressed or implied, with respect to the material contained herein or for any errors or omissions that may have been made. The publisher remains neutral with regard to jurisdictional claims in published maps and institutional affiliations.

This Springer imprint is published by the registered company Springer Nature Singapore Pte Ltd.
The registered company address is: 152 Beach Road, #21-01/04 Gateway East, Singapore 189721, Singapore

Contents

Control Techniques and Their Applications

Robotics and Autonomous Vehicles

About the Editors

Hariharan Muthusamy received a Ph.D. in Mechatronic Engineering (2010) from the University of Malaysia Perlis (UniMAP), Malaysia, a Master of Engineering in Applied Electronics (2006) from the Government College of Technology, India, and a Bachelor of Engineering in Electrical and Electronics Engineering (2002) from Government College of Technology (Affiliated to Bharathiar University), India. He is an Associate Professor in the Department of Electronics Engineering, National Institute of Technology Uttarakhand, India. He has published over 150 papers in refereed journals and conference proceedings. His major research interests include speech signal processing, biomedical signal and image processing, machine learning, deep learning, and optimization algorithms. He has supervised 9 Ph.D. and 4 Masters (research) students in the field of his expertise.

János Botzheim earned his M.Sc. and Ph.D. degrees from the Budapest University of Technology and Economics in 2001 and 2008, respectively. He joined the Department of Automation at Szechenyi Istvan University, Gyor, Hungary in 2007 as a senior lecturer, in 2008 as an assistant professor, and in 2009 as an associate professor. He was a visiting researcher at the Graduate School of System Design at the Tokyo Metropolitan University from September 2010 to March 2011 and from September 2011 to February 2012. He was an associate professor in the Graduate School of System Design at the Tokyo Metropolitan University from April 2012 to March 2017. He was an associate professor in the Department of Mechatronics, Optics, and Mechanical Engineering Informatics at the Budapest University of Technology and Economics from February 2018 to August 2021. He is the Head of the Department of Artificial Intelligence at Eötvös Loránd University, Faculty of Informatics, Budapest, Hungary, since September 2021. His research interest areas are computational intelligence, automatic identification of fuzzy rule-based models and some neural network models, bacterial evolutionary algorithms, memetic algorithms, applications of computational intelligence in robotics, and cognitive robotics. He has about 180 papers in journals and conference proceedings.

Richi Nayak is the Leader of the Applied Data Science Program at the Centre for Data Science and a Professor of Computer science at Queensland University of Technology, Brisbane Australia. She has a driving passion to address pressing societal problems by innovating the Artificial Intelligence field underpinned by fundamental research in machine learning, data mining, and text mining. Her research has resulted in the development of novel solutions to address industry-specific problems in Marketing, K 12 Education, Agriculture, Digital Humanities, and Mining. She has made multiple advances in social media mining, deep neural networks, multi-view learning, matrix/tensor factorization, clustering, and recommender systems. She has authored over 180 high-quality refereed publications. Her research leadership is recognized by multiple best paper awards and nominations at international conferences, QUT Postgraduate Research Supervision awards, and the 2016 Women in Technology (WiT) Infotech Outstanding Achievement Award in Australia. She holds a Ph.D. in Computer Science from the Queensland University of Technology and a Master in Engineering from IIT Roorkee.

Computer Vision

Challenges and Opportunity for Salient Object Detection in COVID-19 Era: A Study

Vivek Kumar Singh and Nitin Kumar

1 Introduction

Humans have the ability to identify visually informative scene regions in the image effortlessly and rapidly based on perceived distinctive features. These filtered regions contain rich information about objects depicted in an image. Salient Object Detection (SOD) aims to highlight important objects or regions and suppress background regions in the image. SOD methods transform an input image into a probability map called saliency map [1] that expresses how much each image element (pixel/region) grabs human attention. An example of salient object detection is illustrated in Fig. 1. Salient Object Detection (SOD) has been widely applied as pre-processing step in computer vision applications such as object detection [4, 5], video summarization [6], and image retrieval [7].

Coronavirus disease (COVID-19) is an infectious disease [8–10] which has posed several challenges to salient object detection, for example, due to use of face mask, face detection performance is decreased. Diffusion of the disease has been occurring from person to person quickly in the world. The disease is called COVID-19 and the virus is denoted as SARS-CoV-2 which is a family of viruses effective for devolving acute respiratory syndrome. COVID-19 common clinical features are fever, dyspnea, cough, myalgia, and headache [11]. The most common diagnosis tool used for diagnosis of COVID-19 is the reverse-transcription polymerase chain reaction (RT-PCR). Further, chest radiological imaging including computed tomography (CT) and X-ray

V. Kumar Singh (✉)
Sharda University, Greater Noida, India
e-mail: vivek.singh1@sharda.ac.in

N. Kumar
National Institute of Technology, Uttarakhand, India
e-mail: nitin@nituk.ac.in

© The Author(s), under exclusive license to Springer Nature Singapore Pte Ltd. 2023
H. Muthusamy et al. (eds.), *Robotics, Control and Computer Vision*, Lecture Notes in Electrical Engineering 1009, https://doi.org/10.1007/978-981-99-0236-1_1

Fig. 1 An example of salient object detection process, **a** input image, **b** saliency map [3], and **c** ground truth

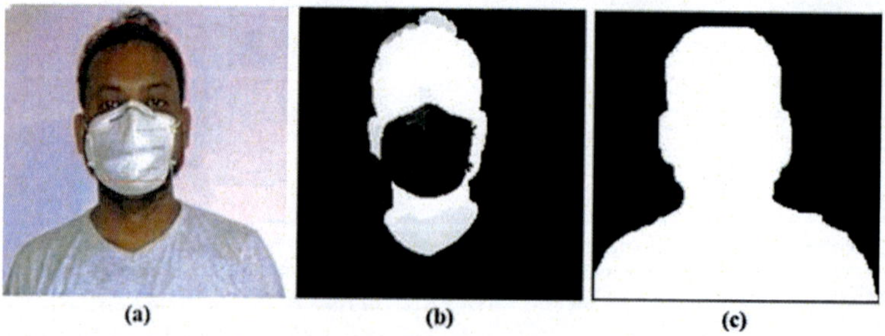

Fig. 2 A motivational example of this study, **a** input image, **b** saliency map obtained from Graph-Based Manifold Ranking (GMR) [31] method, and **c** ground truth

is playing important role in the early diagnosis and treatment of this disease [12]. Researchers are looking for detecting infected patients through medical image processing like X-rays and CT scans [13]. COVID-19 is a pandemic virus that infected many people worldwide and continues spreading from person to person. The disease also affected the lifestyle of humans such as education, office work, transportation, economic actives, etc. Therefore, our main motivation is to look at the impact of the virus on salient object detection performance and the applicability of salient object detection approach to control spreading of the virus. Figure 2 shows a motivational example of this study. In this figure, input image contains a human with face mask, in which saliency map does not highlight the masked region of the face. The purpose of this research work is to analyze the effectiveness of saliency detection on the images generated around the current human life activities. In this study, we propose a dataset which use to validate our suggested challenges in salient object detection due to COVID-19.

The rest of this paper is structured as follows. Section 2 illustrates the related works on salient object detection methods and novel Coronavirus-2019 (COVID-2019). In Sect. 3, a detailed discussion about the challenges and opportunities for salient object detection in the COVID-19 era is presented. Suggested challenges are evaluated and analyzed in Sect. 4. Finally, conclusion and future works are presented in Sect. 5.

2 Related Work

A large number of salient object detection methods have been reported in literature. These methods are broadly categorized into two categories: bottom-up methods and top-down methods. Bottom-up salient object detection methods utilize the appearance contrasts between objects and their surrounding regions in the image. The earliest bio-inspired bottom-up saliency method was proposed by Itti et al. [1]. This method has extracted three low-level visual features such as luminance, color, and orientation and exploits center-surround mechanisms to compute the saliency maps. Achanta et al. [14] proposed a simple and efficient saliency detection approach that computes saliency value of each image pixel by subtracting the Gaussian blurred version of the image from the mean pixel value of the image. Goferman et al. [15] presented four principles, namely, local low-level features, global considerations, visual organizational rules, and high-level factors to compute saliency maps. Perazzi et al. [16] suggested a saliency detection method based on color contrast. Cheng et al. [17] proposed a global contrast-based saliency computation approach which utilizes Histogram-based Contrast (HC) and Region-based Contrast (RC) for saliency estimation. Liu and Yang [18] proposed saliency detection method that exploited color volume and perceptually uniform color differences and combined foreground, center, and background saliency to obtain saliency map. Top-down salient object detection methods calculate the saliency values with the help of high-level priors. Gao et al. [19] computed saliency values of interest points by their mutual information and extracted discriminant features. Yang et al. [20] proposed a novel saliency detection method that jointly learned Conditional Random Field (CRF) for generation of saliency map. Jiang et al. [21] suggested saliency estimation method that effectively integrated shape prior into an iterative energy minimization box. Recently, convolutional neural networks (CNNs) have drawn great attention of computer vision researchers. Wang et al. [22] presented saliency detection method that employed two different deep networks to compute the saliency maps. Wang et al. [23] proposed the PAGE-Net for saliency calculation. Ren et al. [24] suggested the CANet, which has combined high-level semantic and low-level boundary information for salient object detection. Currently, computer vision and machine learning approaches have been rapidly applied for Coronavirus disease-2019 (COVID-19) detection. Ozturk et al. [25] proposed an automatic COVID-19 detection model that exploited deep learning method to detect and classify COVID-19. Waheed et al. [26] proposed an Auxiliary Classifier Generative Adversarial Network (ACGAN) called CovidGAN which has produced synthetic chest X-ray (CXR) images. Fan et al. [27] suggested a novel COVID-19 lung CT infection segmentation network called Inf-Net. Zhou et al. [28] presented a fully automatic, rapid, accurate, and machine-agnostic method for identifying the infection regions on CT scans. Wang et al. [29] suggested a novel noise-robust framework to learn from noisy labels for the segmentation. A summary of the recent research works for object detection during COVID-19 is given in Table 1.

Table 1 Recent research work for object detection during COVID-19

S. no.	Authors	Method	Modality	Remarks
1	Ozturk et al. [25]	Deep learning	Chest X-ray	The model is fully automated, it does not required manual features extraction
2	Waheed et al. [26]	Auxiliary Classifier Generative Adversarial Network (ACGAN)	Chest X-ray	It is a powerful method to generate unseen samples that can be utilized to design effective and robust convolutional neural networks (CNNs)
3	Fan et al. [27]	Deep Network (Inf-Net)	Lung computed tomography (CT) image	The Inf-Net first roughly located an infected region and then exploit the boundaries by means of reverse attention and edge information for accurately identifying the infected region
4	Zhou et al. [28]	A fully automatic, rapid, accurate, and machine-agnostic	CT scans	The segmentation method achieves a good trade-off between the complexity of the deep learning model and the accuracy of the model
5	Wang et al. [29]	A noise-robust framework	CT image	The method aims for learning from noisy labels for COVID-19 pneumonia lesion segmentation from CT images where clean labels are difficult and expensive to acquire

3 Challenges and Opportunity for SOD in COVID-19 Era

In this section, we present study of the impact of COVID-19 situation on identifying the most significant regions at an early stage from the natural images. This study provides scenarios in which images can be changed due to COVID-19 pandemic but the target object is still unchanged for any object detection application. Our work aims at studying this effect to enable the future salient object detection methods addressing such scenarios.

3.1 Challenges

The first challenging scenario is the complexity of the image where the appearance such as color and texture of foreground regions and background regions is similar. This is a difficult scenario for salient object detection methods because several methods exploit color and texture as distinctive features for calculating saliency value to each image element. Therefore, if foreground and background image regions have similar features then the methods may fail to highlight salient regions and suppress background regions. Secondly, saliency detection process is very challenging in real-time images in which the target object is partially hidden by some other objects. This scenario is known as occlusion problem in natural images. The saliency detection methods may fail to identify object in the image which is partially blocked by other objects.

Figure 3 shows various visual challenges of salient object detection in natural images. Similar color and texture of foreground and background regions in the complex natural images are shown in Fig. 3a. An owl is situated in a place where the surrounding location is homogeneous to the owl, the saliency detection task faces problem to identify owl bird from real-time image as shown in Fig. 3a. Partial occlusion problem in real-time images is depicted in Fig. 3b. In a cow body, some regions are blocked by wooden poles which is shown in Fig. 3b, images are taken from PASCAL-S [30] dataset, and in this scene cow is target object to which salient regions are identified, but the methods may detect it partially. Figure 2a illustrates the effect of coronavirus on human real image. In this scene, a man is wearing a white face mask that is not similar to the human face skin. It is a case of partial occlusion where the human face is partially hidden by the face mask.

Moreover, the face mask shows high center-surrounding difference than the targeted object (i.e., man). Hence, the salient object detection methods may identify the face mask as an important object instead of the man. This is a challenge for salient object detection methods to achieve better performance on the visual data generated in COVID-19 era. The COVID-19 pandemic has affected appearance of real-time visual images surrounding human life. For example, nowadays, people are wearing Personal Protective Equipment (PPE) which includes a face mask, gloves, gowns, head cover, shoe cover, etc. to safeguard them from COVID-19. All the images taken

Fig. 3 Visual examples of some challenging scenarios in salient object detection. Appearance similarity between foreground and background is illustrated in (**a**) [30]. Partial occlusion scenario in real-time images is depicted in (**b**) [30]

Fig. 4 Example of some people has appeared together [30]

from public places captured the human face with blockages by face mask. This situation can be considered as an occlusion problem in the natural images. It poses a challenge to computer vision applications and most of them fail to identify hidden face in the presence of a face mask. This is also challenging for salient object detection methods to uniformly highlight the human face. In addition, these PPE can visually appear similar to the surrounding environment in terms of color and texture. Any object identification computer vision application can be easily misguided to identify wrong objects in an image.

Further, COVID-19 has also affected the visual appearance of groups of people due to the following social distancing in public places. On many occasions, people are capturing group images as shown in Fig. 4, image is adopted from PASCAL-S [30] dataset. In this image, all people are together to form an object and salient object detection methods can easily detect it as a salient object. However, today in group images, people are maintaining minimum defined distance which is popularly known as social distancing. Such effects may degrade the performance of salient object detection because the target object is all the people in the image, in contrast, saliency detection methods may detect some people out of all the people who appeared in the image. A summary of these challenges is also given in Table 2.

Table 2 Challenges and opportunities for salient object detection in COVID-19 era

S. no.	Challenges	Reason	Opportunity
1	Low contrast between foreground and background	People are wearing Personal Protective Equipment (PPE) which may be similar with surrounding environment	Need to develop SOD methods which can work better in low contrast situations
2	Occlusion problem with human face in the real-time images	For fighting with COVID-19 humans are wearing face mask which illustrates high contrast between humans face skin and face mask in terms of color and texture	SOD methods which can address the occlusion problem effectively
3	Group of object may not be detected simultaneously	Today's people are not standing very close due to social distancing rule implemented for controlling transmission of coronavirus virus. Therefore, in group images each and very people are considered as individual objects while the significant meaning of the image is to capture all the people present on the location.	SOD methods which can detect multiple objects at a distance
4	Saliency detection methods may be misguided by protected gears to highlight non-salient regions as salient regions	The face mask may become more important object than the human in the image. Whereas the image is captured for keeping the human as target object by photographer	Intelligent SOD methods are required to detect actual salient object in an image
5	Keeping an eye on the student activity in online teaching	It is difficult by an instructor to monitor the students in an online class due to no direct interaction	SOD methods are required which can keep an eye on the student activities

3.2 *Opportunities*

COVID-19 period has emerged as a great opportunity for computer vision researchers to contribute in battling COVID-19 disease. This is also an opportunity for salient object detection methods. For battling with the COVID-19 disease, salient object detection methods are required to focus on the challenges discussed in Sect. 3.1. In this section, we discuss research opportunities and directions for handling the challenges that emerged in COVID-19 era for salient object detection. The low contrast image has a similar appearance of foreground and background regions. Such types of images can be captured during COVID-19 as people are wearing Personal Protective Equipment (PPE) which may have similar color and texture with the surrounding environment. This scenario provides an opportunity to discover visual features which have the discriminative capability to classify foreground and background regions from the input image. The partial occlusion problem may occur in COVID-19 environment as people are wearing a face mask. This effect on the visual scene may influence the performance of salient object detection as partial occlusion is a challenging scenario for saliency detection. Consequently, it is an opportunity for researchers to introduce such saliency detection approaches which can deal with partial occlusion in a better manner.

During COVID-19, people are following social distancing, which affects the visual appearance of people. However, with the social distancing people are scattered on the whole image and it is very difficult to identify all the humans who have appeared for salient object detection. This is an opportunity to find such methodologies which can deal with multiple object detection in a scene. Furthermore, the education system is also facing a big problem during this COVID-19 pandemic. The educational institutions are conducting their classes using online platforms. In such a mode, controlling class behavior is very challenging for the instructor. In this process, the visual data are coming from various sources, hence it is very difficult to identify which visuals are important. This is yet another opportunity to identify salient regions from a different source of visual data. A summary of these opportunities is also given in Table 2.

4 Experimental Result

In this section, we illustrate the evaluation and analysis of different scenarios of salient object detection which may be affected due to COVID-19 on the proposed dataset. This study proposes a dataset which contains 100 natural images, out of which 50 images include face masked humans whereas others consist of unmasked faces in different scenarios.

This dataset contains a variety of images of three people by capturing these images through a mobile camera either from the rear angle or from the front angle with proper illumination. The ground truth is generated manually by one user which provides consistent result with pixel-wise human annotation. The qualitative evaluation and

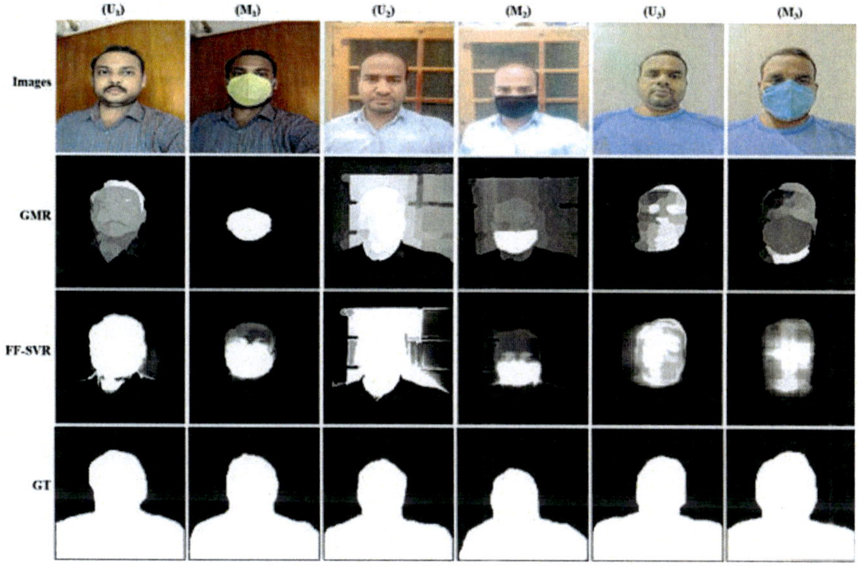

Fig. 5 Qualitative study on samples of images of proposed dataset. First row represents original Images, GMR [31] and FF-SVR [32] saliency maps are depicted in second and third rows, respectively, fourth row shows ground truth (GT)

study is presented under different conditions as shown in Fig. 5. In this figure, U_i and M_i represent unmasked and masked i-th images, respectively. For this study, we have applied existing saliency detection methods such as Graph-Based Manifold RankingGMR [31] and Fusion Framework for Salient Object Detection based on Support Vector Regression (FF-SVR) [32] for generating saliency maps. It can be observed from Fig. 5, when a human visually appears with a face mask then visual attention is distracted by the face mask such as in M_1, only masked region is highlighted while in U_1 whole face is detected. Similarly, in M_2, masked region is located whereas in U_2 entire human face is highlighted. In addition, in M_3 mask region is not detected, while face is identified in U_3. This evaluation and analysis support our suggested challenges for salient object detection in COVID-19.

5 Conclusion and Future Work

COVID-19 pandemic has noticeably affected human lives across the world and the death rate is also alarming. In this study, we have focused on various scenarios of salient object detection which may be affected due to the presence of the COVID-19 pandemic worldwide. Nowadays, people are wearing various modalities such as Personal Protective Equipment (PPE), face masks which change the visual appearance of people in outside places. Such visual changes have put certain challenges in the real-time images, namely, low contrast between foreground and background, partial occlusion and online monitoring, etc. These challenges for salient object detection have also come with certain opportunities for the researchers and practitioners working in this research area. We have evaluated these challenges on the proposed dataset to provide experimental support. In future work, we will explore saliency detection models that can effectively handle the COVID-19 challenges.

References

1. Itti L, Koch C, Niebur E (1998) A model of saliency-based visual attention for rapid scene analysis. IEEE Trans Pattern Anal Mach Intell 20(11):1254–1259
2. Alpert S, Galun M, Basri R, Brandt A (2007) Image segmentation by probabilistic bottom-up aggregation and cue integration. In: IEEE conference on computer vision and pattern recognition, 2007. CVPR\'07 , pp 1–8
3. Singh VK, Kumar N (2019) Saliency bagging: a novel framework for robust salient object detection. Vis Comput 1–19
4. Ren Z, Gao S, Chia L-T, Tsang IW-H (2014) Region-based saliency detection and its application in object recognition. IEEE Trans Circuits Syst Video Technol 5(24):769–779
5. Zhang D, Meng D, Zhao L, Han J (2017) Bridging saliency detection to weakly supervised object detection based on self-paced curriculum learning. arXiv:1703.01290
6. Simakov D, Caspi Y, Shechtman E, Irani M (2008) Summarizing visual data using bidirectional similarity. In: 2008 IEEE conference on computer vision and pattern recognition, pp 1–8
7. Gao Y, Shi M, Tao D, Xu C (2015) Database saliency for fast image retrieval. IEEE Trans Multimed 17(3):359–369
8. Lau H, Khosrawipour V, Kocbach P, Mikolajczyk A, Ichii H, Schubert J, Bania J, Khosrawipour T (2020) Internationally lost COVID-19 cases. J Microbiol Immunol Infect
9. Lippi G, Plebani M, Henry BM (2020) Thrombocytopenia is associated with severe coronavirus disease 2019 (COVID-19) infections: a meta-analysis. Clinica Chimica Acta
10. Zhang J, Yan K, Ye H, Lin J, Zheng J, Cai T (2020) SARS-CoV-2 turned positive in a discharged patient with COVID-19 arouses concern regarding the present standard for discharge. Int J Infect Dis
11. Huang C, Wang Y, Li X, Ren L, Zhao J, Hu Y, Zhang L, Fan G, Xu J, Gu X et al. (2020)Clinical features of patients infected with 2019 novel coronavirus in Wuhan, China. Lancet 395(10223):497–506
12. Zu ZY, Jiang MD, Xu PP, Chen W, Ni QQ, Lu GM, Zhang LJ (2020) Coronavirus disease 2019 (COVID-19): a perspective from China. Radiology 200490
13. Nguyen TT (2020)Artificial intelligence in the battle against coronavirus (COVID-19): a survey and future research directions, vol 10. (Preprint, DOI)
14. Achanta R, Hemami S, Estrad F, Susstrunk S (2009) Frequency-tuned salient region detection. In: 2009 IEEE conference on computer vision and pattern recognition, pp 1597–1604
15. Goferman S, Zelnik-Manor L, Tal A (2011) Context-aware saliency detection. IEEE Trans Pattern Anal Mach Intell 34(10):1915–1926
16. Perazzi F, Krähenbühl P, Pritch Y, Hornung A (2012) Saliency filters: contrast based filtering for salient region detection. In: 2012 IEEE conference on computer vision and pattern recognition, pp 733–740
17. Cheng M-M, Mitra NJ, Huang X, Torr PHS, Hu S-M (2015) Global contrast based salient region detection. IEEE Trans Pattern Anal Mach Intell 37(3):569
18. Liu GH, Yang JY (2019) Exploiting color volume and color difference for salient region detection. IEEE Trans Image Process a Publ IEEE Signal Process Soc 28(1):6

19. Gao D, Han S, Vasconcelos N (2009) Discriminant saliency, the detection of suspicious coincidences, and applications to visual recognition. IEEE Trans Pattern Anal Mach Intell 31(6):989–1005
20. Yang J, Yang M-H (2016) Top-down visual saliency via joint CRF and dictionary learning. IEEE Trans Pattern Anal Mach Intell 39(3):576–588
21. Jiang H, Wang J, Yuan Z, Liu T, Zheng N, Li S (2011) Automatic salient object segmentation based on context and shape prior. BMVC 6(7):9
22. Wang L, Lu H, Ruan X, Yang M-H (2015) Deep networks for saliency detection via local estimation and global search. In: Proceedings of the IEEE conference on computer vision and pattern recognition, pp 3183–3192
23. Wang W, Zhao S, Shen J, Hoi SCH, Borji A (2019) Salient object detection with pyramid attention and salient edges. In: Proceedings of the IEEE conference on computer vision and pattern recognition, pp 1448–1457
24. Ren Q, Lu S, Zhang J, Hu R (2020) Salient object detection by fusing local and global contexts. IEEE Trans Multimed
25. Ozturk T, Talo M, Yildirim EA, Baloglu UB, Yildirim O, Acharya UR (2020) Automated detection of COVID-19 cases using deep neural networks with X-ray images. Comput Biol Med 103792
26. Waheed A, Goyal M, Gupta D, Khanna A, Al-Turjman F, Pinheiro PR (2020) Covidgan: data augmentation using auxiliary classifier gan for improved covid-19 detection. IEEE Access 8:91916–91923
27. Fan D-P, Zhou T, Ji G-P, Zhou Y, Chen G, Fu H, Shen J, Shao L (2020) Inf-Net: automatic COVID-19 lung infection segmentation from CT images. IEEE Trans Med Imag
28. Zhou L, Li Z, Zhou J, Li H, Chen Y, Huang Y, Xie D, Zhao L, Fan M, Hashmi S et al (2020) A rapid, accurate and machine-agnostic segmentation and quantification method for CT-based COVID-19 diagnosis. IEEE Trans Med Imag
29. Wang G, Liu X, Li C, Xu Z, Ruan J, Zhu H, Meng T, Li K, Huang N, Zhang S (2020) A noise-robust framework for automatic segmentation of COVID-19 pneumonia lesions from CT images. IEEE Trans Med Imag
30. Li Y, Hou X, Koch C, Rehg JM, Yuille AL (2014) The secrets of salient object segmentation. In: Proceedings of the IEEE conference on computer vision and pattern recognition, pp 280–287
31. Yang C, Zhang L, Lu H, Ruan X, Yang M-H (2013) Saliency detection via graph-based manifold ranking. In: Proceedings of the IEEE conference on computer vision and pattern recognition, pp 3166–3173
32. Singh VK, Kumar N (2021) A novel fusion framework for salient object detection based on support vector regression. In: Proceedings of the Springer conference on evolving technologies for computing, communication and smart world, pp 437–450

Human Activity Recognition Using Deep Learning

Amrit Raj, Samyak Prajapati, Yash Chaudhari, and Ankit Kumar Rouniyar

1 Introduction

In the current age, the products of the 4th Industrial Revolution are establishing their prevalence in our daily lives and technology has advanced to such a level that going "off-grid" is no longer a viable option. The boom in technology is directly correlated with the boom in the economical position of a nation, and while it has proven apt in ameliorating the quality of life, the general trend is leading us to an over-reliance on technology. This dependence has several pros and cons associated with it, where it all depends on us humans, on how we decide to make use of it. Mobile phones and laptops have now become commonplace items that are at arm's reach for most of us.

Data from such sources can prove valuable in establishing a security-critical surveillance system as proven in the 2013 Boston Marathon Bombings [1] where videos recordings from mobile phones used by citizens aided the investigators in determining the cause of the explosion. With the given abundance of CCTV cameras in nearly every public location, a system designed for activity recognition could prove invaluable in circumventing illegal activities. Such systems could be used for recognizing abnormal and suspicious activities at crowded public locations and aid the on-ground personnel in flagging an individual as needed.

This work has the potential to be extended for applications in areas including assisted living/healthcare to detect activities carried out by patients, or to detect

A. Raj (✉) · S. Prajapati · Y. Chaudhari · A. K. Rouniyar
National Institute of Technology Delhi, New Delhi, India
e-mail: 181210008@nitdelhi.ac.in

S. Prajapati
e-mail: 181210046@nitdelhi.ac.in

Y. Chaudhari
e-mail: 181210062@nitdelhi.ac.in

A. K. Rouniyar
e-mail: 181210011@nitdelhi.ac.in

© The Author(s), under exclusive license to Springer Nature Singapore Pte Ltd. 2023
H. Muthusamy et al. (eds.), *Robotics, Control and Computer Vision*, Lecture Notes in Electrical Engineering 1009, https://doi.org/10.1007/978-981-99-0236-1_2

if a certain person has fallen, and needs active assistance. Systems like these can also be deployed to monitor activities in smart homes, which would then allow the central system to control the lighting and HVAC units depending on the activity being performed.

This paper is organized in the following. Section 2 contains the literature review of related works. Section 3 describes the dataset used and Sect. 4 presents the chosen models, along with the details of the performance metrics that were used. Section 5 consists of the results obtained and finally, Sect. 6 consists of the conclusions and the future works.

2 Literature Review

The works of Mohammadi et al. [2] built their results on CNNs which were pre-trained on "ImageNet" [3] weights and performed transfer learning along with the use of attention mechanisms to achieve an average Top-1 classification accuracy of 76.83% across 8 models. They were also involved in the creation of ensemble models with 4 models that yielded the highest accuracies and achieved an action classification accuracy of 92.67% (Top-1). Geng et al. [4] performed feature extraction on raw video inputs using pre-trained CNNs, and then they performed pattern recognition using an SVM classifier on the extracted features to classify the videos based on acting classes. Bourdev et al. [5] define the term *poselet* as to express a part of one's pose. Their work focuses on the creation of an algorithm to pick or choose the best poselet in the sample space. They proposed a two-layer regression model for detecting people and localizing body components. The first layer would detect the local patterns in the input image as it contains poselet classifiers. The second layer in turn would combine the output of the classifiers in a max-margin framework.

In their research, González et al. [6] proposed an adaption to the Genetic Fuzzy Finite State Machine (GFFSM) method after selecting the three best features from the human activity data using Information Correlation Coefficient (ICC) analysis followed by a wrapper Feature Selection (FS) method. The data used by them was gathered using two triaxial accelerometers on the subject's wrists while performing activities that were going to be recognize at a later stage. While doing the review on Video-based Activity Recognition, Ke et al. [7] have addressed three stages of activity recognition. The first stage is Human Object Segmentation, where they have divided the task into two categories, the static camera segmentation, and the moving camera segmentation, and discussed the same. The second stage is Feature Extraction and Representation, where they have extracted the global features as well as local features, this is because the global features are sensitive to noise, occlusion, and variation of viewpoint. The third stage is Activity Detection and Classification Algorithms, where they have discussed classification algorithms like Dynamic Time

Warping (DTW), K Nearest Neighbor (KNN), Kalman Filter, and Binary tree multi-dimensional indexing. They have also discussed the various applications of human activity recognition, specifically healthcare systems and surveillance systems, and the challenges associated with them.

In their research, Liu et al. [8] proposed to use a set of attributes, directly associated with visual characteristics to represent human actions. They claimed that a representation based on action attributes would be more descriptive and distinct, as compared to the traditional methods.

Ji et al. [9] in their work proposed a 3D CNN model for human action recognition, this model is designed to extract features from the spatial dimensions as well as the temporal dimensions by performing 3D convolutions, as a result capturing the motion information encoded in multiple adjacent frames. They propose regularizing the outputs with high-level features to boost the performance of the model.

3 Data Source

The Stanford 40 Action Classification Dataset [10] was used in this work for training the images. It contains 9532 images across 40 action classes (each class is exhibited in Fig. 1) with around 180–300 images dedicated for each action class. The image collection contained numerous activities which resulted in a colossal number of candidate attributes. In addition, the number of possible interactions between the attributes in terms of co-occurrence statistics. Subsequently, a custom dataset [11] was also created which embodies three YouTube URLs for each action class present in the Stanford 40 dataset. Each URL is a copy-right free and royalty-free "stock" video, with the video length ranging from 15–30 s.

Table 1 depicts the class distribution of images in the original Stanford-40 dataset and videos in the custom dataset.

4 Methodology

4.1 Models Chosen

4.1.1 ResNet-50

ResNet 50 is a deep convolutional neural network (CNN) which is 50 layers in "depth"; it was proposed by He et al. in the paper titled "Deep Residual Learning for Image Recognition" at CVPR [12]. It can alternatively be represented as a Directed Acyclic Graph; this is largely due to the presence of residual blocks, and in turn, skip connections. To make it possible to train deeper networks, Skip connections enable the parameter gradients to propagate more easily from the output layer to the earlier

Fig. 1 Examples of action classes in Stanford 40 dataset

layers of the network. This increased network depth can result in higher accuracies on more difficult tasks. It has publicly available model weights that were trained on the ImageNet dataset and achieved a Top-1 classification accuracy of 75.3% on the ImageNet dataset.

4.1.2 ResNet-101

ResNet 101 is a deep CNN that, as the name suggests, is 101 layers deep. It was also proposed in their paper by He et al. [12]. It is composed of 100 convolutional layers along with a single fully connected layer as its output layer with softmax activation. Being a model of the ResNet family, it makes use of residual blocks (illustrated in Fig. 2), which use skip connections to propagate the output of a previous layer to the "front". As with ResNet 50, it also has publicly available weights that were trained on the ImageNet dataset, achieving a Top-1 classification accuracy of 76.4%.

4.1.3 InceptionV3

Proposed by Szegedy et al. in their paper [13] the model is made up of symmetric and dropout layers, asymmetric building blocks, including convolution layers, average pooling layers, max-pooling layers and fully connected layers, concatenation layers. Throughout the model, Batch Normalization was used to a great extent and applied to activation inputs. The final activation for the output layer is often chosen as softmax

Table 1 Distribution of action classes

Class	Stanford-40 imagery dataset	Custom video dataset
Applauding	184	3
Blowing bubbles	159	3
Brushing teeth	100	3
Cleaning the floor	112	3
Climbing	195	3
Cooking	188	3
Cutting trees	103	3
Cutting vegetables	89	3
Drinking	156	3
Feeding a horse	187	3
Fishing	173	3
Fixing a bike	128	3
Fixing a car	151	3
Gardening	99	3
Holding an umbrella	192	3
Jumping	195	3
Looking through a microscope	91	3
Looking through a telescope	103	3
Phoning	159	3
Playing guitar	189	3
Playing violin	160	3
Pouring liquid	100	3
Pushing a cart	135	3
Reading	145	3
Riding a bike	193	3
Riding a horse	196	3
Rowing a boat	85	3
Running	151	3
Shooting an arrow	114	3
Smoking	141	3
Taking photos	97	3
Texting message	93	3
Throwing Frisby	102	3
Using a computer	130	3
Walking the dog	193	3

(continued)

Table 1 (continued)

Class	Stanford-40 imagery dataset	Custom video dataset
Washing dishes	82	3
Watching TV	123	3
Waving hands	110	3
Writing on a board	83	3
Writing on a book	146	3

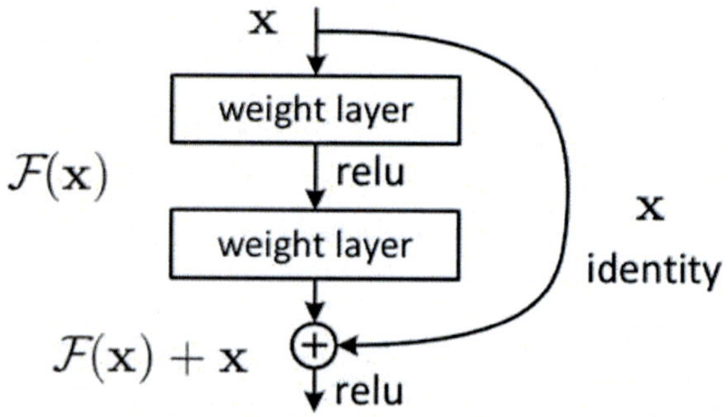

Fig. 2 Skip connections in a residual block

activation for multi-class classification. RMSprop or Stochastic Gradient Descent as popular optimizers for this model due to a large number of trainable parameters. It achieved a Top-1 classification accuracy of 78.8% on the ImageNet dataset.

4.1.4 InceptionResNetV2

This model was proposed by Szegedy et al. [14], the network is 164 layers in "depth" and is a variation of the InceptionV3 model which borrows some ideas from Microsoft's original ResNet works [12, 15]. Residual connections allow for shortcuts in the model and have allowed researchers to successfully train even deeper neural networks, which has led to increased performance when compared to its base, InceptionV3. It achieved a Top-1 classification accuracy of 80.1% on the ImageNet dataset.

4.2 *Workflow*

The images were first augmented with random rotations between 0 and 359 degrees followed by resizing them to 256 $\times$ 256 pixels. The augmented images were then used to train four CNNs, namely ResNet50, ResNet101, InceptionV3, and InceptionResNetV2 using Keras. The models were initialized with "ImageNet" weights

Table 2 Optimized hyperparameters

Model	Learning rate	Momentum	Dropout
ResNet50	1e-3	0.9	–
ResNet101	1e-3	0.9	0.2
Inception V3	1e-3	0.9	–
Inception ResNetV2	1e-4	0.9	0.2

and Stochastic Gradient Descent (SGD) was chosen as the optimizer. The dataset was then divided into a 90:10 train-test split. The metrics were further improved by using different combinations of regularization layers, dropout layers and by hyper-parameter tuning; the final optimized hyperparameters are exhibited in Table 2. To introduce the modality of classification by the use of videos, the trained models were tested by decomposing the videos into individual frames, and then each frame was tested by each model and the predicted class with the highest frequency was chosen as the class exhibited in the video.

A browser-based end-to-end deployment was also created using Streamlit to have a visualizable experience for the end-user of the product. The user can choose from multiple models for detecting action classes, and then the user can decide whether they wish to run on images on a video. In case the user elects to run on a single image, the UI would allow them to upload a single image, and that same image would be used by the selected model to generate a prediction. The prediction would then be printed out below the input image, with a confidence value as well. In case the user instead wishes to detect the most prominent action class of a video, they would be given the option to insert a video URL, which would be downloaded in the background and decomposed into individual frames, the aforementioned steps would then be initiated to run inference for the video and the results would then be printed out below the video. The entire workflow is exhibited as a flowchart in Fig. 3.

4.3 Performance Metrics

The metrics chosen for model evaluation were chosen as Top-1 Accuracy, Precision, Recall, AUROC (Area under ROC Curve), and F1 Score. The mathematical formulas for the metrics are described below as a function of True Positives (TP), True Negatives (TN), False Positives (FP), and False Negatives (FN). The AUROC is calculated by Reimann summation of the curve plotted between the TP Rate and the FP Rate.

$$Accuracy = \frac{TP + FN}{TP + TN + FP + FP} \tag{1}$$

$$Precision = \frac{TP}{TP + FP} \tag{2}$$

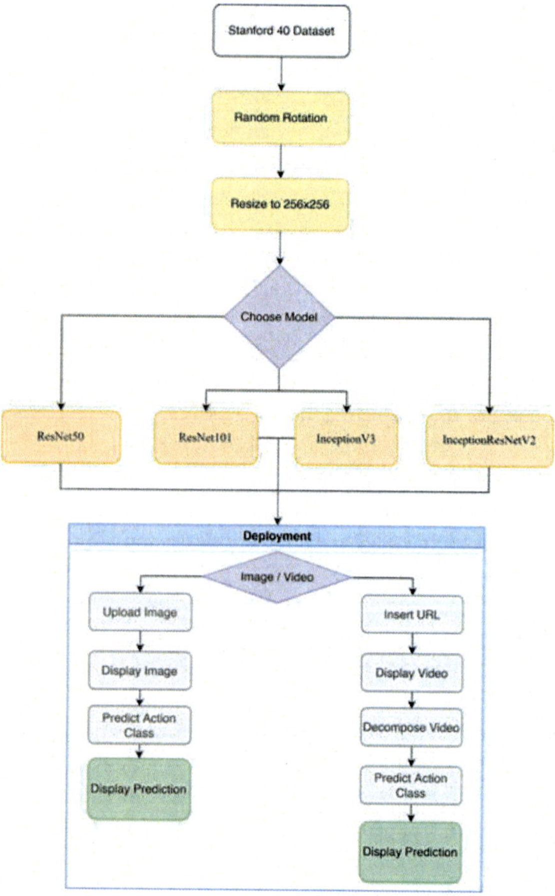

Fig. 3 Process flow of the implemented methodology

$$Recall = \frac{TP}{TP + FN} \tag{3}$$

$$F1\,Score = 2 * \frac{Precision * Recall}{Precision + Recall} \tag{4}$$

5 Results

The performance evaluation metrics achieved after training and testing on Stanford-40 imagery and corresponding videos were tabulated in Tables 3 and 4 respectively. The accuracy mentioned henceforth refers to the Top-1 accuracy.

Table 3 Metrics on Stanford-40 imagery

Model	Accuracy (%)	Precision	Recall	AUC	F1 score
ResNet50	77.55	0.81	0.75	0.96	0.78
ResNet101	80.41	0.84	0.78	0.97	0.81
InceptionV3	79.16	0.82	0.77	0.96	0.79
Inception ResNetV2	77.46	0.85	0.71	0.98	0.77

Table 4 Metrics on Stanford-40 videos

Model	Accuracy (%)	Precision	Recall	AUC	F1 Score
ResNet50	47.50	0.47	0.47	0.73	0.47
ResNet101	54.16	0.54	0.58	0.76	0.56
Inception V3	42.50	0.42	0.42	0.70	0.42
Inception ResNetV2	49.16	0.49	0.49	0.73	0.49

As evident from the results of Table 3, we can see that the models (initialized with Image-Net weights) were able to perform quite well without the use of computationally heavy techniques such as transfer learning.

The lower prediction accuracy in the video classification task exhibits the need for certain "memory" in the neural network for predicting prominent action classes in videos. In such a scenario, a hybrid network with LSTMs and would undoubtedly perform better where the previous prediction has a considerable impact on the current prediction.

6 Conclusions and Future Work

With the availability of computational equipment which enables us to perform such computations in real time, the implications of systems that automatically detect the context of a frame are quite significant. Keeping the current COVID-19 pandemic in mind, computer vision is a field that has progressed incredibly in the span of a few months. Our activity recognition model could easily be trained to differentiate between "Wearing a mask properly", "Wearing a mask improperly" and "Not wearing a mask" and can then be used to flag down violators. This technology, coupled with some hardware, could also be used to create an access control system where only specific categories could be allowed access, such as in a construction site, where many workers tend to skimp off on wearing necessary protective gear. The potential for smart surveillance using this technology is endless, as it can be used to automate the tedious process of monitoring the video feeds of CCTV cameras and automatically flagging down individuals; or it can be used from a statistical background, such as using it in a gymnasium to understand the most popular form of activities that the members prefer, and thus, can develop them further.

The models could further be improved upon by training further with fine-tuning the hyperparameters and making use of transfer learning. Models with 3D CNN layers or hybrid models that incorporate memory-based models such LSTMs could be used for improving the accuracies of video action classification as well. The use of multiple datasets in classification would expand the scope of use case scenarios of, such as the Sports-1 M Dataset [16], which consists of almost one million videos for around 487 sporting activities, and UCF101 Dataset [17], which consists of 13,320 videos for various common actions. Data from mobile sensors such as accelerometer, heart rate sensor, pedometer, barometer, et cetera could also be used in assisting the models in analyzing the conditions of the human body and assessing that in making the prediction. A weighted ensemble model or a cascaded network can also be used for improving overall accuracy in the classification of action categories.

References

1. Hunt for Boston bomber in iPhone era (2013) Financial times. (18 Apr 2013). https://www.ft.com/content/48adc938-a781-11e2-bfcd-00144feabdc0
2. Mohammadi S, Majelan SG, Shokouhi SB (2019) Ensembles of deep neural networks for action recognition in still images. In: 2019 9th international conference on computer and knowledge engineering (ICCKE), Mashhad, Iran, 2019, pp 315–318. https://doi.org/10.1109/ICCKE48569.2019.8965014
3. Deng J, Dong W, Socher R, Li L, Li K, Fei-Fei L (2009) ImageNet: a large-scalemax-pooling hierarchical image database. In: 2009 IEEE conference on computer vision and pattern recognition, Miami, FL, USA, 2009, pp 248–255. https://doi.org/10.1109/CVPR.2009.5206848
4. Geng C, Song JX (2016) Human action recognition based on convolutional neural networks with a convolutional auto-encoder. https://doi.org/10.2991/iccsae-15.2016.173
5. Bourdev L, Malik J (2009) Poselets: Body part detectors trained using 3D human pose annotations. In: 2009 IEEE 12th international conference on computer vision, 2009, pp 1365–1372. https://doi.org/10.1109/ICCV.2009.5459303
6. González S, Sedano J, Villar JR, Corchado E, Herrero L, Baruque B (2015) Features and models for human activity recognition. Neurocomputing. https://doi.org/10.1016/j.neucom.2015.01.082
7. Ke S-R, Thuc H, Lee Y-J, Hwang J-N, Yoo J-H, Choi K-H (2013) A review on video-based human activity recognition. Computers. https://doi.org/10.3390/computers2020088
8. Liu J, Kuipers B, Savarese S (2011) Recognizing human actions by attributes. CVPR 2011. Published. https://doi.org/10.1109/cvpr.2011.5995353
9. Ji S, Xu W, Yang M, Yu K (2013) 3D convolutional neural networks for human action recognition. IEEE Trans Pattern Anal Mach Intell 35(1):221–231. https://doi.org/10.1109/TPAMI.2012.59
10. Yao B, Jiang X, Khosla A, Lin AL, Guibas LJ, Fei-Fei L (2011) Human action recognition by learning bases of action attributes and parts. In: International conference on computer vision (ICCV), Barcelona, Spain. 6–13 Nov 2011
11. Prajapati S, Raj A (2021) djsamyak/DM-Stanford40. GitHub. https://github.com/djsamyak/DM-Stanford40. (Apr 2021)
12. He K, Zhang X, Ren S, Sun J (2016) Deep residual learning for image recognition. In: Proceedings of the IEEE conference on computer vision and pattern recognition, pp 770–778

13. Szegedy C, Vanhoucke V, Ioffe S, Shlens J, Wojna Z (2016) Rethinking the inception architecture for computer vision. In: Proceedings of the IEEE conference on computer vision and pattern recognition, pp 2818–2826
14. Szegedy C, Ioffe S, Vanhoucke V, Alemi A (2017) Inception-v4, inception-resnet and the impact of residual connections on learning. In: Proceedings of the AAAI conference on artificial intelligence, vol 31, no 1. (Feb 2017)
15. He K, Zhang X, Ren S, Sun J (2016). Identity mappings in deep residual networks. In: The European conference on computer vision. Springer, Cham, pp. 630–645. (Oct 2016)
16. Karpathy A, Toderici G, Shetty S, Leung T, Sukthankar R, Fei-Fei L (2014) Large-scale Video Classification with Convolutional Neural Networks. In: Soomro K, Zamir AR, Shah M (eds) UCF101: a dataset of 101 human action classes from videos in the wild, CRCV-TR-12-01, Nov 2012
17. Soomro K, Zamir AR, Shah M (2012) UCF101: a dataset of 101 human action classes from videos in the wild, CRCV-TR-12-01, Nov 2012.

Recovering Images Using Image Inpainting Techniques

Soureesh Patil, Amit Joshi, and Suraj Sawant

1 Introduction

Image inpainting is an actively researched area of deep learning which aims to fill the missing pixels of the image as realistically as possible following the context. This idea is not new and it has been researched for a long time. Approaches to the inpainting tasks can be classified as sequence-based, Convolutional Neural Network (CNN)-based, and Generative Adversarial Network (GAN)-based [1]. Initial approaches used partial differential equations with fluid-dynamics-based approach and Fast Marching method for inpainting [2, 3]. However, these approaches needed manual intervention for creating masks and worked for small damage only. Due to the high availability of data, deep-learning-based approaches can produce better results but realistic image inpainting is still a difficult task. GAN framework served as a base to several inpainting approaches to train the models effectively using adversarial loss function [4]. Context encoders started using GANs for inpainting but had drawbacks for mask sizes and semantic textures. Later models improved on the context encoders to support variable size images and masks to prevent blurry output. Using deep neural networks with established structures like Visual Geometry Group (VGG), learning structural knowledge with shared generators, newer approaches like

S. Patil (✉) · A. Joshi · S. Sawant
Department of Computer Engineering and IT, College of Engineering, Pune (COEP), 411005 Pune, Maharashtra, India
e-mail: patilse18.comp@coep.ac.in

A. Joshi
e-mail: adj.comp@coep.ac.in

S. Sawant
e-mail: sts.comp@coep.ac.in

© The Author(s), under exclusive license to Springer Nature Singapore Pte Ltd. 2023

H. Muthusamy et al. (eds.), *Robotics, Control and Computer Vision*, Lecture Notes in Electrical Engineering 1009, https://doi.org/10.1007/978-981-99-0236-1_3

training generative models to map a latent prior distribution to natural image manifolds are being explored [5–7]. The use of descriptive text is also helpful to generate better semantics [8].

Image inpainting techniques are abundantly available but the choice of the inpainting technique for a particular task depends on various factors like total damaged area, availability of computational resources, memory, and space requirements. Hence, this work provides a comparative analysis of readily available and commonly used techniques, Navier–Stokes and Telea algorithms. The rest of the paper is organized as a literature review in Sect. 2 followed by the proposed methodology in Sect. 3. Section 4 throws light on the experimental setup, achieved results, and its discussions followed by the conclusion in Sect. 5.

2 Literature Review

Pathak et al. proposed context encoders consisting of CNN trained to generate content based on the context of its surroundings. An important contribution of this paper was the "Channel-wise fully connected layer". They achieved state-of-the-art performance for semantic inpainting and the learned features were useful in other computer vision tasks [9]. Context encoders were lacking texture details for predicted pixels. Yang et al. proposed a framework by combining the techniques of neural style transfer and context encoders and obtained enhanced texture details [10]. Many approaches were inefficient in handling diverse-size images. Iizuka et al. proposed a Fully Convolutional Network with Dilated Convolution and local and global discriminators and obtained better texture details for diverse images [11]. Demir et al. demonstrated a combination of PatchGAN and GGAN discriminators. This enhanced local texture details of generated pixels [12]. Yan et al. proposed guidance loss to improve decoded features of the missing region and shift connection layer to enhance global semantic and local texture [13]. Yu et al. proposed contextual attention to obtain information from distant spatial locations. They achieved better training stability by using Wasserstein GAN (WGAN) adversarial loss and weighted L1 loss [14]. Wang et al. proposed the idea of ID-MRF loss term, multi-column structure, and weighted L1 loss following previous trends to obtain high-quality results [15]. Liu et al. proposed the idea of Partial Convolution to obtain state-of-the-art results [16]. Many inpainting methods usually generate blurry images due to usage of L1 loss only. Nazeri et al. proposed an Edge Map of the missing region which contains prior information. They separated the task of image inpainting into edge prediction and image generation to obtain high-quality inpainting [17].

Yu et al. developed DeepFill v2 with Gated Convolution and SN-patch GAN to obtain better inpainting results as compared to other methods [18]. Vitoria et al. incorporated a novel Generator and Discriminator to build on improved WGAN [19]. They produced the ability to recover large regions by learning semantic information. The approaches toward inpainting were able to handle irregular holes but they were not able to generate textures of damaged areas. Guo et al. proposed Fixed-Radius

Nearest Neighbors (FRNN) to solve this issue. Using N blocks-one dilation strategy and residual blocks is effective for smaller irregular holes. However, for larger holes, this method needed to be trained using a large number of parameters [20]. Zeng et al. proposed Pyramid-Context Encoder Network (PEN-Net) based on U-Net to learn contextual semantics from full-resolution input and decode it effectively. This network can be further refined for high-resolution images [21]. Image inpainting results highly depend on input and many models yield unsatisfactory results when the object overlaps with the foreground due to lack of information. Xiong et al. proposed a foreground-aware inpainting system that outperformed other models on complex compositions [22].

Li et al. proposed Spatial Pyramid Dilation (SPD) residual blocks for handling different image and mask sizes. They applied Multi-Scale Self-Attention (MSSA) to enhance coherency and obtained high PSNR scores [23]. For training inpainting models, it is usually assumed that missing region patterns are known. This limits the application scope. Wang et al. proposed Visual Consistency Network (VCNet), a blind inpainting system, which first learns to locate the mask and then fills the missing regions [24]. Liu et al. proposed a coherent semantic attention layer to preserve the contextual structure and modeled the semantic relevance between hole features [25]. Zhao et al. proposed an Unsupervised Cross-space Translation GAN (UCTGAN) model and were able to create visually realistic images. Their new cross-semantic attention layer improved realism and appearance consistency [26]. For GAN-based inpainting tasks, feature normalization helps in training. Most of the methods applied feature normalization without considering its impact on mean and variance shifts. Yu et al. proposed Basic and Learnable Region Normalization methods and obtained better performance than full spatial normalization [27]. Liu et al. proposed Probabilistic Diverse GAN (PDGAN) and achieved diverse inpainting results by modulation of random noise [28]. Liao et al. introduced a joint optimization framework of semantic segmentation and image inpainting by using the Semantic-Wise Attention Propagation (SWAP) module and obtained superior results for complex holes [29]. Zhang et al. proposed a context-aware SPL model for inpainting that uses global semantics to learn local textures [30]. Marinescu et al. proposed a generalizable Bayesian Reconstruction through Generative Models (BRGM) using Bayes' theorem for image inpainting [31]. Although there are a lot of conditional GANs proposed for image inpainting, they underperform when it comes to large missing regions. Zhao et al. proposed a generic Co-Mod-GAN structure to represent conditional and stochastic styles [32].

3 Proposed Methodology

This section explains the OpenCV algorithms used for comparative analysis and custom error masks for producing corrupt images. The two explored areas are.

1. Algorithms
   (a) Telea algorithm.
   (b) Naiver–Stokes algorithm.
2. Custom error masks.

3.1 Algorithms

3.1.1 Telea Algorithm

This algorithm is based on the Fast Marching Method. It inpaints missing pixels proximal to known pixels first, similar to manual heuristic operations. First, one of the invalid boundary pixels is picked and inpainted. This is followed by the selection of all boundary pixels iteratively to inpaint the whole boundary region. Invalid pixels are replaced by the normalized weighted sum of neighboring pixels with more weightage given to closer pixels. Hence, the newly created valid pixels are more influenced by local valid pixels lying on the normal line of the boundary region and contours. After inpainting one pixel, the next invalid pixel is chosen using the Fast Marching Method and slowly propagates toward the center of the unknown region from the image as shown in Fig. 1.

3.1.2 Naiver–Stokes Algorithm

This algorithm is based on fluid dynamics. It involves the solution of Navier–Stokes equation for incompressible fluids. It uses a partial differential equation. It builds on the fact that edges are supposed to be continuous. The algorithm travels along the edges going from valid to invalid region. Using the heuristic principle, it joins the

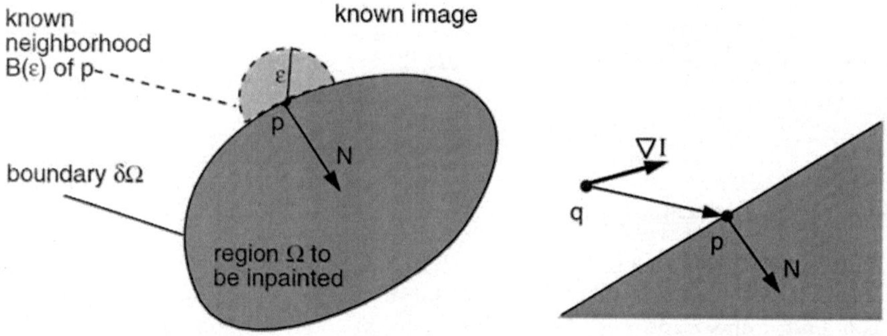

Fig. 1 Inpainting illustration

points with the same intensity to form contours, also known as isophotes. The edges are considered analogous to the incompressible fluid and using the fluid dynamics methods, the isophotes are continued in the unknown region. In the end, color is filled to reduce the minimum variance in the concerned area.

3.2 Custom Error Masks

This work aims to analyze the results on the Oxford Buildings dataset; a medium-sized dataset, consisting of different objects and contexts with custom error masks.

It is emphasized to use manually crafted binary error masks covering smaller damage across different directions. Diagonal, Horizontal, and Vertical masks are aimed to corrupt the image counters on small scale along with respective directions. The center mask is used to simulate a large corrupted area. Custom error masks are shown in Fig. 2.

The effectiveness of the Navier–Stokes algorithm and Telea algorithm is analyzed by measuring established metrics like Peak Signal-to-Noise Ratio (PSNR) and

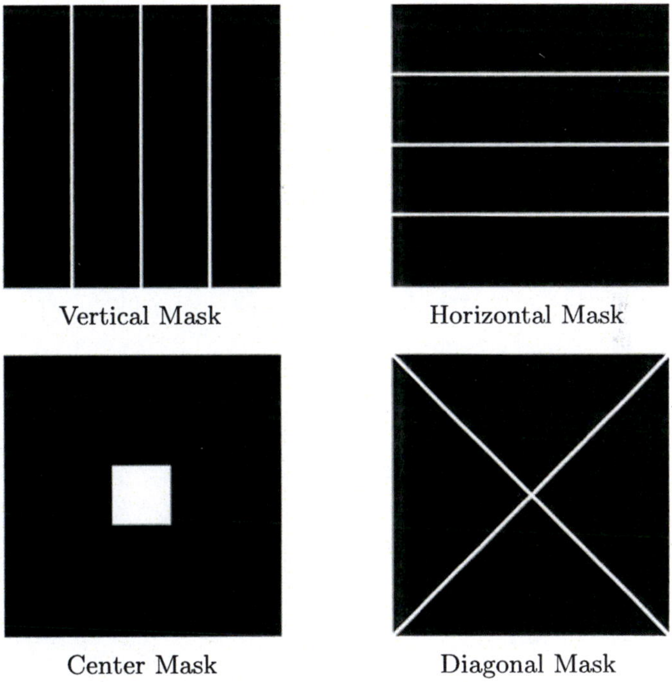

Fig. 2 Custom error masks

Fig. 3 Sample images

Structural Similarity Index Measure (SSIM). Runtime and memory allocated by the algorithms are additionally considered to understand their complexity. The sample images are shown in Fig. 3.

4 Results and Discussion

This section explains the observed results for the two algorithms discussed in this work. The main criterion of evaluation is PSNR and SSIM values observed for both algorithms.

4.1 Experimental Setup

For practical comparison of the two algorithms, this work had the following testing setup specifications:

1. CPU: Intel Core i5-1035G1.
2. RAM: 8GB (3200 MHz).

This work uses Python and OpenCV library for the implementation of sequential approaches. The OpenCV library contains the implementation of the Navier–Stokes method and Telea method of image inpainting. To get the corrupted images, four different crafted binary masks are used.

4.1.1 Dataset

The Oxford Buildings dataset contains 5062 images obtained from querying Flicker by 17 different keywords [33]. It contains 11 different landmarks and the images are of different resolutions. They are preprocessed to 256×256 resolution for uniformity. These preprocessed images are then damaged according to different error masks and provided as input to the inpainting algorithms.

4.2 Performance Considerations

To get the quality assessment of the inpainting results, PSNR and SSIM are used, which are part of the OpenCV library.

4.2.1 PSNR

The PSNR between two images is the peak signal-to-noise ratio measured in decibels. This ratio is generally used in the computing efficiency of compressed images. The higher the PSNR, the better the quality of the reconstructed image. The Mean Squared Error (MSE) represents the cumulative squared error between the compressed and the original image, whereas PSNR represents a measure of the peak error. Lower the value of MSE, lower the error. PSNR is calculated using MSE, followed by an equation containing logarithms and MSE. For colored images, PSNR is computed differently. Images are converted to color spaces of different intensity channels and PSNR is computed on those intensity channels.

4.2.2 SSIM

Structural Similarity Index Measure is a perceptual metric to quantify image quality degradation caused due to various image processing techniques. It computes the structural similarity between two images. It is based on the visible structure between two images and measures the difference between them. The higher value of SSIM indicates strong structural similarity between the two images.

4.3 Discussion

PSNR and SSIM are established metrics to assess image similarities in image processing tasks. Along with that, this work also uses runtime and memory consumption as supplementary metrics. This work has obtained the average values of the metrics on each error mask. Both Navier–Stokes and Telea algorithms performed best at

Table 1 Vertical mask results

	Navier–Stokes	Telea
PSNR	33.8326	34.03554
SSIM	0.97698	0.976927
Memory [KB]	196.70	196.70
Runtime [ms]	3.64	3.547

Table 2 Horizontal mask results

	Navier–Stokes	Telea
PSNR	34.12692	34.23631
SSIM	0.977691	0.977399
Memory [KB]	196.70	196.70
Runtime [ms]	4.5128	3.568

Table 3 Diagonal mask results

	Navier–Stokes	Telea
PSNR	32.24496	32.0982
SSIM	0.966922	0.963651
Memory [KB]	196.70	196.70
Runtime [ms]	9.56	9.709

Table 4 Center mask results

	Navier–Stokes	Telea
PSNR	28.78492	28.90572
SSIM	0.962625	0.963922
Memory [KB]	196.70	196.70
Runtime [ms]	3.679	3.184

horizontal contour recovery with PSNR 34.12692 and 34.23631, respectively. For the central mask, as the larger area containing the most useful semantic information was damaged, the algorithms couldn't recover the images effectively as seen from PSNR values 28.78492 and 28.90572, respectively. Memory consumption in all cases is the same and efficient (196.70 KB). Runtime for diagonal mask shows that it is costly to recover discontinuous area along different contours than continuous areas. Both algorithms have efficient and equal runtime ranging between 3 and 10 ms. The detailed results are summarized in Tables 1, 2, 3, and 4. Sample recovered images for Vertical Mask, Horizontal Mask, Diagonal Mask, and Center Mask are shown in Figs. 4, 5, 6, and 7, respectively.

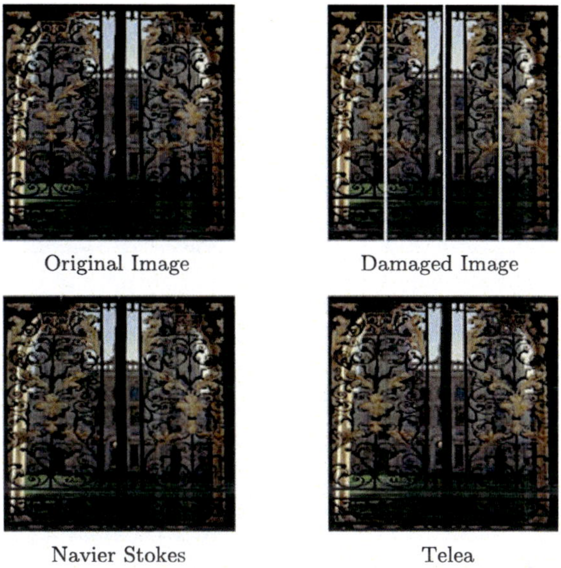

Fig. 4 Vertical mask results

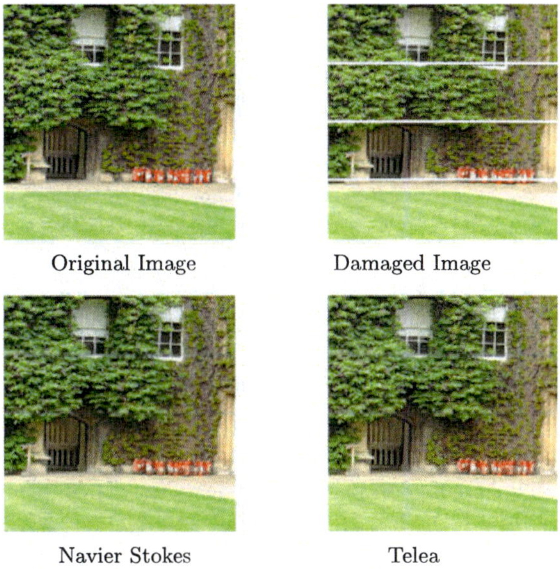

Fig. 5 Horizontal mask results

Fig. 6 Diagonal mask results

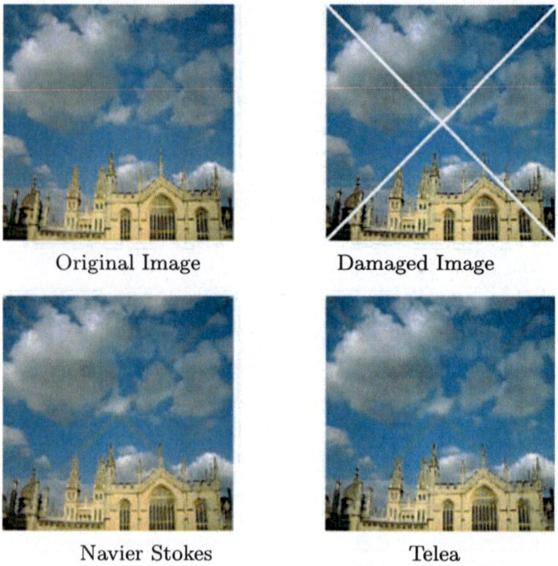

Fig. 7 Center mask results

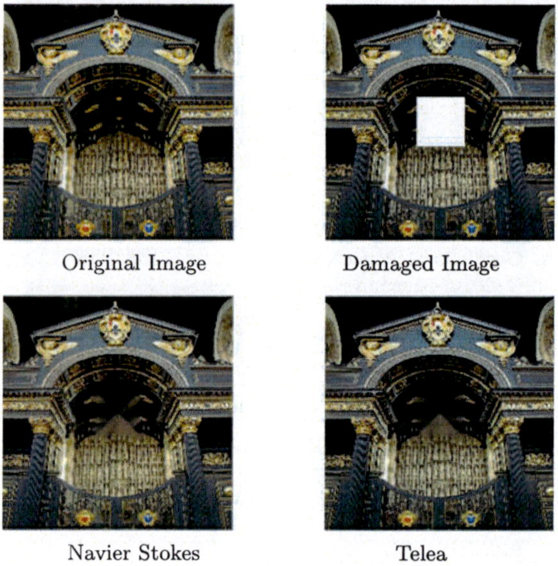

5 Conclusion

Image inpainting problem is an actively researched area and there are many solutions available to this problem. These solutions have a trade-off between complexity and accuracy. The purpose of this work is to appraise new users and researchers of the effectiveness of readily available algorithms. For most common use-cases, the smaller area needs to be inpainted managing time and space complexity. This work shows that PSNR up to 34.23631 and SSIM up to 0.977399 can be achieved with the Telea algorithm. For larger corrupt regions, both methods failed to achieve decent PSNR and SSIM values. Hence, these algorithms are not suitable for recovering larger corrupted regions. Both algorithms are highly efficient in time and space complexities and suitable for small damage recovery. Overall, Telea algorithm performs slightly better than Navier–Stokes algorithm. The future scope of this work aims to consider these algorithms a baseline for further study. The study will be done against CNN-based and GAN-based algorithms which provide better inpainting for complex semantics.

References

1. Elharrouss O, Almaadeed N, Al-Maadeed S, Akbari Y (2020) Image inpainting: a review. Neural Process Lett 51(2):2007–2028
2. Bertalmio M, Bertozzi AL, Sapiro G (2001) Navier-stokes, fluid dynamics, and image and video inpainting. In: Proceedings of the 2001 IEEE computer society conference on computer vision and pattern recognition. CVPR 2001, vol 1, pp I–I. IEEE
3. Telea A (2004) An image inpainting technique based on the fast marching method. J Graph Tools 9(1):23–34
4. Goodfellow I, Pouget-Abadie J, Mirza M, Xu B, Warde-Farley D, Ozair S, Courville A, Bengio Y (2014) Generative adversarial nets. In: Advances in neural information processing systems, vol 27
5. Hui Z, Li J, Wang X, Gao X (2020) Image fine-grained inpainting. arXiv:2002.02609
6. Lahiri A, Jain AK, Agrawal S, Mitra P, Biswas PK (2020) Prior guided gan based semantic inpainting. In: Proceedings of the IEEE/CVF conference on computer vision and pattern recognition, pp 13696–13705
7. Yang J, Qi Z, Shi Y (2020) Learning to incorporate structure knowledge for image inpainting. In: Proceedings of the AAAI conference on artificial intelligence, vol 34, pp 12605–12612
8. Zhang L, Chen Q, Hu B, Jiang S (2020) Text-guided neural image inpainting. In: Proceedings of the 28th ACM international conference on multimedia, pp 1302–1310
9. Pathak D, Krahenbuhl P, Donahue J, Darrell T, Efros AA (2016) Context encoders: feature learning by inpainting. In: Proceedings of the IEEE conference on computer vision and pattern recognition, pp 2536–2544
10. Yang C, Lu X, Lin Z, Shechtman E, Wang O, Li H (2017) High-resolution image inpainting using multi-scale neural patch synthesis. In: Proceedings of the IEEE conference on computer vision and pattern recognition. pp. 6721–6729 (2017)
11. Iizuka S, Simo-Serra E, Ishikawa H (2017) Globally and locally consistent image completion. ACM Trans Graph (ToG) 36(4):1–14
12. Demir U, Unal G (2018) Patch-based image inpainting with generative adversarial networks. arXiv:1803.07422

13. Yan Z, Li X, Li M, Zuo W, Shan S (2018) Shift-net: image inpainting via deep feature rearrangement. In: Proceedings of the European conference on computer vision (ECCV), pp 1–17
14. Yu J, Lin Z, Yang J, Shen X, Lu X, Huang TS (2018) Generative image inpainting with contextual attention. In: Proceedings of the IEEE conference on computer vision and pattern recognition, pp 5505–5514
15. Wang Y, Tao X, Qi X, Shen X, Jia J (2018) Image inpainting via generative multi-column convolutional neural networks. arXiv:1810.08771
16. Liu G, Reda FA, Shih KJ, Wang TC, Tao A, Catanzaro B (2018) Image inpainting for irregular holes using partial convolutions. In: Proceedings of the European conference on computer vision (ECCV), pp 85–100
17. Nazeri K, Ng E, Joseph T, Qureshi FZ, Ebrahimi M (2019) Edgeconnect: generative image inpainting with adversarial edge learning. arXiv:1901.00212
18. Yu J, Lin Z, Yang J, Shen X, Lu X, Huang TS (2019) Free-form image inpainting with gated convolution. In: Proceedings of the IEEE/CVF international conference on computer vision, pp 4471–4480
19. Vitoria P, Sintes J, Ballester C (2018) Semantic image inpainting through improved wasserstein generative adversarial networks. arXiv:1812.01071
20. Guo Z, Chen Z, Yu T, Chen J, Liu S (2019) Progressive image inpainting with full-resolution residual network. In: Proceedings of the 27th ACM international conference on multimedia, pp 2496–2504
21. Zeng Y, Fu J, Chao H, Guo B (2019) Learning pyramid-context encoder network for high-quality image inpainting. In: Proceedings of the IEEE/CVF conference on computer vision and pattern recognition, pp 1486–1494
22. Xiong W, Yu J, Lin Z, Yang J, Lu X, Barnes C, Luo J (2019) Foreground-aware image inpainting. In: Proceedings of the IEEE/CVF conference on computer vision and pattern recognition, pp 5840–5848
23. Li CT, Siu WC, Liu ZS, Wang LW, Lun DPK (2020) Deepgin: deep generative inpainting network for extreme image inpainting. In: European conference on computer vision. Springer, pp 5–22
24. Wang Y, Chen YC, Tao X, Jia J (2020) Vcnet: a robust approach to blind image inpainting. In: Computer vision–ECCV 2020: 16th European conference, Glasgow, UK, August 23–28, 2020, Proceedings, Part XXV 16. Springer, pp 752–768
25. Liu H, Jiang B, Xiao Y, Yang C (2019) Coherent semantic attention for image inpainting. In: Proceedings of the IEEE/CVF international conference on computer vision, pp 4170–4179
26. Zhao L, Mo Q, Lin S, Wang Z, Zuo Z, Chen H, Xing W, Lu D (2020) Uctgan: diverse image inpainting based on unsupervised cross-space translation. In: Proceedings of the IEEE/CVF conference on computer vision and pattern recognition, pp 5741–5750
27. Yu T, Guo Z, Jin X, Wu S, Chen Z, Li W, Zhang Z, Liu S (2020) Region normalization for image inpainting. In: Proceedings of the AAAI conference on artificial intelligence, vol 34, pp 12733–12740
28. Liu H, Wan Z, Huang W, Song Y, Han X, Liao J (2021) Pd-gan: probabilistic diverse gan for image inpainting. In: Proceedings of the IEEE/CVF conference on computer vision and pattern recognition, pp 9371–9381
29. Liao L, Xiao J, Wang Z, Lin CW, Satoh S (2021) Image inpainting guided by coherence priors of semantics and textures. In: Proceedings of the IEEE/CVF conference on computer vision and pattern recognition, pp 6539–6548
30. Zhang W, Zhu J, Tai Y, Wang Y, Chu W, Ni B, Wang C, Yang X (2021) Context-aware image inpainting with learned semantic priors. arXiv:2106.07220
31. Marinescu RV, Moyer D, Golland P (2020) Bayesian image reconstruction using deep generative models. arXiv:2012.04567
32. Zhao S, Cui J, Sheng Y, Dong Y, Liang X, Chang EI, Xu Y (2021) Large scale image completion via co-modulated generative adversarial networks. arXiv:2103.10428
33. Philbin J (2007) Oxford buildings dataset. http://www.robots.ox.ac.uk/~vgg/data/oxbuildings/

Literature Review for Automatic Detection and Classification of Intracranial Brain Hemorrhage Using Computed Tomography Scans

Yuvraj Singh Champawat, Shagun, and Chandra Prakash

1 Introduction

In this study, we have investigated the problem of detection of intracranial brain hemorrhage and the classification of its various subtypes. Intracranial hemorrhage (ICH) is a life-threatening emergency that corresponds to acute bleeding within the skull (cranium) [1]. It is a severe type of stroke that occurs when the brain is deprived of oxygen and blood supply. The most common reasons for the occurrence of intracranial hemorrhage are arteriovenous malformations, hypertension (high blood pressure), and head trauma. Other possible causes include vascular abnormalities, venous infarction, bleeding disorders or treatment with anticoagulant therapy, atherosclerosis (build-up of fatty deposits in the arteries), and smoking or heavy alcohol use. The symptoms of intracranial hemorrhage depend on the affected part of the brain. Generally, symptoms of bleeding within the brain include difficulty in breathing, severe headache, loss of vision, loss of balance, light sensitivity, dizziness, and sudden weakness. Intracranial hemorrhage constitutes a major threat and can be fatal. Rapid bleeding into intracranial compartments can even cause sudden death. According to recent medical surveys, brain hemorrhage has become one of the main causes of death and many disabilities. As per the various studies done in India, the diffusion of stroke ranges from 334 to 424/100,000 in urban areas and 84 to 262/100,000 in rural areas [2] (Fig. 1).

Y. S. Champawat (✉) · Shagun · C. Prakash
Department of Computer Science and Engineering, National Institute of Technology Delhi, New Delhi, India
e-mail: 181210064@nitdelhi.ac.in

Shagun
e-mail: 181210049@nitdelhi.ac.in

C. Prakash
e-mail: cprakash@nitdelhi.ac.in

© The Author(s), under exclusive license to Springer Nature Singapore Pte Ltd. 2023

H. Muthusamy et al. (eds.), *Robotics, Control and Computer Vision*, Lecture Notes in Electrical Engineering 1009, https://doi.org/10.1007/978-981-99-0236-1_4

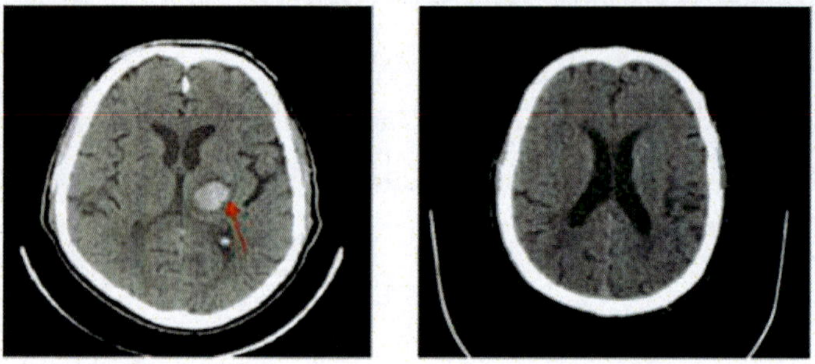

Fig. 1 Sample images of CT scan with intracranial hemorrhage (marked with red arrow) and healthy brain

Intracranial Brain Hemorrhage comprises five types, named as, epidural hemorrhage, subdural hemorrhage, subarachnoid hemorrhage, intraventricular hemorrhage, and intraparenchymal hemorrhage [1].

- Epidural Hemorrhage: It is a type of hemorrhage in which the blood accumulates between the thick outer membrane, that is, the dura mater, and the skull. The main cause of such hemorrhage is when a skull fracture or injury tears the underlying blood vessels.
- Subdural Hemorrhage: It is a type of hemorrhage in which the blood accumulates within the skull but outside the tissue of the brain. It causes when any brain injury bursts the outer blood vessels on the skull head. It sometimes does not show symptoms and needs no treatment.
- Subarachnoid Hemorrhage: It is a type of hemorrhage in which the blood accumulates in the space surrounding the brain. It is mainly caused when any blood vessel presents on the surface of the brain's outer tissue bursts. It is a severe type of stroke and needs immediate treatment.
- Intraventricular Hemorrhage: It is a type of hemorrhage in which the blood accumulates into the brain's ventricular system. It mainly occurs due to a lack of oxygen in the brain or traumatic birth. It also has a high mortality rate, especially among newborn babies.
- Intraparenchymal Hemorrhage: It is a type of hemorrhage in which the blood accumulates within the brain parenchyma region, that is, the tissue region of the brain. It mainly occurs due to sudden trauma, tumors, rupture of inner brain arteries or veins, or birth disorders (Fig. 2).

It is well known that India is facing a shortage of both trained medical staff and medical facilities. As per statistics presented in Thayyil and Jeeja [4], India comprises approx. 17% of the total world population but contributes to about 20% of the total world disease burden. About 70% of the total population of the country resides in rural areas but approx. 74% of the total trained medical staff lives in urban areas, leaving behind 26% for the majority of the population. As per a survey conducted in

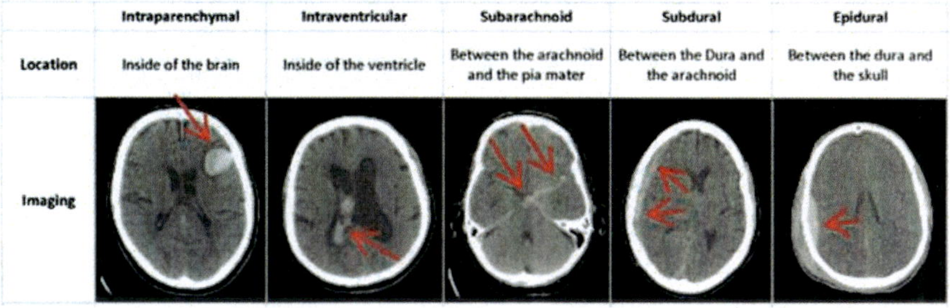

Fig. 2 Types of hemorrhages: (From Left to Right) Intraparenchymal, Intraventricular, Subarachnoid, Subdural, Epidural. Source [3]

March 2018, a shortfall in health facilities at different levels is about: 18% at the Sub-Centre level, 22% at the PHC level, and 30% at the CHC level [5]. Thus, there is a lot of burden on the existing medical staff. The professional medical staff works day and night for the well-being of society. Examples of this have been seen in the past two years during the COVID-19 pandemic. The advancements in science and technology, particularly in the field of artificial intelligence should be implemented and used in such a way that it helps and supports our medical workforce. AI-assisted tools and chatbots, AI-powered robots, and various computer-aided diagnostic systems should be promoted more and more. Real-time automatic diagnosis of severe health issues like intracranial brain hemorrhage will definitely prove a milestone in medical history. It will save thousands of patients per year who lost their lives due to late treatment and improper diagnosis of hemorrhage.

The rest of this paper is organized as follows: Sect. 2 describes the existing methods of diagnosis of ICH and comparison between CT scan images and MRI images for diagnosis purpose. Section 3 describes how machine learning and deep learning techniques can assist in the detection of ICH and also presents the summary of some previously done works, the comparison table, and analysis based on the obtained table. Section 4 describes some limitations of this study, presents the future research work for related to the field and lastly, and concludes our paper.

2 Methods for Diagnosis of Brain Hemorrhage

Intracranial Brain Hemorrhage is a severe type of stroke that can affect the functioning of brain cells and thus can lead to critical symptoms and can eventually lead to the death of a patient. Fast and effective treatment is generally required in case of an ICH emergency. In some cases, major surgeries are also required to save the life of a patient. Diagnosis of ICH is done by either CT scan or Magnetic Resonance Imaging (MRI) [6, 7]. Neurologists and Radiologists require images of the inner regions of the brain, in order to locate and confirm the presence of hemorrhage. Further, they perform the volumetric analysis of ICH on the basis of the spread of blood over brain tissues. This is an important step of treatment because performing this provides information about location, position, volume, and subtype of hemorrhage. Generally, a CT scan is done first, and then if further clear and detailed images are required then MRI is done. Due to the better image quality of MRI sometimes, it is being assumed that MRI should be preferred over CT scan for diagnosis, but this is not always true. CT scans have many advantages over MRI. Imaging in case of CT scan is fast, generally takes 10–15 min while MRI might take 35–45 min and in case of an emergency, the patient might not have that much time and need instant treatment. Moreover, a CT scan can also be performed if the patient is taking a drip but MRI cannot be done in that case. CT scan machines are easily available as compared to MRI machines and performing CT scans is also less costly. MRI scan cannot be per- formed in case if a patient is having any metallic or electrical implant in the body. Also, in MRI the body of the patient is completely passed into the machine thus it might lead to a state of unconsciousness. Sometimes, patients might not fit into the MRI scanning machine due to their weight. Generally, it is recommended to the patient to stay still in the MRI machine but sometimes it might not be feasible for the patient due to old age or pain. MRI also has some advantages over CT scan like the dose of harmful X-rays is high in case of CT scan while MRI works on the magnetic and electrical power. Frequent CT scans can increase the risk of cancer to the patient. The quality of images and information provided by MRI scans is much better as compared to CT scan images.

Thus, it can be seen that both types of diagnostic imaging processes have their own pros and cons. It has been observed that the image quality of a CT scan is sufficient enough to provide details and information about brain hemorrhage so that doctors can start initial treatment. Head CT scan images can even show the acute hemorrhage or abnormality present in brain tissues. That's why doctors prefer CT scans over MRI for the accurate diagnosis of brain hemorrhage. If frequent imaging reports are required or radiologists need further details of inner brain tissues then MRI is done. Due to these reasons, we have chosen Computed Tomography (CT) scan for the diagnosis of Intracranial Brain Hemorrhage as our work.

3 Machine Learning for Diagnosis of Brain Hemorrhage

Intracranial Brain Hemorrhage is a very serious health problem that requires immediate and intensive medical treatment. The delay in proper treatment might lead to the death of the patient. The diagnosis of ICH using CT scans is a very complex process and generally requires a very experienced radiologist. Sometimes it is not possible to have an experienced radiologist available all the time. Which leads to a lack of treatment. Moreover, the volumetric analysis of ICH using CT scan images is a very complex and error-prone process. In the case of complex ICH, it becomes very difficult to estimate the volume of the Hemorrhage. Thus, a rapid and accurate alternative method of diagnosis is necessary for the treatment process achieving success over

ICH. The advancements in the field of machine learning and deep learning, particularly computer vision, attracts the research community to propose computer-aided, rapid, and accurate mechanisms for the automatic diagnosis of various diseases. As the diagnosis of hemorrhage depends on the images obtained from CT scan or MRI, a self-learning algorithm can be trained to obtain a model that can learn the patterns from the normal and abnormal images. On the basis of these learned patterns the model can detect the traces of disease present in medical images. In recent years, a lot of work has been done in the field of diagnosis using machine learning [3, 8–18]. Some of these are, detection of pneumonia and COVID-19 using X-ray images of chest, classification of brain tumor into benign and malignant, detection of breast cancer, treatment of dead cells related skin infections, detection of degenerative diseases like Parkinson and Alzheimer, in Diabetic Retinopathy, assisting doctors for prescribing medicines and ICU calls, detection of stage of Diabetes and many more.

The detection and classification of ICH using machine learning techniques generally follows the pipeline presented in Fig. 3. The first stage of the pipeline is Data collection or Data acquisition, in this stage the medical images along with proper metadata of patients are collected from different hospitals or radiology centres. These images are later used for training and testing of models. The following step is the Data preparation step, which includes various data pre-processing techniques applied on the medical images to make them ready for the input to model. This is an important step as in this step noise and extra, unwanted information are removed from images and various data augmentation techniques are applied. Next stage is Dataset partition, this stage includes dividing the dataset into training, validation, and test sets. Following is the Training stage, this is the most important stage in the pipeline as it includes feature extraction, feature selection and classification on the basis of features obtained. The performance of the model is highly dependent on the methods that are being adopted for feature extraction and classification in this stage. Lastly, the trained model is being tested on the test dataset images and performance and generalizability of the model is evaluated on the basis of various parameters like accuracy, recall, precision, F1-score, AUC, sensitivity, specificity, etc. [19]. The brief description of some of the most commonly used performance metrics is as follows.

- Accuracy: It is defined as the ratio of the sum of true positives and true negatives to the total number of data instances available.

$$\text{Accuracy} = (\text{TP} + \text{TN})/(\text{TP} + \text{FP} + \text{TN} + \text{FN}) \tag{1}$$

- Recall: It is defined as the ratio of true positives to the sum of true positives and false negatives.

$$\text{Recall} = (\text{TP})/(\text{TP} + \text{FN}) \tag{2}$$

- Precision: It is defined as the ratio of true positives to the sum of true positives and true positives.

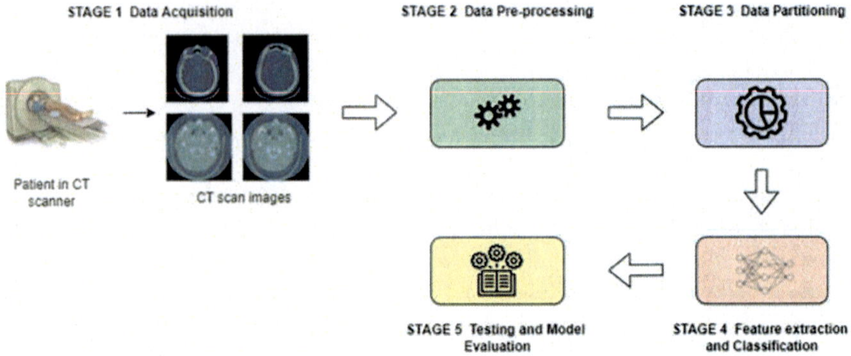

Fig. 3 The block diagram represents general pipeline for the diagnosis of brain hemorrhage

$$\text{Precision} = (\text{TP})/(\text{TP} + \text{FP}) \tag{3}$$

- Sensitivity: It is defined as the ability of the model to predict true positives from the total given labels for each class. In binary classification, sensitivity is similar to recall. In medical diagnosis, high sensitivity is preferred because if a patient is having hemorrhage but classified as healthy, that is, no hemorrhage present then it might lead to a big trouble.
- Specificity: It is defined as the ability of the model to predict true negatives from total given labels for each class. In binary classification, specificity is similar to precision.
- F1-score: It is defined as the measure of model's accuracy on the complete dataset. Mathematically it is being calculated using values of precision and recall.

$$\text{F1} - \text{score} = 2 * (\text{Precision} * \text{Recall})/(\text{Precision} + \text{Recall}) \tag{4}$$

- Area Under Curve (AUC): It is defined as the probability of predicting a random positive class data instance correctly. It can take values between 0 and 1. When the value of AUC is 0 it represents that 100% predictions made are wrong and when it is 1 then it represents that 100% predictions made are correct.
- Log Loss: In classification tasks, this metric is based on the probabilities calculated for different classes. Lower the value of log loss means better the predictions are made by model. For multilabel classification, we can assign weights to the probabilities of different classes. The weighted log loss is preferred because it efficiently deals with the class imbalance issues.

$$Log\ Loss = -\frac{1}{N}\sum_{i=1}^{N} y_i * log(p_i) + (1 - y_i) * log(1 - p_i) \tag{5}$$

where TP stands for True Positives, TN stands for True Negatives, FP stands for False Positives, FN stands for False Negatives, y stands for the true label of a data instance and p stands for predicted label of a data instance (Fig. 3).

Depending on stage 4, feature extraction and classification, we have divided the approaches for building pipeline into four types:

- Both feature extraction and classification based on machine-learning techniques and algorithms.
- Feature extraction based on deep learning models and classification using machine learning algorithms.
- Both feature extraction and classification based on deep learning techniques and algorithms.
- Classification using IoT-powered techniques or segmentation-based algorithms.

3.1 Both Feature Extraction and Classification Based on Machine Learning Techniques and Algorithms

In this approach, after applying suitable data pre-processing methods to input images, the useful features are extracted using different standard manual methods and then traditional machine learning-based classifiers like SVM, Random Forest, KNN, etc. are trained on the obtained features (Fig. 4).

Shahangian and Pourghassem [20], implemented a pipeline for the segmentation of the hematoma region for its area evaluation and classification into subtypes. This pipeline includes pre-processing techniques, skull removal methods, brain ventricles removal technique, morphological filtering processes, segmentation of ICH region, feature extraction, quantifiable feature selection using genetic algorithm, and lastly, classification of ICH into subtypes. The skull and brain ventricles were removed by applying a check on the intensity values of the CT scan. Then a median filter was applied to remove noise and the largest area object had been selected from the binary image to get only the brain region. ICH segmentation was performed by applying a threshold to pixel intensities. For the classification purpose, a KNN algorithm and a multilayer perceptron (MLP) model with a tan sigmoid activated output layer were trained. MLP model outperformed KNN.

Liu et al. [7], dealt differently with the nasal cavity and encephalic region CT scans. From Fig. 5 we can observe that both types of CT scans have different textures, thus, the method working efficiently with brain regions might not work well with the nasal

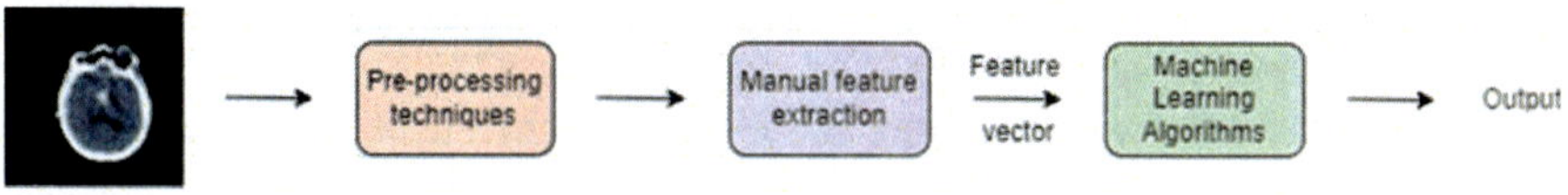

Fig. 4 This flow diagram represents the pipeline for both feature extraction and classification based on machine learning techniques and algorithms

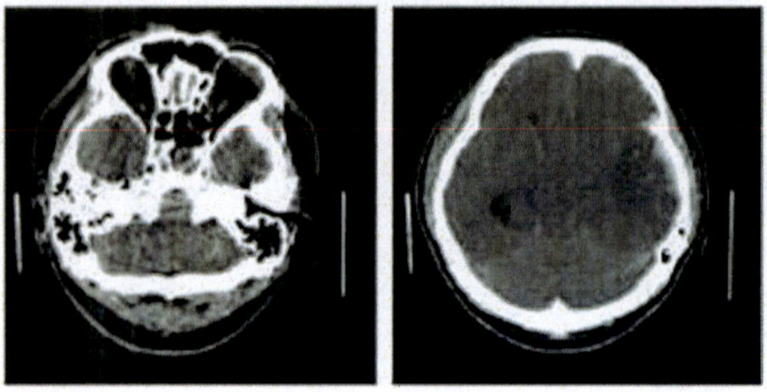

Fig. 5 Nasal Cavity (left side image) and Encephalic Region (right side image). Source [13]

cavity. Both are separated on the basis of texture analysis using Wavelet transform. Skull removal and gray matter removal methods were applied to the encephalic region to get segmented hemorrhages. Then 12 different features corresponding to intensity distribution and texture descriptions were extracted. Entropy calculation was employed to select good features and a Support Vector Machine (SVM) classifier was trained to distinguish abnormal slices (slice consisting of ICH) from normal slices.

Al-Ayyoub et al. [21], proposed a pipeline that includes skull removal, segmentation of ICH, morphological methods, extraction of the region of interest, feature extraction, and classification. For the segmentation purpose, Otsu's method was applied followed by the opening transformation technique. Region of Interest is obtained by applying the region growing algorithm on the output obtained after segmentation. Finally, features based on the size, shape, and position of hemorrhage ROI were extracted. The SVM, Multinomial logistics regression (MLR), Multilayer perceptron model, Decision tree, and Bayesian network classifiers were trained independently on features. The MLR classifier outperforms others.

3.2 *Feature Extraction Based on Deep Learning Models and Classification Using Machine Learning Algorithms*

In this approach, after applying suitable data pre-processing methods to input images, the pre-trained Convolutional Neural Networks (CNN) are imported and are trained end-to-end in order to extract features from images. The traditional machine learning algorithms applied on the top of these CNN models are then trained for performing classification using the obtained features (Fig. 6).

Salehinejad et al. [8], stacked three windows of CT scan images to get 3- channel input for 2D-CNN models. They have used pre-trained SE-ResNeXt-50 and SE-ResNeXt-101 models as the backbone for extracting features from images and have applied traditional machine-learning algorithms like LightGBM, CatBoost, and XGBoost for classification. To utilize the interdependency among slices of a CT scan they applied a sliding window module. For testing the generalizability of the models, they tested them on a private external validation dataset. This is an important step,

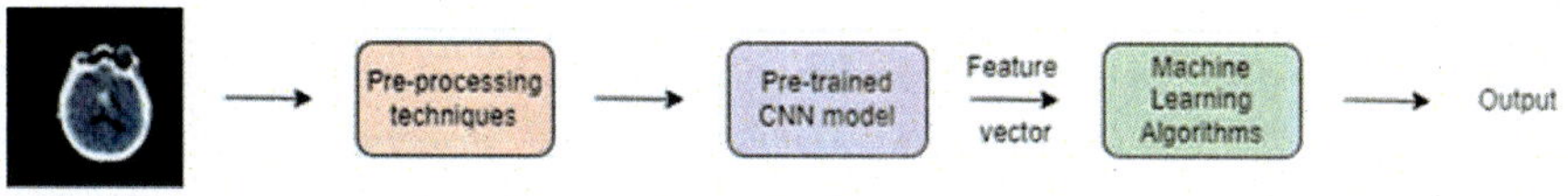

Fig. 6 This flow diagram represents the pipeline for feature extraction using pre-trained convolutional neural network (CNN) model and classification based on machine learning algorithms

especially in the case of medical images. Testing the models on a dataset consisting of temporally and geographically different images indicates the generalization power of models.

Sage and Badura [9], have applied regions of interest, that is, brain region cropping and skull removal methods before giving image input to the ResNet-50 model. They performed brain region cropping by determining the largest binary object from the CT scan image after applying Otsu Algorithm. The skull removal method was applied by reducing the values of pixels having the highest intensities to zero. Two-branch architecture was used to train the classification model. In the first branch, three different windows were stacked and in the second branch, three consecutive subdural windows were stacked to get a 3-channel image. SVM and Random Forest were applied on top of the ResNet-50 network for predicting the class.

3.3 *Both Feature Extraction and Classification Based on Deep Learning Techniques and Algorithms*

In this approach, after applying suitable data pre-processing methods to input images, the pre-trained Convolutional Neural Networks (CNN) are imported and then a transfer learning protocol is followed to train these models for performing classification. The features obtained from the pre-output layer of these models can also be used to train Bi-LSTM network layers in order to utilize the spatial interdependence among slices of CT scan (Fig. 7).

He et al. [10], developed a classification model using pre-trained CNN models like SE-ResNeXt50 and EfficientNet-B3 as the backbone. They have used weighted

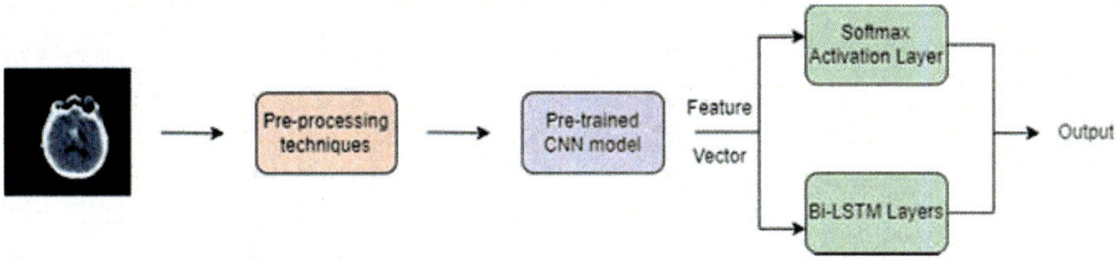

Fig. 7 This flow diagram represents the pipeline for feature extraction using pre-trained convolutional neural network (CNN) model and classification using softmax activation function layer or Bi-LSTM layers as output layers

multi-label logarithmic loss for the training of models. For improving the performance, they employed K-fold cross-validation (K = 10 in their case) and pseudo-label technique. Using the pseudo-label technique, 52,260 new images were added to the training dataset which was originally present as unlabeled data in the RSNA dataset.

Anaya and Beckinghausen [11], proposed a multi-label classification model for classifying ICH into its subtypes. The features were extracted using pre-trained MobileNet and ResNet-50 networks. On the basis of experimental results, the authors concluded that it is most difficult to detect epidural hemorrhage using a CT scan. This is probably due to the presence of an epidural hematoma near the skull region of the head.

Juan Sebastian Castro et al. [12], proposed a binary classification model for detecting hemorrhage in CT scans. The brain region from CT scans was extracted from the background and then a single window (WW = 80; WL = 50) was applied to get the brain parenchyma region. They have used pre-trained VGG-16 and a customized CNN model as the backbone for the classification model. The training was performed using two protocols, one is slices randomized and another is subject randomized.

Lewicki et al. [13], presented a multi-label classification model for the detection and classification of ICH into its subtypes. Due to heavy negative bias and high-class imbalance among positive classes in the RSNA dataset [22], class weights were applied to loss function and recall/precision tuning was performed. A batch of 3-channel CT scan images produced by stacking three different windows was fed as input to the ResNet-50 model for training purposes.

Patel et al. [14], used a private dataset to train the combination of CNN and Bi-LSTM networks for predicting the probabilities corresponding to each class. Initially, features of the CT scan images were extracted using CNN and then the output spatial vectors of consecutive slices were together given as input to Bi-LSTM layers. The Bi-LSTM network was applied to utilize the interdependency among slices of a CT scan. Rotation and Random Shifting augmenting techniques were also applied. The authors also specified the importance of pre-training of CNN models before applying end-to-end training for fine-tuning.

Nguyun et al. [15], trained a CNN and Bi-LSTM combinational network on the RSNA dataset and used the CQ500 dataset [23] for external validation. They have applied various types of augmenting techniques to improve the generalizability of models. To deal with the class imbalance problem in the RSNA dataset they have applied weighted binary cross-entropy loss for training. They have used ResNet-50 and SE-ResNeXT-50 models as the feature extractors.

Burduja et al. [3], proposed a slice-based classification model using the ResNeXt-101 network for feature extraction and Bi-LSTM layers on top. The Res- NeXt-101 network outputs a 2048-seized feature vector for each image. Then, PCA was applied to reduce the dimensions of this feature vector to a 120-sized vector. This reduced feature vector was given as input to recurrent neural networks. The outputs of RNN were concatenated to the prediction probabilities obtained as outputs from Res-NeXt-101. These concatenated feature vectors were used to train the final output softmax-activated layer. They have also compared performances of ResNeXt-101 and EfficientNet-B4 and concluded that ResNeXt-101 gives better results. The authors also presented the importance of using spatial dependency of slices of a CT scan. By utilizing this characteristic, the numbers of false positives and false negatives can be reduced. The GRAD-CAM saliency maps were also presented for the approx. visualization of the ICH region.

Hoon et al. [16], proposed a multi-label classification model for the detection of ICH and classification into its subtypes. The model consisted of a combination of pre-trained Xception network and Bi-LSTM layers. The sigmoid-activated layer was applied as the output classifying layer. The major positive point of this work is that several image augmenting techniques were applied especially on the Epidural Hemorrhage subclass to deal with the class imbalance issue in the RSNA dataset [22].

3.4 Classification Using IoT-Powered Techniques or Segmentation-Based Algorithms

In this approach, after applying suitable data pre-processing methods to input images, various medical image segmentation algorithms are applied to get the segmented image of ICH. Features are then extracted from the obtained segmented images using either a manual feature extraction process or pre-trained CNN by fine-tuning the model. Later on, these features are used to train the machine learning algorithms or CNN models for the classification purpose. Internet-of-Things (IoT) powered techniques can also be used for getting processed images in electrical format. These electrical signals act as feature vectors of images which are then used to train classifiers (Fig. 8).

Sage and Badura [9], presented the comparison between various ICH segmentation algorithms. Majorly, three techniques have been used for the segmentation of ICH, named as, Thresholding technique, Region Growing technique, and Clustering techniques. The authors have implemented and compared the proposed multilevel segmentation approach (MLSA), watershed method, and EM method on the basis

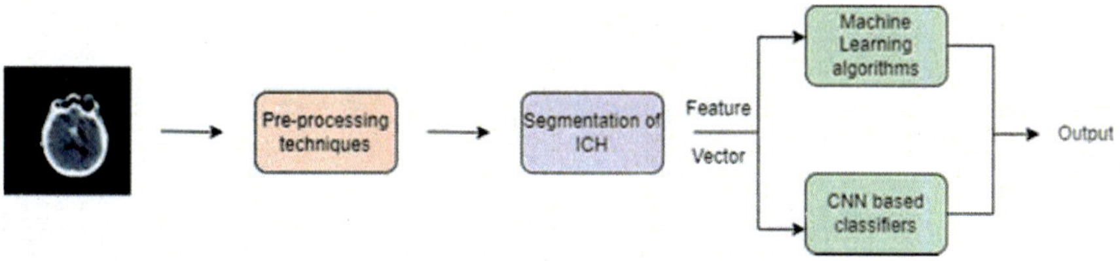

Fig. 8 This flow diagram represents the pipeline for classification of ICH using feature vectors that are extracted from segmented ICH image. For the classification purpose, any classifying model can be applied

of the time taken to process a single image and average PCC values. The MLSA technique has performed better than other methods.

Vincy Davis et al. [12], presented a model for the diagnosis and classification of ICH. The model includes the conversion of CT scan image into the grayscale image then resizing and edge detection were applied. After that several morphological techniques like opening and closing transformations and boundary smoothing methods were applied. Segmentation of ICH was performed using Watershed Algorithm. The paper also presents the importance of the Watershed algorithm in extracting hematoma regions. An ANN model was trained using features extracted from Gray Level Co-occurrence Matrix (GLCM) method.

Patel et al. [14], proposed a CNN model inspired by U-Net for the segmentation of the ICH region in the CT scan image. The model was trained on the ground truth labeled images. The segmented hematoma was classified into its sub-types. They have applied several data augmentation techniques for achieving better- generalized outcomes. It also discussed possible reasons for achieving better results for subtypes and not-so-good results for other subtypes.

Balasooriya et al. [24], presented a pipeline for the diagnosis of ICH using image segmentation of hemorrhage region in CT scan using watershed algorithm. As the first step, the input images were converted to greyscale and were reduced to 2- dimensional images. Then various morphological techniques were applied for removing noises and disturbances from CT scan image, preparing it for segmentation purpose. Features extracted manually from segmented images were used to train artificial neural network (ANN).

Chen et al. [25], presented a smart Internet-of-Things (IoT) based technique for classification of ICH using machine learning algorithms. In the setup, a Wi-Fi sensor was placed in between CT scan machine and Arduino board. Two types of sensors were applied, for converting the CT scan images into electrical signals and saving them to the server. A complementary metal oxide semiconductor (CMOS) sensor was used to convert medical images into electrical signals and ESP8266 Wi-Fi module was used for posting data to server. The electrical signals obtained were used to train the Support vector machine (SVM) and Feedforward neural network (FNN) model for classification. A mobile application was also developed for testing CT scan images and generating reports in real time (Table 1).

3.5 Research Challenges Related to the Field and Suggestions for Future Scope

With reference to Table 1, we can infer some research challenges and can provide some suggestions for future scope related to the field, which have to be taken care of while implementing a model for detection and classification of intracranial brain hemorrhage into its subtypes. The following are some measures.

Table 1 Table presents the comparison of reviewed papers on the basis of common parameters primarily related to the implementation of work. The parameters included are application of the paper, dataset used in the paper, windowing policies adopted for the CT scans to convert them into 3-channel images, pre-processing techniques applied before feature extraction and training of models, saliency or heat maps presenting the presence of ICH in CT scan, performance metrics included in work, strong points related to the methods adopted by authors and review comments for the presented work

Author [Year]	Application of paper	Dataset	Window policy	Pre-processing	Saliency map visualization	Performance metrics	Strong points of methods adopted	Review comments
Liu et al. [7]	Splitting of CT scan images into nasal cavity and encephalic region. Classification of ICH into its subtypes	Private dataset	–	Skull removal; gray Matter removal; wavelet transforms	–	Accuracy = 80%, Recall = 88%	1. Presented pre-processing methods for discarding abnormal slices 2. Wavelet and haralick texture-based model for splitting of CT scan images	1. Dataset not made publicly available 2. Applicable only on encephalic region images 3. Poor feature extraction and selection methods
Balasooriya et al. [24]	Detection of ICH in CT scan using segmentation of region of interest	Private dataset	–	Opening and closing transformation; watershed algorithm; background noise removal techniques	–	Accuracy = 80%, Recall = 88%	1. Implemented various pre-processing techniques on images before segmentation	1. Small private dataset was used 2. Manual feature extraction was done

(continued)

Table 1 (continued)

Author [Year]	Application of paper	Dataset	Window policy	Pre-processing	Saliency map visualization	Performance metrics	Strong points of methods adopted	Review comments
Saini and Banga [6]	Segmentation of ICH region and Detection of abnormal slices of CT scan	Private dataset	–	–	–	Highest Accuracy = 97.1% using MLSA method, Highest Precision = 94.69% using K-means, Highest Recall = 90.07% using K-means and FCM	1. Presented a comparative analysis of various segmentation methods	1. No pre-processing techniques applied 2. No presentation of classifier algorithm was given 3. Proposed segmentation method is also unclear

(continued)

Table 1 (continued)

Author [Year]	Application of paper	Dataset	Window policy	Pre-processing	Saliency map visualization	Performance metrics	Strong points of methods adopted	Review comments
Shahangian and Pourghassem [20]	Segmentation of ICH region and Classification of ICH into subtypes	Private dataset	–	Skull removal; Brain Ventricles removal; Median Filter; Soft tissue Edema removal;	–	Highest Accuracy = 93.3% using Multilayer Perceptron model, For segmentation, highest accuracy obtained is for epidural ICH = 96.22%	1. Implemented pre-processing techniques on images before feature extraction 2. Various segmentation techniques were implemented and comparative analysis was presented	1. Dataset is not made publicly available. Small dataset 2. No window policy applied 3. Only three subtypes (Epidural, Intracerebral and Subdural Hematoma) are classified 4. The method proposed for segmentation is based on pixel intensity division. This method is not so promising and might not work better in case of complex CT scans

(continued)

Table 1 (continued)

Author [Year]	Application of paper	Dataset	Window policy	Pre-processing	Saliency map visualization	Performance metrics	Strong points of methods adopted	Review comments
Al-Ayyoub et al. [21]	Segmentation of ICH region and Classification of ICH into its subtypes	Private dataset	–	Skull removal; Segmentation using Otsu's method; Opening operation; region growing	–	Accuracy for detection of Hemorrhage = 100% Accuracy for classification of ICH into subtypes = 92%	1. Texture-based ICH segmentation was applied 2. Various morphological techniques and region of interest extraction techniques are presented	1. Dataset not made publicly available 2. Poor feature extraction and selection methods 3. Only three subtypes (Epidural, Intra-parenchymal and Subdural Hematoma) are classified

(continued)

Table 1 (continued)

Author [Year]	Application of paper	Dataset	Window policy	Pre-processing	Saliency map visualization	Performance metrics	Strong points of methods adopted	Review comments
Davis and Devane [26]	Segmentation of ICH region and Classification of ICH into subtypes	Private dataset	–	Edge Detection; Opening and closing operations; Median Filter; Watershed Algorithm (Segmentation)	–	Error in detection of ICH = 0.47838	1. Segmentation using watershed algorithm is presented	1. Small dataset (just 35 images) 2. No window policy 3. Only two subtypes (Intracerebral and Subdural Hematoma) are classified 4. Poor feature extraction and selection methods
Majumdar et al. [27]	Segmentation of ICH region and Classification of ICH into its subtypes	Private dataset	–	Data augmentation	–	Sensitivity = 81% Specificity = 98%	1. Described model for segmentation of ICH region 2. Presented analysis for false negatives in diagnosis	1. No proper pre-processing applied 2. Small private dataset (just 134 CT scans)

(continued)

Table 1 (continued)

Author [Year]	Application of paper	Dataset	Window policy	Pre-processing	Saliency map visualization	Performance metrics	Strong points of methods adopted	Review comments
Anaya and Beckinghausen [11]	Detection and classification of ICH into subtypes	RSNA	–	–	–	Accuracy = 76%, Recall = 93%	1. Presented a detailed analysis of obtained results 2. Stated importance of 3D - CNN for classification	1. No window policy 2. No pre-processing done 3. Small dataset (only 5000 images from RSNA were used)
Castro et al. [12]	Detection of ICH in CT scan	CQ500	Brain window (WW = 80; WL = 50)	Background Removal; Anisotropic filter	–	Accuracy = 98%, Recall = 97% F1-score = 98%	1. Two protocol training: - Slices randomized and Subject randomized	1. Small Dataset 2. Only detection of ICH, No classification into subtypes
Patel et al. [14]	Detection of ICH in CT scan using spatial interdependency among slices of CT scan	Private dataset	–	Data augmentation	–	Highest AUC = 0.96	1. Used spatial interdependency by using Bi-LSTM network	1. Not made dataset publicly available 2. No pre-processing and visualization techniques applied

(continued)

Table 1 (continued)

Author [Year]	Application of paper	Dataset	Window policy	Pre-processing	Saliency map visualization	Performance metrics	Strong points of methods adopted	Review comments
He et al. [10]	Detection and Classification of ICH into subtypes	RSNA	–	Data augmentation	–	Weighted mean log loss = 0.0548	1. Applied K-fold cross-validation which improves the performance of model	1. No window policy applied 2. No pre-processing done
Lewicki et al. [13]	Detection and Classification of ICH into subtypes	RSNA	Brain window (WW = 80; WL = 40); Subdural window (WW = 200; WL = 80); Bone window (WW = 2800;WL = 600)	–	–	Highest Accuracy = 93.3% Average per-class Recall = 76%	1. All the performance metrics are presented as per-class measures which helps in better analysis for diagnosis among subtypes of ICH	1. No pre-processing done 2. No visualization of ICH presented 3. Only one classifier is trained

(continued)

Table 1 (continued)

Author [Year]	Application of paper	Dataset	Window policy	Pre-processing	Saliency map visualization	Performance metrics	Strong points of methods adopted	Review comments
Sage and Badura [9]	Detection and Classification of ICH into subtypes	RSNA	Brain window (WW = 80; WL = 40); Subdural window (WW = 200; WL = 100) Bone window (WW = 2800;WL = 600)	Brain region Cropping Skull Removal	–	Highest Accuracy reported for: Intraventricular = 96.7% Intraparenchymal = 93.3% Subdural = 89.1% Epidural = 76.9% Subarachnoid = 89.7%	1. Pre-processing techniques were applied on images before training phase 2. Made use of spatial inter-dependency among slices of a CT scan	1. Not used spatial inter-dependency among slices 2. No saliency map visualization 3. Only subset of RSNA was used

(continued)

Table 1 (continued)

Author [Year]	Application of paper	Dataset	Window policy	Pre-processing	Saliency map visualization	Performance metrics	Strong points of methods adopted	Review comments
Nguyun et al. [15]	Detection and Classification of ICH into subtypes using Spatial Interdependency among slices of CT scan	RSNA; CQ500 (external validation)	Brain window (WW = 80; WL = 40); Subdural window(WW = 215; WL = 75) Bone window (WW = 2800;WL = 600) Brain window (WW = 80; WL = 40); Subdural window (WW = 200; WL = 80); Soft Tissue window (WW = 380; WL = 40)	Data augmentation	–	Weighted mean log loss: For SE ResNext-50 = 0.05218 For ResNet-50 = 0.05289	1. Used interdependency among slices by applying Bi-LSTM network 2. Tested models on CQ500 dataset for external validation	1. No pre-processing and visualization techniques applied

(continued)

Table 1 (continued)

Author [Year]	Application of paper	Dataset	Window policy	Pre-processing	Saliency map visualization	Performance metrics	Strong points of methods adopted	Review comments
Burduja et al. [3]	Detection and Classification of ICH into subtypes using Spatial Interdependency among slices of CT scan	RSNA		Data augmentation	GRAD-CAM heat maps presented	Weighted mean log loss = 0.04989	1. Used interdependency among slices by applying Bi-LSTM network 2. Presented saliency maps	1. No pre-processing techniques applied
Hoon et al. [16]	Detection and Classification of ICH in CT scan using Spatial Interdependency among slices of CT scans	RSNA	Brain window (WW = 80; WL = 40); Subdural window (WW = 200; WL = 80); Bone window (WW = 1800;WL = 400)	Data augmentation; Data balancing	–	Weighted mean log loss = 0.07528	1. Addressed problem of class imbalance in RSNA dataset and presented data balancing techniques	1. No pre-processing and visualization techniques applied 2. Number of labels are shown as number of images in dataset which is not correct

(continued)

Table 1 (continued)

Author [Year]	Application of paper	Dataset	Window policy	Pre-processing	Saliency map visualization	Performance metrics	Strong points of methods adopted	Review comments
Chen et al. [25]	Detection and Classification of ICH in CT scan using Internet of Things based system	Private dataset	–	–	–	Accuracy for SVM = 80.67% Accuracy for Feedforward Neural Network = 86.7%	1. Presented the importance and use of IoT-based devices for diagnosis of diseases 2. Implemented an end-to-end mobile application for the real time use	1. Dataset used is small and not made publicly available 2. No pre-processing techniques were applied 3. More better classifiers could be used for achieving better results
Salehinejad et al. [8]	Detection and Classification of ICH into subtypes using Spatial Interdependency among slices of CT scan	RSNA; private dataset (external validation)	Brain window (WW = 80; WL = 40); Subdural window (WW = 200; WL = 80); Soft Tissue window (WW = 380; WL = 40)	–	GEAD-CAM and GRAD-CAM+ + heat maps presented	RSNA: AUC = 98.4%, Sensitivity = 98.8%, Specificity = 98.0% External Validation: AUC = 95.4%, Sensitivity = 91.3%, Specificity = 94.1%	1. Tested models on external validation dataset which proves better generaliz-ability of model 2. Presented saliency maps	1. No pre-processing done 2. Haven't made private dataset public

- Adequate pre-processing techniques must be applied on the image data before the feature extraction and classification phase because pre-processing techniques help in removing noise and not so required information from image data, and increases the quality of the image data leading to better feature extraction.
- Saliency or Heat maps of the CT scan image showing the location of the ICH region must be presented. This might help the radiologists in locating the acute ICH region and also proves the credibility of the classification model that it is considering the ICH part in real and not classifying on the basis of some external bias.
- The adjacent slices of a CT scan have almost similar texture composition and have similar characteristics. Thus, while training the classification model one should use this spatial interdependence among slices of CT scan. This leads to better results comparatively.
- Using only a single window of CT scan image cannot help much in the diagnosis of ICH because there might be a case when hemorrhage is present in the bone region of the brain and if only soft tissue windows are being considered then it might not provide clear insights of the hemorrhage. Thus, a combination of different windows of a CT scan image should be preferred for diagnosis.
- If any researcher is preparing or collecting their own private dataset then they should make the dataset publicly available with proper metadata. This motivates the research community to further work in the field.
- Most of the papers published before the year 2019 have either used their private datasets or have used the CQ500 dataset. But almost all works that are being done after 2019 have used RSNA dataset because the RSNA ICH Detection Challenge

RSNA Intracranial Hemorrhage Detection Challenge [28] was launched in the year 2019. Generally, it has been observed that the number of images in both private datasets and CQ500 is much less than the number of images present in RSNA dataset. Thus, models trained on the RSNA dataset can be considered more promising on the grounds of generalizability.

- To the best knowledge of the authors, as of now, there is no publicly available dataset for the segmentation and extraction of ICH from CT scan images for its volumetric analysis. For the treatment of ICH, its volumetric analysis is considered a crucial step and due to lack of publicly available dataset, it becomes difficult for new researchers to work in this field.
- For the purpose of segmentation of the ICH region, the proposed algorithm should be robust to the quality of the input CT scan images. The algorithm should not be trained on any particular types of CT scans like encephalic regions only. It should be able to locate and extract the hemorrhage regions of all subtypes present in all types of CT scan images.

4 Conclusion and Future Work

This study aims to investigate the problem of detection of Intracranial Brain Hemorrhage and classification into its subtypes. Intracranial Hemorrhage (ICH) is a life-threatening emergency that corresponds to acute bleeding within the skull (cranium). Thousands of people die every year due to the lack of instant treatment of ICH. We have shown the significance of machine learning and deep learning, in the field of diagnosis of ICH. Along with the general insights of intracranial hemorrhage and its subtypes, the paper described the existing methods of diagnosis using CT scan and MRI. Our study also explains how AI/ML techniques can be used for the detection and extraction of the ICH region. In the review process of previously done works, the paper consists of a state-of-art ranging from data handling to feature extraction and classification. All these stages in the pipeline were explored and analyzed individually. The works are compared on the basis of various dimensions like application of work, the dataset used, data pre-processing steps included, heat maps presented, AI/ML techniques employed and classifiers used, etc.

We have compared different previously done studies in the field of detection and classification of Intracranial Brain Hemorrhage on the basis of some common parameters. But there are some limitations of this study that need to be addressed in future work. Firstly, we have majorly reviewed works which are using deep learning techniques. This is because it has been observed that the performance of deep learning models is generally much better than that of traditional machine learning methods and algorithms. Almost all studies related to this field done in recent years have employed only deep learning-based CNN models for classification. Secondly, we assumed that the reader is having some prior knowledge about the implementation details of various algorithms and methods presented in this study. That is why we have not shown the working details or theoretical information about these algorithms. Thirdly, some specific parameters like hyperparameters values (batch size, learning rate, number of nodes or layers in customized networks, epochs, kernel size, etc.), number of images in the datasets, information about data splitting, and results of the reviewed works have not been presented. This is because these parameters were differently implemented in different studies and thus, cannot be directly compared. Lastly, we have not implemented any codes for the confirmation of the results claimed in the reviewed studies. Also, we do not guarantee the qualitative results of these studies in real-time applications for the diagnosis of ICH.

Further, for the future works to be done, it would be suggested for aiming to implement several pipelines for the detection and classification of ICH using CT scans. In these pipelines, one can implement different pre-processing techniques like skull removal methods, head cropping methods, enhancing the medical image quality by applying CLAHE, Gamma correction or Histogram equalization, etc., and different image data augmentation techniques. Then compare the results obtained from these pipelines to get the best pre-processing and augmenting techniques to be followed for achieving the best results. For the classification purpose, it would be suggested to use pre-trained CNN models for feature extraction and Bi-LSTM network layers on top to use the interdependency among slices in a CT scan. As classifiers, one can train both traditional machine learning algorithms like SVM, KNN, XGBoost, and Random Forest on top of CNN models and softmax activated final output layer. Along with the combination of CNN and Bi-LSTM, one can also use the 3D-CNN model for classification purposes. Saliency heat maps for the visualization of the location of the Hemorrhage region in the CT scan image should be presented in work.

This study has certain limitations, but it also provides insightful information about the problem and suggests several solutions to overcome the current challenges related to the field. We think it will motivate other researchers who would like to contribute in the future to this field. The role of radiologists and neurologists cannot be replaced by the intelligence of machine learning and deep learning models. We have presented support to the healthcare workforce. The primary aim for this study is to bridge the gap between AI/ML experts and trained medical staff so that they cooperate proactively with each other. In the near future, it is hoped that AI/ML techniques will be accurate and reliable enough to be used in the diagnosis of Intracranial Brain Hemorrhage in real-time so that we can together win over this life-threatening disease.

References

1. Brain Bleed, Hemorrhage (Intracranial Hemorrhage) (2021). https://my.clevelandclinic.org/health/diseases/14480-brain-bleed-hemorrhage-intracranial-hemorrhage. Accessed 10 Dec 2021
2. Pandian JD, Sudhanb P (2013) Stroke epidemiology and stroke care services in India. Elsevier Public Health Emergency Collection. https://doi.org/10.5853/jos.2013.15.3.128
3. Burduja M., Ionescu RT, Verga N (2020) Accurate and efficient intracranial hemorrhage detection and subtype classification in 3D CT scans with convolutional and long short term memory neural networks. Sensors MDPI 20:5611.
4. Thayyil J, Jeeja MC (2013) Issues of creating a new cadre of doctors for rural India. Int J Med Public Health 3(1). (Jan–Mar 2013)
5. Kumar A, Nayar RK, Koyac SF (2020) COVID-19: Challenges and its consequences for rural health care in India. Elsevier Public Health Emergency Collection. https://doi.org/10.1016/j.puhip.2020.100009
6. Saini S, Prof Banga VK (2013) A review: hemorrhage intracranial segmentation in Ct brain images. Int J Eng Res Technol (IJERT) 2(10), ISSN: 2278-0181. (Oct 2013)
7. Liu R., Tan CL, Leong TY, Lee CK, Pang BC, Lim CCT, Qi T, Tang S, Zhang Z (2008) Hemorrhage slices detection in brain CT images. In: IEEE 19th international conference on pattern recognition. https://doi.org/10.1109/ICPR.2008.4761745
8. Salehinejad H, Kitamura J, Ditkofsky N, Lin A, Bharatha A, Suthiphosuwan S, Lin H, Wilson JR, Mamdani M, Colak E (2021) A real-world demonstration of machine learning generalizability: intracranial hemorrhage detection on head CT. Scientific Reports. Article number: 17051
9. Sage A, Badura P (2020) Intracranial hemorrhage detection in head CT using double-branch convolutional neural network, support vector machine, and random forest. Appl Sci MDPI 10:7577. https://doi.org/10.3390/app10217577
10. He J (2020) Automated detection of intracranial hemorrhage on head computed tomography with deep learning. In: ICBET 2020: proceedings of the 2020 10th international conference on biomedical engineering and technology, pp 117–121

11. Anaya E, Beckinghausen M (2019) A deep learning approach to classifying intracranial hemorrhages. In: CS230: deep learning, Fall 2019, Stanford University, CA
12. Castro JS, Chabert S, Saavedra C, Salas R (2019) Convolutional neural networks for detection of intracranial hemorrhage in CT images. In: Proceedings of the 4th congress on robotics and neuroscience 2564
13. Lewicki T, Kumar M, Hong R, Wu W (2020) Intracranial hemorrhage detection in CT scans using deep learning. In: 2020 IEEE sixth international conference on big data computing service and applications (Big Data Service), pp 169–172. https://doi.org/10.1109/BigdataService49289.2020.00033.
14. Patel A., Van De Leemput SC, Prokop M, Ginneken BV, Manniesing R (2019) Image level training and prediction: intracranial hemorrhage identification in 3D non-contrast CT. IEEE Access. Digital Object Identifier https://doi.org/10.1109/ACCESS.2019.2927792. (2019).
15. Nguyen NT, Tran DQ, Nguyen NT, Nguyen HQ (2020) A CNN-LSTM architecture for detection of intracranial hemorrhage on CT scans. In: Medical imaging with deep learning 2020. arXiv:2005.10992v3 [cs.CV]
16. Hoon K, Chung H, Lee H, Lee J (2020) Feasible study on intracranial hemorrhage detection and classification using a CNN-LSTM network. In: 42nd annual international conference of the IEEE engineering in medicine & Biology Society (EMBC). DOI: https://doi.org/10.1109/EMBC44109.2020.9176162. (2020).
17. Chilamkurthy S, Ghosh R, Tanamala S, Biviji M, Campeau NG, Venugopal VK (2018) Deep learning algorithms for detection of critical findings in head CT scans: a retrospective study. Lancet J 392(10162):2388–2396
18. Anam C, Budi WS, Haryanto F, Fujibuchi T, Dougherty G (2019) A novel multiple-windows blending of CT images in red-green-blue (RGB) color space: phantom's study. Sci Vis 11(5):56–69. https://doi.org/10.26583/sv.11.5.06
19. Shervin M (2019) 20 Popular machine learning metrics. Part 1: classification & regression evaluation metrics. Towards data science. Accessed: 21 Dec 2021
20. Shahangian B, Pourghassem H (2013) Automatic brain hemorrhage segmentation and classification in CT scan images. In: IEEE 8th Iranian conference on machine vision and image processing (MVIP). https://doi.org/10.1109/IranianMVIP.2013.6780031
21. Al- Ayyoub M, Alawad D, Al-Darabsah K, Inad AJ (2013) Automatic detection and classification of brain hemorrhages. WSEAS Trans Comput 12(10). (Oct 2013)
22. RSNA Intracranial Hemorrhage Detection Identify acute intracranial hemorrhage and its subtypes. Competition on Kaggle By RSNA. https://www.kaggle.com/c/rsnaintracranial hemorrhage-detection/data. Accessed 10 Dec 2021
23. CQ500 Head CT scan dataset. http://headctstudy.qure.ai/dataset. Accessed 10 Dec 2021
24. Balasooriya U, Perera MUS (2012) Intelligent brain hemorrhage diagnosis using artificial neural networks. In: 2012 IEEE business, engineering & industrial applications colloquium (BEIAC). https://doi.org/10.1109/BEIAC.2012.6226036
25. Chen H, Khan S, Kou B, Nazir S, Liu W, Hussain A (2020) A smart machine learning model for the detection of brain hemorrhage diagnosis based internet of things in smart cities. Hindawi Complex. 2020, Article ID 3047869:10. https://doi.org/10.1155/2020/3047869
26. Davis V, and Dr Devane S (2017) Diagnosis & classification of brain hemorrhage. In: IEEE international conference on advances in computing, communication and control (ICAC3). https://doi.org/10.1109/ICAC3.2017.8318764
27. Majumdar A, Brattain L, Telfer B, Farris C, Scalera J (2018) Detecting intracranial hemorrhage with deep learning. In: Annual International Conference of the IEEE Engineering in Medicine and Biology Society (EMBC) Jul 2018, pp 583– 587. https://doi.org/10.1109/EMBC.2018.8512336
28. RSNA Intracranial Hemorrhage Detection Challenge (2019). https://www.rsna.org/education/ai-resources-and-training/ai-image-challenge/rsna-intracranial-hemorrhagedetection-challenge-2019. Accessed 10 Dec 2021

A Pilot Study for Profiling Diabetic Foot Ulceration Using Machine Learning Techniques

Irena Tigga, Chandra Prakash, and Dhiraj

1 Introduction

Diabetes is a disorder of metabolism, in which there is Glucose in the bloodstream in enormous amounts and the body is not able to convert this glucose into energy. Diabetes is of two kinds—Type-1 Diabetes and Type-2 Diabetes. Type-1 Diabetes is a condition in which one's immune system destroys insulin-making cells present in the pancreas. In type-1 Diabetes there is very less production of insulin or no production of insulin. Type-2 Diabetes is a condition where the production of insulin is normal but the insulin receptors of the cells lose their sensitivity. Both conditions lead to the accumulation of glucose in enormous amounts. Ideally glucose should be consumed by the cell to produce energy. This condition is known as Hyperglycemia where the glucose value is greater than 140 mg/dl. The study focuses on Type-2 Diabetes which is considered to be difficult to cure. Long-term complications of diabetes include nephropathy leading to renal failure, retinopathy with potential loss of vision, peripheral neuropathy with risk of foot ulcers, amputations, Charcot disease, and autonomic neuropathy causing gastrointestinal, genitourinary, and cardiovascular symptoms. According to the International Diabetes Federation's statistics, there are over 425 million people worldwide who have Diabetes Mellitus. Heart disease, stroke, renal failure, blindness, and diabetic foot ulceration (DFU) are all serious consequences of diabetes mellitus (DM) [1]. The focus of this study is on complications regarding foot. Diabetic Foot is considered to be the most serious and

I. Tigga (✉) · C. Prakash
National Institute of Technology Delhi, Delhi 110040, India
e-mail: 202211009@nitdelhi.ac.in

C. Prakash
e-mail: cprakash@nitdelhi.ac.in

Dhiraj
Central Electronics Engineering Research Institute (CSIR-CEERI), Pilani 333031, India
e-mail: dhiraj@ceeri.res.in

© The Author(s), under exclusive license to Springer Nature Singapore Pte Ltd. 2023
H. Muthusamy et al. (eds.), *Robotics, Control and Computer Vision*, Lecture Notes in Electrical Engineering 1009, https://doi.org/10.1007/978-981-99-0236-1_5

costly complication of Diabetic Mellitus. Many people have to undergo amputation. This can be avoided if we can detect it at an early stage and provide the patient with the required treatment. Foot ulceration is very common in Diabetic Mellitus. As the tissue doesn't get enough energy and oxygen to heal wounds, this complicates the situation and later leads to amputation.

In diabetic individuals, foot ulcers are a severe life-threatening condition that is a primary key to amputation. The mortality rate is high, and ulcers that have healed sometimes resurface also. Amputation of a leg occurs in around 1 million diabetes patients each year [2]. Failure to identify the severity stage and establish a correct methodology for treatment planning could be the cause of amputation. The medical, economic, and social implications of these foot problems are significant.

Diabetic foot ulcers are currently diagnosed manually by clinicians. Approaches for early detection of diabetic foot includes: Dermatologic and Musculoskeletal; Vascular and Neurological Assessment. Dermatologic and Musculoskeletal Assessment includes examination regarding the change in skin color, temperature, and edema. Musculoskeletal assessment includes deformities that increase plantar pressure leading to skin breakdown. Vascular Assessment is done to check the blood flow in foot arteries and veins, Peripheral vascular disease (PVD) is a significant complication of diabetes and can produce changes in blood flow that will induce a change in skin temperature. Neurological Assessment includes pressure assessment with nylon filament Semmes–Weinstein monofilament test; vibration testing with a 128-Hz tuning fork; testing for pinprick sensation and ankle reflex assessment.

Thermograpy is a thriving technique which is used for various medical applications to diagnose various diseases [3]. In the manual pathway, an evaluation based on the patient's medical history, a comprehensive examination of the ulcer, and various medical tests such as X-rays, MRIs, and CT scans are used in the judgments on early identification and ulcer progression suppression. Diabetic foot ulcers cause swelling ankles and feet in patients. As a result, manual evaluations with medical equipment might be painful and inconvenient. Early diagnosis of people at risk of DFU may allow for earlier care to avoid foot ulcers, amputation, and death. Thermography is a non-invasive imaging technique that is used to detect thermal changes in diabetic feet [1, 4]. Several researches [1, 4] have proposed thermogram-based approaches for identifying persons at risk of DFU by recognizing a specific heat distribution in an infrared image.

As Diabetic foot leads to uneven temperature distribution in foot which is due to artery damage, this leads to increased interest in study of thermogram image of diabetic foot. Experts found Diagnosing Diabetic foot with the help of Plantar Thermogram is way more convenient, contactless, fast and non-intrusive. It helps to visualize plantar temperature distribution [5]. Earlier studies considered many patterns like—butterfly, whole high, inverse butterfly, inner high, whole low, forefoot low, tip toe low and anomaly [4]. However it has not been fully elucidated to what extent the individual variation of plantar thermogram pattern can show different trends between controlled and diabetic groups. Later novel classification done by the concept of foot angiosome was introduced. Classification divides plantar region into

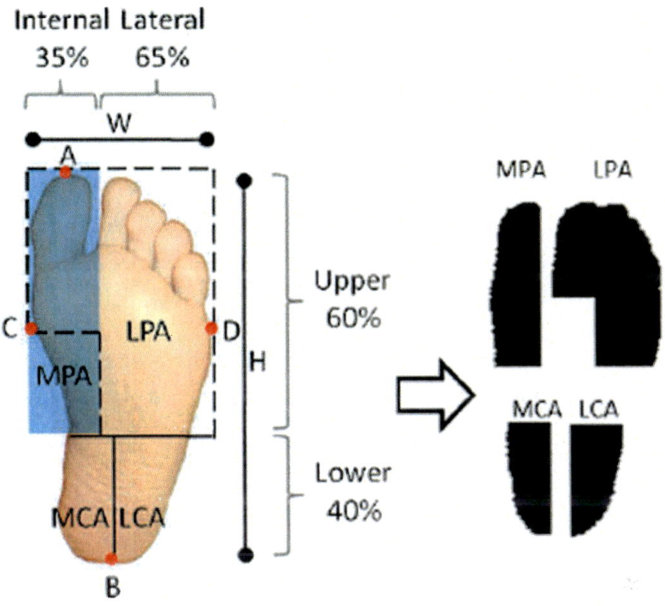

Fig. 1 Proportional foot divisions into plantar angiosomes [2]

four parts: MPA (medial plantar artery), LPA (lateral plantar artery), MCA (medial calcaneal artery), and LCA (lateral calcaneal artery) [1, 2]. As shown in Fig. 1.

The past work performed in DFU in the context of the application of Machine Learning and Deep Learning is mainly done over the thermogram image data where feature selection and extraction is being performed by deep learning models. Various deep learning models are provided with pre-trained data in order to get high accuracy [6]. Francisco et al. in 2017, Thermoregulation of healthy individuals, overweight–obese, and diabetic is discussed. In this paper, conventional foot assessment methods, infrared.

Muhammad et al. proposed computer-aided diagnosis of the diabetic foot using infrared Thermography. Different techniques for thermal image analysis are presented in this paper. Among them, asymmetric temperature analysis is a commonly used technique as it is simple to implement and yielded satisfactory results in previous studies. In 2019, Dineal et al., create Database of Plantar thermograms. It also discusses various challenges to capture and analyze thermogram data and provides a database which is composed of 334 individual thermograms from 122 diabetic subjects and 45 non-diabetic subjects. Each thermogram includes four extra images corresponding to the plantar angiosomes, and each image is accompanied by its temperature.

Many techniques had been used for processing thermogram patterns like spatial patterns, segmentation, active contour models, edge detection, and diffuse clustering [2]. Later further work was done on Image classification using Deep Learning where models like GoogLeNet and AlexNet performance was compared with ANN and

SVM. There are some issues that need to be addressed when DL is used these include the dataset size, the appropriate labeling of the samples, the segmentation and selection of Regions of Interest (ROIs), the use of pre-trained structures in the mode of transfer learning, or the design of a proper new learning-structure from scratch, among others [7]. Proper feature selection and appropriate hyper parameter adjustment can provide high-accuracy classification results using traditional ML techniques. In this study feature extraction, feature ranking, and Machine learning (ML) methods are explored.

This study provides a comparative analysis of various ML techniques when performed on the thermogram database for the DFU profile of the subjects [2]. Grid search provides the best hyper parameter for the models and helps to get high accuracy for Random Forest and SVM.

2 Methodology Proposed

This section presents the methodology proposed for the Profiling diabetic foot ulceration Using Machine Learning Techniques for Rehabilitation Fig. 2. Illustrates the methodology used in this pilot study.

In this methodology, firstly the dataset which is in the form of excel containing information for each individual (DM patients and the CG people) is preprocessed. In this dataset preprocessing includes the identification and treatment of missing values and encoding categorical data. The next step is Data Analysis and feature extraction, where data has been analyzed using various pandas libraries in order to figure out features correlation and relevance. The next step includes the application of ML models on processed data and thereafter applying Hyper parameter optimization in

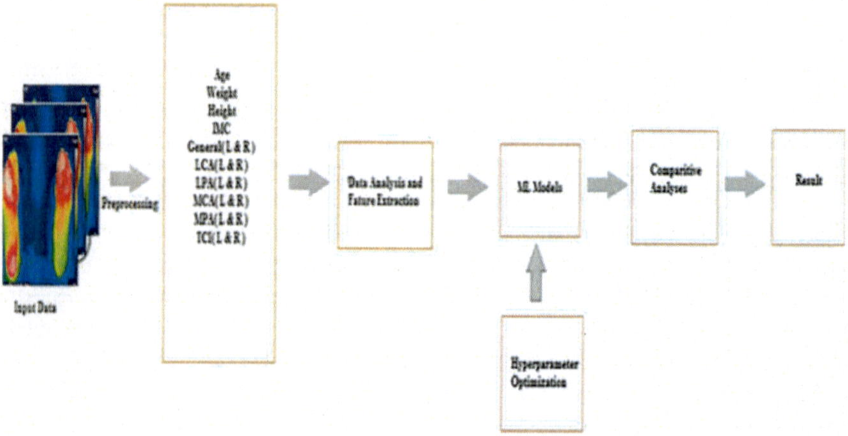

Fig. 2 Methodology used for the Profiling diabetic foot ulceration Using Machine Learning Techniques

order to get optimum results. Further steps describe the various comparative analyses done using different ML models and different ratios of training and test sets. Results give a clear understanding for various features and their role in different ML models, and thereby concluding which ML model provides optimum result with which set of features. This analysis points to important features that can be checked for abnormal temperature change in those regions which can be focused in order to avoid DFU at an early stage.

2.1 Dataset Used

Thermogram database [2] is used which contain features like Age Weight, Height, IMC, R_General, R_LCA, R_LPA, R_MCA, R_MPA, R_TCI, L_General, L_LCA, L_LPA, L_MCA, L_MPA, L_TCI, Result. The database is composed of 167 plantar thermograms, which were obtained from 122 diabetic subjects (referring here as DM-diabetes mellitus) and 45 non-diabetic subjects (referring as CG-Controlled Group).

The subjects were recruited from the General Hospital of the North, the General Hospital of the South, the BIOCARE clinic and the National Institute of Astrophysics, Optics and Electronics (INAOE) over a period of 3 years (from 2012 to 2014) [2]. There was much research done under capturing correct and accurate thermograms, posture and angle at which Thermogram took matters and In order to obtain accurate and useful thermograms for clinical practice, the recommendations of the International Academy of Clinical Thermology were followed [8]. The dataset consists of data in two formats; One format consists of a thermogram image where they provide csv containing temperature at each pixel this record has been maintained for each subject.

Dataset includes information about following:

- Gender,
- Age,
- Weight,
- Height,
- IMC (stands for BMI in french),
- R_General and L_General (general temperature of the foot. R_ represents the Right foot and L_ represents the Left foot),
- R_LCA and L_LCA (temperature value in celsius for the lateral calcaneal artery),
- R_LPA and L_LPA (temperature value in celsius for the lateral plantar artery),
- R_MCA and L_MCA (temperature value in celsius for the medial calcaneal artery)
- R_MPA and L_MPA (temperature value in celsius for the medial plantar artery)
- R_TCI and L_TCI (based on the mean differences between corresponding angiosomes of the foot from a diabetic subject)

Based on Fig. 3, output, Age, Weight, Height, IMC, R_General, R_MCA, R_MPA, L_MCA are selected as features for the study.

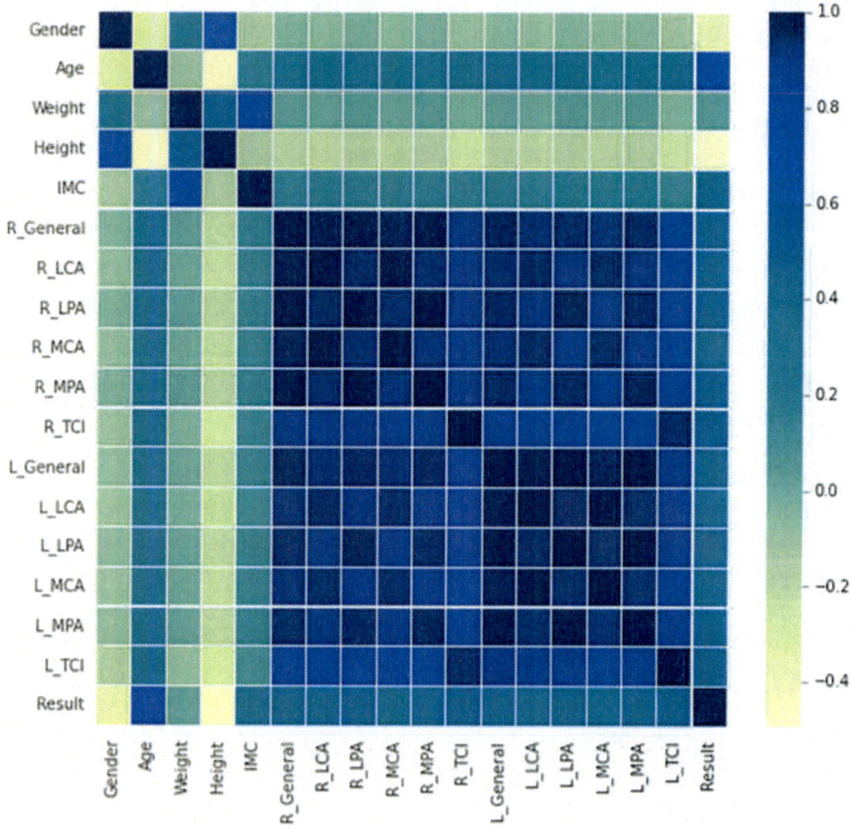

Fig. 3 Correlation matrix of features

2.2 *Machine Learning Used*

In this study, k-nearest neighbor, naïve bayes, decision tree, random forest, logistic Regression, Support vector machine (SVM), and ada boost methods are explored on the dataset for the independent feature analysis.

KNN: K nearest neighbors is a simple algorithm that stores all available cases and classifies new cases by a majority vote of its k neighbors. Various Distance functions can be used–Euclidean, Manhattan, Minkowski, and Hamming distance. The first three functions are used for continuous functions and the fourth one (Hamming) for categorical variables.

Naive Bayes Classifier: A Naïve Bayes Classifier assumes that the presence of a particular feature in a class is unrelated to the presence of any other feature. The model is easy to build and particularly useful for very large data sets. Bayes theorem provides a way of calculating posterior probability P(c|x) from P(c), P(x), and P(x|c).

Decision Tree: It is a type of supervised learning algorithm that is mostly used for classification problems. It works for both categorical and continuous dependent variables. Entropy and information gain are the building blocks of decision trees (Entropy is a metric for calculating uncertainty. Information gain is a measure of how uncertainty in the target variable is reduced, given a set of independent variables.) In this the population is divided into two or more homogeneous sets and is done based on the most significant attributes/ independent variables to make as distinct groups as possible. It is very important in the application of ML in mission-critical industries such as health: its ability to offer interpretable predictions to some degree that can also be introspective easily by humans. The decision tree of the dataset used is shown in Fig. 4.

Random Forest: It is a trademark term for an ensemble of decision trees. To classify a new object based on attributes, each tree gives a classification and we say the tree "votes" for that class. The forest chooses the classification having the most votes (over all the trees in the forest). The random forest employs the bagging method to generate the required prediction. Systematically generates a subset of data and attributes.

Logistic Regression: It is a linear regression model but the logistic regression uses a more complex cost function, this cost function can be defined as the "Sigmoid

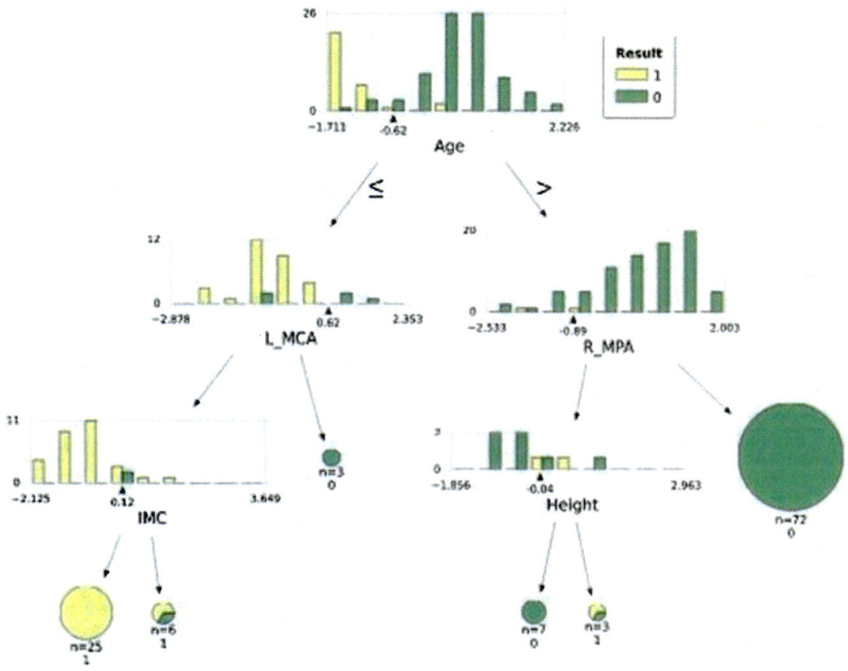

Fig. 4 Visualization of decision tree

function" or "logistic function". The sigmoid function/logistic function is a function that resembles an "S" shaped curve when plotted on a graph. It takes values between 0 and 1 and "squishes" them towards the margins at the top and bottom, labeling them as 0 or 1.

Support Vector Machine (SVM): It is a supervised machine learning algorithm that can be used for both classification and regression challenges. But it's mostly used for classification. Here in SVM we plot each data item as a point in n-dimensional space with the value of each feature being the value of a particular coordinate. Then, we perform classification by finding the hyper-plane that differentiates the two classes very well. Support Vectors are simply the coordinates of individual observation. The SVM algorithm has a technique called the kernel trick. The SVM kernel is a function that takes low dimensional input space and transforms it to a higher dimensional space.

AdaBoost: In the case of AdaBoost, higher points are assigned to the data points which are miss-classified or incorrectly predicted by the previous model. This means each successive model will get a weighted input. Later all the models are aggregated to develop the final model. The individual models are known as the weak learners.

The result of these models has been discussed in detail in the next section.

3 Result

The tenfold cross-validation procedure is used to evaluate each algorithm with a 70% and 30% training and testing data split, configured with the same random seed to ensure that the same splits to the training data are performed and that each algorithm is evaluated in precisely the same way. The result is illustrated in Table 1.

Figure 5 illustrates the accuracy score spread throughout each cross validation fold for each algorithm using a box and whisker plot. For machine learning approaches, grid search is used to find the best possible set of parameters. Table 1 shows that

Table 1 Accuracy for normal and k = tenfold cross-validation with parameters

ML techniques	Accuracy	Accuracy with k = tenfold	Hyper parameters setting
KNN	95.68	93.49	Euclidean distance, neighbors = 16
Naïve Bayes	76.47	93.41	–
Decision tree	94.77	93.41	Gini, max.depth = 11
Random forest	97.39	95.18	Gini, max.depth = 7
Logistic regression	95.65	93.45	–
SVM	93.93	95.22	C = 1, degree = 3, gamma = 0.1, kernel = sigmoid
Ada boost	98.25	94.63	Gini, max.depth = 2

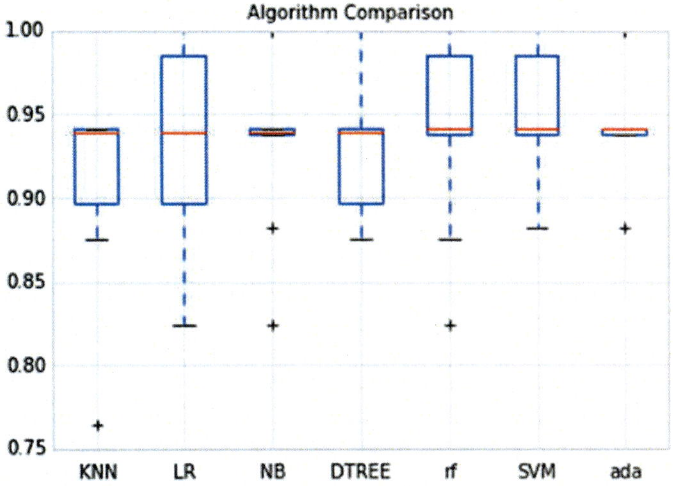

Fig. 5 Algorithm comparison using tenfold cross validation

the Random Forest and SVM classification methods provide higher classification accuracy for tenfold cross validation.

In this pilot study, five cases have been considered on the dataset used with training and testing split. The hypothesis is that machine learning accuracy and dataset features are not correlated.

Case 1 consists of 10% training and the rest 90% as testing data. In case 2, it is 30 and 70% respectively. This ratio is 50% for both training and testing in case 3. Case 4 comprises 70% training and 30% testing followed by 90 and 10% for case 5 as shown in Table 2.

Table 3 shows the results of five cases over k-Nearest Neighbor, Naïve Bayes, Decision Tree, Random Forest, Logistic Regression, Support Vector Machine (SVM), and Adaboost methods. Random Forest and Logistic Regression are able to classify the DFU profile even with only 10% of data split as the training set. Naïve Bayes and Logistic Regression are also able to perform well in Case 2. When the split was 50% for both training and testing, the Decision Tree accuracy was 96.42. This suggests that the accuracy of the DFU profiling may be independent of the dataset size because features are prominent identifiers.

Figure 6 shows the support feature for the machine learning technique and which is not relevant for the respective split ratio with five cases. Age is a major factor

Table 2 Cases consider for result analysis with respect to different training and test spits

Case 1		Case 2		Case 3		Case 4		Case 5	
Train data	Test data	Train data	Test data	Train data	Test data	Train data	Test data	Train data	Test data
10	90	30	90	50	50	70	30	90	10

Table 3 Accuracy for five cases considered in the study over machine learning techniques

ML technique	Case 1	Case 2	Case 3	Case 4	Case 5
KNN	72.8	93.16	94.04	88.25	94.11
Naïve Bayes	86.09	95.72	92.85	76.47	88.23
Decision tree	92.71	92.3	96.42	94.11	88.23
Random forest	94.03	92.3	95.23	94.11	100
Logistic regression	94.03	95.72	92.85	92.15	100
SVM	92.05	94.87	91.66	92.15	91.66
Ada boost	92.71	88.88	94.04	92.15	100

that is a prominent feature in DFU profiling. This is in correlation with the standard factor responsible for the diabetic foot. It can be concluded that Height, weight, IMC(BMI), and R_general are major factors contributing to the accuracy of the model for profiling.

Table 4 presents the detailed effect of features on accuracy for five cases with respect to the machine learning techniques used. This analysis can help in localizing the foot regions which are more sensitive toward ulcer formation.

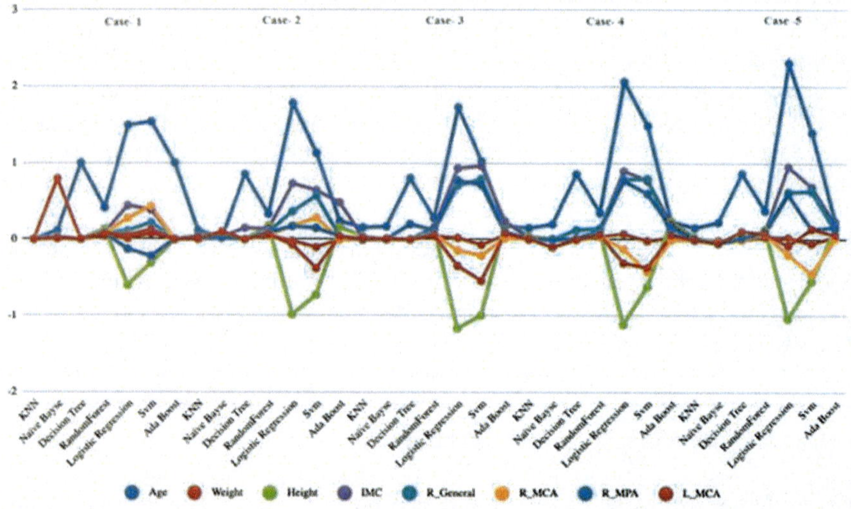

Fig. 6 Feature importance across various ML techniques under five different cases

Table 4 Effect of features on accuracy for five cases with respect to the machine learning techniques used

	ML technique	Age	Weight	Height	IMC	R_General	R_MCA	R_MPA	L_MCA
Case 1	KNN	0	0	0	0	0	0	0	0
	Naïve Bayes	0.11	0.79	0.03	0.029	0.012	0.011	0.01	0.008
	Decision tree	1	0	0	0	0	0	0	0
	Random forest	0.41323	0.04004	0.13328	0.09129	0.09398	0.0756	0.07495	0.07667
	Logistic regression	1.49027	0.00953	−0.60241	0.43996	0.10717	0.27533	−0.129	0.05483
	SVM	1.53529	0.07203	−0.31452	0.3811	0.21512	0.43094	−0.2202	0.1254
	Ada boost	1	0	0	0	0	0	0	0
Case 2	KNN	0.11257	0	0.04311	0.03832	0.00479	0	0.00359	0.00599
	Naïve Bayes	0.022	0.049	0.03	0.03	0.025	0.0175	0.016	0.1
	Decision tree	0.85303	0	0	0.14697	0	0	0	0
	Random forest	0.33165	0.04995	0.17966	0.13545	0.09977	0.0493	0.09017	0.06405
	Logistic regression	1.77988	−0.01645	−0.98581	0.72805	0.36352	0.1799	0.17141	−0.07007
	SVM	1.13259	−0.1013	−0.72769	0.64185	0.56792	0.28346	0.14939	−0.37955
	Ada boost	0.24	0	0.16	0.48	0.04	0	0.04	0.04
Case 3	KNN	0.15569	0.01198	0.04072	0.02994	−0.00599	−0.00359	−0.0012	0.00958
	Naïve Bayes	0.17	0.01	0	0	0	0	0	0.01
	Decision tree	0.80013	0	0	0	0	0	0.19987	0
	Random forest	0.27614	0.04561	0.16	0.15484	0.11733	0.06546	0.12208	0.05853
	Logistic regression	1.724	0.02271	−1.16841	0.93483	0.79002	−0.20745	0.71247	−0.54154
	SVM	1.02314	−0.06888	−0.99766	0.96458	0.79002	−0.20745	0.71247	−0.54154
	Ada boost	0.18	0.06	0.22	0.24	0.1	0.02	0.1	0.08

(continued)

Table 4 (continued)

	ML technique	Age	Weight	Height	IMC	R_General	R_MCA	R_MPA	L_MCA
Case 4	KNN	0.1497	0.0012	0.04671	0.02275	0.00838	0.01198	0.00719	0.00838
	Naïve Bayes	0.2	0.01	0.005	0	0	−0.01	−0.01	−0.1
	Decision tree	0.85736	0	0	0	0.12439	0	0.01825	0
	Random forest	0.34974	0.04069	0.12047	0.11784	0.14628	0.04633	0.11341	0.06525
	Logistic regression	2.06001	0.0744	−1.11712	0.89576	0.78607	−0.12007	0.76832	−0.30817
	SVM	1.48522	−0.01443	−0.61837	0.77172	0.79232	−0.42062	0.5987	−0.37048
	Ada boost	0.24	0.02	0.24	0.2	0.08	0	0.16	0.06
Case 5	KNN	0.15329	-0.0012	0.03234	-0.00359	0.01317	0.0012	0.01317	−0.0012
	Naïve Bayes	0.225	−0.01	−0.04	−0.04	−0.04	−0.04	−0.05	−0.05
	Decision tree	0.86431	0	0	0	0	0	0.0278	0.10789
	Random forest	0.37902	0.03398	0.13559	0.10975	0.09956	0.07276	0.08811	0.08122
	Logistic regression	2.30622	0.03049	−1.0383	0.95525	0.62142	−0.19123	0.59201	−0.07688
	SVM	1.3981	−0.04961	−0.54835	0.68762	0.63167	−0.44485	0.14139	0.14171
	Ada boost	0.24	0.04	0.16	0.2	0.12	0.02	0.16	0.06

4 Discussion

By the above analysis we can infer that taking high ratios for training data sets leads to overfitting of various classifiers. And taking low training data as compared to test data leads some classification to predict with high accuracy but internally they have used only one feature or very less feature for classification. For example while performing Decision Tree Classification over Plantar Thermogram database when the split is set to be 10% for training and 90% for testing. The classifier gives 100% accuracy when applying grid search over it, this was because age was the only feature used for the creation of a decision tree. But as we know we cannot classify straight forward by only one feature, also the feature age which is very common information and doesn't have much significance with the medical problem of diabetic foot.

Naive Bayes accuracy falls as we increase ratio for training dataset in the split, this shows that this model is not suitable for classifying this dataset. The split ratio of 70:30 is best situated as mostly all ML models are performing well. Random forest and SVM performs better for Plantar Thermogram Database.

The earlier finding worked on image data which purely work on pattern, which is a complex process as DFU leads to foot deformation and the pattern isn't fixed. This analysis can help in localizing the foot regions which are more sensitive towards ulcer formation. This can be inferred from the feature importance for each ML model and also provide a clear idea of which features are more important to be recorded and which data split ratio helps to get optimum result.

5 Conclusion

This study shows that detecting diabetic foot in an early stage with the help of Thermogram data is a very good approach. The Thermogram data consists of the temperature of four angiosome regions of the plantar area along with personal details like age, weight, etc. The paper comes up with the conclusion that using diabetic foot thermogram data as input for machine learning technique in order to classify the Diabetes Mellitus Group and Control Group and keeping the 70:30 ratio for training the dataset gives a balanced result which is free from overfitting and underfitting. Results indicate that Random Forest is performing better and all the features have positive importance in the Random Forest machine learning technique. The accuracy and performance of the model can further be increased by hyper parameter tuning methods.

References

1. Cajacuri LAV (2014) Early diagnostic of diabetic foot using thermal images. HAL 11 Jul 2014
2. Hernandez-Contreras DA, Peregrina-Barreto H, de Jesus Rangel-Magdaleno J, Renero-Carrillo F-J (2019) Plantar thermogram database for the study of diabetic foot complications. IEEE Access. (4 Nov 2019)
3. Lahiri BB, Bagavathiappan S, Jaya kumar T, Philip J (2012) Medical applications of infrared thermography: a review. Infrared Phys Technol 55(4). (July 2012)
4. Mori T, Nagase T, Takehara K, Oe M, Ohashi Y, Amemiya A, Noguchi H, Ueki K, Kadowaki T, Sanada H (2013) Morphological pattern classification system for plantar thermography of patients with diabetes. J Diabetes Sci Technol 7(5). (September 2013)
5. Adam M, Ng EYK, Tan JH, Heng ML, Tong JWK, Acharya UR (2017) Computer aided diagnosis of diabetic foot using infrared thermography: a review. (25 Oct 2017)
6. Gamage C, Wijesinghe I, Perera I (2019) Automatic scoring of diabetic foot ulcers through deep CNN based feature extraction with low rank matrix factorization. In: 2019 IEEE 19th international conference on bioinformatics and bioengineering (BIBE). https://doi.org/10.1109/bibe.2019.00069
7. Cruz-Vega I, Hernandez-Contreras D, Peregrina-Barreto H, de Jesus Rangel-Magdaleno J, and Ramirez-Cortes JM (2020) Deep learning classification for diabetic foot thermograms. (Mar 2020)
8. International Academy of Clinical Thermology (2002) Thermography guidelines: standards and protocols in clinical thermographic imaging. Redwood City, CA, USA
9. Peregrina-Barreto H, Morales-Hernandez LA, Rangel-Magdaleno JJ, Avina-Cervantes JG, Ramirez-Cortes JM, Morales-Caporal R (2014) Quantitative estimation of temperature variations in plantar angiosomes: a study case for diabetic foot. In: Computational and mathematical methods in medicine, vol 2014
10. Renero-C FJ (2017) The thermoregulation of healthy individuals, overweight–obese, and diabetic from the plantar skin thermogram: a clue to predict the diabetic foot, vol 8

A Deep Learning Approach for Gaussian Noise-Level Quantification

Rajni Kant Yadav, Maheep Singh, and Sandeep Chand Kumain

1 Introduction

Image noise removal has been an active topic of research in the domain of image processing. Noise in image processing is a random variation of brightness or color in images that do not portray the true information of the image. The presence of noise in an image alters the true value of pixels and causes a loss of information which is a disadvantage to image processing. A few common types of noises that can be found in images are Gaussian noise, Salt and Pepper noise, Speckle noise and more [1–3]. Noise may be introduced in the image during capturing, transmission, or due to electrical faults in the capturing device [4–6].

Noise reduction techniques have been a domain of extensive study over the last few years. Most of these studies are focused on additive white gaussian noise as it is one of the most common types of noise present in an image. It is hard to object that these techniques have proven to be very helpful in Digital Image Processing (DIP) [7]. However, these techniques are based on the assumption that the images to be processed are noisy. The possibility of the image being noise-free is being ignored. Almost all of the aforementioned methods suffer in determining if the images are corrupted by noise and therefore another processing overhead where the noisy images have to be sorted out manually in advance. Therefore, noise quantification also becomes a necessary step in image denoising. The development of a Gaussian noise quantification model is the main interest of the author(s). In this research article,

R. K. Yadav (✉) · M. Singh · S. C. Kumain
Department of Computer Science and Engineering, National Institute of Technology, Srinagar, Uttarakhand, India
e-mail: mt20cse004@nituk.ac.in

M. Singh
e-mail: maheepsingh@nituk.ac.in

S. C. Kumain
e-mail: sandeep.chandphd2021@nituk.ac.in

© The Author(s), under exclusive license to Springer Nature Singapore Pte Ltd. 2023
H. Muthusamy et al. (eds.), *Robotics, Control and Computer Vision*, Lecture Notes in Electrical Engineering 1009, https://doi.org/10.1007/978-981-99-0236-1_6

the author(s) are presenting a Convolutional Neural Network (CNN) model which is inspired by LeNet and AlexNet architecture [8, 9]. The proposed model will help to identify and apply the appropriate algorithm based on the amount of noise available in the image.

The paper is further organized as follows. Section 2 is a brief review of related work. Section 3 introduces the proposed model. Section 4 delineates the experimental results. Finally, Sect. 5 concludes the work and talks about the future scope of the work.

2 Related Work

A lot of work has been done in image noise reduction so far. An image-denoising collaborative filtering method using sparse 3D transform domain is proposed by Dabov et al. [10]. The Non-Local Mean (NLM) technique such as Block Matching and 3D filtering (BM3D) [11] is one of the powerful image-denoising techniques used by the researchers. Another Prefiltered Rotationally Invariant Non-Local Means 3D (PRI-NLM3D) technique is proposed by Manjon et al. [12]. This NLM-based denoising technique provided good accuracy scores in terms of Peak Signal-to-Noise Ratio (PSNR), Structural Similarity Index Measure (SSIM), and Universal Image Quality Index (UQI) measures for denoising Magnetic Resonance(MR) images. Gondara et al. [13] proposed a deep learning approach for medical image denoising based on a convolutional autoencoder. This model was compared with NLM and median filter and yielded better SSIM scores for a small training sample of 300.

However, less work has been done on identifying the type and amount of noise present in the image. Quantification of the noise is no less important than noise reduction. For identifying the type of noise present in an image a voting-based deep CNN model is proposed by Kumain et al. [14]. This model is only giving information about the type of noise present in an image. For quantifying the Gaussian noise present in an image Chauh et al. [15] proposed a deep learning approach based on CNN. This CNN method quantified the Gaussian noise into ten classes with the noise levels of $\sigma = 10, 20, 30, 40, 50, 60, 70, 80$, and 90 to corrupt the image, and achieved an accuracy of 74.7%. A noise classifier based on CNN was proposed by Khaw et al. [16] utilizing the Stochastic Gradient Descent (SGD) optimization technique. The noisy image was fed as input to the model and based on the distinctive features extracted by the sequence of convolutional and pooling layers. The CNN classification methods have yielded excellent results in several domains such as Handwritten Character Recognition or Character Classification [17], Vehicle, Logo Recognition [18], Face Classification [19], and Bank Notes Series identification [20].

In the real-world scenario, if the clean image is unavailable and only the noisy image is available then the performance parameters such as PSNR and SSIM fail to work. So, quantification of noise level becomes necessary. The author(s) have proposed a CNN model for Gaussian noise quantification. The next section describes the same.

3 Proposed Model

In this section, the architecture of the proposed model for the quantification of Gaussian noise has been discussed. The model architecture is shown in Fig. 1.

The author(s) developed a noise quantification model which is based on the deep learning technique. The proposed model is inspired by the LeNet and AlextNet architecture [8, 9]. The proposed architecture is addressing the multiclass classification problem and classifies the input image into 11 different classes where the 10 classes represent images corrupted by 10 different levels of Gaussian noise with a mean zero and 1 class represents a noise-free image. The specification of the dataset utilized for the model training and testing as per the prepared dataset discussed in the experimental analysis section.

CNN has been used to develop the classifier model. Input images are resized to $256 \times 256 \times 3$ and perturbed by Gaussian noise levels of variance 0.01, 0.02, 0.03, 0.04, 0.05, 0.1, 0.2, 0.3, 0.4, and 0.5. The noisy image is fed to the first Convolutional (Conv2D) layer. The filter values are adjusted through backpropagation repeatedly to get the optimum set of values for the best classification accuracy. The model has a series of four alternating Conv2D and MaxPool layers. MaxPool layer is used to subsample the feature maps. Further, the author(s) utilized the dropout layer to reduce the overfitting of the model. In the model development process, the author(s) utilized Rectified linear unit (ReLU) as an activation function and Adam as an optimizer. The

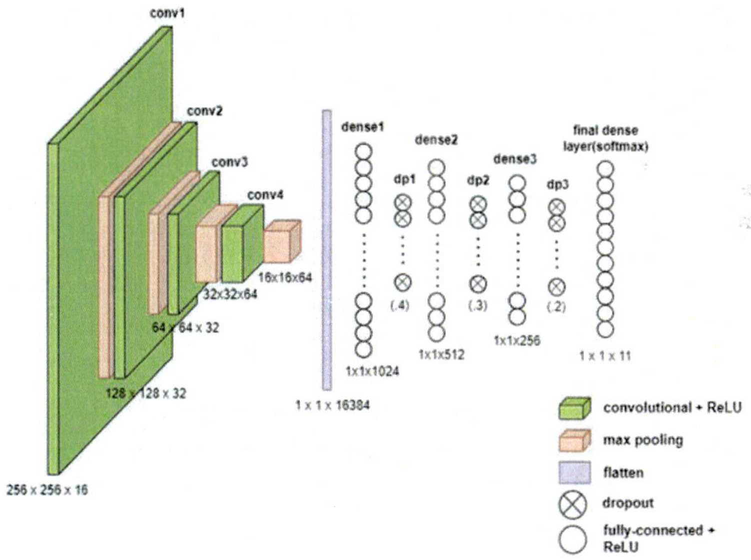

Fig. 1 Architecture of the proposed model depicting the CNN layers used

Table 1 Model Description with Parameters of each CNN layer

S.no.	Operation	Kernal size	Stride	Parameters
1	Input Image (256, 256,3)	–	–	–
2	Convolution + ReLU	3*3 @ 16	1	448
3	Max – Pooling	2*2	2	0
4	Convolution + ReLU	3*3 @ 32	1	4640
5	Max – Pooling	2*2	2	0
6	Convolution + ReLU	3*3 @ 32	1	9248
7	Max – Pooling	2*2	2	0
8	Convolution + ReLU	3*3 @ 64	1	18496
9	Max – Pooling	2*2	2	0
10	Flatten			0
11	Dense Layer 1 (1024) + ReLU			16778240
12	Dropout (.40)			0
13	Dense Layer 2 (512) + ReLU			524800
14	Dropout (.30)			0
15	Dense Layer 3 (256) + ReLU			131328
16	Dropout (.20)			0
17	Dense Layer 4 (11) + Softmax			2827
Total Trainable Parameter: 17,470,027				

softmax function is utilized in the final dense layer to predict probabilities for each class of image. The class with the highest probability will be chosen. The model summary with the number of trainable parameters is as per Table 1.

4 Experimental Results and Analysis

This section comprises the steps utilized for dataset preparation and the classification results. Section 4.1 talks about the dataset preparation. Sect. 4.2 is about the performance parameters used for evaluation, and Sect. 4.3 describes the experimental results.

4.1 Dataset Preparation

In the dataset preparation process, due to the non-availability of the specific dataset, the noisy dataset was prepared by incorporating the Gaussian noise at different levels of variance with 0 mean. First, the 2000 images are taken randomly from the MSRA10K [21] dataset, and further, noise is incorporated. For model training, 70% of the sample is utilized. The remaining data was split into validation and testing set with 15% data in each. Along with the noise-free class, a total of 11 classes were created. A brief description of the dataset is as per Table 2.

4.2 Performance Parameters

For overall evaluation, the classification report [22] and confusion matrix [23] have been used. The key terms for this are as follows.

(a) Precision: Precision is the measure that out of the total predicted positives for a class, how many are actually positive. The equation for precision is as follows:

$$precision = \frac{True\ Positive}{True\ Positive + False\ Positive} \tag{1}$$

(b) Recall: Recall is the measure of how many positives were correctly classified, out of the total number of positives for a particular class. The equation for recall is as follows:

$$recall = \frac{True\ Positive}{True\ Positive + False\ Negative} \tag{2}$$

(c) F1-Score: It is the weighted harmonic average between Precision and Recall. The best score is represented by 1 and the worst score is represented by 0.

Table 2 Description of the noisy dataset

S.no	Description
1	Total number of clean images are: 2000
2	Noise variance level: 0.01, 0.02, 0.03, 0.04, 0.05, 0.1, 0.2, 0.3, 0.4, 0.5
3	Total training sample images: $1400 * 11 = 15400$
4	Total validation sample images: $300 * 11 = 3300$
5	Total testing sample images: $1400 * 11 = 3300$

(d) Confusion Matrix: It is a matrix that represents the result in the form of a table. The diagonals represent true positives for the corresponding class. The rows represent actual values and the columns represent predicted values.

4.3 Experimental Results

During the training of the model, the ModelCheckpoint and EarlyStopping [24] functionality of *keras*(python) library were used to get the model with the best validation accuracy. Since it is difficult to estimate the exact number of epochs, EarlyStopping was utilized and the patience level was set to 50 during the model training. Figures 2 and 3 are showing the model training/validation accuracy and training and validation loss, respectively.

After analyzing the accuracy and loss graph for training and validation, the effect of the dropout layer during the model development process was observed. The dropout layer is useful for solving the problem of overfitting. However, it can be seen that there are fluctuations in the accuracy and loss graph. Nevertheless, the author(s) have been able to save the best model using the ModelCheckpoint [24] feature of *keras*(python) library. The best model was achieved at epoch 116. Due to a patience level of 50, the training process automatically ended in epoch 166. The best-saved model was used to yield the accuracy of the model on the test set.

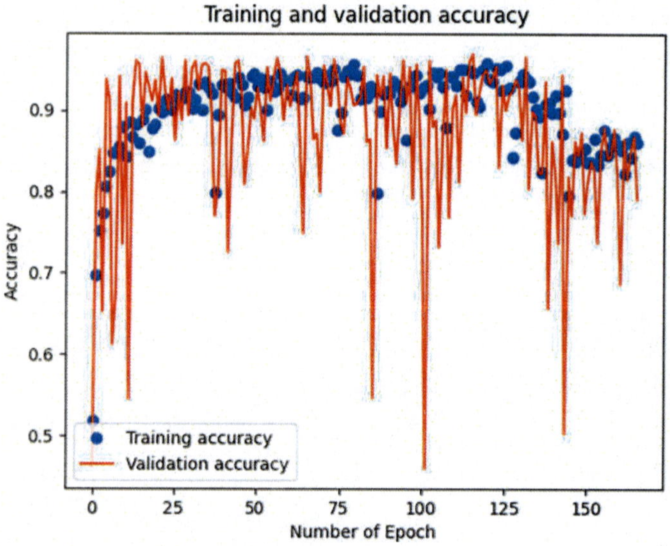

Fig. 2 Accuracy graph for training and validation phases

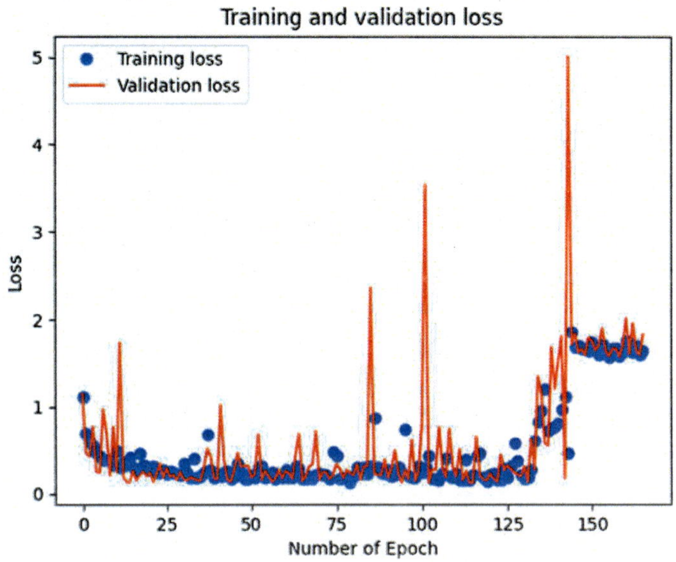

Fig. 3 Loss graph for training and validation phases

Class	0	1	2	3	4	5	6	7	8	9	10
0	299	0	1	0	0	0	0	0	0	0	0
1	0	298	0	2	0	0	0	0	0	0	0
2	2	0	285	0	13	0	0	0	0	0	0
3	0	0	0	299	0	1	0	0	0	0	0
4	0	0	1	0	293	0	6	0	0	0	0
5	0	0	0	2	0	296	0	2	0	0	0
6	0	0	0	0	5	0	273	0	22	0	0
7	0	0	0	0	0	7	0	292	0	1	0
8	0	0	0	0	0	0	12	0	288	0	0
9	0	0	0	0	0	0	0	41	0	259	0
10	0	0	0	0	0	0	0	0	0	0	300

Fig. 4 Confusion matrix for the proposed model

The experimental results as per the performance parameters discussed above are shown in Figs. 4 and 5. The proposed model has shown better results compared to Chau et al. [13]. The author(s) in the aforementioned paper used standard images from the USC-SIPI dataset [25]. Noise levels of $\sigma = 10, 20, 30, 40, 50, 60, 70, 80$, and 90 were used. This paper achieved an accuracy of 74.7%, whereas the quantification model present in this paper achieved 96% accuracy, which is much higher.

Noise Level Variance	Class	Precision	Recall	F1-Score	Support
0.01	0	0.99	1.00	1.00	300
0.10	1	1.00	0.99	1.00	300
0.02	2	0.99	0.95	0.97	300
0.20	3	0.99	1.00	0.99	300
0.03	4	0.94	0.98	0.96	300
0.30	5	0.97	0.99	0.98	300
0.04	6	0.94	0.91	0.92	300
0.40	7	0.87	0.97	0.92	300
0.05	8	0.93	0.96	0.94	300
0.50	9	1.00	0.86	0.92	300
Noise Free	10	1.00	1.00	1.00	300
	Accuracy			**0.96**	**3300**

Fig. 5 Classification report for the proposed model

5 Conclusion and Future Work

In this research article, the author(s) have proposed a model for quantification of the Gaussian noise present in the image. Using quantitative parameters such as SSIM, PSNR, it is not possible to identify the strength of noise reduction, if the clean image is not available. This quantification model can help evaluate a denoising model in terms of the amount of noise it has denoised when there is no clean image available. The author(s) here have addressed only 11 classes for this quantification task. The work can be further extended by incorporating more levels of noise and developing a generalized model which will address other types of noise as well.

References

1. Ambulkar S, Golar P (2014) A review of decision based impulse noise removing algorithms. Int J Eng Res Appl 4:54–59
2. Verma R, Ali J (2013) A comparative study of various types of image noise and efficient noise removal techniques. Int J Adv Res Comput Sci Softw Eng 3(10)
3. Singh M, Govil MC, Pilli ES, Vipparthi SK (2019) SOD-CED: salient object detection for noisy images using convolution encoder-decoder. IET Comput Vision 13(6):578–587
4. Hosseini H, Hessar F, Marvasti F (2015) Real-time impulse noise suppression from images using an efficient weighted-average filtering. IEEE Signal Process Lett 22:1050–1054
5. Bovik A (2000) Handbook of image and video processing, 2nd ed. Elsevier Academic Press
6. Kumain SC, Singh M, Singh N, Kumar K (2018) An efficient Gaussian noise reduction technique for noisy images using optimized filter approach. In: 2018 first international conference on secure cyber computing and communication (ICSCCC), pp 243–248
7. Chang SG, Bin Y, Vetterli M (2000) Adaptive wavelet thresholding for image denoising and compression. IEEE Trans Image Process 9:1532–1546

8. LeCun Y, Bottou L, Bengio Y, Haffner P (1998) Gradient-based learning applied to document recognition. Proc IEEE 86(11):2278–2324
9. Krizhevsky A, Sutskever I, Hinton GE (2012) Imagenet classification with deep convolutional neural networks. Adv Neural Inf Process Syst 25:1097–1105
10. Dabov K, Foi A, Katkovnik V, Egiazarian K (2007) Image denoising by sparse 3-D transform-domain collaborative filtering. IEEE Trans Image Process 16(8):2080–2095
11. Bhujle HV, Vadavadagi BH (2019) NLM based magnetic resonance image denoising-a review. Biomed Signal Process Control 47:252–261
12. Manjón JV, Coupé P, Buades A, Collins DL, Robles M (2012) New methods for MRI denoising based on sparseness and self-similarity. Med Image Anal 16(1):18–27
13. Gondara L (2016) Medical image denoising using convolutional denoising autoencoders. In: 2016 IEEE 16th international conference on data mining workshops (ICDMW), pp 241–246
14. Kumain SC, Kumar K (2021) VBNC: voting based noise classification framework using deep CNN. In: Conference proceedings of ICDLAIR2019, pp 357–363
15. Chuah JH, Khaw HY, Soon FC, Chow CO (2017) Detection of Gaussian noise and its level using deep convolutional neural network. In: TENCON 2017-2017 IEEE region 10 conference, pp 2447–2450. (Nov 2017)
16. Khaw HY, Soon FC, Chuah JH, Chow CO (2017) Image noise types recognition using convolutional neural network with principal components analysis. IET Image Proc 11(12):1238–1245
17. LeCun Y, Boser BE, Denker JS, Henderson D, Howard RE, Hubbard WE, Jackel LD (1990) Handwritten digit recognition with a back-propagation network. In: Advances in neural information processing systems, pp 396–404
18. Huang Y, Wu R, Sun Y, Wang W, Ding X (2015) Vehicle logo recognition system based on convolutional neural networks with a pretraining strategy. IEEE Trans Intell Transp Syst 16(4):1951–1960
19. Lawrence S, Giles CL, Tsoi AC, Back AD (1997) Face recognition: a convolutional neural-network approach. IEEE Trans Neural Netw 8(1):98–113
20. Feng BY, Ren M, Zhang XY, Suen CY (2014) Automatic recognition of serial numbers in bank notes. Pattern Recogn 47(8):2621–2634
21. MSRA10K Dataset. https://mmcheng.net/msra10k/. Accessed 01 Dec 2021
22. Classification Report. https://muthu.co/understanding-the-classification-report-in-sklearn/. Accessed 01 Dec 2021
23. Confusion Matrix. https://www.geeksforgeeks.org/confusion-matrix-machine-learning/. Accessed 05 Dec 2021
24. Callbacks API. https://keras.io/api/callbacks/. Accessed 05 Dec 2021
25. USC-SIPI Dataset. https://sipi.usc.edu/database/. Accessed 05 Dec 2021

Performance Evaluation of Single Sample Ear Recognition Methods

Ayush Raj Srivastava and Nitin Kumar

1 Introduction

Biometrics [1] are physical or behavioral characteristics that can uniquely identify a human being. Physical biometrics include face, eye, retina, ear, fingerprint, palmprint, periocular, footprint, etc. Behavioral biometrics include voice matching, signature and handwriting, etc. There have been several applications [1] of biometrics in diverse areas such as ID cards, surveillance, authentication, security in banks, airports and corpse identification. Ear [2] is a recent biometric which has drawn the attention of the research community. This biometric possesses certain characteristics which distinguish it from other biometrics, e.g. less amount of information is required than the face, where the person is standing in a profile manner facing the camera, face recognition do not perform satisfactorily. Further, no user cooperation is required for ear recognition as required by other biometrics such as iris and fingerprint.

The ear is one of those biometrics whose permanence attribute is very high. Unlike our face which changes considerably throughout our life, the ear experiences very less changes. Further, it is fairly collectible and in the post-covid scenario, it can be considered as a safer biometric since the face and hands are covered with masks or gloves. It can be more acceptable if we do not bother a user for more number of samples. In a real-world scenario, the problem of ear recognition becomes more complex when only a single training sample is available. Under these circumstances, one sample per person (OSPP) [3] architecture is used. This methodology has been highlighted in the research community over all the problem domains such as face recognition [3, 4], ear recognition [5] and other biometrics. The reason OSPP is popular is the preparation of the dataset; specifically, the collection of the sample from the source is very easy. However, recognition becomes more complex due to the lack of samples. Hence, the model cannot be trained in the best possible manner.

A. R. Srivastava · N. Kumar (✉)
NIT Uttarakhand, Srinagar Uttarkhand 246174, India
e-mail: nitin@nituk.ac.in

© The Author(s), under exclusive license to Springer Nature Singapore Pte Ltd. 2023
H. Muthusamy et al. (eds.), *Robotics, Control and Computer Vision*, Lecture Notes in Electrical Engineering 1009, https://doi.org/10.1007/978-981-99-0236-1_7

There are several methods suggested in the literature by researchers for addressing OSPP for different biometric traits. Some of the popular methods include Principal Component Analysis (PCA), Kernel PCA, Wavelet transformation, Fourier transformation with frequency component masking and wavelet transformation using subbands. These methods have been employed for different biometrics and under different experimental settings. However, it is not clear which method performs best for ear recognition under a single training sample. Hence, there is a need to compare the performance of the aforementioned methods for ear recognition. In this paper, the performance of all the aforementioned methods is compared on three standard publicly available datasets *viz.*, Indian Institute of Technology-Delhi (IIT-D) [6], Mathematical Analysis of Images (AMI) [7] and Annotated Web Ears (AWE) [8].

The rest of the paper is organized as follows: Sect. 2 reviews the methods available in the literature briefly. Section 3 describes the single sample ear recognition methods whose performance is compared in this paper. Experimental setup and results are given in Sect. 4. Finally, conclusion and future work are given in Sect. 5.

2 Related Work

PCA method was used for ear recognition by Zhang et al. [9] in 2008. This method extracted local as well as global features. Linear Support Vector Machine (SVM) was used for classification. Later in 2009, Long et al. [10] proposed using wavelet transformations for ear recognition. The proposed method was better than PCA and Linear Discriminant Analysis(LDA) [11] previously implemented. In 2011, Zhou et al. [12] used the color Scale Invariant Feature Transform (SIFT) method for representing the local features. In the same year, Wang et al. [13] employed an ensemble of the local binary pattern (LBP), direct LDA (linear discriminant analysis) and waterlet transformation methods for recognizing ears. The method was able to give accuracy up to 90% depending upon the feature dimension given as input. A robust method for ear recognition was introduced in 2012 by Yuan et al. [14]. They proposed an ensemble method of PCA, LDA and random projection for feature extraction and a sparse classifier for classification. The proposed was able to recognize partially occluded image samples. In 2014, Taertulakarn et al. [15] proposed ear recognition based on Gaussian curvature-based geometric invariance. The method was particularly robust against geometric transformations. In the same year, an advanced form of wavelet transformation along with discrete cosine transformation was introduced by Ying et al. [16]. The wavelet used weighted distance which highlighted the contribution of low-frequency components in an image.

In 2016, Ling et al. [17] used Deep Neural Network for ear recognition. The proposed method also took advantage of CUDA cores for training the model. The final model was quite accurate against hair-, pin- and glass-occluded ear image. The same year, the **One Sample Per Person** (OSPP) problem for ear biometric was tackled by Long et al. [18]. This method used an adaptive multi-keypoint descriptor sparse representation classifier. This method was occlusion-resistant and better than contemporary methods. The recognition time was a little high in the band of 10–12 s. In 2017, Emersic et al. [8] introduced an extensive survey of methods of ear recognition. In this paper, different divisions were suggested for recognition approaches depending on the technique used for feature extraction *viz.*, holistic, geometric, local and hybrid. Holistic approaches describe the ear with global properties. In this approach, the ear sample is analyzed as a whole and local variations are not taken into consideration. Methods using geometrical characteristics of the ear for feature representation are known as geometric approaches. Geometric characteristics of the ear include the location of specific ear parts, shape of ear, etc. Local approaches describe local parts or the local appearance of the ear and use these features for the purpose of recognition. Hybrid approaches involve those techniques which cannot be categorized into other categories or are an ensemble of different category methods. The paper also introduced a very diverse ear dataset called Annotated Web Ears (AWE) which has been used in this paper also.

In 2018, the deep transfer learning method was proposed as a deep learning technique for ear biometric recognition by Ali et al. [19] over a pretrained CNN model called ALexNet. The methodology involved using a state-of-the-art training function called *Stochastic Gradient Descent with Momentum (SGDM)* and a momentum of 0.9. Another deep learning-based method was suggested in 2019 by Natchapon et al. [20]. In this method, a CNN architecture was employed for frontal-facing ear recognition. It was more acceptable due to the fact that the creation of the face dataset simultaneously created the ear dataset. In the same year, Matthew et al. [21] proposed a variation of wavelet transformation and successive PCA for **single sample** ear recognition. In 2020, Ibrahim et al. [22] introduced a variation of Support Vector Machine (SVM) for ear biometric recognition called Learning Distance Metric via DAG Support Vector Machine. In 2021, deep unsupervised active learning methodology was proposed by Yacine et al. [23]. The labels were predicted by the model as it was unsupervised. Conditional deep convolutional generative adversarial network (cDCGAN) was used to color the grayscale image which further increased the accuracy of recognition.

3 Methodology

3.1 PCA

Principal Component Analysis, or PCA [11], is a method used to reduce the dimensions of samples. It extracts those features which contain more variation in the intensity values. Its popularity owes to the fact that although the size of data is reduced, still it is an unsupervised method. Reducing the number of variables of a dataset naturally comes at the expense of accuracy, but the trick in dimensionality reduction is to trade a little accuracy for simplicity, because smaller datasets are easier to explore and visualize and make analyzing data much easier and faster for machine learning algorithms without extraneous variables to process. So in a nutshell, the idea of PCA is simple—reduce the number of variables of a dataset, while preserving as much information as possible. For the very basic method, the image is directly fed to PCA which is used for dimensionality as well as noise reduction. The resulting components are known as eigenears [5]. These eigenears then constitute a feature vector which is given as input to the SVM model for classification in our research work.

3.2 KPCA

PCA is a linear method which means that it can only be applied to datasets which are linearly separable. It does an excellent job for datasets, which are linearly separable. But, if we use it for non-linear datasets, we might get a result which may not be the optimal dimensionality reduction. Kernel PCA [9] uses a kernel function to project the dataset into a higher dimensional feature space, where the data is linearly separable. Hence, using the kernel, the originally linear operations of PCA are performed in a reproducing kernel Hilbert space.

Most frequently used kernels include cosine, linear, polynomial, radial basis function (rbf), sigmoid as well as pre-computed kernels. Depending upon the type of dataset on which these kernels are applied, different kernels may have different projection efficiency. Thus, the accuracy depends solely on the kernel used in the case of KPCA.

3.3 Fourier

Fourier analysis [24] is named after Jean Baptiste Joseph Fourier (1768–1830), a French mathematician and physicist. Joseph Fourier, while studying the propagation of heat in the early 1800s, introduced the idea of a harmonic series that can describe any periodic motion regardless of its complexity. Fourier Transform is a mathematical process that relates the measured signal to its frequency content. It is used for analyzing the signals. It involves the decomposition of the signals in the frequency domain in terms of sinusoidal or cosinusoidal components. Fourier transform of a function of time is a complex-valued function of frequency, whose magnitude (absolute value) represents the amount of that frequency present in the original function, and whose argument is the phase offset of the basic sinusoid in that frequency. The Fourier transform is not limited to functions of time, but the domain of the original function is commonly referred to as the time domain.

When the image is transformed, there are usually bright areas signifying the edges or high-frequency components and dull areas signifying noise or low-frequency components [25]. In the proposed methodology, the high- as well as low-frequency components are sequentially masked, and the inverse of the masked frequency profile is converted back to a spatial domain using "Inverse" of Fourier transformation. The spatial domain image is then fed to PCA for significant dimensional projection and then to SVM for classification.

3.4 *Wavelet*

The edge is the most important high-frequency information of a digital image. The traditional filter eliminates the noise effectively. But it will make the image blurry. So it is aimed to protect the edge of the image when reducing the noise in an image. The wavelet analysis method is a time-frequency analysis method which selects the appropriate adaptive frequency band based on the characteristics of the signal. Then the frequency band matches the spectrum which improves the time-frequency resolution. The wavelet analysis method has an obvious effect on the removal of noise in the signal.

In this paper, for directly applying the wavelet transformation [10] as well as for further wavelet analysis, the "Discrete" Meyer class of wavelets is used. According to the features of the multi-scale edge of the wavelet, we analyze the de-noising method of the Meyer wavelet transform which is based on a soft and hard threshold. "Discrete" Meyer is a comparatively simpler wavelet as compared to other classes of wavelets. It has only 2 variables, namely the scaling function and the wavelet function.

After wavelet analysis of the samples, unlike the Fourier method where the transformed image had to be converted back to the spatial domain for further processing and classification, the processed feature vector is directly fed into PCA for dimensionality reduction. It is a distinguishing feature of wavelet and Fourier transformations where the former transformation preserves the locality of features but the latter takes a holistic approach to conversion to the frequency domain. Feature vector from PCA is input to SVM for classification.

3.5 *Wavelet Using Subbands*

In this method, a little more sophisticated wavelet called the "Biorthogonal 1.1" wavelet is used. In this family of wavelets, the scaling and wavelet functions of discrete Meyer wavelets is extended by introducing a decomposition and reconstruction parameter to both of the wavelet parameters. A biorthogonal wavelet is used to transform the image in the frequency domain. Further, it divides the image into subbands [21] depending on the frequency components as low-low (LL), low-high (LH), high-low (HL) and high-high (HH). Here, the LL subband is the approximate image and c, whereas the LH, HL and HH subbands inherently include the edge information of horizontal, vertical and diagonal directions, respectively (Fig. 1).

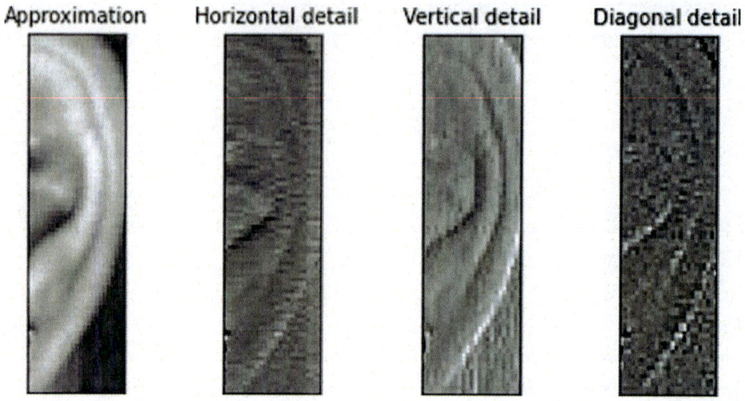

Fig. 1 Subbands of image of Biorthogonal wavelet transformation

In this method, a mean image is derived from the HH and LL subband. The HH band contains diagonal details and LL is the approximate image. This mean the image is then fed to PCA and SVM classifier for the purpose of classification.

4 Experimental Results

In this section, we compare the performance of ear recognition methods in a single sample scenario. These methods include PCA [11], KPCA [9], Wavelet transformation [10], Fourier transformation with frequency masking [25] and Wavelet transformation using subbands [21]. The performance of these methods is compared in terms of average classification accuracy by varying the number of reduced dimensions and repeating the experiments 25 times. The experiments have been performed on three publicly available datasets *viz.*, IIT-D [6], AMI [7] and AWE [8]. A summary of these datasets is given in Table 1.

The KPCA method has been implemented with five different kernels *viz.*, linear, polynomial, radial basis function (RBF), cosine and sigmoid. However, the results with polynomial, RBF and sigmoid are not encouraging. Hence, the results with the remaining two kernels, i.e. linear and cosine have been shown in this paper.

Table 1 Summary of datasets used for experiments

Dataset	Subjects	Images	Yaw	Occlusion	Accessories	Image size	Ethnicity
IIT-D	221	793	None	None	Yes	50 × 180	Asian
AMI	100	700	Mild	Mild	None	492 × 702	White
AWE	100	1000	Severe	Severe	Yes	Varying	Variation

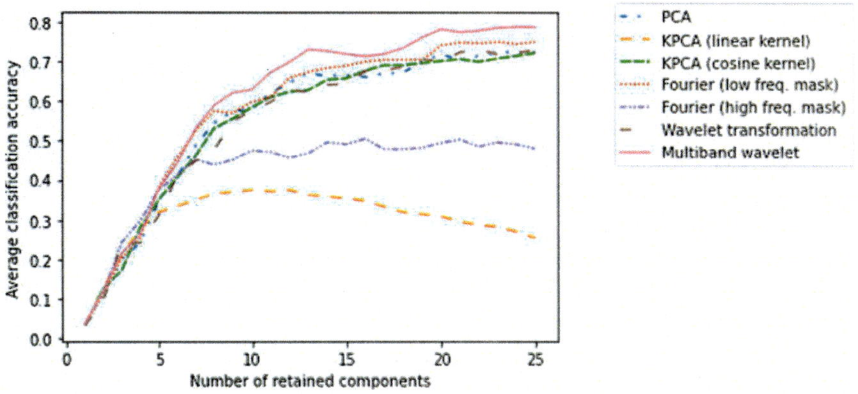

Fig. 2 Average classification accuracy on IIT Delhi ear dataset

Further, in the Fourier transformation-based method, frequency masking has been done sequentially for low as well as high-frequency components and the results for both have been shown in this paper. Now, we discuss the results obtained on individual datasets.

The average classification accuracy on the IIT Delhi ear dataset for all the compared methods is shown in Fig. 2. It can be readily observed from Fig. 2 that KPCA with linear kernel and Fourier transformation with high-frequency mask give poor performance. The accuracy of these methods does not increase even when the number of reduced features is increased. For the remaining methods, the classification accuracy lies between 71.4 and 79.8% with 25 components. The highest accuracy is given by multiband wavelet transformation with 8 or more features. The reason for the higher accuracy of even the most basic methods like PCA is due to the fact that IIT-D database samples are pre-processed. The ear region is tightly cropped and there is almost no noise or occlusion. So the performance of all the methods is generally on the higher side. In the accuracy plot, it is evident that when using Fourier transformation, a low mask is giving accuracy near 74%, whereas the high-frequency masking method is yielding a maximum accuracy of 50%. This signifies that in the ear image, data is concentrated in the high-frequency components or the edges.

The classification accuracy on the AMI dataset is shown in Fig. 3. Here also, the least performance is reported by KPCA with linear kernel and Fourier transform with a high-frequency components mask. However, the highest accuracy of other methods has a large deviation from 45% to approximately 80%. The highest accuracy is reported by multiband wavelet transform which is marginally higher than that of the IIT Delhi dataset. But PCA and KPCA with cosine kernel have shown a large drop in performance. This is due to the fact that the AMI dataset contains ear images with occlusion and larger image sizes with redundant features.

The classification accuracy on the AWE ear dataset is shown in Fig. 4. On this dataset also, the least performance is reported by KPCA with linear kernel and Fourier transform with high-frequency components mask. However, the highest accuracy

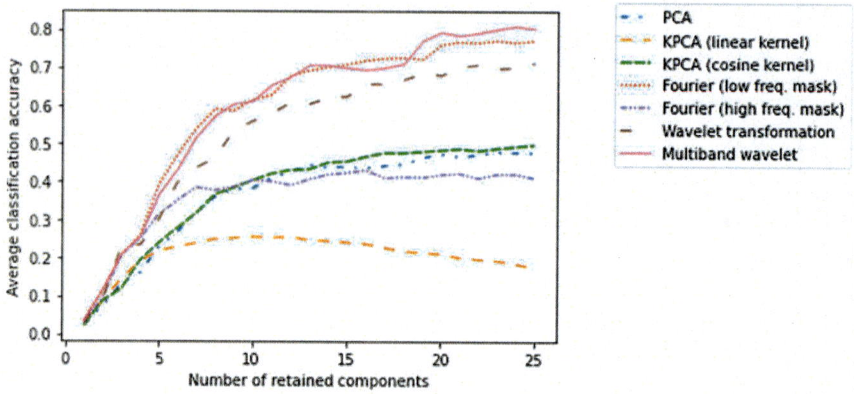

Fig. 3 Average classification accuracy on AMI ear dataset

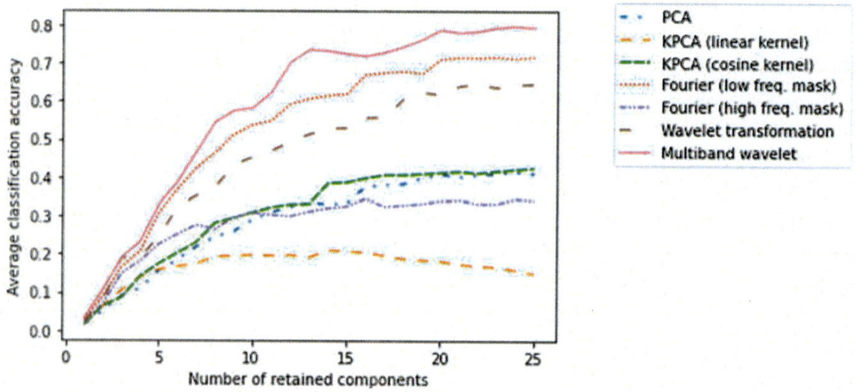

Fig. 4 Average classification accuracy on AWE ear dataset

of other methods has a large deviation from 40% to approximately 78%. It is also observed that the classification accuracy saturates after 15 components. Further, PCA and KPCA report drop in performance in comparison to the AMI dataset. This is due to the high diversity of ear images such as yaw, high occlusion and variation in ethnicity. The highest classification accuracy is again reported by multiband wavelet transformation.

A summary of the highest average classification accuracy reported by five compared methods on the three datasets after 25 iterations is given in Table 2. It is apparent from Table 2 that the Wavelet transformation with multiband gives the highest as well as most consistent accuracy of the three datasets. The variation in performance by all the compared methods is the least on the IIT Delhi ear dataset and largest on AWE dataset. These results also support the characteristics of individual datasets

Table 2 Highest classification accuracy of compared methods on three datasets

Method/dataset	PCA (%)	KPCA (%)	Wavelet (%)	Fourier (%)	Wavelet with multiband (%)
IIT-D	71.59	71.03	71.69	74.15	**79.88**
AMI	45.58	48.23	69.94	78.52	**80.42**
AWE	41.29	45.21	65.12	71.21	**79.47**

in terms of pre-processed images, and the presence of variations such as occlusion, noise contents and yaw movement. These observations can be listed succinctly as follows:

- The highest and consistent performance on the three datasets is given by wavelet transformation with multiband.
- The worst performance is reported by KPCA with linear kernel and Fourier transform with high-frequency components mask.
- PCA and KPCA with cosine kernel show large deviations across different datasets.
- The variation in performance by all the compared methods is the least on the IIT Delhi ear dataset and largest on AWE dataset.

5 Conclusion and Future Work

Ear recognition has emerged as an attractive research area in the past two decades. This problem becomes more challenging when there is only one sample per person available for training. In literature, there have been several methods which have been suggested for ear recognition under different experimental settings. In this paper, we have attempted to investigate which method performs best for single sample ear recognition. We have compared the performance of five methods on three publicly available datasets. It has been found that the wavelet subband-based method performs best on all three datasets. In future work, it can be explored how the deep learning-based methods can be exploited for single sample ear recognition.

References

1. Jain A, Bolle R, Pankanti S (1996) Introduction to biometrics. In: Jain AK, Bolle R, Pankanti S (eds.) Biometrics. Springer, Boston, MA. https://doi.org/10.1007/0-306-47044-6_1
2. Yuan L, Mu Z, Xu Z (2005) Using ear biometrics for personal recognition. In: Li SZ, Sun Z, Tan T, Pankanti S, Chollet G, Zhang D (eds) Advances in biometric person authentication. IWBRS 2005. Lecture notes in computer science, vol 3781. Springer, Berlin, Heidelberg. https://doi.org/10.1007/11569947_28

3. Kumar N, Garg V (2017) Single sample face recognition in the last decade: a survey. Int J Pattern Recognit Artif Intell. https://doi.org/10.1142/S0218001419560093
4. Zhao W, Chellappa R, Phillips P, Rosenfeld A (2003) Face recognition: a literature survey. ACM Comput Surv 35(4):399–458. https://doi.org/10.1145/954339.954342
5. Kumar N (2020) A novel three phase approach for single sample ear recognition. In: Boonyopakorn P, Meesad P, Sodsee S, Unger H (eds) Recent advances in information and communication technology 2019. Advances in intelligent systems and computing, vol 936, Springer, Cham. https://doi.org/10.1007/978-3-030-19861-9_8
6. Kumar A, Wu C (2012) Automated human identification using ear imaging. Pattern Recognit 41(5)
7. AMI Ear database. https://ctim.ulpgc.es/research_works/ami_ear_database/
8. Emeršič Z, Štruc V, Peer P (2017) Ear recognition: more than a survey. Neurocomputing 255:26–39. ISSN 0925-2312. https://doi.org/10.1016/j.neucom.2016.08.139. https://www.sciencedirect.com/science/article/pii/S092523121730543X
9. Zhang H, Mu Z (2008) Ear recognition method based on fusion features of global and local features. In: 2008 international conference on wavelet analysis and pattern recognition, pp 347–351. https://doi.org/10.1109/ICWAPR.2008.4635802.
10. Long Z, Chun M (2009) Combining wavelet transform and Orthogonal Centroid Algorithm for ear recognition. In: 2009 2nd IEEE international conference on computer science and information technology, pp 228–231 (2009). https://doi.org/10.1109/ICCSIT.2009.5234392
11. Kaçar Ü, Kirci M, Güeş E, İnan T (2015) A comparison of PCA, LDA and DCVA in ear biometrics classification using SVM. In: 2015 23nd signal processing and communications applications conference (SIU), pp 1260–1263. https://doi.org/10.1109/SIU.2015.7130067
12. Zhou J, Cadavid S, Mottaleb M (2011) Exploiting color SIFT features for 2D ear recognition. In: 2011 18th IEEE international conference on image processing, pp 553–556. https://doi.org/10.1109/ICIP.2011.6116405
13. Wang Z, Yan X (2011) Multi-scale feature extraction algorithm of ear image. In: 2011 International conference on electric information and control engineering, pp 528–531. https://doi.org/10.1109/ICEICE.2011.5777641
14. Yuan L, Li C, Mu Z (2012) Ear recognition under partial occlusion based on sparse representation. In: 2012 international conference on system science and engineering (ICSSE), pp 349–352. https://doi.org/10.1109/ICSSE.2012.6257205
15. Taertulakarn S, Tosranon P, Pintavirooj C (2014) Gaussian curvature-based geometric invariance for ear recognition. In: The 7th 2014 biomedical engineering international conference, pp 1–4. https://doi.org/10.1109/BMEiCON.2014.7017396
16. Ying T, Debin Z, Baihuan Z (2014) Ear recognition based on weighted wavelet transform and DCT. In: The 26th Chinese control and decision conference (2014 CCDC), pp 4410–4414. https://doi.org/10.1109/CCDC.2014.6852957
17. Tian L, Mu Z (2016) Ear recognition based on deep convolutional network. In: 2016 9th international congress on image and signal processing, biomedical engineering and informatics (CISP-BMEI), pp 437–441. https://doi.org/10.1109/CISP-BMEI.2016.7852751
18. Chen L, Mu Z (2016) Partial data ear recognition from one sample per person. IEEE Trans Hum Mach Syst 46(6):799–809. https://doi.org/10.1109/THMS.2016.2598763
19. Almisreb A, Jamil N, Din N (2018) Utilizing alexnet deep transfer learning for ear recognition. In: 2018 Fourth international conference on information retrieval and knowledge management (CAMP), pp 1–5. https://doi.org/10.1109/INFRKM.2018.8464769
20. Petaitiemthong N, Chuenpet P, Auephanwiriyakul S, Theera-Umpon N (2019) Person identification from ear images using convolutional neural networks. In: 2019 9th IEEE international conference on control system, computing and engineering (ICCSCE), pp 148–151. https://doi.org/10.1109/ICCSCE47578.2019.9068569
21. Zarachoff M, Sheikh-Akbari A, Monekosso D (2019) Single image ear recognition using wavelet-based multi-band PCA. In: 2019 27th European signal processing conference (EUSIPCO), pp 1–4. https://doi.org/10.23919/EUSIPCO.2019.8903090

22. Omara I, Ma G, Song E (2020) LDM-DAGSVM: learning distance metric via DAG support vector machine for ear recognition problem. In: 2020 IEEE international joint conference on biometrics (IJCB), pp 1–9. https://doi.org/10.1109/IJCB48548.2020.9304871
23. Khaldi Y, Benzaoui A, Ouahabi A, Jacques S, Ahmed A (2021) Ear recognition based on deep unsupervised active learning. IEEE Sens J 21(18):20704–20713. (15 Sept 2021). https://doi.org/10.1109/JSEN.2021.3100151
24. Gonzalez R, Woods R (2006) Digital Image Processing, 3rd edn. Prentice-Hall Inc, USA
25. Frejlichowski D (2011) Application of the polar-fourier greyscale descriptor to the problem of identification of persons based on ear images. In: Image processing and communications challenges, vol 3. Springer, Berlin, Heidelberg, pp 5–12

AI-Based Real-Time Monitoring for Social Distancing Against COVID-19 Pandemic

Alok Negi, Krishan Kumar, Prachi Chauhan, Parul Saini, Shamal Kashid, and Ashray Saini

1 Introduction

The COVID-19 epidemic has impacted the lives of millions of people worldwide, and the crisis's consequences are still being felt. The COVID-19 catastrophe has been dubbed the worst economic disaster since the great depression. It is a sobering reminder of long-standing imbalances in our societies. The daily struggles of the COVID-19 pandemic are constantly compared to living in any war environment [1]. The long-term social and economic effects of the COVID-19 epidemic are uncertain, but many people are concerned that lockdown-related education cuts affected 1.6 billion students globally, resulting in a loss of 0.3–0.9 years of education. According to World Bank statistics, five months global shutdown could result in 10 trillion dollars in lost wages over dollars lifetimes. Economic shocks from the pandemic are highly probable to increase school dropout rates, and nearly two-thirds of the households surveyed lead to a decline in agricultural and non-agricultural income (the latter

A. Negi (✉) · K. Kumar · P. Saini · S. Kashid · A. Saini
Department of Computer Science and Engineering, National Institute of Technology, Srinagar (Garhwal) 246174, Uttarakhand, India
e-mail: aloknegi.phd2020@nituk.ac.in

K. Kumar
e-mail: kkberwal@nituk.ac.in

P. Saini
e-mail: parulsaini.phd2020@nituk.ac.in

S. Kashid
e-mail: kashid.shamalphd2021@nituk.ac.in

A. Saini
e-mail: mt20cse001@nituk.ac.in

P. Chauhan
Department of Information Technology, Govind Ballabh Pant University of Agriculture and Technology, Pantnagar 263153, Uttarakhand, India

© The Author(s), under exclusive license to Springer Nature Singapore Pte Ltd. 2023

H. Muthusamy et al. (eds.), *Robotics, Control and Computer Vision*, Lecture Notes in Electrical Engineering 1009, https://doi.org/10.1007/978-981-99-0236-1_8

being more severe), as well as a large majority (94%) reporting reduced remittances received, which is consistent with international reports from the first months just after infection [2]. In all, almost three-fourths' people reported an unambiguous reduction in income. Our evidence validates worries regarding pandemics' adverse negative externalities. When a pandemic hit, it sent most employees scurrying home, resulting in income disparity and hurting employment prospects for all of those with only a high school diploma while having little effect on those with graduate degrees.

The COVID-19 pandemic's trajectory shows the changing environment regulating both coronavirus transmission and its socio-economic consequences. Since mid-March, a second severe wave of diseases has resulted in lockdowns, the second set of stringent measures established following the original epidemic in the spring. Although COVID-19 transmission behaviors and consequences differed between rural and urban regions, there were significant implications on rural incomes and livelihoods. COVID-19's effects on revenue, food security, and dietary variety are progressively appearing as global trends and local variances. Vaccination rates are already increasing, and people are looking forward to a safer post-pandemic future. However, specific essential actions are required after vaccines, such as masks, which are essential to prevent transmission and save lives. Social distancing, avoiding crowded, confined, and close-contact situations, proper ventilation, washing hands, concealing sneezes and coughs, and much more used to be parts of a complete "Do it all!" strategy [3].

Coronaviruses can be disseminated when persons with the infection have close, constant contact with those who are not infected. It generally involves staying more than 15 minutes approximately two meters from an infected individual, such as conversing with someone for illustration. The more people come into proximity with the droplets from coughs and sneezes of an infected individual, the more susceptible you are to get the virus. It necessitates the use of a new measurement notion. These measures are sometimes referred to as "social distancing" that include activities like temporarily prohibiting socializing in public areas like entertainment or sporting events, restricting the usage of non-essential public transportation, or encouraging more work at home. In general, social distancing is an attempt to prevent coronavirus transmission in big gatherings such as meetings, movie theaters, weddings, and public transportation. Schools, universities, malls, and movie theaters are now shuttered across the country to emphasize the need for social distance. People are being encouraged to work from home and have as little interaction with others as possible.

Wearing a mask and keeping a six-foot distance to prevent the disease from spreading as advised by the WHO. It is indeed essential for citizens to have a specified level of social interaction for better mental wellness. As a result, distinct stages of artificial intelligence (AI) can be followed depending on the disease's spread [4]. Therefore, we developed an AI-based model for real-time monitoring of individuals for social distancing that uses YOLOv3 person identification, VGG-16-based face mask classifier, Dual Shot Face Detector-based face detection, and DBSCAN clustering. The main objectives and contribution of this paper are as follows:

- To use real-time video streams to monitor persons who are breaking the rules of Social Distancing.
- Building a data-driven framework to assist governments in establishing a secure de- and re-confinement planning schema for their respective regions.
- To assist in navigating future waves of viral transmission and other unforeseeable negative consequences.
- To create a decision-making tool that can be used not only for the present epidemic but also for future pandemics, which we all know are coming, especially as we witness the repercussions of global climate change.
- To prevent the transmission of new infection waves by shifting from a reactive to a proactive approach.

The remaining part of the paper is laid out as follows: Section 2 describes the related work followed by the proposed methodology in Sect. 3. Section 4 describes results and discussion. Section 5 brings the paper to a conclusion and outlines future research.

2 Related Work

In this crucial time, social disengagement is one of humanity's most urgent calls. In this way, countries are preventing infection and reducing infection, and flattening the infection-to-community curve. The "lockdown," as it is known, will essentially lower the viral load and the number of infected cases that need to be treated. Masks can help prevent the infection from spreading from the person who is wearing it to others. COVID-19 [5] is not protected by masks alone; they must be used with physical separation and hand cleanliness.

In the COVID-19 pandemic, identifying persons who use face masks is complex, and detection of facemask-with high accuracy has practical applications in COVID-19 epidemic prevention. As a consequence, Qin et al. [6] proposed a four-step technique for identifying facemask-wearing conditions: image pre-processing, facial recognition and cropping, image super-resolution, and facemask-wearing scenario detection. For face image classification, the approach integrated an SR network with a classification network (SRCNet). The input images were processed with facial detection and cropping, SR, and facemask-wearing condition identification to recognize the facemask-wearing scenario. Finally, SRCNet achieved a 98.70% accuracy and outperformed conventional end-to-end image classification methods by over 1.5% in kappa.

For face mask identification, Loey et al. [7] introduced a hybrid model that combined deep and conventional machine learning. There were two sections to the model. The initial step was to extract features using Resnet50, one of the most popular deep supervised learning models. The second section dealt with the identification of face masks using traditional machine learning techniques. Conventional machine learning techniques such as the Support Vector Machine (SVM), decision trees, and collaborative algorithms were investigated.

In order to work and travel securely during the COVID-19 outbreak, Xiao et al. [8] created a deep learning-based security detection technique that relied on machine vision rather than manual monitoring. To identify unlawful actions of workers without masks in workplaces and highly populated locations, convolutional neural network VGG-19 modifies the original 3 FC layers with 1 Flatten layer and 2 FC layers, as well as the original Softmax classifier with two labeled Softmax classification layers, Masked workers (Mask) and unmasked workers were subjected to training and testing (Un-mask). The upgraded network model's precision for identifying whether or not to wear a mask has grown by 10.91% and 9.08%, respectively, while its recall rate has enhanced by 11.4% and 8.39%.

Hussai et al. [9] deployed deep learning to classify and recognize face emotions in real-time. They classified seven face expressions using VGG-16. The suggested model was trained using the KDEF dataset and has an accuracy of 88%. The use of masks is an essential part of the covid-19 prevention process. Due to embedded devices' limited memory and computational capability, real-time surveillance of persons wearing masks or not is complicated. Roy et al. [10] tested several prominent object detection methods on the Moxa3K benchmark dataset to address these issues, including YOLOv3 YOLOv3Tiny, SSD, and Faster R-CNN. As a good combination of accuracy and real-time inference, the YOLOv3 small model gave an excellent mAP of 56.27% with an FPS of 138. The backbone of YOLOv3 is Darknet-53 in [11] applied the YOLOv3 algorithm to detect faces. The accuracy of the proposed technique was 93.9%. It was developed using the CelebA and WIDER FACE datasets, which contain over 600,000 shots.

Din et al. [12] presented a new GAN-based network that can automatically delete masks covering the facial region and recreate the vision by filling in the empty hole. Nieto-Rodrguez et al. [13] recommended that ICDSC participants engage with a system that divides faces into two categories: those with surgical masks and those without. The system establishes a per-person ID through tracking, resulting in only one warning for a mask-less face over several frames in a video. The system can achieve five frames per second with several faces in VGA images on a standard laptop. The tracking method significantly reduces the number of false positives. The system's output includes confidence values for both mask and non-mask face detections.

3 Proposed Work

This work aims to use real-time video streams to track persons who are breaking the rules of social distancing. Furthermore, a VGG16-based Face Mask Classifier model is trained and deployed to recognize people who are not wearing a face mask. For detecting prospective intruders, the suggested technique also employs YOLOv3 and DBSCAN clustering. The detailed flow is drawn in Fig. 1.

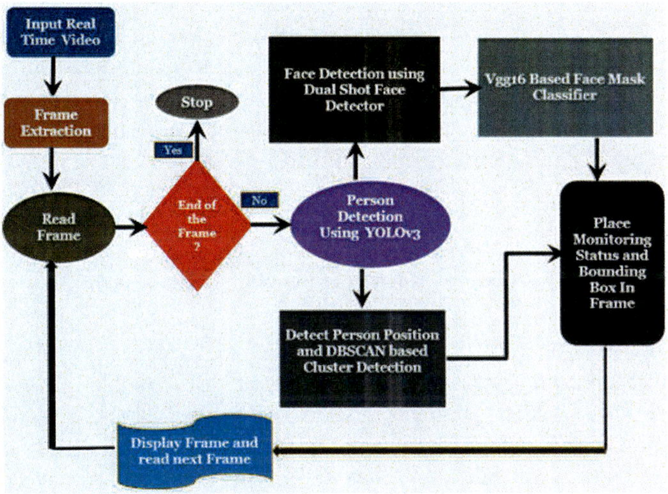

Fig. 1 Proposed model for social distancing

Firstly the frames are extracted from the real-time video and passed to the YOLOv3 model for person detection. Further, faces are detected from the frame using a Duel shot face detector, and a vgg-16 based face mask detection classifier is used to check whether a person is wearing a mask or not. Person position is also detected with DBSCAN clustering for cluster detection. Then bounding box and monitoring status is placed into the frame, and finally, frames are displayed. This process is done for each frame until the end of the frame.

3.1 Person Detection Using YOLOv3

Real-time object detection model YOLOv3 (You Only Look Once) is used for the person detection, which is pretrained on the COCO dataset. Yolov3 used a better hybrid architecture of YOLOv2, Residual networks, and Darknet-53 for the feature extraction. Inside each residual block, the network is created using a bottleneck structure (1 × 1 followed by 3 × 3 convolution layers) and a skip connection. Due to ResNet, the performance of the network will not be harmed by overlaying layers. Furthermore, the mass of fine-grained features is not lost because the more profound layers receive more information directly from the shallower layers.

The model made use of the Darknet-53 architecture, which was designed with a 53-layer network for feature extraction training. The detection head for the training object detector was then stacked with 53 more layers, giving YOLOv3 a total of 106 layers of the fully convolutional underlying architecture. Instead of stacking the prediction layers at the last layers as before, YOLOv3 added them to the side network.

YOLOv3's most significant feature is that it detects at three distinct scales. Three distinct scale detectors were created using the features from the last three residual blocks. 1 × 1 kernel is applied on each detection layer responsible for predicting the bounding box for feature map of each grid cell. 416 × 416 resolution is used in this work to get the bounding box on a person.

3.2 *Face Mask Classifier using VGG16*

On the SMFD dataset, the VGG-16 model is used as a face mask classifier to determine whether a person is wearing a mask or not. In VGG-16, the first two convolutional layers have 64 filters with 3 × 3 sizes to generate 224 × 224 × 64 volume. The next layer is the pooling layer, which reduces the height and width of volume 224 × 224 × 64 to 112 × 112 × 64. Then again, there are more conv layers with 128 filters, and 112 × 112 × 128 will be the new dimension. After that, a pooling layer is applied, resulting in a new dimension of 56 × 56 × 128. Then VGG-16 has two convolutional layers with 256 filters followed by pooling layer, three convolutional layers with 512 filters followed by pooling layer, and three convolutional layers with 512 filters followed by a pooling layer. Finally, Vgg16 has a 7 × 7 × 512 into a Fully connected layer (FC) with 4096 hidden units and a softmax output of one of 1000 classes. As shown in Figure 2, the three fully connected layers of the original VGG16 are replaced with two dense layers with 128 and 2 hidden nodes, respectively. The softmax activation function is utilized to create a second dense layer for the final output.

3.3 *Face Detection using Dual Shot Face Detector (DSFD)*

Across low resolution or covered images, the MTCNN and Haar-Cascades face detectors are ineffective; hence DSFD is utilized in this study for a wide range of orientations to detect the face. Cv2 and face detection library are used for DSFD with Confidence threshold (0.5) and IOU threshold (0.3). After applying the model, it will return a tensor with (N, 5) shape where N is the no of faces and xmin, ymin, xmax, ymax, detection confidence values.

4 Result and Analysis

For this work, training is performed on Google Colab using python script for only 30 epochs. Adam optimizer with Batch size 32 is used for it. There are total 14,780,610 parameter out of which 65,922 are trainable and remaining 14,714,688 non-trainable parameter. Real-time video with 25 fps is used for this work.

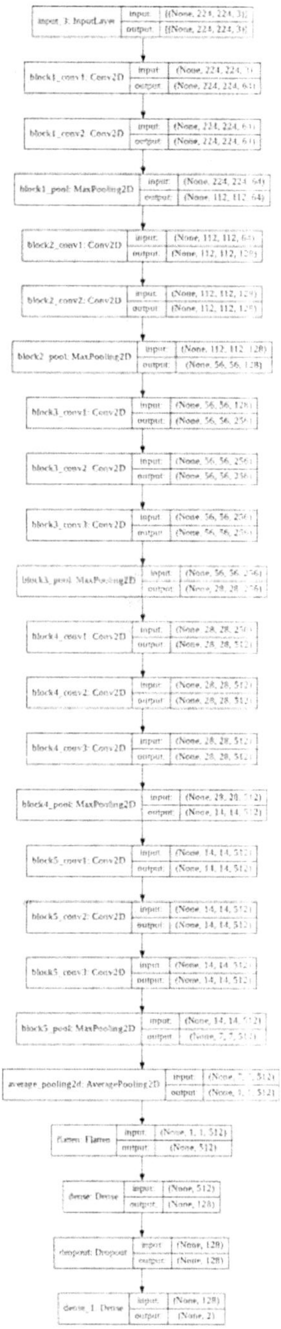

Fig. 2 Layered architecture of VGG-16

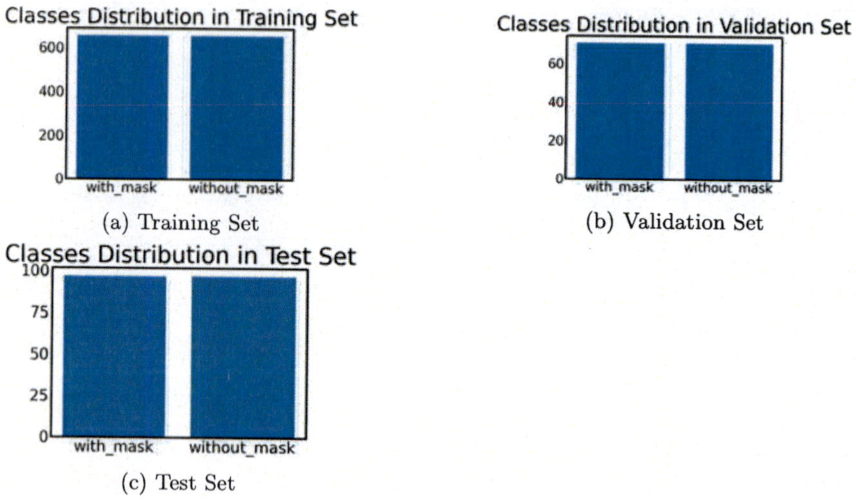

Fig. 3 Distribution of dataset

4.1 *Dataset Description*

A simulated Masked Face Dataset (SMFD) is used for the face mask classifier. The dataset contains a total of 1651 images as shown in Figure 3. The training set has a total of 1315 images for both masked and without masks. The validation and test set contains 142 and 194 images, respectively.

4.2 *Data Preprocessing and Augmentation*

Data-augmentation can help to increase the number of images (creating image variations) and provide the images in batch to the model. The images are not replicated in batches, and they also help to avoid model overfitting. Images are resized into $224 \times 224 \times 3$ due to different sizes and to decrease the scale. ImageDataGenerator is used for the augmentation with rescale (1./255), zoom range (0.2), shear range (0.2), and horizontal flip (true) parameters. Figure 4 shows the random transformation of the images using data augmentation.

4.3 *Performance Metrics*

The performance analysis for the proposed work is performed on the basis of Accuracy curve, Loss curve, Precision, Recall, F1 score, and Confusion matrix. Equations 1, 2, 3, 4, and 5 show the mathematics behind the each metrics.

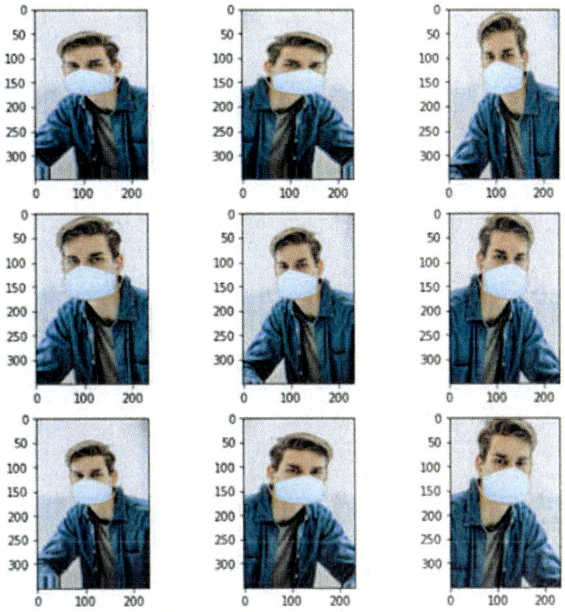

Fig. 4 Random transformation using data augmentation

$$Accuracy = (TP + TN)/(FN + TP + TN + FP) \tag{1}$$

Categorical cross entropy as shown in Eq. 2 is used as a metric for this work. A perfect classifier gets the logloss of 0.

$$logloss = -1/N \sum_{i=1}^{N} \sum_{j=1}^{M} y_{ij} log(p_{ij}) \tag{2}$$

A Classification report is used to measure the quality of predictions from a classification algorithm. The report shows the main classification metrics precision, recall, and f1-score on a per-class basis. There are four ways to check if the predictions are right or wrong:

- TN(True Negative): when a case was negative and predicted negative
- TP(True Positive): when a case was positive and predicted positive
- FN(False Negative): when a case was positive but predicted negative
- FP(False Positive): when a case was negative but predicted positive

Precision is the ability of a classifier not to label an instance positive that is negative. It is defined as the ratio of true positives to the sum of true and false positives for each class.

$$Precision = TP/(TP + FP) \tag{3}$$

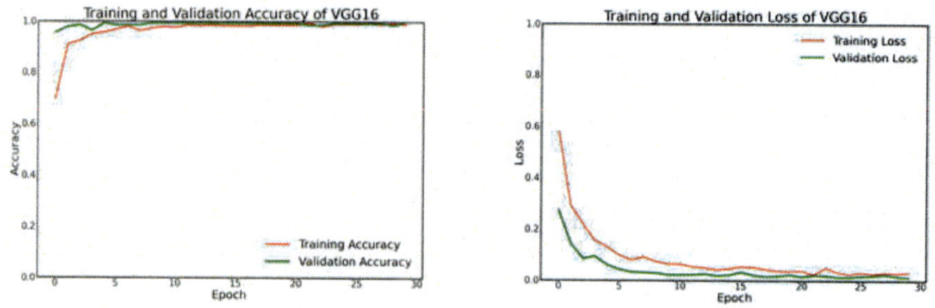

(a) Accuracy Curve

(b) Loss Curve

Fig. 5 Accuracy and loss curve of VGG16

Recall is the ability of a classifier to find all positive instances. For each class, it is defined as the ratio of true positives to the sum of true positives and false negatives.

$$Recall = TP/(FN + TP) \tag{4}$$

$$f1Score = 2 * (Precision * Recall)/(Precision + Recall) \tag{5}$$

The proposed work recorded the training accuracy of 99.32% with a loss score of 0.02, while the accuracy for the validation set recorded 100% with 0.01 loss as shown in Fig. 5a and b. Our proposed model achieved 98.97% accuracy with 0.02 loss for the test set.

Confusion Matrix is shown in Figs. 6 and 7 for validation and testing set with normalization and without normalization respectively. True Negatives, False Positives, False Negatives, True Positives values are recorded 71, 0, 0, 71 for the validation set while 97, 0, 2, 95 for the test set, respectively. So our model achieved the 100% precision, recall, and F1 score for the validation set. In the validation set, both the classes (with mask, without mask) recorded 100% precision, recall, f1-score with a support value of 71 each for a total of 142 images as shown in Table 1. Support is the number of actual occurrences of the class in the specified dataset. Imbalanced support in the training data may indicate structural weaknesses in the reported scores of the classifier and could indicate the need for stratified sampling or rebalancing. Overall, 100%, 97.94%, and 98.96% precision, recall, and F1 score are recorded for the test set. Further, in the test set with mask class recorded the 98%, 100%, 99% precision, recall, f1-score with support value of 97 while without mask class recorded the 100%, 98%, and 99% precision, recall, f1-score with a support value of 97 for total 194 images as shown in Table 2. Sample images obtained for the real-time videos using the proposed work are displayed in Fig. 8.

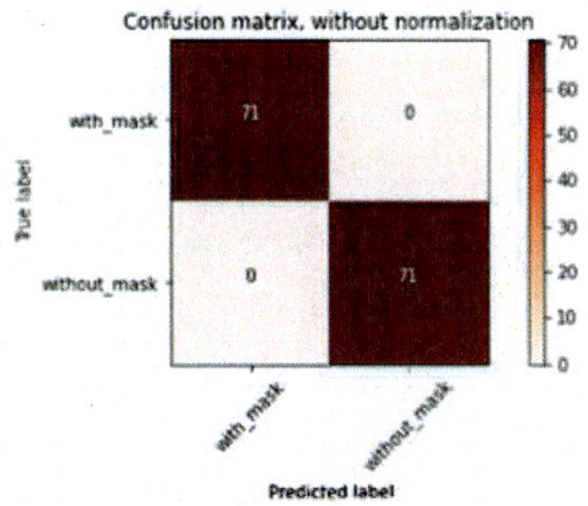

(a) Without Normalization

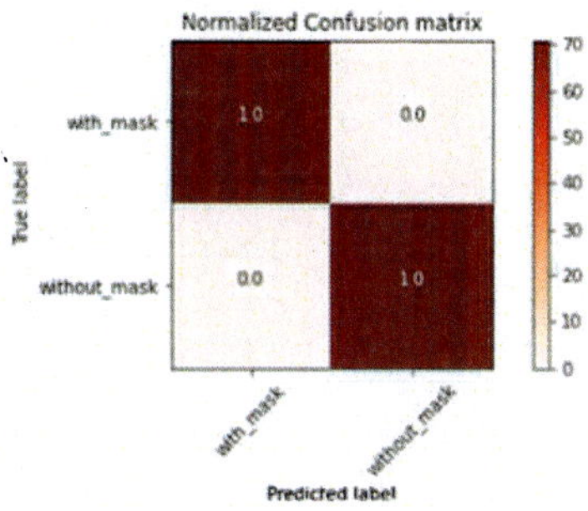

(b) With Normalization

Fig. 6 Confusion matrix for validation set

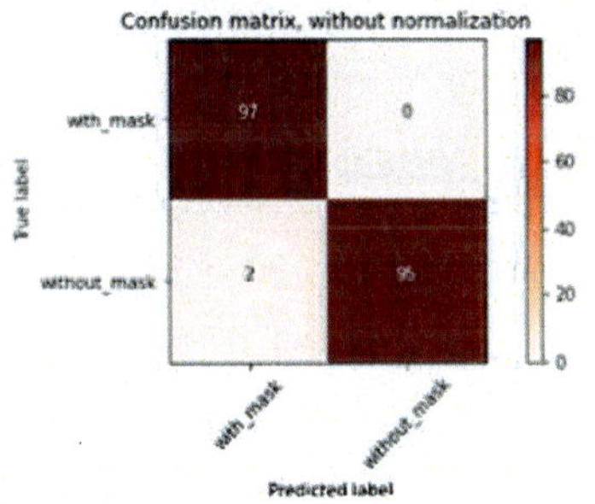

(a) Without Normalization

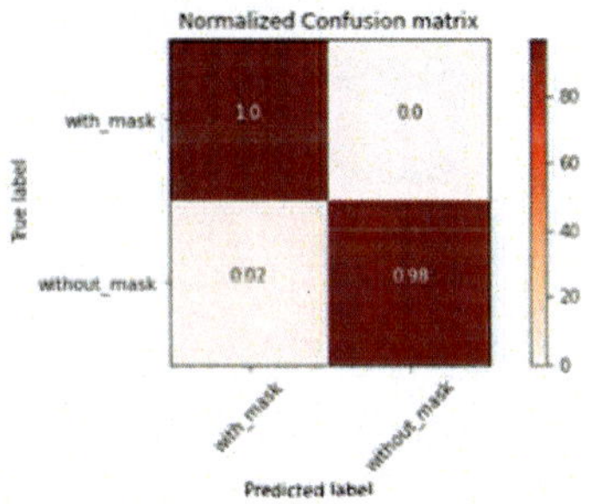

(b) With Normalization

Fig. 7 Confusion matrix for test set

Table 1 Classification report for validation set (In Percent)

Category	Precision	Recall	f1 score
Overall	100.00	100.00	100.00
With mask	100.00	100.00	100.00
Without mask	100.00	100.00	100.00

Table 2 Classification report for test set (In Percent)

Category	Precision	Recall	f1 score
Overall	100.00	97.94	98.96
With mask	98.00	100.00	99.00
Without mask	100.00	98.00	99.00

Fig. 8 Results obtained using proposed work

4.4 Comparison with Related Works

We compared the proposed work with some other state-of-the-arts models and found more excellent and nearer results. Starting from the [7], the author used the same dataset for the face mask detection classifier and obtained 94 and 98.7% accuracy using ensemble classifier, 96 and 95.64% using decision trees classifier, 100 and 99.49% using SVM classifier. Similarly, the work proposed in [14] recorded 98.59 and 98.97% accuracy for validation and test set using VGG-16. Nagrath et al. [15] recorded 92.64% accuracy and 93% f1 score. The work proposed in [15] obtained 93% accuracy. Zhang et al. [16] recorded 84.10 mAP for the face mask detection. Our work has recorded 99.32%, 100%, and 98.97% accuracy for training, validation, and test set, respectively, in just 30 epochs.

The proposed work yielded promising results in only 30 epochs, but it may be expanded to include more standard datasets such as RMFD, LFW, and others. For blurred faces caused by quick movement or noise during capture, blurring augmentation (Motion blur, Average blur, Gaussian blur, etc.) might be utilized.

5 Conclusion

The proposed work can enhance real-time public health governance, decision-making, and related data insights around the world—not only for the virus we currently face but also for the pandemics we will inevitably face in the future. In this work, AI-based real-time monitoring of people for social distancing is implemented using YOLOv3 person detection, VGG-16-based face mask classifier, Dual Shot Face Detector-based face detection, and DBSCAN clustering. The proposed work achieved 99.32%, 100%, and 98.97% accuracy for training, validation, and test set. The proposed study may be expanded using more advanced neural networks (Yolov5, VGG19, Resnet, Densenet, etc.) and standard dataset such as RMFD, LFW, etc. A successful solution would assist governments and companies in making quick and confident decisions about proper confinement strategy for their region while also reducing the number of lives and livelihoods lost.

References

1. Sohrabi C, Alsafi Z, O'neill N, Khan M, Kerwan A, Al-Jabir A, Iosifidis C, Agha, R (2020) World Health Organization declares global emergency: a review of the 2019 novel coronavirus (COVID-19). Int J Surg 76:71–76
2. Altmann DM, Douek DC, Boyton RJ (2020) What policy makers need to know about COVID-19 protective immunity. The Lancet 395(10236):1527–1529
3. Gandhi M, Rutherford GW (2020) Facial masking for Covid-19-potential for "variolation" as we await a vaccine. N Engl J Med 383(18):e101
4. Alimadadi A, Aryal S, Manandhar I, Munroe PB, Joe B, Cheng X (2020) Artificial intelligence and machine learning to fight COVID-19. Physiol Genomics 52(4):200–202
5. Hazarika BB, Gupta D (2020) Modelling and forecasting of COVID-19 spread using wavelet-coupled random vector functional link networks. Appl Soft Comput 96:106626
6. Qin B, Li D (2020) Identifying facemask-wearing condition using image super-resolution with classification network to prevent COVID-19. Sensors 20(18):5236
7. Loey M, Manogaran G, Taha MHN, Khalifa NEM (2021) A hybrid deep transfer learning model with machine learning methods for face mask detection in the era of the COVID-19 pandemic. Measurement 167:108288
8. Xiao J, Wang J, Cao S, Li B (2020) Application of a novel and improved VGG-19 network in the detection of workers wearing masks. J Phys Conf Ser 1518(1):012041. (IOP Publishing, Apr 2020)
9. Hussain SA, Al Balushi ASA (2020) A real time face emotion classification and recognition using deep learning model. J Phys Conf Ser 1432(1):012087. (IOP Publishing)
10. Roy B, Nandy S, Ghosh D, Dutta D, Biswas P, Das T (2020) MOXA: a deep learning based unmanned approach for real-time monitoring of people wearing medical masks. Trans Indian Natl Acad Eng 5(3):509–518
11. Li C, Wang R, Li J, Fei L (2020) Face detection based on YOLOv3. In: Recent trends in intelligent computing, communication and devices, pp 277–284. Springer, Singapore
12. Din NU, Javed K, Bae S, Yi J (2020) A novel GAN-based network for unmasking of masked face. IEEE Access 8:44276–44287
13. Nieto-Rodríguez A, Mucientes M, Brea VM (2015) Mask and maskless face classification system to detect breach protocols in the operating room. In: Proceedings of the 9th international conference on distributed smart cameras, pp 207–208. (Sept 2015)

14. Negi A, Kumar K, Chauhan P, Rajput R (2021) Deep neural architecture for face mask detection on simulated masked face dataset against covid-19 pandemic. In: 2021 International Conference on Computing, Communication, and Intelligent Systems (ICCCIS). IEEE, pp 595–600
15. Nagrath P, Jain R, Madan A, Arora R, Kataria P, Hemanth J (2021) SSDMNV2: a real time DNN-based face mask detection system using single shot multibox detector and MobileNetV2. Sustain Cities Soc 66:102692
16. Zhang J, Han F, Chun Y, Chen W (2021) A novel detection framework about conditions of wearing face mask for helping control the spread of covid-19. IEEE Access 9:42975–42984

Human Activity Recognition in Video Sequences Based on the Integration of Optical Flow and Appearance of Human Objects

Arati Kushwaha and Ashish Khare

1 Introduction

Human activity recognition has emerged as a pivotal research problem in recent years due to its potential applications in several intelligent automated monitoring applications such as intelligent surveillance, robot vision, automated healthcare monitoring, entertainment, video analytics, security and military applications, etc. Video data is booming very fast due to the advancements in multimedia technology such as smartphones, drones, movies, and surveillance cameras in the modern era. So it has become essential to predict and monitor semantic video contents automatically. Therefore, human activity recognition systems have become an innovative solution to such automated monitoring of visual systems and encouraged the adoption and usability of intelligent monitoring visual applications [1, 2]. Vision-based activity recognition often becomes more difficult for real-world applications when the irregular motion of non-stationary cameras records activity videos. Such videos have a complex background, varying illumination conditions, different poses, orientations, and scaling of objects. Therefore, activity recognition involves parsing the complex video sequences and learning complex activity patterns. Therefore, the extraction of compelling features plays a vital role in activity recognition.

Over the last decade, various handcrafted feature descriptors were proposed, such as single feature descriptors and a combination of multiple feature descriptors [1, 3, 4], and some encoding schemes with mid-level representation such as Bag-of-Words (BoW) [5] and Fisher Vector [6] have been considered for activity recognition task using several machine learning algorithms. Since realistic videos have a dynamic range of varying details. Human activity recognition in realistic videos is still a challenging and open problem for research. For accurate recognition of human activity, there is a need for an excellent and discriminative feature descriptor that selects

A. Kushwaha · A. Khare (✉)
Department of Electronics and Communication, University of Allahabad, Allahabad, India
e-mail: khare@allduniv.ac.in

© The Author(s), under exclusive license to Springer Nature Singapore Pte Ltd. 2023
H. Muthusamy et al. (eds.), *Robotics, Control and Computer Vision*, Lecture Notes in Electrical Engineering 1009, https://doi.org/10.1007/978-981-99-0236-1_9

relevant visual data and reduces unnecessary visual content [7]. This fact motivated us to design a novel framework for human activity recognition for motion activities recorded in realistic and multi-view environments. This work used the integration of multiple feature representation techniques to represent human activities recorded by static and moving cameras with varying scales, poses, orientations of human objects, and changing illumination conditions. In the proposed approach, first, we performed object segmentation by the method proposed by Kushwaha et al. [8, 9] to capture the moving human objects (to compute human appearance in the subsequent frames) [10]. Then, computed the magnitude and orientation information of moving objects using the optical flow technique [11], followed by the histogram of oriented gradient [12] of optical flow features to capture the dynamic pattern of human activities [13]. The final feature vector is constructed by a fusion of local-oriented gradients of magnitude and orientation, which is then processed by a multi-class support vector machine to compute the class scores of each activity category. The proposed method's effectiveness is empirically justified by conducting several extensive experiments. Therefore, to analyze the proposed framework, we considered three publically available datasets that are IXMAS [14], UT Interaction [15], and CASIA [16], and the results of the proposed method were compared with several state-of-the-art methods. The recognition result demonstrates the usefulness of the proposed method over considered state-of-the-art methods.

The rest of the paper is organized as follows: Sect. 2 has a detailed literature review study. Section 3 consists of details of the proposed framework. The experimental result and detailed discussion are given in Sect. 4. We concluded the proposed work in Sect. 5.

2 Literature Review

With the increase in video recording cameras in different firms like visual surveillance, film crews, drones, robotics, and smartphones, computer vision scientists have increased their interest in developing an automated monitoring system. Therefore, video-based human activity recognition (HAR) has become one of the most critical research problems over the last few decades in different computer vision applications, such as security monitoring, gaming entertainment, smart indoor security, intelligent visual surveillance, military applications, healthcare, robot vision, and daily life activity monitoring. The process of capturing and recognizing human activities is affirmed to be cumbersome and challenging due to the high degree of freedom of human body motion with unpredictable appearance visibility such as personal style and activity length, clothing, and object appearance in different viewpoints and scales. Feature extraction techniques always play a crucial role in accurately recognizing human activities. Researchers in this field used one of two types of feature extraction techniques: (1) self-learning techniques from raw data based on deep learning approaches, and (2) traditional handcrafted feature descriptor-based techniques. Traditional handcrafted feature descriptor-based techniques are problem-specific and based on feature descriptors designed by experts. Although deep learning-based methods perform well for activity recognition [17]. However, they are highly computationally complex and require extensive sample data and powerful machines to process that data [18]. These methods give rise to the over-fitting problem with small-scale datasets. In the past few years, a large number of feature descriptors has been used for activity recognition tasks such as Local Binary Pattern (LBP), Local Ternary Pattern (LTP), Histogram of Oriented Gradient (HOG), Scale Invariant Feature Transform (SIFT), and Space–Time Interest Points (STIP) [1–4, 11, 12]. Kushwaha et al. [1] proposed an approach for activity recognition tasks to integrate multiple features viz. multiclass LBP, HOG, and DWT to represent complex human activities uniquely. Ladjailia et al. [2] proposed an algorithm for human activity recognition tasks based on motion information followed by a KNN machine learning classifier for activity recognition. Al-Faris et al. [10] proposed a human activity recognition task based on appearance and motion information. They combined motion history image and local motion vector to represent the human activities followed by multiclass KNN classifier for classification of activities. Kushwaha and Khare [19] had proposed an approach for human activity recognition by utilizing local ternary patterns and histograms of oriented gradients followed by a machine learning classifier to compute class scores of activities. In Kushwaha and Wolf [20], the authors had proposed a human activity recognition system for motion activities. They computed optical flow vectors followed by HOG descriptor to represent dynamic motion patterns followed by multiclass support vector machine for activity recognition. Yeffet et al. [21] developed an algorithm for activity recognition in which they used Local Ternary Pattern (LTP) to represent human action followed by an SVM classifier to compute class scores. Nigam and Khare [22] developed an algorithm for human activity recognition to integrate uniform binary patterns and moment invariants followed by a binary SVM classifier to recognize human activities. Seemanthini and Manjunath [23] proposed a framework for human action recognition. They first used the segmentation technique to extract objects of interest, followed by a HOG descriptor to represent complex human actions, followed by an SVM classifier to compute class scores of action classes. From the detailed study of literature on human activity recognition, feature representation plays a crucial role in achieving good performance and is application dependent. So there is a need to design an efficient and discriminative feature descriptor. In the present work, we have proposed a feature representation technique for human activity recognition for video sequences based on the integration of the appearance of the object of interest and the motion information of the moving object.

3 The Proposed Method

The ultimate goal of this work is to present a framework for the recognition of human activity based on supervised learning, which is recorded for real-world applications by single and multi-camera. We designed a novel feature descriptor to represent complex motion activities in this work. The general framework of the proposed work is shown in Fig. 1. Since excellent and discriminative feature descriptors always play a crucial role in the activity recognition task, we first segmented the moving objects from complex video data to capture the objects of interest and reduce the unnecessary background content from the video clips. Then, we used the optical flow technique [11] to compute the magnitude (motion or velocity vectors) and orientation (direction) information of each moving pixel of an object further to avoid the noise and background content [8]. Then magnitude and orientation information is further used to compute the histogram of oriented gradients (HOG) [12] because it captures the dynamic pattern of complex motion activities more discriminatively. At last, the unique dynamic pattern of magnitude and orientation information captured by the histogram of oriented gradients is further integrated using the feature fusion strategy (concatenation) to construct the final feature vector. We have taken velocity and direction information to construct the final feature vector to avoid inter and intra-class variations and redundant information that may confuse the classifier on training. The sample data of different activity categories may have the same magnitude (velocity) but not the direction [8, 9]. A multiclass support vector machine then processes the final constructed feature vector to compute the class scores of activities [24]. The proposed work consists of the following steps:

i. The object segmentation technique proposed by Kushwaha et al. [8] separates the complex background and computes human appearance in the subsequent video frames.
ii. The optical flow technique [11] has been used to compute the magnitude (velocity vector) and orientation (direction) of each moving pixel and to eliminate background noise.
iii. Along with the temporal axis, we integrated optical flow vectors with a histogram of oriented gradients (HOG) to compute dynamically oriented histograms of optical flow sequences.
iv. Finally, a local-oriented histogram of the velocity vector and orientation information is integrated using a feature fusion strategy to construct the final feature vector.
v. We used a one-vs-one multiclass support vector machine to compute the class scores of human activities [24].

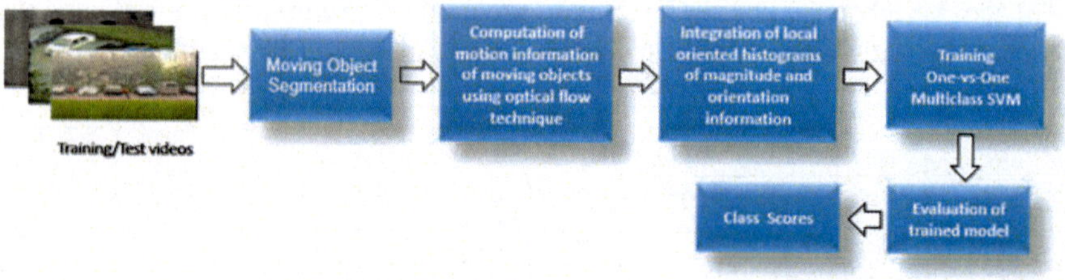

Fig. 1 Schematic diagram of the proposed human activity recognition algorithm

3.1 The Algorithm

The proposed algorithm of human activity recognition

Input: $n \in$ # activity classes in the database

Output: class score of activity categories

A. *for* $i = 1$ *to* n

$f(f_{i,1}, f_{i,2}, f_{i,3}, f_{i,4}, \ldots\ldots f_{m,n}) \leftarrow Object\_Segmentation(v_i)$

$m \leftarrow$ # *frame sequences in n activity categories*

for $j = 1$ *to* m

$[Magnitude, Orientation] \leftarrow Optical\_flow(f_{i,j})$

$[Dynamic\ oriented\ histograms\ of\ motion\ vectors] \leftarrow HOG(Optical\_flow\_vectors)$

$[\xi_{i,j}] \leftarrow feature\_fusion(dynamic\ oriented\ histograms\ of\ motion\ vectors)$

end for

end for

B. SVM model $\leftarrow Multiclass\_SVM(\xi)$

C. For test video clips, repeat step A

E. Evaluation of trained SVM model with test data

F. Generate class scores of each activity category

---

4 Experimental Result and Discussion

To prove the empirical justification of the proposed framework, we conducted several extensive experiments on three publically available datasets, namely IXMAS [14], UT Interaction [15], and CASIA [16]. IXMAS [14] is a multi-view human activity dataset that contains 13 activity categories of daily life which are do nothing, check watch, crossing arms, scratching head, sitting down, getting up, turning around, walking, waving, punching, kicking, pointing, picking up, throwing over the head, and throwing from the bottom up. This dataset consists of low-resolution video clips recorded by five different cameras from different views. UT interaction [15] dataset consists of six activity categories: shaking hand, hugging, pointing, kicking, pushing, and punching recorded by a static camera. This dataset was created with challenges like wider area, aerial view, and complex human–human interaction activities. CASIA [16] is a realistic and multi-view human activity dataset recorded by outdoor video cameras from different viewing angles. This dataset consists of two types of activities: (i) Eight activity categories were performed by single person. These activities are running, walking, jumping, bending, fainting, wandering, crouching, and punching a car and (ii) Two or more persons perform seven high-level activities: fighting, robbing, overtaking, following, meeting and parting, following and gathering, and meeting and gathering. This dataset has many challenges, such as complex background, varying illumination conditions, and different clothing appearances. Sample frames of the considered dataset for this experimentation are shown in Fig. 2. The effectiveness of the proposed method is proven by comparing its results with the results of other existing state-of-the-art methods [19–23, 25]. To analyze the result of the proposed method, we considered classification accuracy as a performance measure which is mathematically defined as [19, 20]

$$\text{Classification accuracy} = (C_A/T_A) \times 100 \tag{1}$$

where C_A is the number of correct activity sequences and T_A is the number of activity sequences taken to be tested, and the result of the proposed method and other existing methods considered for comparison [19–23, 25] on IXMAS, UT interaction, and CASIA datasets is presented in Table 1.

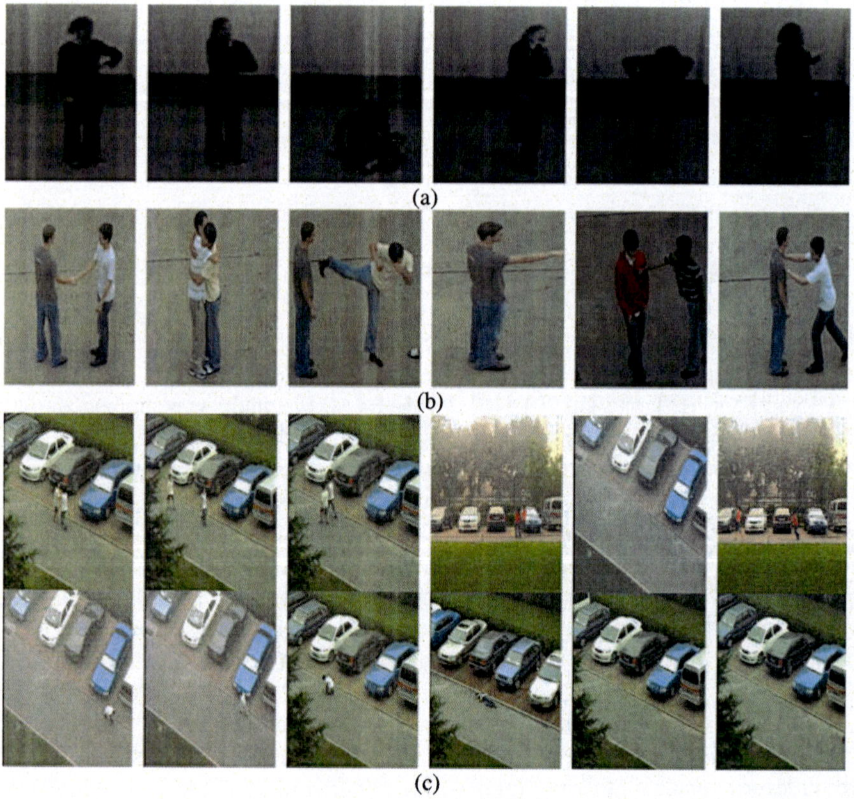

Fig. 2 Sample frames of the considered datasets. **a** IXMAS [14], **b** UT Interaction [15], and **c** CASIA [16]

Table 1 Performance of the proposed method with other state-of-art methods

Method	Accuracy (%) for IXMAS	Accuracy (%) for UT interaction	Accuracy (%) for CASIA (single person)	Accuracy (%) for CASIA (interaction)
Kushwaha and Khare [19]	93.19	100	95.04	93.00
Kushwaha et al. [20]	88.21	99.31	97.95	94.33
Yeffet and Wolf [21]	76.32	99.05%	91.87	95.66%
Nigam and Khare [22]	40.89	86.19%	38.77%	27.00%
Seemanthini and Manjunath [23]	54.31	80.92%	44.92	30.20%
Aly and Sayed [25]	82.76	90.00%	35.71%	57.14%
The proposed Method	93.35	99.11%	97.39%	96.53%

As illustrated in Table 1, it can be observed that the proposed method achieves the highest classification value for the IXMAS dataset (99.19%), CASIA (interaction) (96.35%), second-highest for UT Interaction (99.11%), and CASIA (single person) (97.35%). Although for UT Interaction, Kushwaha et al. [19] achieve the highest accuracy value (100%). For CASIA (single person), Kushwaha et al. [20] achieve the highest accuracy (97.95%), but both accuracy values are comparable to the result of the proposed method; therefore, the overall performance of the proposed method is good. The reason behind excellent accuracy is that the proposed method can extract more discriminant features and provide exemplary performance in low resolution with multi-view and realistic data by the proposed feature descriptor. The efficient object segmentation technique in the proposed method followed by motion information and histogram of oriented gradients gives another reason for excellent accuracy. From Table 1, one can see that the proposed method gives better results for low-resolution data recorded by different views, i.e. for human–human interaction and human-object interaction with the capability to deal with challenges like varying illumination conditions, presence of complex background and camera motion, and variation in scales, poses, and orientations. The recognition results demonstrate the usefulness of the proposed method for real-world applications, e.g. surveillance systems having complex activities and outdoor scenes recorded from different viewing angles.

5 Conclusion

This paper presents human activity recognition framework for motion activities in a realistic and multi-view environment. In this work, we designed a novel feature representation technique based on integrating the object's appearance of interest and the object's motion information. Therefore, we used the object segmentation technique to extract the human object and the optical flow technique to compute the velocity (magnitude) and orientation (direction) information of moving human objects. We considered velocity and direction information to avoid variations in intra-class activities because samples of different activity categories may have the same velocity but not the orientation. The histogram of orientated gradients computation then follows the magnitude and orientation information to compute the dynamic pattern of human activities, which gives a relative distribution of information of each activity category uniquely and in a more discriminative way. The final feature vectors are constructed by integrating local-oriented histogram of optical flow vectors using feature fusion strategy followed by multiclass support vector machine to compute the class score of human activities. The effectiveness of the proposed method is established by conducting several experiments on three different publically available datasets that are IXMAS, UT Interaction, and CASIA. The result of the proposed method was analyzed by comparing its result with several existing state-of-the-art methods. The result of the proposed method demonstrates the outperformance of the method over the other state-of-the-art methods.

Acknowledgements This work was supported by the Science and Engineering Research Board (SERB), Department of Science and Technology (DST), New Delhi, India, under Grant No. CRG/2020/001982.

References

1. Kushwaha A, Khare A, Srivastava P (2021) On integration of multiple features for human activity recognition in video sequences. Multimedia Tools Appl 1–28
2. Ladjailia A, Bouchrika I, Merouani HF, Harrati N, Mahfouf Z (2020) Human activity recognition via optical flow: decomposing activities into basic actions. Neural Comput Appl 32(21):16387–16400
3. Khare M, Binh NT, Srivastava RK (2014) Human object classification using dual tree complex wavelet transform and Zernike moment. In: Transactions on large-scale data and knowledge-centered systems, vol XVI. Springer, Berlin, Heidelberg, pp 87–101
4. Srivastava P, Khare A (2018) Utilizing multiscale local binary pattern for content-based image retrieval. Multimedia Tools Appl 77(10):12377–12403
5. Sivic J, Zisserman A (2003) Video Google: a text retrieval approach to object matching in videos. In: Proceedings ninth IEEE international conference on computer vision, Nice, France, vol 1, pp 1470–1477. https://doi.org/10.1109/ICCV.2003.1238663
6. Sánchez J, Perronnin F, Mensink T, Verbeek J (2013) Image classification with the fisher vector: theory and practice. Int J Comput Vis 105(3):222–245
7. Souly N, Shah M (2016) Visual saliency detection using group lasso regularization in videos of natural scenes. Int J Comput Vis 117(1):93–110

8. Kushwaha A, Khare A, Prakash O, Khare M (2020) Dense optical flow based background subtraction technique for object segmentation in moving camera environment. IET Image Proc 14(14):3393–3404
9. Kushwaha A, Prakash O, Srivastava RK, Khare A (2019) Dense flow-based video object segmentation in dynamic scenario. In: Recent trends in communication, computing, and electronics. Springer, Singapore, pp 271–278
10. Al-Faris M, Chiverton J, Yang L, Ndzi D (2017) Appearance and motion information based human activity recognition. In: IET 3rd international conference on intelligent signal processing (ISP 2017). IET, pp 1–6
11. Farnebäck G (2003) Two-frame motion estimation based on polynomial expansion. In: Scandinavian conference on image analysis. Springer, Berlin, Heidelberg, pp 363–370
12. Dalal N, Triggs B (2005) Histograms of oriented gradients for human detection. In: Proceedings of IEEE computer society conference on computer vision and pattern recognition, vol 1, pp 886–893
13. Li X (2007) HMM based action recognition using oriented histograms of optical flow field. Electron Lett 43(10):560–561
14. Kim SJ, Kim SW, Sandhan T, Choi JY (2014) View invariant action recognition using generalized 4D features. Pattern Recogn Lett 49:40–47
15. Ryoo MS, Aggarwal JK (2009) Spatio-temporal relationship match: video structure comparison for recognition of complex human activities. In: 2009 IEEE 12th international conference on computer vision. IEEE, pp 1593–1600
16. Wang Y, Huang K, Tan T (2007) Human activity recognition based on r transform. In: 2007 IEEE conference on computer vision and pattern recognition, pp 1–8
17. Singh R, Dhillon JK, Kushwaha AK, Srivastava R (2019) Depth based enlarged temporal dimension of 3D deep convolutional network for activity recognition. Multimedia Tools Appl 78(21):30599–30614
18. Simonyan K, Zisserman A (2014) Very deep convolutional networks for large-scale image recognition. arXiv preprint arXiv:1409.1556
19. Kushwaha A, Khare A (2021) Human activity recognition by utilizing local ternary pattern and histogram of oriented gradients. In: Proceedings of international conference on big data, machine learning and their applications. Springer, Singapore, pp 315–324
20. Kushwaha A, Khare A, Khare M (2021) Human activity recognition algorithm in video sequences based on integration of magnitude and orientation information of optical flow. Int J Image Graph 22:2250009
21. Yeffet L, Wolf L (2009) Local trinary patterns for human action recognition. In: 2009 IEEE 12th international conference on computer vision, pp 492–497
22. Nigam S, Khare A (2016) Integration of moment invariants and uniform local binary patterns for human activity recognition in video sequences. Multimedia Tools Appl 75(24):17303–17332
23. Seemanthini K, Manjunath SS (2018) Human detection and tracking using HOG for action recognition. Procedia Comput Sci 132:1317–1326
24. Boser BE, Guyon IM, Vapnik VN (1992) A training algorithm for optimal margin classifiers. In: Proceedings of the fifth annual workshop on Computational learning theory, pp 144–152
25. Aly S, Sayed A (2019) Human action recognition using bag of global and local Zernike moment features. Multimedia Tools Appl 78(17):24923–24953

Multi-agent Task Assignment Using Swap-Based Particle Swarm Optimization for Surveillance and Disaster Management

Mukund Subhash Ghole, Arabinda Ghosh, and Anjan Kumar Ray

1 Introduction

Natural calamities such as earthquakes, hurricanes, floods, volcanic eruptions, or man-made disasters result in significant losses in lives and property and cause degradation in various sectors. The disruptions lead to a huge socioeconomic burden for the affected areas, such as in 2017 the Hurricane Harvey in the USA caused $125 billion losses [1], estimated annual losses due to bush-fire in Australia is approximately $400 million [2]. These events are usually unpredictable in nature which plead us to take emergency measures to prevent, preserve, and save lives and property. These measures are called disaster responses, which include search and rescue missions [3] or surveillance and monitoring operations [4]. These responses can be smartly and swiftly handled by an intelligent multi-agent system (MAS). MAS is a combination of two or more agents that agree to work on a common objective, through coordination. These agents require partial autonomy to make certain decisions on their own and the capability to interact among peers. Applications of MAS can be found in RoboCup [5], coastal patrolling [6], traffic management [7], etc.

In this paper, a MAS is considered based on its popularity and the potential for solving different real-life problems. The objective of this work is to use this collaborative work framework of the MAS in rescue or surveillance operations in disaster-affected areas, for example, recently in Uttarakhand, India, a flash flood due to glacier burst has triggered a large-scale rescue operation [8]. Here, two areas of interest are considered to mimic disaster-affected smart cities. The first one is Gangtok, Sikkim, India which frequently experiences earthquakes [9], and the other

M. S. Ghole (✉) · A. Ghosh · A. K. Ray
Department of Electrical and Electronics Engineering, National Institute of Technology Sikkim, Ravangla 737139, Sikkim, India
e-mail: mukund.ghole@gmail.com

A. Ghosh
e-mail: phee16023@nitsikkim.ac.in

© The Author(s), under exclusive license to Springer Nature Singapore Pte Ltd. 2023
H. Muthusamy et al. (eds.), *Robotics, Control and Computer Vision*, Lecture Notes in Electrical Engineering 1009, https://doi.org/10.1007/978-981-99-0236-1_10

is Marina Beach, Chennai, India, which was affected by Tsunami [10]. An emergency environment is created by invoking tasks inside the areas of interest. The objective of the MAS system is to go on to complete these tasks satisfying different real-life constraints. The distribution of these tasks among these agents is a challenging problem [11, 12]. Thus, a design of an effective procedure for task assignment (TA) is required.

TA is a process to assign tasks using the available resources (in our case, agents) in such a way that the agent system effectively performs the tasks. TA plays an important role in different real-life problems such as surveillance [13], disaster management [14], intelligent parcel delivery system [15], and waste collection management [16]. Since these agents are assigned the tasks, the sequence in which these tasks are completed has a great impact on the total resources used by the agent system. In this paper, the agents are deployed from a base camp to complete some tasks and return to the starting position. This process can be represented as a traveling salesman problem (TSP). TSP is a problem in which an agent has to complete all the tasks by visiting them only once and finally coming back to starting point once all the tasks are completed in the most efficient route. Now, to solve the TSP problem and reduce the resource consumption of the agent system, a swap-based particle swarm optimization (PSO) paradigm is proposed.

PSO is a meta-heuristic algorithm that optimizes a problem by iteratively improving a probable solution [17]. It solves the problem by having a population of probable solutions called particles. Each particle moves in the search space influenced by its local best position and best position among all the particles in the search space. In the proposed method, the objective of PSO is to optimize the assigned task sequence of individual agent to reduce the resource consumption. A variant of PSO named swap-based PSO is used. Various applications of the swap-based PSO algorithm are post-earthquake scenario problem [4], intelligent welding robots path optimization [18], flexible job scheduling problem [19], team formation problem [20], in partial shading of solar panels [21], vehicle routing problem [22], etc. This has motivated the authors to use the swap-based PSO algorithm in this paper. The key contributions of this work are highlighted as follows:

1. A task assignment approach for a multi-agent system is developed which is suitable for surveillance and disaster management. It is assumed that all service requests (tasks) appear at the same time in the form of respective GPS coordinates.
2. A two-stage approach for the assignment of tasks is proposed. At first, the tasks are distributed for each agent based on available resources such as proximity of resources and task completion overhead. This breaks down the problem as a traveling salesman problem for each agent.
3. Then the assigned tasks for an agent are further optimized for the sequence of executions by a proposed swap-based particle swarm optimization.
4. Extensive results are presented to demonstrate the feasibility of the proposed method. It is demonstrated on the Google Maps considering the real coordinates of two different locations (M. G. Marg at Gangtok, India which had experi-

enced earthquakes [9] and Marina Beach at Chennai, India which was affected by Tsunami [10]).

This paper is arranged in the following way, the proposed method is presented in Sect. 2. The results are given in Sect. 3, followed by the conclusion and future direction of this work in Sect. 4.

2 Proposed Methodology

In this work, a task assignment approach for a multi-agent system with sequence optimization is considered. The following assumptions are considered in the proposed method:

1. Point-based agents and tasks are considered for the simulations.
2. All tasks are appearing at the same time.
3. An obstacle-free environment is considered.
4. Stationary tasks are considered.
5. A homogeneous agent system is considered where all the agents in the agent system are of the same specifications.

Now, let us consider an MAS of N number of agents and M number of tasks (service requests) in a workspace where

$$A_i \in A, \forall i = 1, 2, \ldots, N \tag{1}$$

$$T_j \in T, \forall j = 1, 2, \ldots, M \tag{2}$$

where A_i is the position of the agent i and T_j is the position of the task j. Next, the stage I of the proposed method is presented.

2.1 Stage I: Assignment of the Task to MAS

The assignment of tasks to agents is dependent on key factors like availability of resources, the proximity of resources, task completion overhead, etc. In the proposed method, the assignment of tasks will begin with calculating the distance between the agents and tasks considering these key factors and represented as

$$d_{ijr} = \left\| A_i, T_{jr} \right\| + dia_i, \ \forall A_i, \ \forall T_{jr} \tag{3}$$

where jr is the index of M_r number of unassigned tasks, T_{jr} is the position of the unassigned tasks, and dia_i is the distance overhead of the agent i. Initially, $M_r = M, \ dia_i = 0, \ T_{jr} = T_j$.

Now, the job is to find the closest agent to each task (Eq. 4) and the closest task for each agent (Eq. 5):

$$d_{ta} = arg\ \underset{min\ i}{min}(d_{ijr}),\ \forall\, T_{jr},\ i = 1, 2, \ldots, N \tag{4}$$

$$d_{at} = arg\ \underset{min\ jr}{min}\ (d_{ijr}),\ \forall\, A_i,\ \forall\, T_{jr} \tag{5}$$

Now, an agent i will be assigned a task j iff,

$$i = d_{at} \text{ and } j = d_{ta} \tag{6}$$

Now, when a task is assigned to an agent, the corresponding agent will be denoted as $A_{(assigned,i)}$ and the corresponding task will be denoted as $T_{(assigned,jr)}$. Let, there be a binary matrix C_i of agent i such that

$$\begin{aligned} C_i(T_l, T_q) &= 1, \text{ if agent } i \text{ goes to task } T_q \text{ from task } T_l \\ &= 0, \text{ otherwise} \end{aligned} \tag{7}$$

C_i is a square matrix of dimension $M_{(assigned,i)} + 1$ where $M_{(assigned,i)}$ is the number of tasks that are assigned to agent i. Initially, agent i must go only to its first assigned task T_l from its starting point such that

$$\sum_{l=1}^{M_{(assigned,i)}} C_i(A_i, T_l) = 1 \tag{8}$$

Now, we update the remaining tasks and position of agents to the already assigned tasks as follows:

$$A_{(assigned,i)} = T_{(assigned,jr)} \tag{9}$$

Let us consider that p number of tasks are assigned in this time step. Therefore, $M_r = M_r - p$. Now, the distance overhead will be calculated as

$$dia_i = \left\| A_{(assigned,i)}, T_{(assigned,jr)} \right\| + dia_i + \delta_i \tag{10}$$

where δ_i is the ith agent task completion cost. Now, if the ith agent goes to task $T_{(assigned,jr)}$ from task T_l, the binary matrix C_i will be updated as

$$C_i(T_l, T_{(assigned,jr)}) = 1 \tag{11}$$

Repeat the process from Eqs. (3) to (11), until $M_r = 0$, i.e., till all the tasks are assigned.

After the assignment process, the agent will be at the end of its assigned task. Now, the agent has to come back to its starting position which is represented as

$$
\begin{aligned}
A_{(assigned,i)} &= A_{(\text{starting point},i)} \\
C_i(T_{M_{(assigned,i)}}, A_{(\text{starting point},i)}) &= 1 \\
dia_i &= \left\| T_{M_{(assigned,i)}}, A_{(\text{starting point},i)} \right\| + dia_i
\end{aligned} \tag{12}
$$

$A_{(\text{starting point},i)}$ is the starting position of the agent i and $T_{M_{(assigned,i)}}$ is the last task assigned to agent i. Now, this problem can be presented as TSP.

2.2 Representation of the Proposed Method as a TSP

Consider there are $M_{(assigned,i)}$ tasks and there is an agent i which has to visit each task only once and finally come back to the starting point. The objective of TSP is to complete these tasks in such a way that the total cost incurred by the agent after completing all the tasks and returning back to its starting point is minimum. Thus, the objective of the agent i is defined as

$$O - min\,[dia_i] \tag{13}$$

$$\text{subject to:} \quad \sum_{s=1,s\neq l}^{M_{(assigned,i)}} C_i(T_l, T_s) = 1, \ \forall\, l \tag{14}$$

$$\sum_{l=1,l\neq s}^{M_{(assigned,i)}} C_i(T_l, T_s) = 1, \ \forall\, s \tag{15}$$

where Eq. (14) represents that the agent goes to any one of the task T_s (excluding T_l) from T_l and Eq. (15) represents that the agent comes from any one of the task T_l (excluding T_s) to T_s.

One of the objectives of this work is to minimize Eq. (13) (henceforth will be called as path cost) using the constraints given in Eqs. (14) and (15). This leads to the stage II of the proposed method presented in the next section.

2.3 Stage II: Optimizing Task Sequence Using Swap-Based PSO

PSO is a meta-heuristic optimization method, inspired by the social behavior of organisms such as flock of birds or school of fish where it tries to optimize an objective function iteratively by having a population of probable solutions called particles. Each particle is influenced by its current solution, its own best known solution, and the best known solution among the population. To optimize the task sequence of each agent, a swap-based PSO technique [23] is proposed in this paper.

2.3.1 Initialization of Swap Operations

In the original PSO, each particle starts with an initial position from a defined search space. In the proposed method, each particle will start with a sequence of tasks assigned to a particular agent. Let's consider that there is an agent i having the sequence of tasks assigned as discussed in the previous subsection. Let this agent have K number of PSO particles (henceforth will be called as particles) with each particle k containing the random sequence of the tasks assigned to agent i, so this kth particle is defined as

$$Z_k = (T_1, T_2, \ldots, T_{M(assigned,i)}), \forall\, k = 1, 2, \ldots, K \tag{16}$$

2.3.2 Swap Operator

Swap operator ($SO(T_i, T_j)$) is a process of exchanging the position of Task T_i and the Task T_j when applied on the kth particle solution (Z_k). Therefore, the new solution is

$$Z_k^{new} = Z_k + SO(T_i, T_j) \tag{17}$$

"+" sign indicates that the swap operator ($SO(T_i, T_j)$) is acting on Z_k to obtain Z_k^{new}. For instance, let Z_k be (1,3,2,4) and then $SO(1, 2)$ acts on Z_k to get Z_k^{new} as (3, 1, 2, 4).

2.3.3 Swap Sequence

The swap sequence is defined as the collection of swap operators of particle k and it is denoted as

$$SS_k = (SO_1 + SO_2 + \cdots + SO_{(M(assigned,i)-1)}) \tag{18}$$

Furthermore, a consensus of swap sequence is formed by merging multiple swap operators such that

$$SS_{total} = SS_1 \oplus SS_2 \oplus \ldots \tag{19}$$

2.3.4 Generation of Swap Sequence

Let a normal solution of kth particle be Z_k and a target solution of kth particle be $Z_k(tgt)$. The swap sequence that should operate on Z_k to get $Z_k(tgt)$ is defined as

$SS(Z_k, Z_k(tgt))$, e.g., let Z_k = (2, 3, 1, 4) and let $Z_k(tgt)$ = (1, 2, 3, 4). The swap sequence that is generated will be $SS_k(Z_k, Z_k(tgt))$ = $(SO(1,3)+SO(2,3))$. So, here first $SO(1,3)$ will act on Z_k to get Z_k = (1, 3, 2, 4) and then $SO(2,3)$ will act to get Z_k = (1, 2, 3, 4) which is $Z_k(tgt)$.

2.3.5 Velocity Update

The velocity of each particle will be updated as

$$V_k(t+1) = \omega(t) \times V_k(t) \oplus N_{pk}(t) \times P(t) \oplus N_{gk}(t) \times G(t) \quad (20)$$

t is the current iteration; V_k is the velocity of the particle k which is actually the consensus swap sequence of particle k; $\omega(t)$ is the inertia weight update for iteration t; $N_{pk}(t)$ and $N_{gk}(t)$ are the number of swap operators to be allowed to operate on $Z_k(t)$ for both $P(t)$ and $G(t)$, respectively; and $P(t)$ and $G(t)$ represent the swap sequence that is generated by comparing $Z_k(t)$ with $Z_k^{pb}(t)$ and $Z_k^{gb}(t)$, respectively.

$$N_{pk}(t) = \text{ceil}(\alpha \times N_k(t)); \ \alpha = rand[0, 1] \quad (21)$$

$$P(t) = SS(Z_k(t), Z_k^{pb}(t)) \quad (22)$$

$$N_{gk}(t) = \text{ceil}(\beta \times N_k(t)); \ \beta = rand[0, 1] \quad (23)$$

$$G(t) = SS(Z_k(t), Z_k^{gb}(t)) \quad (24)$$

$N_k(t)$ is the total number of swap operators required to generate the swap sequences $P(t)$ and $G(t)$ separately.

2.3.6 Position Update of Particles

At the end of each iteration, consensus swap sequence $V_k(t+1)$ is applied on $Z_k(t)$ to obtain $Z_k(t+1)$ as

$$Z_k(t+1) = Z_k(t) + V_k(t+1) \quad (25)$$

With $Z_k(t+1)$ the path cost $(dia_k(t+1))$ is calculated. The personal best solution of each particle Z_k^{pb} will be updated iff

$$dia_k(t+1) < dia_k^{pb}(t) \quad (26)$$

The global best solution Z_k^{gb} will be updated iff

$$dia_k(t+1) < dia_k^{gb}(t) \quad (27)$$

The process will be repeated from Eqs. (20) to (27) for each particle $k \in K$ until $t = iter_{max}$. The final solution and path cost of an agent i will be Z_k^{gb} and dia_k^{gb}, respectively.

3 Results

Extensive simulations are done on Google Maps. Two locations are considered for simulations, one is Marg at Gangtok, India [24] and the other is Marina Beach at Chennai, India [25]. The process of calculating the aerial distance between two GPS coordinates is given in Appendix 5. For each location, two PSO variants are considered. These are

1. Biased swap-based PSO, in this variation the agents which are assigned the tasks in TA, their task execution sequence is optimized by taking each agent's task sequence and the respective path cost from the TA process as initial global best task sequence and initial global best path cost for the particles.
2. Unbiased swap-based PSO, in which random task sequence is considered as the initial global best sequence and initial global best path cost for the particles.

In the proposed method, the following parameter values are considered: number of PSO particles are 20 for all variations; $iter_{max}$ are 100 and 50 for M. G. Marg, Gangtok and Marina Beach at Chennai, respectively; number of tasks are 100 and 50 for M. G. Marg, Gangtok and Marina Beach, Chennai, respectively; and number of agents are 20. The values of ω are updated using the methods and values presented in [26]. The result of the proposed task assignment method is demonstrated in Fig. 1 and the results of the agents' assignment by the proposed biased swap-based PSO method are shown in Fig. 2 for M. G. Marg, Gangtok. In both the figures, the movements of agent 11, 12, and 20 are shown. By the proposed task assignment process as shown in Fig. 1, agent 11 is assigned with tasks 17, 52, 47, and 84 which incurred a path cost of 142.503 units; agent 12 is assigned with tasks 38, 80, and 18 which incurred a path cost of 130.814 units; and agent 20 is assigned with tasks 73, 79, 94, 44, 26, 15, 58, 56, 82, 36, and 53 which incurred a path cost of 187.868 units. On the contrary, in Fig. 2, agent 11 is assigned with tasks 17, 52, 84, and 47 which incurred a path cost of 140.228 units; agent 12 is assigned with tasks 38, 18, and 80 which incurred a path cost of 130.814 units; and agent 20 is assigned tasks 73, 79, 94, 44, 53, 15, 58, 56, 82, 36, and 26 which incurred a path cost of 177.622 units. Similar observations of improvements are also noted for Marina Beach, Chennai, India as demonstrated through Fig. 3 (stage I or task assignment results) and Fig. 4 (unbiased swap-based PSO results) using the movements of agents 1, 14, and 16. This shows that with the inclusion of swap-based PSO algorithm, the cost incurred at the individual agent level and the total cost incurred by the MAS have improved.

The analysis is extended to demonstrate the effects of the maximum number of iterations on total path cost by all agents. Three variations in the maximum number of iterations are considered here. For each variation, the proposed method is simulated

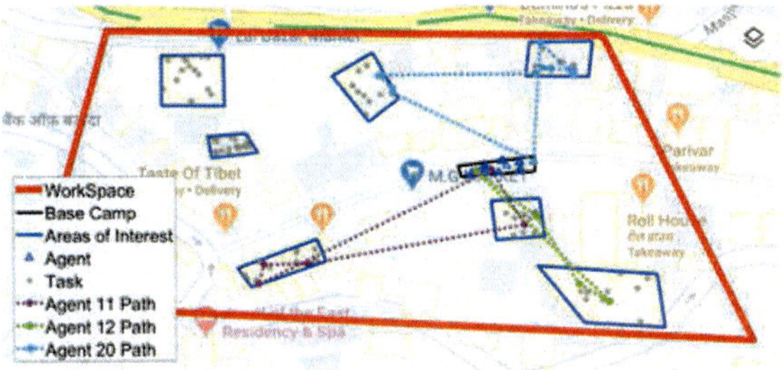

Fig. 1 Task assignment results for agents at M. G. Marg, Gangtok, India. Here, the assignments of agents 11, 12, and 20 are shown

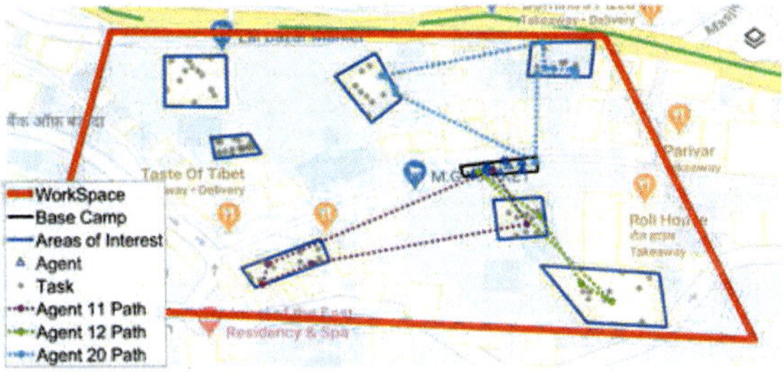

Fig. 2 Results of agents' assignments using biased swap-based PSO for agents 11, 12, and 20 at M. G. Marg, Gangtok, India

Fig. 3 Task assignment results for agents at Marina Beach, Chennai, India. Here, the assignments of agents 1, 14, and 16 are shown

Fig. 4 Results of agents' assignments using unbiased swap-based PSO for agents 1, 14, and 16 at Marina Beach, Chennai, India

Table 1 Effect of variations in the total number of iterations in PSO algorithm for both PSO variations for M. G. Marg, Gangtok

M. G. Marg, Gangtok, India			
TA	Iterations	Biased swap-based PSO	Unbiased swap-based PSO
2867.949	50	2844.294	2843.297
	100	2840.402	2840.663
	150	2840.663	2841.323

Table 2 Effect of variations in the total number of iterations in PSO algorithm for both PSO variations for Marina Beach, Chennai, India

Marina Beach, Chennai, India			
TA	Iterations	Biased swap based PSO	Unbiased swap based PSO
33897.434	50	33881.104	33881.134
	100	33884.206	33885.079
	150	33881.133	33881.104

10 times and the best results among 10 runs are presented here. In Table 1, the variations are shown for M. G. Marg, Gangtok and Marina Beach, Chennai for both variants of the PSO algorithm. It is observed from Table 1 that there is a good improvement in the results of path cost for 100 and 150 iterations as compared to 50 iterations in biased swap-based PSO mode. Thus, for the task assignment problem at M. G. Marg, Gangtok, the increasing number of iterations are improving the total path cost by all agents. However, in case of Marina Beach, Chennai, the increasing number of iterations have negligible effect on the total path cost by all agents as shown in Table 2.

4 Conclusion

In this work, a multi-agent task assignment procedure is developed for simultaneous tasks for disaster management and/or surveillance of different areas of interest. A two-stage task assignment approach supported by a swap-based PSO algorithm ensures that each agent receives an optimized execution sequence of tasks. The proposed method is implemented on the Google Maps using GPS coordinates of Marina Beach, Chennai, India and M. G. Marg, Gangtok, India. Results show that each agent attends respective tasks and returns to the base successfully. It is also observed that the swap-based PSO has improved the execution sequence of tasks along with the total task assignment cost in comparison to the stand-alone task assignment process. In future, the authors would like to extend this work to consider the proximity of resources (e.g., fuel), the priority of tasks, and dynamic task assignment where tasks will appear randomly or sequentially.

5 Aerial Distance Between Two GPS Coordinates

The proposed method is implemented on Google Maps. To calculate the aerial distance between two GPS coordinates, the law of cosines model is used, where it is assumed that earth is spherical [27] and this model considers all the GPS coordinates at the mean sea level. The following process is used for the calculation of aerial distance. Let $GPS_1^o = (lat_1^o, lon_1^o)$, where lat_1^o and lon_1^o are in decimal degree. Then,

$$GPS_1^r = GPS_1^o \times \left(\frac{\pi}{180}\right)$$
$$GPS_2^r = GPS_2^o \times \left(\frac{\pi}{180}\right)$$

Let $A = sin(lat_1^r) \times sin(lat_2^r)$ and $B = cos(lat_1^r) \times cos(lat_2^r) \times cos(lon_2^r - lon_1^r)$, then aerial distance $= cos^{-1}(A + B) \times$ earth radius. To obtain accuracy in few meters, cos^{-1} needs to be accurate up to 10 decimal places or in double format.

References

1. Sun W, Bocchini P, Davison BD (2020) Applications of artificial intelligence for disaster management. Nat Haz 103(3):2631–2689
2. Jyoteeshkumar RP, Sharples JJ, Lewis SC, Perkins-Kirkpatrick SE (2021) Modulating influence of drought on the synergy between heatwaves and dead fine fuel moisture content of bushfire fuels in the Southeast Australian region. Weather Clim Extremes 31:100300
3. Malaschuk O, Dyumin A (2020) Intelligent multi-agent system for rescue missions. In: Advanced technologies in robotics and intelligent systems. Springer, pp 89–97
4. Zhu M, Du X, Zhang X, Luo H, Wang G (2019) Multi-UAV rapid-assessment task-assignment problem in a post-earthquake scenario. IEEE Access 7:74542–74557
5. Asada M, Stone P, Veloso M, Lee D, Nardi D (2019) RoboCup: a treasure trove of rich diversity for research issues and interdisciplinary connections [TC spotlight]. IEEE Robot Autom Mag 26:99–102
6. Turner IL, Harley MD, Drummond CD (2016) UAVs for coastal surveying. Coast Eng 114:19–24
7. Hamidi H, Kamankesh A (2018) An approach to intelligent traffic management system using a multi-agent system. Int J Intell Transp Syst Res 16(2):112–124
8. BBC News, Uttarakhand Dam Disaster: race to rescue 150 people missing in India. https://www.bbc.com/news/world-asia-india-55975743. Accessed 4 Feb 2022
9. Baruah S, Bramha A, Sharma S, Baruah S (2019) Strong ground motion parameters of the 18 September 2011 Sikkim Earthquake Mw = 6.9 and its analysis: a recent seismic hazard scenario. Nat Haz 97(3):1001–1023
10. Satpathy KK (2005) Impact of Tsunami on Meiofauna of Marina Beach, Chennai, India. Curr Sci-Bangalore 89(10):1646

11. Ghole MS, Ghosh A, Singha A, Das C, Ray AK (2021) Self organizing map-based strategic placement and task assignment for a multi-agent system. In: Advances in intelligent systems and computing. Springer, pp 387–399
12. Ghole MS, Ray AK (2020) A neural network based strategic placement and task assignment for a multi-agent system. In: Lecture notes in electrical engineering. Springer, pp 555–564
13. Gu J, Su T, Wang Q, Du X, Guizani M (2018) Multiple moving targets surveillance based on a cooperative network for multi-UAV. IEEE Commun Mag 56(4):82–89
14. Li P, Miyazaki T, Wang K, Guo S, Zhuang W (2017) Vehicle-assist resilient information and network system for disaster management. IEEE Trans Emerg Top Comput 5(3):438–448
15. Wang F, Wang F, Ma X, Liu J (2019) Demystifying the crowd intelligence in last mile parcel delivery for smart cities. IEEE Netw 33(2):23–29
16. Shao S, Xu SX, Huang GQ (2020) Variable neighborhood search and Tabu search for auction-based waste collection synchronization. Transp Res Part B: Methodol 133:1–20
17. Kennedy J, Eberhart R (1995) Particle swarm optimization. In: Proceedings of ICNN'95-international conference on neural networks. IEEE, pp 1942–1948
18. Yifei T, Meng Z, Jingwei L, Dongbo L, Yulin W (2018) Research on intelligent welding robot path optimization based on GA and PSO algorithms. IEEE Access 6:65397–65404
19. Gu XL, Huang M, Liang X (2020) A discrete particle swarm optimization algorithm with adaptive inertia weight for solving multiobjective flexible job-shop scheduling problem. IEEE Access 8:33125–33136
20. El-Ashmawi WH, Ali AF, Tawhid MA (2019) An improved particle swarm optimization with a new swap operator for team formation problem. J Indus Eng Int 15(1):53–71
21. Li H, Yang D, Su W, Lu J, Yu X (2019) An overall distribution particle swarm optimization MPPT algorithm for photovoltaic system under partial shading. IEEE Trans Indus Electron 66(1):265–275
22. El-Hajj R, Guibadj RN, Moukrim A, Serairi M (2020) A PSO based Algorithm with an Efficient Optimal Split Procedure for the Multiperiod Vehicle Routing Problem with Profit. Annals of Operations Research 291(1):281–316
23. Liu X, Su J, Han Y (2007) An improved particle swarm optimization for traveling salesman problem. In: International conference on intelligent computing, pp 803–812
24. MG Marg, Gangtok, India, lat 27.32860 (deg) and lon 88.61230 (deg), (Google Earth). Accessed 4 Feb 2022
25. Marina Beach, Chennai, India, lat 13.056327 (deg) and lon 80.283403 (deg), (Google Earth). Accessed 4 Feb 2022
26. Huang X, Li C, Chen H, An D (2020) Task scheduling in cloud computing using particle swarm optimization with time varying inertia weight strategies. Clust Comput 23(2):1137–1147
27. Calculate distance, bearing and more between Latitude/Longitude points. https://www.movable-type.co.uk/scripts/latlong.html. Accessed 4 Feb 2022

Facemask Detection and Maintaining Safe Distance Using AI and ML to Prevent COVID-19—A Study

Ankita Mishra, Piyali Paul, Koyel Mondal, and Sanjay Chakraborty

1 Introduction

COVID-19 was initially reported in Wuhan, China, and then it has been unrolled to the whole world. The rapid spread of the coronavirus has resulted in 4 million global deaths by Oct 21, 2021. COVID-19 is becoming a headache for everyone. Everyone is afraid of this disease. The COVID-19 pandemic has created a difficult scenario for the entire world; as a result, everyone is taking drastic measures to stem the spread of coronavirus. Coronavirus spread can be kept away by maintaining distance and wearing masks to prevent the transmission of the virus from one person to another.

In a nutshell, the performances of this study are mentioned as follows:

- This paper makes an extensive study on some recent research works to detect facemasks worn by people and check safe distances through machine learning and deep learning techniques along with the concept of image processing.
- Performances among several state-of-the-art methods are investigated and compared.
- Discusses the benefits and applications of these recent studies.

The rest of the paper is organized as follows. Section 2 discusses some state-of-the-art methods proposed for handling the COVID-19 spread. This section discusses various facemasks and social distancing approaches where machine learning, deep learning, and image processing play vital roles. Then a brief comparison section among some popular methods in this domain is discussed and analyzed in Sect. 3.

A. Mishra · P. Paul · K. Mondal
Department of CSE, JIS University, Kolkata, India

S. Chakraborty (✉)
Department of Computer Science and Engineering, Techno International New Town, Kolkata, India
e-mail: sanjay.chakraborty@tint.edu.in

© The Author(s), under exclusive license to Springer Nature Singapore Pte Ltd. 2023
H. Muthusamy et al. (eds.), *Robotics, Control and Computer Vision*, Lecture Notes in Electrical Engineering 1009, https://doi.org/10.1007/978-981-99-0236-1_11

Eventually, we conclude the paper in Sect. 4 by highlighting some of the future research scopes.

2 Related Study

2.1 *Facemask and Safe Distance Detection Using SVM and CNN*

The paper [1] presents a clear picture of recent studies about machine learning and also artificial intelligence to handle all sides of COVID-19 trouble in various ranges like molecular, clinical, and societal applications. In [2] this paper, a hybrid deep learning and machine learning model is applied for detecting facemasks. The first part is designed for feature extraction using ResNet-50 and for the classification process of facemasks using decision trees, support vector machine (SVM), and the ensemble method. In [3] this paper, the main goal is to identify the crowd. To adopt live video, Raspberry Pi is used along with an RPi camera. Then the video is prepared frame-by-frame. To identify people, vehicles in the video image processing are used. Moreover, TensorFlow and OpenCV take an important role to do so. A model has been established in [4] this paper to identify masked and physical distances among construction workers in order to protect their safety during the COVID-19 epidemic. Among the several models to achieve 99.8% accuracy for recognizing facemask, a fast region-based CNN inception ResNet version2 network is chosen. The goal of [5] this work is to develop RetinaFaceMask, a unique facemask detector that can detect facemasks and contribute to public healthcare. In [6] this paper, Image classification studies the performance of the convolution neural network (CNN). In [7] this paper, MobileNet which is a new model architecture that is based on depth-wise dividable convolutions is suggested. In [8] this book, neural networks are discussed broadly and how they can be used extensively to predict diseases. A facemask detection model based on computer vision and deep learning has been proposed in this research [9]. This model can be used in conjunction with a computer or laptop camera to determine whether people are wearing masks on their faces. In [10] this paper, the main goal of this study is to gain more about social distancing and facemask detection. Normally object detection is taking place for social distancing and faces are being used to identify masks on faces. OpenCV is generally used for all this. Opencv Darknet is responsible for target tracking.

2.2 *Facemask and Safe Distance Detection Using CNN Along with YOLO Models and Internet of Things (IoT)*

In [11] this paper, using image processing and deep learning, real-time social distance is calculated. The YOLO location model is used here. YOLO has three tuning boundaries. In the first place, the edges are marked. At that stage of the jump, the box is sorted which then determines the center of the bounding box. The edge is pre-prepared to provide three results, which are assurance, the bounding box, and the centroid of each person. In [12] this paper, a system has been proposed that uses computer vision and the MobileNet V2 architecture for the benefit of the environment and automatically monitors public places to prevent the spread of the COVID-19 virus. In [13] this paper, the intention is to construct a system to detect if a person is wearing a mask or not and notify the resembling authority in a smart city network with the help of CCTV cameras and features extraction from images CNN. In this research paper [14], pre-trained deep neural network models like ResNet Classifier, DSFD, and YOLOv3 bounding box have been utilized to identify individuals and masks and conclude two things: Social distance can abate the expansion of the coronavirus. In this paper [15], an integrated real-time facemask and social distance infraction detection system has been built where objects are identified using YOLO v4. In [16] this paper, recent technology has been used such as computer vision and deep learning. It uses the MobileNetV2 architecture for facemask detection and uses the Euclidean distance formula for distance computing. In [17] this paper, a system has been suggested that monitors human activity using deep learning techniques, assuring human safety in public places. In [18] this paper, the explicit study is based on the conclusions of previous literary work with social distance and related technical predictions. In [19] this paper, a summarized preface to social distance and masks is presented that is the main resource in this present scenario. In [20] this paper, the action can able to differentiate the type of social distance and categorize them as the norm of social distance. In addition, it shows labels according to object identification. The classifier has been applied to live video streams and photos. By observing the distance between two people, it can be confirmed that one person is maintaining social. In [21] this paper, a model has been proposed that can able to detect social distance and facemask by using YOLOv2, YOLOv3, and YOLOv4. Social distance and facemask detection is performed using the Darknet model YOLOV4, from video collected by a camera or user-provided images and videos, identifying whether people follow social distances and whether wearing a mask. In [22] a red line will report this paper, the deep learning and YOLO methods are used to reduce the caliber of coronavirus epidemics by assessing the distance between humans, and any couple failing to comply with the regulations.

2.3 Facemask and Safe Distance Detection Using CNN Along with YOLO Models

In the paper [23], the training and testing of the commonly used deep pre-trained CNN models (DenseNet, InceptionV3, MobileNet, MobileNetV2, ResNet-50, VGG-16, and VGG-19) using the Facemask dataset are simulated. In [24] this paper, the OpenCV is used to gather live input video feeds from webcams and to feed them into deep learning models. Using a complex neural network to classify the various object classes discernible in the video gives us objects that are interested in such as people and a closed box around them and then comparing distances. In [25] this paper, a comparative study of various methods of CNN and machine learning techniques for the detection and identification of a person wearing a facemask to prevent the spread of COVID-19 is given. In [26] this work, a deep learning-based approach for detecting masks has been introduced by using a combination of single and two-stage detectors and then a transfer learning is applied to pre-trained models to measure the accuracy and robustness of the system. In [27] this paper, they manufactured the PWMFD with 9205 high-quality masked face photos and developed SE-YOLOv3, a quick and accurate mask detector with a channel attention mechanism that improved the backbone network's feature extraction capability. The findings show that this Yolo can provide state-of-the-art performance in object identification and classification while requiring significantly less inference time [28]. The fundamental goal of this [29] is to summarize the critical roles of AI-driven approaches (machine learning, deep learning, and so on) and AI-empowered imaging techniques in analyzing, predicting, and diagnosing COVID-19 disease. Various machine learning and deep learning models are developed in the paper [30] to predict the PPIs between the SARS-CoV-2 virus and human proteins, which are then confirmed using biological tests.

3 Comparison and Analysis Among State-of-the-Art Approaches

In this section, we have compared among several state-of-the-art methods. The comparisons are done in terms of used tools and techniques and recognition accuracy.

In Table 1, a comparison of various mask detection techniques is discussed. YOLO categories like YOLO v2, YOLO v3, and YOLO v4 are used to detect objects for facemask identification and to maintain social distance. Along with YOLO, CNN is also used. We can say after analyzing the accuracy graph in Fig. 1

YOLO is a very effective technology to give effective accuracy in terms of this kind of work. In Table 2, a comparison of different social distance-maintaining strategies is noted. All of these papers use a method of deep learning strategies to monitor public activity, ensuring the safety of people in public places. These papers use deep learning algorithms like MobileNetV2, SDBox, etc. All other technologies like Computer Vision and MobileNet V2 are used. We can say that deep learning is

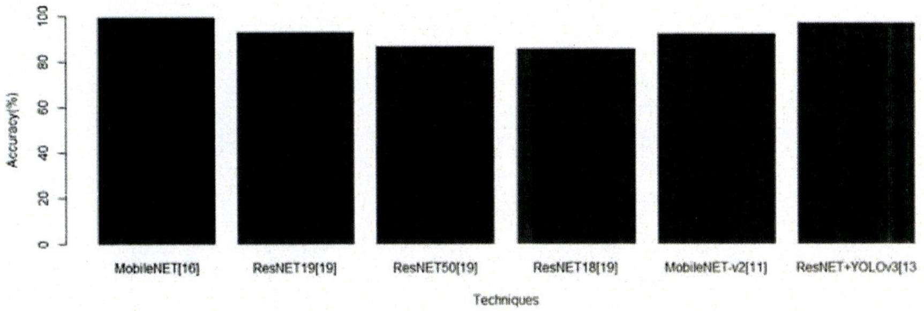

Fig. 1 First comparison among various techniques based on accuracy

another important way to get better accuracy in this type of work after analyzing Fig. 2.

In Table 3, we have considered some popular papers that make a comparison on various social distance-maintaining techniques mask detection techniques. These papers use CNN, AI, Deep Learning, YOLO, MobileNetV2, and so on. We can say these technologies are very much effective to give an accurate result considering the above accuracy graph in Fig. 3.

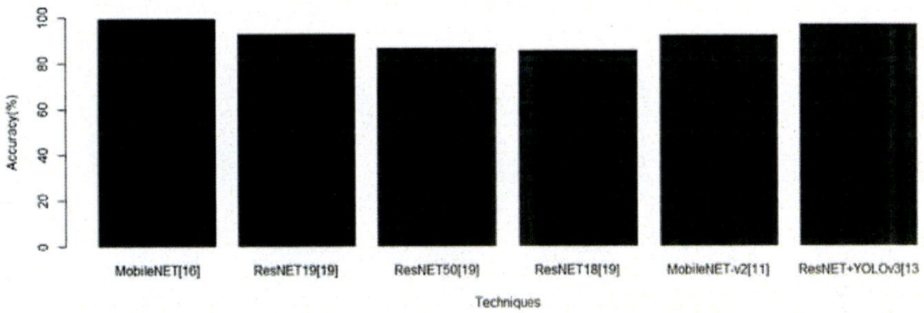

Fig. 2 Second comparison among various techniques based on accuracy

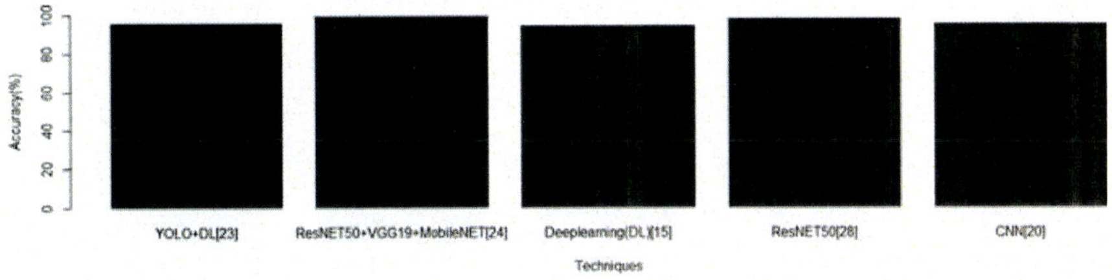

Fig. 3 Third comparison among various techniques based on accuracy

Table 1 First comparison of various techniques

	Real-time facemask and social distancing violation detection system [15]	Real-time facemask detection method [27]	Application of Yolo on mask detection task [28]	Detection and identification of facemask [25]	Social deprivation with protective mask detector [21]
Techniques used	YOLO v4	PWMFD + 9205 and SE-YOLO version3, Darknet53	YOLO NN	CNNs, Google FaceNET, and YOLOv3	YOLO v2
Depth of the models	6120 images × 8000 iterations	13 × 13, 26 × 26, and 52 × 52	Total images: 3145, with mask: 2546, no mask: 508 incorrect mask: 91	–	–
Number of trainable parameters	• Total steps: 8000 • BS: 64 • Mini BS: 64 • Momentum: 0.949 • Decay: 0.0005 • Prior learning rate: 0.001	Four (4) trainable parameters are used in the proposed model	Batch Size = 8, Width = 512, Height = 512, no. of channels = 3 Momentum = 0.9 Decay = 0.0005 Angle = 0 Saturation = 1.5 Exposure = 1.5 Hue = 0.1, LR = 0.001, 200 epochs	–	–
Accuracy	94.5%	93%	91%	CNNs, Google FaceNET, and YOLOv3-98% HGL method with CNN-90% of frontal and 87% of side The SVM classifier-98.64%	95%

Table 2 Second comparison of various techniques

	Social distancing and face mask detection using deep learning [17]	SocialdistancingNet-19 [20]	DL-based safe distance and facemask detection [19]	DL-based safer distancing and facemask detection [12]	Detection using DL and computer vision [14]
Techniques used	Deep learning technique	Deep learning and socialdistancingNet-19 model	Deep learning algorithms like mobilenetV2 and SDBox	CNN, transfer learning, Single Shot Detector (SSD), and MobileNet V2 architecture	ResNet, DSFD, and YOLO version3 along DBSCAN clustering
Model testing and usage	MobileNet use as the backbone The model is tested with images	During pandemics, this method might be utilized in CCTV surveillance to keep an eye on individuals In crowded venues such as train stations, bus stops, markets, streets, mall entrances, schools, colleges, and so on	Dual OpenCV for facemask detection, TensorFlow, and MobileNetV2 model	This method can be employed in temples, shopping malls, metro stations, and airports, among other places	In this study, a detector is used to detect the faces of the participants
Depth of the models	–	Convolution + batch normalization + ReLU 112 × 112 × 64 Max pooling + Convolution + batch normalization + ReLU 56 × 56 × 64 32 global average pooling and 10 fully connected + Softmax	MobileNetV2 supports two-dimensional three-channel image Input image size: 256 × 256	Convolution layer-1 3 × 3 × (3 × (classes + 4)) Convolution layer-2 3 × 3 × (6 × (classes + 4))	–

(continued)

Table 2 (continued)

	Social distancing and face mask detection using deep learning [17]	SocialdistancingNet-19 [20]	DL-based safe distance and facemask detection [19]	DL-based safer distancing and facemask detection [12]	Detection using DL and computer vision [14]
Number of trainable parameters	–	The network input sizes, anchoring box, and feature extraction network are the three tuning parameters in YOLO	Learning rate = 0.0001, EPOCHS = 50 and batch size BS = 32	Adam optimizing, learning rate = $1e^{-4}$, epochs = 20 and BS = 32	–
Accuracy	99.22%	SocialdistancingNet-19–92.8% ResNet: 50–86.5% ResNet: 18–85.3%	–	92% Precision-0.917 Recall-0.917	Between 96.73 and 100%

Table 3 Third comparison of various techniques

	Social Distancing Detection [22]	Real-time artificial intelligence-based facemask detection and safer distancing [23]	Safe distancing and facemask detection through CCTV [16]	Facemask detection using DL [26]	COVID-19 Face Mask Detection [9]
Techniques used	RCNN, SSD, and YOLO	CNN (DenseNet, Inception version3, MobileNet, MobileNet version2, ResNet-50, VGG-16, and VGG-19)	Computer vision + DL Kaggle datasets, and RMFD	ResNet version50	OpenCV, tensor flow, Keras, and CNN
Model testing & usage	This system is suitable for many public ambiances like restaurants, schools, offices, stations, etc.	Public area, a station, a corporate setting, a road, a retail mall, or a test center, where accuracy is critical. Smart city is another application	Detection of distance between two persons	Face masks and social alienation can be detected using this technology in public areas such as schools, airports, and markets	1315 samples Training phase: testing phase = 80:20
Depth of the models	Input image 3 × 3 and 45 frames/sec	InceptionV3: 48 deep layers with 1 × 1, 3 × 3, and 5 × 5 convolutions ResNet-50: 48 convolutional layers with 1 max pooling and also 1 average pooling layer	Input image: 3835, 196 with mask, and 1916 without mask	25,876 input images, 23,858 masked images, and 2018 non-masked images Pool size equal to 5 × 5, a dense ReLU layer of 128 neurons, a dropout of 0.5	1,315 input images, 658 face masks, and 657 without masks images Convolution layer contains 100 kernels Max pooling layer of size 2 × 2
Number of trainable parameters	–	Learning rate:0.0001, epochs: 20–40, and batch size: 32	–	Learning rate-0.03, SGD, momentum = 0.9, and batch size = 64, epochs: 60	Epochs:10, learning rate: 0.0002, and batch size: 32
Accuracy	95.5%	99%	94.2%	Precision-98.86% Recall-98.22%	95%

4 Conclusion

In this study, we have investigated the performance issues of various facemask and safe distance techniques using AI and ML to prevent COVID-19. This paper summarizes some recent popular methods and their proposed approaches along with their applications to stop the spread of this disease. The main challenges of these approaches in face mask detection come from the diversity of in-the-wild scenarios, which include non-mask occlusion, various types of masks, different face orientations, and small or blurred faces. In this paper, the readers can find an extensive set of comparison studies among those methods with respect to some popular parameters that helps them to do their future research works in this domain and can provide a social impact to stop the further spread of this deadly disease. In the future, more new tools and technologies-based research works can be considered to enrich this kind of survey work in this domain. This extensive review will enable the researchers to open the mind to explore possible optimized applications in this field as well as beyond this area.

References

1. Bullock J, Luccioni A, Pham KH, Lam CSN, Luengo-Oroz M (2020) Mapping the landscape of artificial intelligence applications against COVID-19. J Artif Intell Res 69:807–845
2. Loey M, Manogaran G, Taha MHN, Khalifa NEM (2021) A hybrid deep transfer learning model with machine learning methods for facemask detection in the era of the COVID-19 pandemic. Measurement 167:108288
3. Dhanush Reddy KN (2021) Social distance monitoring and facemask detection system for Covid-19 pandemic. Turk J Comput Math Educ (TURCOMAT) 12(12):2200–2206
4. Razavi M, Alikhani H, Janfaza V, Sadeghi B, Alikhani E (2021) An automatic system to monitor the physical distance and facemask wearing of construction workers in a Covid-19 pandemic. arXiv preprint arXiv:2101.01373
5. Jiang M, Fan X, Yan H (2020) Retina facemask: a facemask detector. arXiv preprint arXiv:2005.03950, 2
6. Lubis R. Machine learning (convolutional neural networks) for facemask detection in image and video. Binus University Repository. https://core.ac.uk/reader/328808130
7. Howard AG, Zhu M, Chen B, Kalenichenko D, Wang W, Weyand T, Adam H (2017) Mobilenets: efficient convolutional neural networks for mobile vision applications. arXiv preprint arXiv:1704.04861
8. Nielsen MA (2015) Neural networks and deep learning, vol 25. Determination Press, San Francisco, CA
9. Maurya P, Nayak S, Vijayvargiya S, Patidar M (2021) COVID-19 facemask detection. In: 2nd international conference on advanced research in science, engineering & technology, Paris, France, pp 29–34
10. Bhutada S, Nirupama NS, Mounika M, Revathi M (2021) Social distancing and mask detector based on computer vision using deep learning methods. Int J Res Biosci, Agricult Technol 2(9):81–87
11. Murugan KS, Kavinraj G, Mohanaprasanth K, Ragul KB (2021) Real-time social distance maintaining using image processing and deep learning. J Phys: Conf Ser 1916(1):012190. IOP Publishing

12. Yadav S (2020) Deep learning-based safe social distancing and facemask detection in public areas for covid-19 safety guidelines adherence. Int J Res Appl Sci Eng Technol 8(7):1368–1375
13. Rahman MM, Manik MMH, Islam MM, Mahmud S, Kim JH (2020) An automated system to limit COVID-19 using facial mask detection in the smart city network. In: 2020 IEEE international IOT, electronics and mechatronics conference (IEMTRONICS). IEEE, pp 1–5
14. Shete I (2020) Social distancing and facemask detection using deep learning and computer vision (Doctoral dissertation, Dublin, National College of Ireland). http://norma.ncirl.ie/4419/1/ishashete.pdf
15. Bhambani K, Jain T, Sultanpure KA (2020) Real-time facemask and social distancing violation detection system using YOLO. In: 2020 IEEE Bangalore humanitarian technology conference (B-HTC). IEEE, pp 1–6
16. Savita S (2021) Social distancing and facemask detection from CCTV camera. Int J Eng Res Technol (IJERT) 10(8)
17. Krishna KP, Harshita S (2020) Social distancing and facemask detection using deep learning. In: 10th international conference, IACC 2020, Panaji, Goa, India, December 5–6, 2020, Revised Selected Papers, Part I, vol 1367. Springer Nature
18. Pandiyan P. Social distance monitoring and facemask detection using deep neural network.
19. Bala MMS (2021) A deep learning technique to predict social distance and facemask. Turk J Comput Math Educ (TURCOMAT) 12(12):1849–1853
20. Keniya R, Mehendale N (2020) Real-time social distancing detector using Socialdistancingnet-19 deep learning network. https://doi.org/10.2139/ssrn.3669311, available at SSRN 3669311
21. Babu DCR, Jyothir Vijaya Lakshmi K, Saisri KM, Anjum SR (2021) Social deprivation with protective mask detector. J Eng Sci 12(7):219–226
22. Patil NS, Rani K, Rangappa S, Jain V (2021) Social distancing detection. Int J Res Eng Sci 9(9):50–56
23. Teboulbi S, Messaoud S, Hajjaji MA, Mtibaa A (2021) Real-time implementation of AI-based facemask detection and social distancing measuring system for COVID-19 prevention. Sci Program 1–21
24. Yadav N, Sule N, Yadav S, Kullur S (2021) Social distancing detector using deep learning. Int Res J Eng Technol 8(5):3699–3703
25. Jenitta J, Shrusti BK, Vidya DY, Sinnur VS, Varma S (2021) Survey on detection and identification of facemask. Int J Sci Res Eng Trends 7(2):985–988
26. Sethi S, Kathuria M, Kaushik T (2021) Facemask detection using deep learning: an approach to reduce risk of Coronavirus spread. J Biomed Inform 120:103848
27. Jiang X, Gao T, Zhu Z, Zhao Y (2021) Real-time facemask detection method based on YOLOv3. Electronics 10(7):837
28. Liu R, Ren Z (2021) Application of Yolo on mask detection task. In: 2021 IEEE 13th international conference on computer research and development (ICCRD). IEEE, pp 130–136
29. Chakraborty S, Dey L (2021) The implementation of AI and AI-empowered imaging systems to fight against COVID-19—a review. Smart Healthc Syst Des: Secur Privacy Aspects 301
30. Dey L, Chakraborty S, Mukhopadhyay A (2020) Machine learning techniques for sequence-based prediction of viral–host interactions between SARS-CoV-2 and human proteins. Biomed Journal 43(5):438–450

A Machine Learning Framework for Breast Cancer Detection and Classification

Bagesh Kumar, Pradumna Tamkute, Kumar Saurabh, Amritansh Mishra, Shubham Kumar, Aayush Talesara, and O. P. Vyas

1 Introduction

Medical science and researchers with the implementation of neural network and computer-based techniques have come up with approaches where early detection of breast cancer is possible in which time plays a vital role that when the tumor is detected it is possible to detect it in initial stage and stop cancer cells to grow further as soon as detection of the tumor is found.

Breast cancer starts in the cells of the breasts and spreads throughout the body. Women are more likely than men to develop breast cancer. A mass in the breast, blood extravasation from the nipple, and changes in the consistency or structure of the breast or nipple are all the signs of cancer of the breast which is also known

B. Kumar · P. Tamkute · K. Saurabh · A. Mishra (✉) · S. Kumar · A. Talesara · O. P. Vyas
Indian Institute of Information Technology, Allahabad, India
e-mail: iit2018142@iiita.ac.in

B. Kumar
e-mail: pse2016001@iiita.ac.in

P. Tamkute
e-mail: iec2017059@iiita.ac.in

K. Saurabh
e-mail: pwc2017001@iiita.ac.in

S. Kumar
e-mail: iit2018146@iiita.ac.in

A. Talesara
e-mail: iit2018150@iiita.ac.in

O. P. Vyas
e-mail: opvyas@iiita.ac.in

P. Tamkute · K. Saurabh · A. Mishra · S. Kumar · A. Talesara · O. P. Vyas
Springer Heidelberg, Tiergartenstr. 17, 69121 Heidelberg, Germany

© The Author(s), under exclusive license to Springer Nature Singapore Pte Ltd. 2023

H. Muthusamy et al. (eds.), *Robotics, Control and Computer Vision*, Lecture Notes in Electrical Engineering 1009, https://doi.org/10.1007/978-981-99-0236-1_12

as lobular carcinoma. Breast cancer develops as a result of an aberrant swelling in the cells of the breast, known as a tumor. A tumor can be benign, pre-malignant, or malignant. Cancer affects people's lives in many ways, it also affects the mental state of people, hence early diagnosis is a very crucial thing.

Main objectives of this paper would be predicting the nature of the cancer using Support Vector Machine (SVM) for binary classification as follows.

1.1 Benign Tumor

These are not deadly and not harmful as it shows some abnormal growth or sometimes few changes in the tissue of the breast which is not cancerous. These are basically lump in breast and are non-cancerous and not prone to impact at deadly level. They are basically a lump in the breast which looks scary but they are not cancerous. Benign breast conditions are not harmful.

1.2 Malignant Tumor

Malignant cancer is dangerous. These cells grow and then spread to other parts of the body. When these cells grow, they accumulate together. Malignant conditions are dangerous and need to be identified as soon as possible.

The steps involved in the methodology we used are exploratory data analysis (EDA) and data preprocessing. An SVM is built for the prediction of the nature of tumor, optimizing the SVM classifier, and comparison with other classification models. In this model for the EDA section correlation matrix, scatter plots have been used. SVM is the basic methodology behind the main model. We have made use of cross validation and hyperparameter tuning. In the comparison section, SVM is compared with five other algorithms with the help of scikit-learn pipelining for smooth succession and to avoid loss.

For training Wisconsin breast cancer dataset gathered and maintained by the University of California is used which we have discussed in detail in "Dataset" discussion section.

2 Literature Survey

Dataset is obtained from the Wisconsin breast cancer dataset.

In Ref. [1], "Applying best machine learning Algorithm for classification of Breast cancer", authors did the comparative paper of Random Forest Naive Bayes, SVM, and K-NN to select apparently the most optimal solution. As a result of this paper,

SVM showed highest accuracy of 97%. This paper was done for the understanding of comparative performance of algorithms.

This paper does not have an optimization done on some particular ML algorithm, it can successfully improve the performance.

In Ref. [2], "Using Machine Learning algorithms for breast cancer risk prediction and diagnosis", authors applied various machine learning algorithms on breast cancer data. Here scope for improvement would be fine-tuning of the parameters and data standardization. Ensuring traceability of the data and optimizing the downstream data flow.

In Ref. [3], "An Enhanced Breast Cancer Diagnosis Scheme based on Two-Step-SVM Technique", authors have used a hybrid support vector machine (SVM) and the two-step clustering technique to separate the incoming tumors, and the two-step algorithm and SVM were coupled and used to identify the hidden patterns of malignant and benign tumors. When tested on the UCI-WBC dataset, the proposed hybrid approach improves accuracy by 99.1%. In future paper to improve diagnostic accuracy, an optimization approach can be coupled with the SVM two-step clustering methodology. Also in [21] authors have used SVM classifier with statistical parameters such as entropy, mean, RMS, etc. and have achieved 80% accuracy. Similarly on [22] they have used contrast stretching to increase the contrast of the image. The segmentation of mammogram images has been playing an important role to improve the detection and diagnosis of breast cancer.

3 Dataset

This dataset is the Wisconsin Diagnostic Breast Cancer dataset [4] gathered by the University of California, Irvine machine learning repository. The dataset contains 357(62.74) B-type breast cancer and 212(37.25) m-type breast cancer cases, where B and M denote benign and malignant. The dataset consists of 32 vertical lines with the first column as exclusive ID number; the second column being the diagnosis result (M or B); and after that there are standard deviations, average (mean), and the average of the ten worst measurements. Missing values were not noticed. The exclusive ID numbers of the specimens and the accompanying diagnosis (M and B which actually denote malignancy and benignity) are saved in the first two progressions of the dataset. The columns from three through thirty-two contain thirty actual value attributes generated from digitized capture of cell nucleus which we can use to establish a machine learning configuration to determine the nature of the tumor may it be malignant or benign. A digital image of a fine needle aspiration biopsy of the tumor was used to extract the characteristics. These features give a description of the nuclei of the cell. We have obtained this dataset from Kaggle.

4 Methodology

We have distributed the analysis of this model into five subdivisions.

4.1 Exploratory Data Analysis

So far in the paper we can get a visceral idea regarding the dataset we are working with, now we are going to do a detailed paper of the features and the data values.

Exploratory data analysis (EDA) is a critical course of action which follows feature engineering and data acquisition and it is supposed to be completed prior to any kind of modeling. This is because a data scientist's ability to comprehend the nature of the facts is critical, by not assuming things prior to the analysis. Data research results are incredibly valuable in determining the arrangement and distribution of data, as well as the presence of extreme boundary type points and interrelationships within the data collection.

4.1.1 The Purpose of EDA

1. To better understand data by using summary statistics and visualizations.
2. Find clues about the data's tendencies and quality, as well as formulate assumptions and hypotheses for our analysis.
3. Having a general view of our data is very critical for the data preprocessing to be successful. We can benefit from the use of simple qualitative elucidation to recognize the characteristics of a dataset and emphasize whichever data points that we can view as outliers or noise.

4.1.2 Summary Statistics

Summary statistics are used to summarize significant aspects of a dataset into simple quantitative measures. Standard deviation (SD), mean, and correlation are some of the most commonly used measures. Since our data can be unevenly distributed, we performed the data distortion operation. The distortion result indicates whether the distortion is -ive (left) or +ve (right). Points near the zero have less distortions. Due to the unique grouping of malignant and benign cancer kinds in these the graphs show that "radius mean", "area mean", "concave points mean", "concavity mean", and "perimeter mean" are beneficial in predicting cancer type. It's also important to mention that the parameters "area worst" and "perimeter worst" could be valuable at some point.

4.1.3 Visualization

The process of protruding data, or the chunks of data, in the abstract visuals is known as visualization. Data exploration is used in many various aspects of the data mining process, including preprocessing, modeling, and interpretation of results.

4.1.4 Uni-modal Data Visualization

Determining whichever characteristics are of utmost use in prediction of the nature of the breast cancer tumor is one of the very important purposes of visualization of the dataset over here. The other is to look for broad trends that can help us choose models and hyperparameters. Then, in order to analyze each attribute of our dataset separately, we attempted three distinct methodologies:

1. Density plot.
2. Histogram.
3. Box and Whisper plot.

Histogram is a popular method of depicting numerical data. A histogram resembles a bar graph when the values of the variable are clustered into a fixed number of intervals. The histogram separates the data parameters among bins and sums up the number of observations for each. By examining the geometry of the bins, we can assess whether a feature has a "Gaussian", "skewed", or even "exponential distribution". It can also help us in identifying any potential outliers (Figs. 1 and 2).

From the above plots, the observation can be made that the parameters "perimeter", "radius", "area", "concavity", and "compactness" may have a distribution that is exponential. It's also possible that the "texture", "smoothness", and "symmetry" features have Gaussian or distributions that look like Gaussian distribution. This information has significance because many ML approaches consider the variables of type input to show a Gaussian univariate distribution. Multimodal Data Visualizations (Fig. 3).

(1) Correlation matrix.
(2) Scatter plots.

We can see that mean value parameters between 1 and 0.75 have a strong positive association. The radius and parameter mean values have a strong positive association with the mean area of the tissue nucleus. Concavity and area, concavity and perimeter, and other parameters have a slight positive correlation ("r" belongs to the range of 0.5–0.75). Similarly, the attribute values "texture", "radius", and "parameter mean" have a high negative association with fractal dimension (Fig. 4).

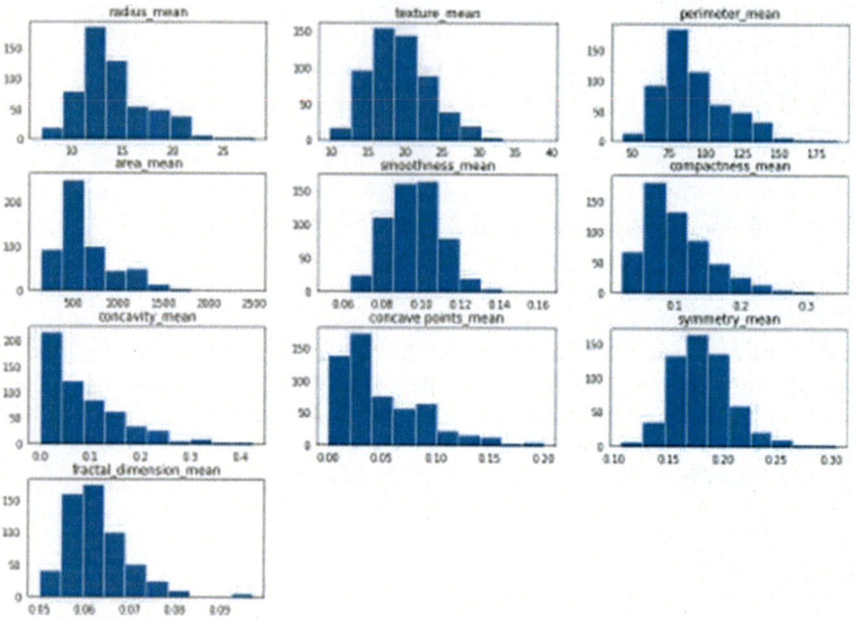

Fig. 1 Histogram of type suffix columns

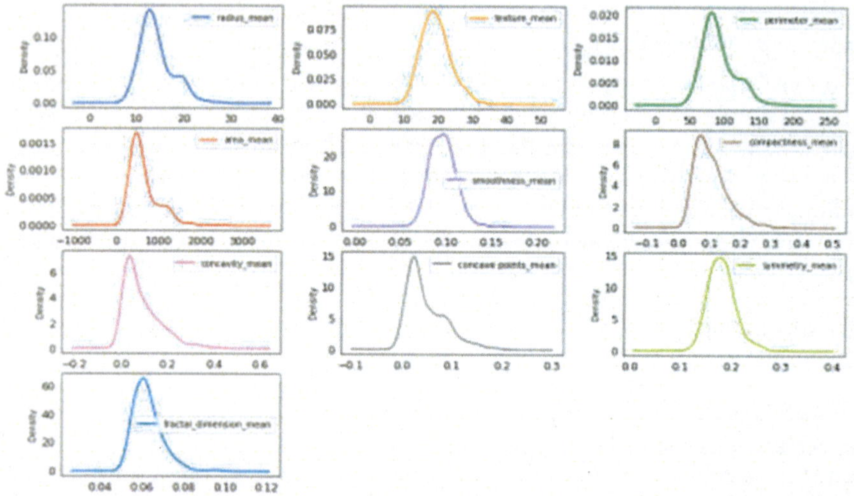

Fig. 2 Density plots mean suffix columns

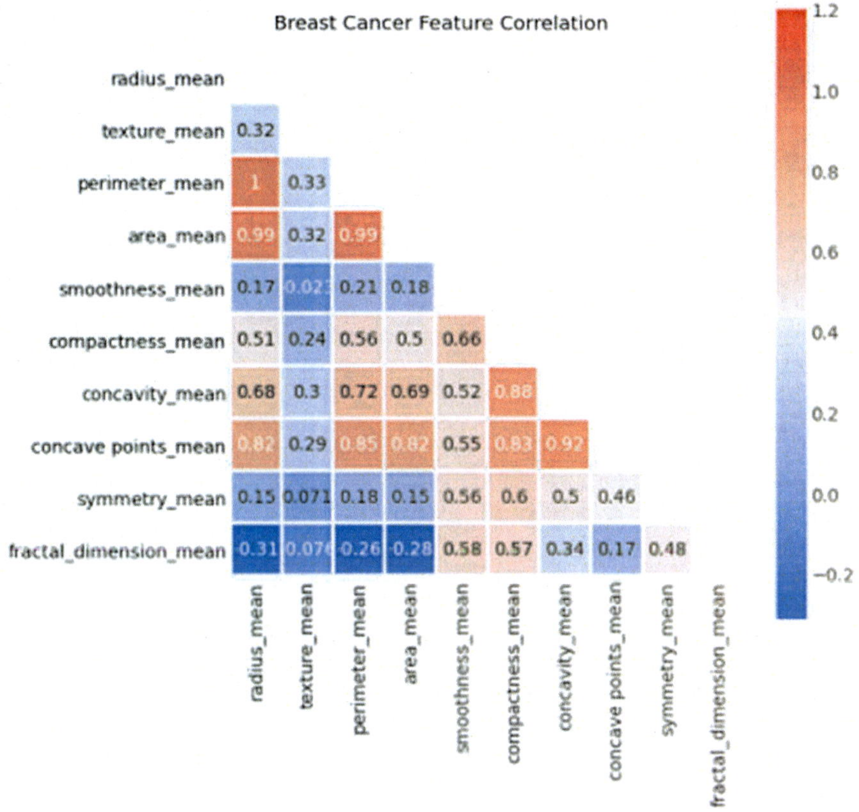

Fig. 3 Correlation matrix

4.1.5 Conclusion of EDA

We can utilize the average values of "area", "cell radius", "compactness", "perimeter", "concavity", and "concave regions" to classify cancer. The presence of malignant tumors is associated with higher values of these parameters. Texture, smoothness, symmetry, and factual dimension mean values do not indicate a preference for one diagnosis over another. There are no obvious significant outliers in any of the histograms that need to be cleaned up.

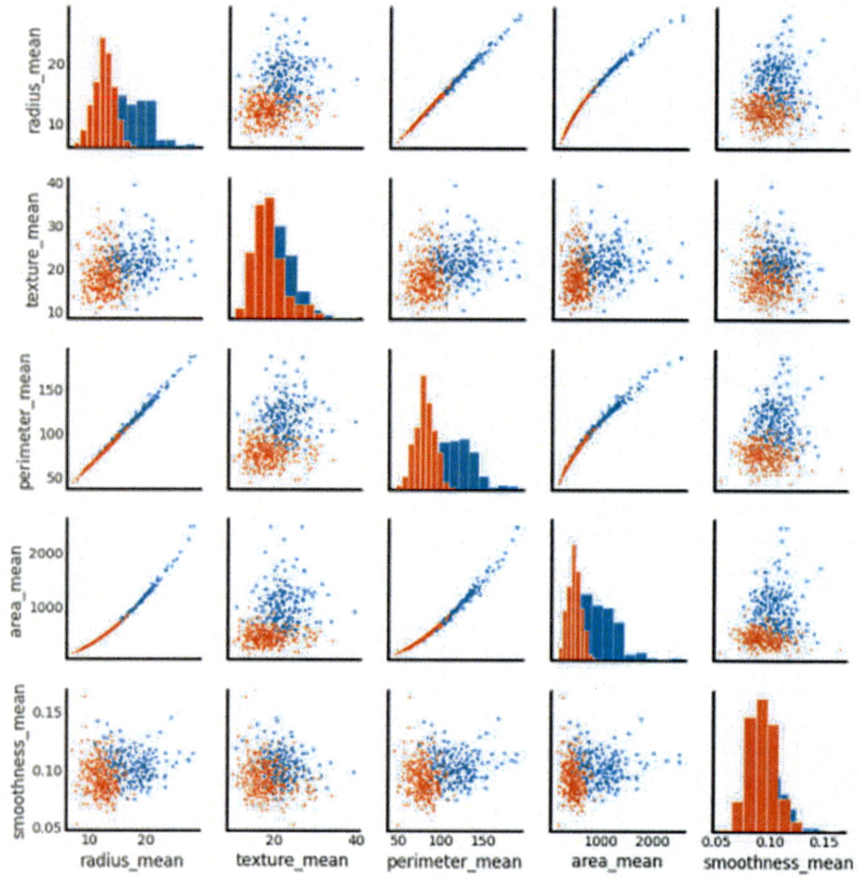

Fig. 4 Scatter plots

4.2 Data Preprocessing

Every predictive analysis paper involves preprocessing of data. Formatting our data in a manner that the nature of the challenge is optimally revealed to the machine learning methodologies will be beneficial. This is a smart idea most of the times.

Following tasks are involved in the preprocessing of data:

1. Categorical data is given numerical values.
2. Missing values are dealt with.
3. Normalization of the attributes (in such a way the systems performance has a negligible influence of small-scale features).

So in this EDA part, data was studied for learning more about how the data was distributed and how the qualities were related to one another. We saw a few things that piqued our interest. In this section, we utilize feature selection, feature extraction, and transformation to minimize dimensionality in high-dimension data. Our goal here is to identify the data's most predictive attributes and filter them to improve the analytics model's predictive capability.

NumPy was used to assign the 30 characteristics to an array X, and the class names were converted to integers from their original textual format (M and B). Malignant tumors are now designated as category 1 and benign tumors as category 0, respectively. Thereafter, we encode the class labels (diagnosis) in the array y, as shown by invoking the transform method of LabelEncoder on two dummy variables.

4.2.1 Assessing Model Accuracy

Splitting of data into train and test sets. Using separate training and testing datasets is the simplest way to measure the effectuation of the machine learning classifier.

We've divided the data into two sets: a testing set and a training set (70% practise, 30% assessment). The algorithm is taught in the first section, forecasts are made in the other section, and the forecasts are compared to the anticipated results in the third section. The length of the split is dependent on the length and details of our dataset, but it's typical to use 67% of it for practice and 33% for assessment. The 80:20 split is also pretty common.

4.2.2 Feature Standardization

In standardization methodology, Gaussian attributes with different means and standard deviations are transformed to a standard Gaussian distribution with a mean of zero and a standard deviation of one [5].

The raw data has different distributions, as shown in exploratory data analysis. It has an influence over almost all ML methodologies. When features are on the same scale, most machine learning and optimization methods perform substantially better.

Let's put the same techniques to the test on a standardized dataset. We are going to use sklearn to scale and modify the data so that each attribute has a mean of zero and a standard deviation of one (Figs. 5 and 6).

4.2.3 Feature Decomposition Using Principal Component Analysis (PCA)

When working with only two dimensions, because a number of attribute couples partition the dataset similarly, that makes it logical to include some of the feature extraction techniques to try to use as many attributes while retaining maximum feasible data. The PCA method will be employed. We now have a reduced dimensional subspace (in this case, from 3D to 2D) in which the data is "most spread" along the new attribute axes after the application of the linear PCA modification.

4.2.4 Deciding Count of Principal Components to Preserve

In order to determine how many main components should be preserved, we typically use scree plot for the summarization of the findings of a principal components analysis. A scree plot shows how much variation from the data each principal component captures.

If the first two or three PCs have captured the majority of the data, the others can be ignored without losing anything critical. A scree plot depicts the amount of variation that each PC extracts from the data. The y-axis represents eigenvalues, which represent the degree of variance. Select the main components to maintain using a scree plot. A good curve is one that is steep, but not too steep. We got the below scree plot.

Here we can observe that the most visible update in incline in the scree plot comes after PC2, which is the scree plot's "elbow". As a result, based on this scree plot, it may be argued that the first three components should be preserved. It's conventional to choose an attribute subsection which has the closest association to the class

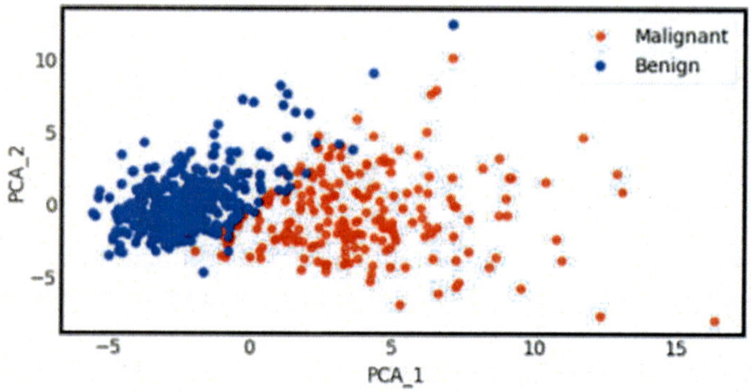

Fig. 5 Feature decomposition using principal component analysis (PCA)

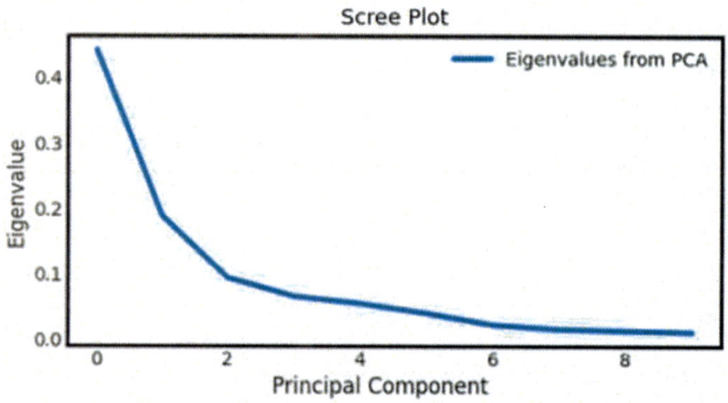

Fig. 6 Deciding count of principal components to preserve

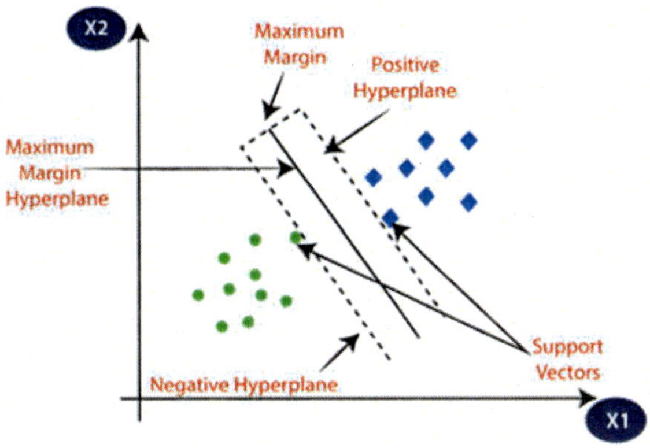

Fig. 7 Support vector machine (Javatpoint)

designation. For providing us with an unprejudiced approximation of our model's true execution, feature selection must be evaluated as part of a complete modeling process.

4.3 Build a Model to Predict Whether Breast Cell Tissue is Malignant or Benign Using Support Vector Machine

The SVM classification algorithm's aim is to identify a hyperplane in an n-D space (with n as the number of attributes) that differentiates among data points. There are several hyperplanes to choose from in order to distinguish between the two types of data points. Here we have a goal to find the plane with the largest margin, or the distance between data points in both groups [6]. The goal of increasing the margin distance is to indicate the presence of subsequent data points (Fig. 7).

4.3.1 Hyperplanes and Support Vectors

Hyperplanes are judgment boundaries that make sorting of the datum. Datums along both sides of the hyperplane may be assigned to separate categories. In case of only two input attributes, the hyperplane is just an irrelevant strip. Whereas in case of input count of attributes being three, the hyperplane becomes a 2D plane. It becomes hard to imagine when the number of features is more than three, hence we can say that the hyperplane's dimension is determined by the count of attributes. In this paper, first, we split the dataset between train and test dataset in 70:30 proportion, i.e., 70 of the data is used to train the model and 30 dataset is used for testing. We analyzed and built a model based on this dataset to determine if such a particular set of manifestations will evolve to form breast cancer. Support vector machine (SVM) is a binary classifier, it looks for a hyperplane which leaves the biggest feasible fragment of points that lie on exactly the side and belong to that particular class, which at the same time amplifies the distance between the hyperplane and every class [7]. SVMs are one of the most recent approaches of machine learning techniques which we can apply in the area of prognosis of carcinoma. In the starting part of SVM does the designation of input vectors into an attribute field which belongs to a higher dimension and recognizes the hyperplane that it does the segregation of the data entries among two sub-classes. The minimal spacing in between judgment hyperplane and the occurrences closest to the border must be kept to as low as possible. The final classifier achieves substantial generalizability and so we can use it for the efficient categorization of new specimens [7].

4.3.2 Important Parameters Under Consideration

Kernel SVMs have critical parameters as follows:

1. The kernel selection from linear, radial basis function (RBF), or polynomial.
2. C : the regularization parameter.
3. Parameters that are particular to the kernel.

Gamma and C parameters have an impact on the model's complexity, with large values of either producing a more complicated model. As a result, good values for such two variables are generally firmly associated, and gamma and C are something that can be regulated in coordination. After performing support vector classification on our model, we got an accuracy of 95%. Now so as to improve the model we have to use few of the techniques.

4.3.3 Cross Validation

We can't certainly say that the model which is trained on the training data would function with accuracy on practical data for every case in machine learning. To tackle this issue, it should be assured that the model we have can give us the accurate result from the data, and has low resultant noise. The cross-validation approach is a go to strategy for this sort of thing. In the cross-validation method, we split the data among various subsections and the Ml model is trained on one subsection of the dataset and the other subsection is used for the re-evaluation.

4.3.4 K-fold Cross Validation

In this technique, the dataset is divided into k subsections, on all of them but one, subsections are trained and then the trained model is evaluated on one subsection. Here a unique subsection designated for testing reasons each time is re-replicated k times.

Here in this model we first checked the model with three-fold cross validation, i.e., K = 3. With this we got an accuracy of 97%. This assessment was done while taking all the available parameters into consideration. Now we will try to cut down parameters. We used three parameters that fit the model better than the other parameters. By doing that we again got the accuracy of 97%. Hence, we can conclude that a small number of features can also give us the model with similar performance. Hence, we need to focus a little on feature selection now. Let's have a detailed discussion on model accuracy.

4.3.5 Receiver Operating Curve

A receiver operating characteristic curve (ROC curve) is a graph that defines the degree of accuracy of the categorization model that works across every categorization criteria. In this plot, y- and x-axes are as follows:

1. Rate of True Positives.
2. Rate of False Positives.

The True Positive Rate (TPR), known as recall or sensitivity, is defined as follows:
TPR = (TP)/(TP + TN).

The False Positive Rate (FPR) can be defined as follows:
FPR = (FP)/(FP + TN),

where FP is false positive, TP is true positive, and TN is true negative. The True Negative Rate (TNR) which is also known as specificity is defined as
TNR = (TN)/(FP + TN).

4.3.6 ROC Plotting

TPR versus FPR at various categorization parameters is plotted on a ROC curve. So as we lower the classification threshold, more items are classified as positive, the number of False Positives and True Positives both increase as a result of the earlier [8]. A typical ROC curve is depicted in the diagram below. A logistic regression can be analyzed multiple times with distinct categorization criteria to analyze the points on an ROC curve. However, this method seems to be insufficient and inefficient but providentially, we have fast, sorting-based method called AUC which give us what we need (Fig. 8).

4.3.7 Area Under the ROC Curve

"Area under the ROC Curve" has short form "AUC". AUC in short assesses full 2D area beneath whole ROC curve from (0, 0) to (1, 0). The AUC value lies between

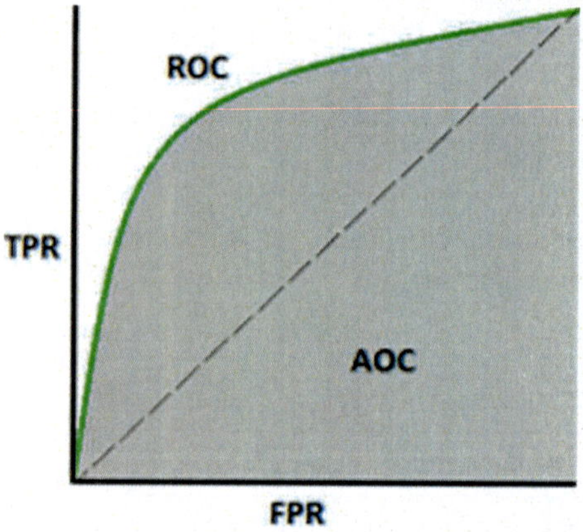

Fig. 8 ROC-AUC curve image

0 and 1. The model with all the incorrect predictions has AUC = 0.0, whereas the model with all predictions correct has AUC = 1.0. The following are two reasons for AUC to be desirable:

1. The AUC stays impervious to scaling. It measures how effectively predictions are sorted instead of measuring absolute values.
2. The categorization boundary has no ramification to AUC. It evaluates the system's classification performance autonomous of the categorization level employed.

4.3.8 Observation

Confusion matrix for the model's current performance is shown in Fig. 9. Now here we have "1" and "0" as the two probable expected classes. Benign equals to 0 which indicates absence of cancer cells and malignant equals to 1 which identifies existence of cancer cells. A total of 174 predictions were made by the classifier. The classifier accurately guessed "yes" and "no" 113 out of 174 cases. In actuality, 64 of the total patients in the data have cancer, whereas the remaining 107 do not. So here we have calculated rates based on confusion matrix:

Accuracy is calculated as:
(TP + TN)/(TP + TN + FP + FN) = (57 + 106)/171 = 0.95.

Rate of Miscellaneous:
(FP + FN)/(TP + TN + FP + FN) = (1 + 7)/171 = 0.05 (0.05 = 1 − 0.95).
True Positive Rate (Sensitivity): Ratio of number of times it predicts yes and is actually yes, with total number of yes it predicted.
TP/actual yes = 57/64 = 0.89.

False Positive Rate:
FP/actual no = 1/107 = 0.01.

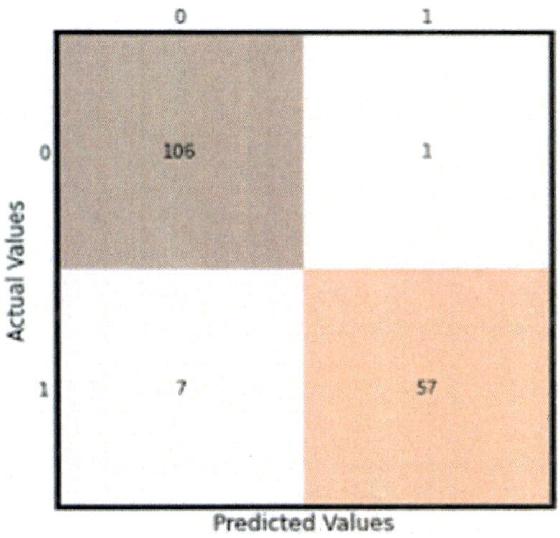

	precision	recall	f1-score	support
0	0.94	0.99	0.96	107
1	0.98	0.89	0.93	64
accuracy			0.95	171
macro avg	0.96	0.94	0.95	171
weighted avg	0.95	0.95	0.95	171

Fig. 9 Confusion matrix

Prevalence:
Actual yes/total = 64/171.

Precision:
TP/(TP + FP) = 57/58 = 0.98.

True Negative Rate:
TN/(actual no) = 106/107.
Now, here we have the ROC curve for this model: In this ROC, we can interpret that, points that are present on diagonal, they have 0.5 probability of being either 0 (no) or 1 (yes). Hence classification model is not really making a difference, hence the decision is being made at random (Fig. 10).

TPR is greater than FPR in the areas over diagonal, and the model suggests that this region outperforms randomness. Let us suppose FPR = 0.01 and TPR = 0.99. In this case, the chance of true positive subsection is (TRP/(TPR+FPR)), i.e., 99%. Besides, suppose F.P.R. stays constant, it is clear that the classification model performs better as we go vertically higher and higher from the diagonal.

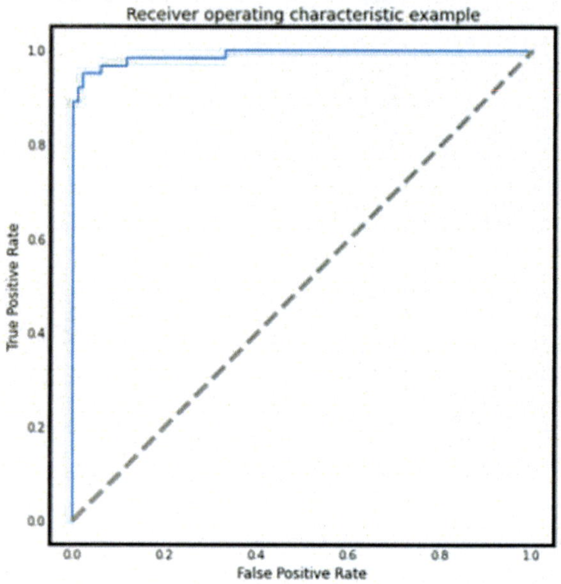

Fig. 10 Confusion matrix

4.4 *Optimizing the SVM Classifier*

To tune their behavior to a specific environment, machine learning models are parameterized. Because models might include a lot of parameters, finding the ideal combination is a search problem. Now in this section, we have used scikit-learn to adjust the SVM classification model's parameters.

Now in this section first we tried applying k-fold cross validation with k = 5, hence we got the accuracy of 96%.

Results are shown in Fig. 11.

4.4.1 Hyperparameter Tuning

A mathematical model containing a number of parameters that must be learned from data is termed as a machine learning model [11]. There are, however, some factors known as hyperparameters that cannot be learned directly. Before the actual training begins, humans frequently choose them based on intuition or trial and error. These factors demonstrate their value by enhancing the model's performance, such as its complexity or learning rate. Models can include a large number of hyperparameters which make determining the best combination of parameters a search problem.

```
The classifier accuracy score is 0.96

              precision    recall  f1-score   support

           0       0.95      0.99      0.97       107
           1       0.98      0.91      0.94        64

    accuracy                           0.96       171
   macro avg       0.96      0.95      0.96       171
weighted avg       0.96      0.96      0.96       171
```

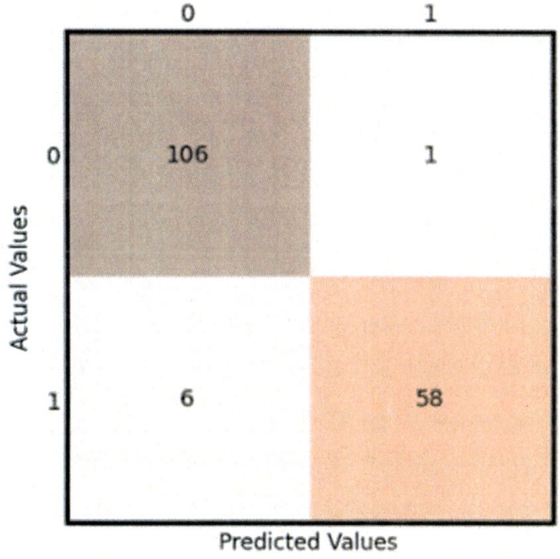

Fig. 11 Classifier matrix

SVM parameters that can be tuned are as follows:

1. Type of kernel.
2. C and gamma parameters.

It is very important to pick the right kernel type because if the transformation is incorrect, the model's outcomes can be very less accurate. We should always check if our data is linear and, if so, we utilize linear SVM (linear kernel). By default the kernel type of SVM is set as RBF (radial basis function), whereas C value is set to 1. Now in scikit-learn library we have following techniques for hyperparameter tuning:

1. GridSearchCV

GridSearchCV uses a dictionary to specify the parameters that can be used to train a model. The grid of parameters is defined as a dictionary, with the keys being the parameters and the values being the test settings. There is one shortcoming to this method [12]. GridSearchCV will go through all of the intermediate hyperparameter combinations, making grid search computationally quite expensive.

2. RandomizedSearchCV

RandomizedSearchCV only runs through a predetermined number of hyperparameter settings, therefore RandomizedSearchCV overcomes the shortcomings of GridSearchCV. It moves randomly throughout the grid to discover the optimal collection of hyperparameters. This method eliminates the need for extra computation [19]. Through this process we got an accuracy of 98% with parameters that suited best for our model as C : 0.1, gamma : 0.001 and the kernel being "linear". The result can be seen in Fig. 11.

4.4.2 Epilogue of This Section

We can successfully classify the malignant and benign breast cancer tumor with the use of the support vector machine methodology. Hyperparameter tuning can give us considerable improvement in the accuracy of the model. Performance of the SVM can be improved as compared to default SVC, when all the parameters are scaled so that the mean is zero and standard deviation is set at one (Figs. 12 and 13).

4.5 Comparison with Other Classification Models

4.5.1 Automate the ML Process Using Pipelines

Now before we jump to comparison we first made the process a little more convenient. We created machine learning pipelines. In a machine learning paper, there are regular workflows that we should automate. Pipelines in the scikit-learn library in Python assist in explicitly defining and automating these operations.

1. Pipelines are useful for resolving issues like data leaks in your test harness.
2. Pipeline is a Python scikit-learn facility for automating workflows of machine learning.
3. Enabling a linear succession of data alterations so as to link together for pipelining.

In this section, we perform the following sub-tasks :

1. Create a validation dataset and separate it from the rest of the data.
2. Set up a ten-fold cross validation for the test system.

3. Create five different classification models.
4. Choose the most appropriate model based on its performance.

Validation Set: While tuning model hyperparameters, a specimen of feature is utilized for offering an impartial analysis for the model trained on a train dataset. As we incorporate the competence on the validation dataset in the system structure, the evaluation gets further skewed. We make use of a validation set for verification of the system, although we only have to use it on high frequency analysis. This information is used for fine-tuning the model hyperparameters [13]. As a result, the model views this information on occasion but never ever learns from it, i.e., it is never used as a training dataset. We adjust higher level hyperparameters on the basis of usefulness of the validation set. From this, our validation set has a contingent repercussion on this model. The dataset is crucial during the model's "development" stage, therefore this process makes sense. So here we separated the validation dataset from the rest of the total dataset. Now it is time to create five classification models with the help

```
The classifier accuracy score is 0.96
```

	precision	recall	f1-score	support
0	0.95	0.99	0.97	107
1	0.98	0.91	0.94	64
accuracy			0.96	171
macro avg	0.96	0.95	0.96	171
weighted avg	0.96	0.96	0.96	171

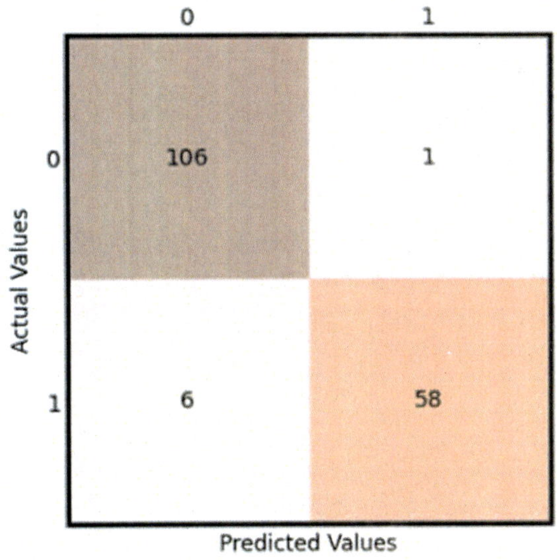

Fig. 12 Classifier matrix

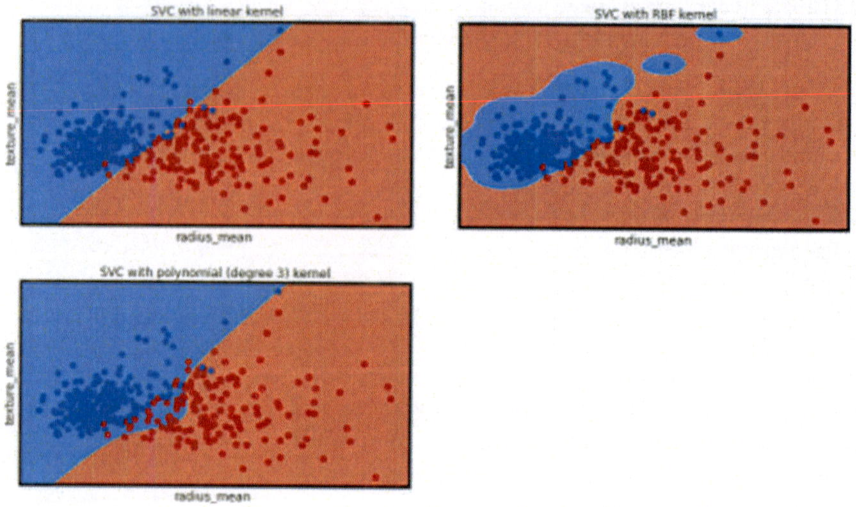

Fig. 13 The decision boundaries of linear, RBF, third-degree polynomial classifiers

of scikit-learn for comparison against each other. Six classification models that we used are as follows:

1. Logistic Regression (LR).
2. Linear Discriminant Analysis (LDA).
3. K-Nearest Neighbor Classification (K-NN).
4. Decision Tree Classifier (CART).
5. Gaussian Naive Bayes Classifier (GaussianNB).
6. Support Vector Classification (SVC) or SVM.

Logistic Regression: Logistic regression is a ML technique which we use for solving categorization problems. It is a predictive analytic approach as it is established on the probability notation [14]. Except for the part where complexity of the cost functions comes into picture the linear regression model has a lot of similarities with logistic regression model. These cost functions are known as the "Sigmoid function" or "logistic function" rather than "a linear function".

Linear Discriminant Analysis (LDA): Linear Discriminant Analysis (LDA), also known as Normal Discriminant Analysis (NDA), or Discriminant Function Analysis (DFA) is a dimensionality reduction approach often used for supervised classification issues. It is used to represent group differences, like separating two or more classes. It is used to paper higher dimensional features onto a lower dimensional space [15].

K-Nearest Neighbor Classification (K-NN): The K-nearest neighbor methodology has its establishments in the supervised learning technique and is one of the most basic ML techniques. In the K-NN method, it is considered that the existing cases and new case/data have huge similarity, hence new case is placed in the category that has highest number of similar existing cases like this one [16]. The K-NN approach saves all available data then does the classification of new data points on the basis of their similarity with current data. This suggests that new data can be put right away in a precise group with use of the K-NN method.

Decision Tree Classifier (CART): The Decision Tree is a supervised learning approach that we can use to overcome regression and categorization difficulties, however we often make its use for categorization [17]. Dataset attributes are shown by internal nodes whereas decision rules are represented by branches, and outcome is represented by each leaf node in a tree-structured classifier.

Gaussian Naive Bayes Classifier: "Gaussian Naive Bayes" is a Naive Bayesian variation which permits serialized input which follows the Gaussian normal distribution. The Bayes theorem is the foundation of the Naive Bayes categorization methods, which are supervised machine learning classification algorithms. It is a straightforward categorization method that works efficiently. When input complexity is substantial, they become advantageous [18]. The Naive Bayesian Classifier may also be of use in solving complex categorization issues.

Support Vector Classification: As we have discussed about this classifier in detail earlier, it uses support vectors and is very efficient.
Here we used k = 10 for cross validation. We got the following accuracy results for the respective model. We figured out that both LDA and logistic regression should be investigated ahead. We have only average efficiency figures. Checking at the accuracy values calculated throughout cross validation folds and seeing how they are distributed is usually considered a good idea. Whiskers and box charts can help us visualize this. These results show that the SVM gives out the most distribution, hence suggesting low variance. Result given by SVM is strikingly lower than expected. Now we repeated this process with the standardized dataset.

ScaledLR: 0.974936 (0.015813).
ScaledLDA: 0.954744 (0.018784).
ScaledKNN: 0.957372 (0.033665).
ScaledCART: 0.937244 (0.032017).
ScaledNB: 0.937115 (0.039261).
ScaledSVM: 0.967436 (0.027483).

These results indicated that SVM accuracy had been increased, and it is giving the highest accuracy achieved so far. SVM, LDA, and LR have shown good results, hence with tuning they can produce better results (Fig. 14).

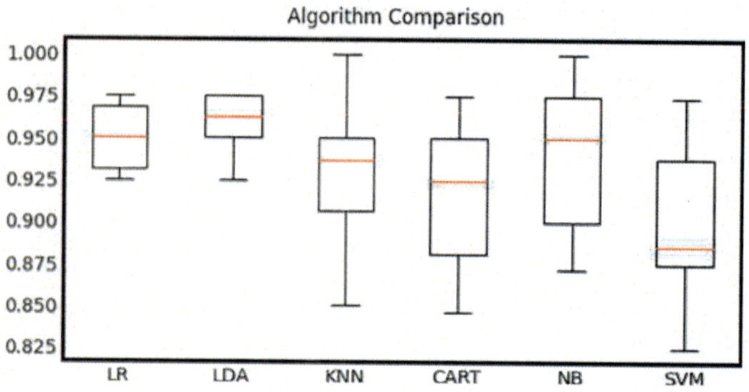

Fig. 14 Comparison between different models

4.5.2 Tuning Algorithm

Now, we have tuned the parameters for SVC and K-NN classifiers. We have used GridSearchCV in tuning these parameters. The GridSearchCV method receives predetermined hyperparameter values. This is achieved by creating a dictionary in which every hyperparameter is listed along with the possible values.

SVM: On hyperparameter tuning SVC we got the following result and parameters as best suited.

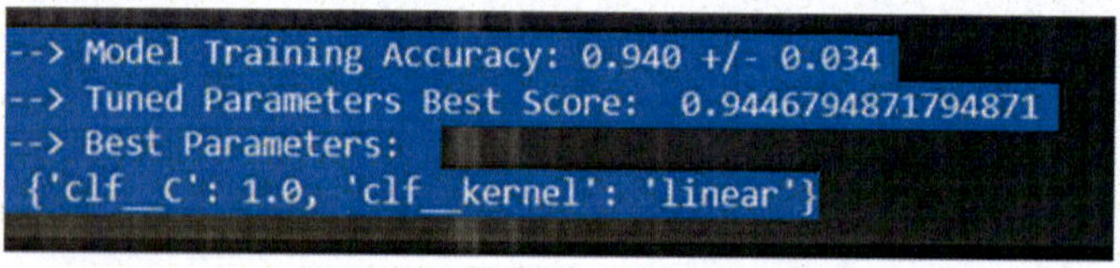

K-NN: Now, in K-NN-classifier parameters "k" and distance metric function can be tuned. By this process we got following results, and best suited parameters.

Hence, from this we have gathered that the SVM performs better. Hence, we decided that the SVM is the best suited model for this machine learning problem.

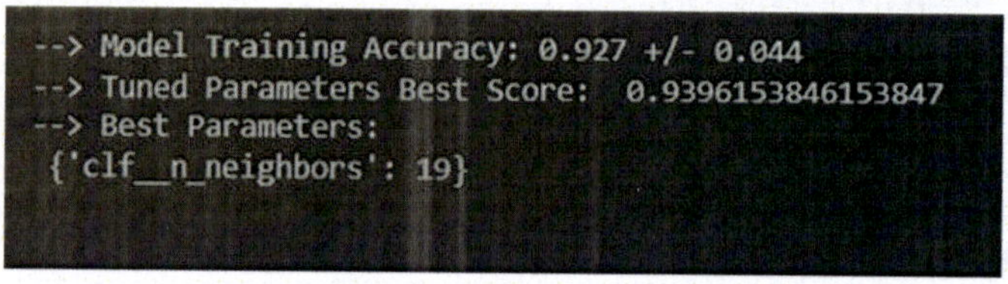

5 Result

We have finalized that SVM is the best performing model for our classification problem, now after running the model individually we got the impressive results on the test dataset as follows.

The result shows that SVM gets an accuracy of 97%, which is the best among the five and hence we have used that.

```
[[114   2]
 [  3  52]]
              precision    recall  f1-score   support

           B       0.97      0.98      0.98       116
           M       0.96      0.95      0.95        55

    accuracy                           0.97       171
   macro avg       0.97      0.96      0.97       171
weighted avg       0.97      0.97      0.97       171
```

6 Conclusion and Future Work

Here we conclude that with the help of hyperparameter tuning and data standardization, SVM best classifies the malignant and benign cancers from given data among the algorithms we considered. In this paper, we have created a model with optimized model based on support vector machine (SVM), and compared its performance with five other methodologies which are logistic regression (LR), linear discriminant analysis (LDA), K-nearest neighbor classification (K-NN), decision tree classifier (CART), and Gaussian Naive Bayes Classifier (GaussianNB). SVM has proved its superiority over others that we can see its evaluation matrix. In the future, there are some possibilities to the direction in which this research can lead. One of them being the possibility of change in type of database; in this paper, we have worked on Wisconsin breast cancer biopsy dataset. This dataset is derived from biopsy results of various breast cancer potential patients. Different results can be obtained from a dataset of X-ray data. Furthermore, complex algorithms can be designed with the help of deep learning methods. Larger datasets can be used to train the model.

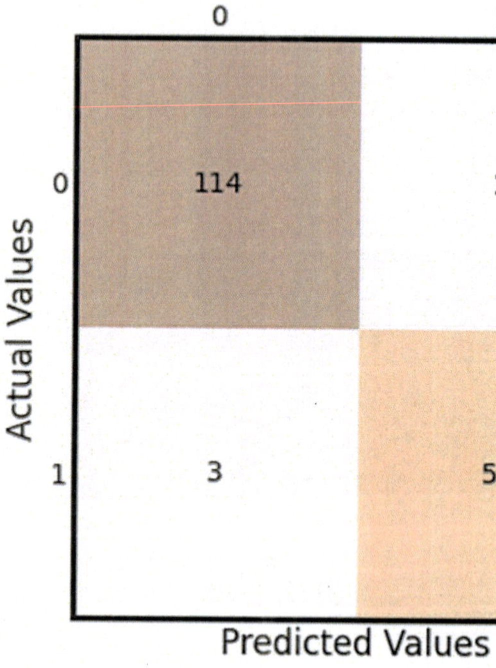

References

1. Khourdifi Y, Bahaj M (2018) Applying best machine learning algorithms for breast cancer prediction and classification. In: 2018 international conference on electronics, control, optimization and computer science (ICECOCS), pp 1—5. https://doi.org/10.1109/ICECOCS.2018.8610632
2. Bharat A, Pooja N, Reddy RA (2018) Using Machine Learning algorithms for breast cancer risk prediction and diagnosis. In: 2018 3rd international conference on circuits, control, communication and computing (I4C), pp 1–4. https://doi.org/10.1109/CIMCA.2018.8739696
3. Osman Ahmed Hamza (2017) An enhanced breast cancer diagnosis scheme based on two-step-SVM technique. Int J Adv Comput Sci Appl 8(4):158–165
4. Dr. Wolberg WH General Surgery Department University of Wisconsin, Clinical Sciences Center. "Breast Cancer Wisconsin (Diagnostic) Data Set" Retrieved from https://www.kaggle.com/uciml/breast-cancer-wisconsin-data
5. Gupta T (2021) Machine learning—Geeksforgeeks. https://www.geeksforgeeks.org/machine-learning/
6. Support Vector Machine (SVM) Algorithm -Javatpoint. https://www.javatpoint.com/machine-learning-support-vector-machine-algorithm
7. Gandhi R (2018) SVM Introduction to Machine Learning algorithms Rohit Gandhi—Datascience. https://towardsdatascience.com/support-vector-machine-introduction-to-machine-learning-algorithms-934a444fca47
8. Unknown (2020) Classification: ROC curve and AUC—google developer website. https://developers.google.com/machine-learning/crash-course/classification/roc-and-auc
9. Narkhede S (2018) Understanding AUC—ROC curve—towards data science. https://towardsdatascience.com/understanding-auc-roc-curve-68b2303cc9c5

10. Czako Z (2018) SVM and Kernel SVM by Czako zoltan—towards data science. https://towardsdatascience.com/svm-and-kernel-svm-fed02bef1200
11. Singh T (2020) Hyperparameter tuning—Geeksforgeeks. https://www.geeksforgeeks.org/hyperparameter-tuning/
12. Tyagikartik, 2021. "SVM Hyperparameter Tuning using GridSearchCV - Geeksforgeeks". Available at https://www.geeksforgeeks.org/svm-hyperparameter-tuning-using-gridsearchcv-ml/
13. Shah T (2017) About train, validation and test sets in machine learning—towards data science. https://towardsdatascience.com/train-validation-and-test-sets-72cb40cba9e7
14. Pant A (2019) Introduction to logistic regression—towards data science. https://towardsdatascience.com/introduction-to-logistic-regression-66248243c148
15. Raman_257 (2021) ML—linear discriminant analysis—Geeksforgeeks. https://www.geeksforgeeks.org/ml-linear-discriminant-analysis/
16. Unknown KNN algorithm for machine learning—Javatpoint. https://www.javatpoint.com/k-nearest-neighbor-algorithm-for-machine-learning
17. Majumder P (2020) Gaussian Naive Bayes, machine learning—Opengenus. https://iq.opengenus.org/gaussian-naive-bayes/
18. Unknown Decision tree classification algorithm—Javatpoint. https://www.javatpoint.com/machine-learning-decision-tree-classification-algorithm
19. Hussain M (2020) Hyperparameter tuning with GridSearchCV—MyGreatlearning. https://www.mygreatlearning.com/blog/gridsearchcv/
20. Gardezi SJS, Elazab A, Lei B (2019) Wang T breast cancer detection and diagnosis using mammographic data: systematic review. J Med Internet Res 21(7):e14464
21. Chanda PB, Sarkar SK (2018) Detection and classification technique of breast cancer using multi Kernal SVM classifier approach. In: 2018 IEEE applied signal processing conference (ASPCON), pp 320–325. https://doi.org/10.1109/ASPCON.2018.8748810
22. Rejani YI, Dr. Selvi ST (2009) Early Detection of breast cancer using SVM classifier technique. Int J Comput Sci Eng 1

Vision Transformers for Breast Cancer Classification from Thermal Images

Lalit S. Garia and M. Hariharan

1 Introduction

Breast cancer is the topmost eminent cause of death among women, resulting in the rising of breast cancer cases worldwide [1], affecting annual breast cancer screening necessary for early detection, and reducing the mortality rate. India positions third highest in cancer cases side by side with China and United States and is increasing by 4.5–5% every year. In India, the death rate for breast cancer is 1.7 times higher than maternal fatality [2]. Thermal imaging is a physiological imaging used as an adjunctive modality and has become an appreciable area of research. Breast thermography is non-contact and non-invasive on the strength of using no radiation and avoiding painful breast compression [3]. Expert radiologists and pathologists are required to diagnose breast cancer, which is time-consuming, and they draw their conclusion formulated on various visual features monitored which may vary from person to person. Computer-aided diagnosis (CAD) systems can support experts to reach decisions automatically. These techniques can also minimize inter-observer variations to implement the diagnosis process replicable. Deep learning algorithms have performed much the same as human experts on object detection and image classification tasks [4]. The convolutional neural network (CNN) is the best-used deep learning model to grasp complex discriminative features among image classes. Different architectures of CNNs such as VGG-16 [5] have presented exceptional results in the past few years on the very large ImageNet dataset. Also, CNNs are utilized on medical images to produce futuristic results.

L. S. Garia · M. Hariharan (✉)
Department of Electronics Engineering, National Institute of Technology, Srinagar (Garhwal), Uttarakhand 246174, India
e-mail: hariharanm@nituk.ac.in

L. S. Garia
ECE Department, BTKIT Dwarahat, Almora, Uttarakhand 263653, India

© The Author(s), under exclusive license to Springer Nature Singapore Pte Ltd. 2023
H. Muthusamy et al. (eds.), *Robotics, Control and Computer Vision*, Lecture Notes in Electrical Engineering 1009, https://doi.org/10.1007/978-981-99-0236-1_13

The transformer architecture [6] is already the superior model in natural language processing (NLP). Inspired by the progress of the self-attention-based deep neural networks of Transformer models in NLP, the Vision Transformer (ViT) [7] architecture is introduced for the image classification application. Input image is split into patches and every embedded patch is treated as a word in NLP during the training of these models. ViT uses self-attention modules to acquire the relation between these embedded patches. Here in, we move forward in applying Transformers in thermal image analysis and examine the potential application of self-attention-based architectures in breast thermal images (thermograms) classification. Specifically, we inspected ViT base model with different patch sizes on the basis of their performance during fine-tuning for our specific task based on the thermogram dataset. The outcomes display the high potential of ViT models in breast thermal image classification. We believe that this is the first study to explore the performance of ViT architectures on the classification of breast thermograms.

2 Related Work

This section presents the review of some of the significant works on breast cancer detection/diagnosis using thermal images, image processing, machine learning, and deep learning.

Zuluaga-Gomez et al. [8] performed a study of the impact of data pre-processing, data augmentation, and database size on a proposed set of CNN models. Tree Parzen Estimator was used for CNN hyperparameters fine-tuning. 57 patients database from DMR-IR database [9] were used, and the CNN models obtained 92% of accuracy and F1-score, which outperformed various state-of-the-art architectures namely Inception, ResNet50, and SeResNet50. The results also confirmed that a CNN model using data-augmentation techniques attained similar performance metrics compared with a CNN that uses a 50% bigger database.

Kakileti et al. [10] explored several CNN architectures for semantic segmentation. Hotspots in the thermal image were detected using naive patch-based classifiers to several variations of the encoder-decoder architecture. 180 subjects were used (private database) and results revealed that encoder-decoder architectures performed better than patch-based classifiers in spite of small thermal image datasets in terms of accuracy.

Torres-Galvan et al. [11] used the DMR-IR database to classify breast thermograms using transfer learning. Pre-trained architectures: GoogLeNet, AlexNet, ResNet50, ResNet101, InceptionV3, VGG-16, and VGG-19 were used. Images were resized to a fixed size of 227×227 or 224×224 pixels, and 173 patients database was randomly split into 70% for training and 30% for validation. The learning rate of 1×10^{-4} and 5 epochs were used for all deep Neural Networks. VGG-16 outperformed with a balanced accuracy of 91.18%, specificity of 82.35%, and sensitivity of 100%.

Fernandez-ovies et al. [12] used 216 patients (41 sick patients and 175 healthy patients) from the DMR-IR dataset (dynamic thermogram) and divided them into 500 healthy and 500 sick patients with breast thermal images with 80% allocation for training and testing (80–20 split) and 20% for validation. Various CNN models such as ResNet18, ResNet34, ResNet50, ResNet152, VGG-16, and VGG-19 were used. The results showed that ResNet50 and ResNet34 produced the highest validation accuracy rate of 100% for breast cancer detection.

Mishra et al. [13] used DCNN on 160 abnormal and 521 healthy breast thermograms of DMR-IR Database. After the conversion of color to grayscale, thermal images were pre-processed, segmented, and then classified using DCNN with SGD optimizer and a learning rate of 0.01. An accuracy of 95.8%, with specificity and sensitivity levels at 76.3 and 99.5%, respectively, resulted.

From the previous works, it can be observed that the researchers have explored and applied different deep convolutional neural network models for the classification of normal and abnormal breast thermograms using the self-collected breast thermal images and the images from the DMR-IR database. The number of images used in different research works was also different. The accuracies were obtained between 90 and 100%. Most of the self-collected datasets are not available for research purpose and current public datasets consist of only two classes of breast thermograms (healthy/normal and abnormal/sick). Though considerable research works have been published in the literature using deep learning models, researchers are continuously working on improving the efficiency of the algorithms, reducing the time complexity of the deep learning models, and improving the detection accuracy. In this paper, Vision Transformer (ViTs)-based solution is proposed for the classification of normal and abnormal breast thermograms.

3 Vision Transformer

Transformer is a powerful and popular model in the Nature Language Processing field. Transformers are networks that operate on sequences of data (a set of words in NLP). These sets of words are tokenized first and then applied as input to the transformers. The fundamental idea behind the transformer is self-attention [6] (a quadratic operation), where there is a connection of each word to every other word in an NLP model. The attention method permits the model to concentrate on the "important" feature of the next input. The Vision Transformer (ViT) [7] applies the same idea of Transformer in NLP. The idea behind Vision Transformer is to utilize the encoding part to implement classification. The input image is divided into many small patches and then flattened into a linear shape. Each patch of the image has been converted to a grid of pixel values and fed to the Transformer encoder, and a learnable class token is also passed into the encoder for classification. Table 1 displays the three ViT models (ViT-Base, ViT-Large, and ViT-Huge) proposed in Dosovitskiy et al. [7].

In this paper, the train from the scratch approach of ViT is applied to the task. To demonstrate the result of splitting image into patches, a random breast thermal

Table 1 Vision transformer models [7]

Model	Layers	Hidden size D	MLP size	Heads	Parameters (M)
ViT-base	12	768	3072	12	86
ViT-large	24	1024	4096	16	307
ViT-huge	32	1280	5120	16	632

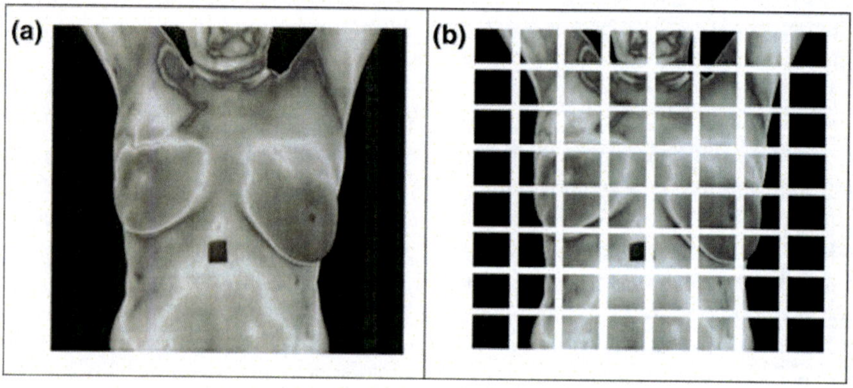

Fig. 1 **a** Original thermogram [9]. **b** Patches

image is chosen and performed patching on it. Figure 1 shows splitting an image into several 32 × 32 patches. The network structure for Vision Transformer is shown in Fig. 2.

4 Results and Analysis

Breast thermograms are used from the Research Data Base (DMR) [9] for this work. Thermograms of healthy and sick patients were acquired using a FLIRSC-620 IR camera having a resolution of 640 × 480 pixels with static and dynamic protocols. The dataset consists of images of individuals aged between 29 and 85 years old. In this work, static thermograms are used as tabulated. 90–10 data split is used for training and testing purposes (Fig. 3 and Table 2).

In order to measure the performance of the ViT, six performance indices are measured as follows:

$$\text{Accuracy (ACC)} = \frac{\text{TP} + \text{TN}}{\text{TP} + \text{FP} + \text{TN} + \text{FN}} \tag{1}$$

$$\text{Sensitivity/Recall (SE)} = \frac{\text{TP}}{\text{TP} + \text{FN}} \tag{2}$$

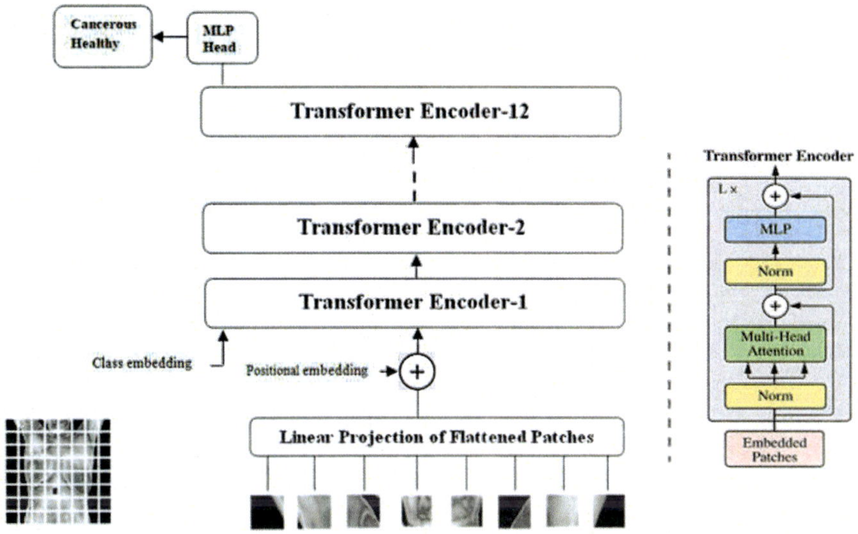

Fig. 2 Vision transformer structure [7]

Fig. 3 Number of thermograms

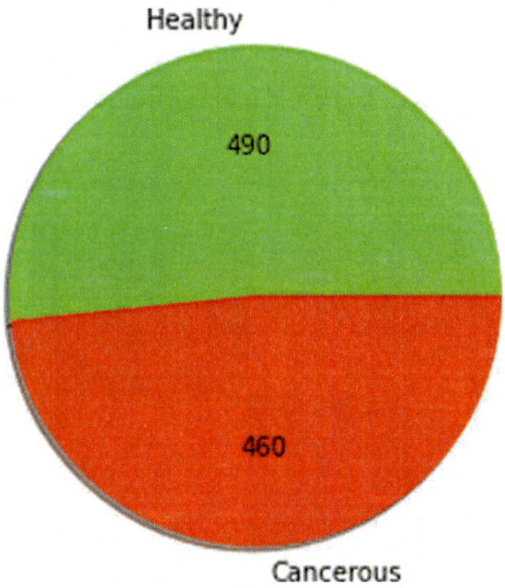

Table 2 Dataset distribution

	Cancerous	Healthy
Train	414	441
Test	46	49
Total	**460**	**490**

$$\text{Specificity (SP)} = \frac{\text{TN}}{\text{TN} + \text{FP}} \tag{3}$$

$$\text{Positive Predictive Value (PPV)/Precision (PRE)} = \frac{\text{TP}}{\text{TP} + \text{FP}} \tag{4}$$

$$\text{Negative Predictive Value (NPV)} = \frac{\text{TN}}{\text{TN} + \text{FN}} \tag{5}$$

$$\text{F1-score } (\text{F}_1) = 2\frac{\text{PRE.SE}}{\text{PRE} + \text{SE}} \tag{6}$$

ViT-Base has 12 encoder layers having 12 heads for multi-head attention. This network has 768 embedded size and 3072 MLP size. In the present study,16 × 16 and 32 × 32 size image patches are given to the input of ViT-B (ViT-B/16 and ViT-B/32). Adam optimizer is used with learning rate 1e-2 for training. 10% of test data is used for validation purposes.

A confusion matrix is drawn for each classifier (Fig. 4). In the present study, the positive and negative cases were allotted to cancerous and non-cancerous patients, respectively. Hence, TP and TN symbolize the number of correctly diagnosed cancerous and non-cancerous patients, respectively. FP and FN represent the number of incorrectly diagnosed cancerous and non-cancerous patients, respectively. Results are tabulated in Table 3.

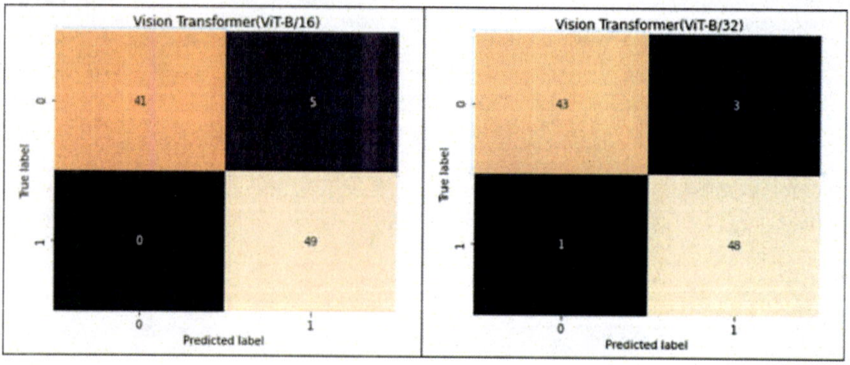

Fig. 4 Confusion matrix

Table 3 Performance evaluation

Model	Patch size	No. of patches	Acc %	SE	SP	PPV	NPV	F1
ViT-B/16	16 × 16	256	94.73	0.89	1.00	1.00	0.91	0.94
ViT-B/32	32 × 32	64	95.78	0.93	0.98	0.98	0.94	0.95

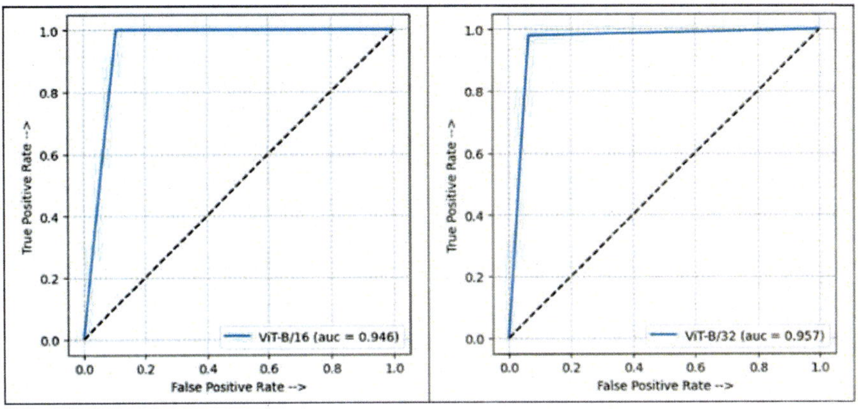

Fig. 5 ROC curve and AUC

Further, the area under the ROC curve (AUC) [14] is calculated to show the overall performance of the ViTs (Fig. 5). F1-score is calculated when the False Negatives and False Positives are important [15].

The proposed ViT model yielded a maximum accuracy of 95.78% for 32 × 32 patches and 94.73% for 16 × 16 patches using the distribution of 90% training and 10% testing. It is also observed from Fig. 5 that the proposed ViT model yielded a maximum AUC of 0.957 for 32 × 32 patches and 0.946 for 16 × 16 patches using the distribution of 90% training and 10% testing. The results of the proposed model cannot be compared directly with the existing works in the literature due to different numbers of images/subjects used, different deep learning models used, transfer learning or learning from scratch techniques used, and acquisition protocols used (dynamic/static). Some of the significant works published in the literature using the DMR dataset with different deep learning models are reported in Table 4. This table clearly indicates that the CNN models proposed by researchers achieved accuracies between 91.8 and 100% either using transfer learning [11] or trained from the scratch which includes ResNet18, ResNet34, ResNet50, SeResNet50, VGG-16, and Inception models. The proposed ViT model yielded a maximum accuracy of 95.78% and a maximum AUC of 0.957 for 32 × 32 patches using the distribution of 90% training and 10% testing. Considering ViT models demand a large-scale dataset for training, and the size of DMR data is relatively small, 90% of the images were used for training and the rest 10% of images were used for testing in this work.

5 Conclusion and Future Scope

Medical images diverge from natural images as they have originally higher resolutions along with smaller regions of interest. As a result, neural network architectures that perform well for natural images probably not be appropriate for medical image analysis.

Table 4 Comparison between different methods using DMR dataset

Authors	Thermograms		Data split (train-test)	Deep learning models used	Acquisition protocol	Epochs, learning rate	Acc%
	H	C					
Gomez et al. [8]	380	740	70–30	ResNet50, SeResNet50, Inception	Dynamic	40, --	92
Galvan et al. [11]	141	32	70–30	VGG-16	Static	5, 1×10^{-4}	91.8
Fernandez et al. [12]	500	500	80–20	ResNet18, ResNet34, ResNet50, VGG-16	Dynamic	--, --	100
Mishra et al. [13]	521	160	-- – --	DCNN	Dynamic	50, 1×10^{-2}	95.8
Present work	490	460	90–10	ViT-B/16, ViT-B/32	Static	50, 1×10^{-2}	95.78

The Vision Transformer model works effectively, and it may require more data to classify the right class. The self-attention mechanism is very powerful not only in the field of NLP, but also in Computer Vision. Splitting the image into many patches helps the model to learn the image better, when sending these patches into the transformer encoder, the self-attention mechanism is applied. It will look for the most significant feature for each class and predict a new input image based on the significant part. The outcomes are compared with the corresponding performance of the CNNs and demonstrate that attention-based ViT models score comparable achievement with CNN methods (95.78% accuracy).

Improving the performance of Vision Transformer is a challenging task. This work can also be extended and modified for low-resolution breast thermal images captured using a mobile camera. The results presented in this analysis reveal new ways to utilize self-attention-based architectures as a substitute for CNNs in different medical image analysis tasks.

References

1. Jemal A, Bray F, Center MM, Ferlay J, Ward E, Forman D (2011) Global cancer statistics. Cancer J Clin 61(2):69–90
2. Pandey N (2018) [World Cancer Day] Why does India have the third highest number of cancer cases among women? https://yourstory.com/2018/02/world-cancer-day-why-does-india-have-the-third-highest-number-ofcancer-cases-among-women/amp
3. Borchartt TB, Conci A, de Lima RCF, Resmini R, Sanchez A (2013) Breast thermography from an image processing view point: a survey. Int. J Signal Process 93(10):2785–2803
4. Yamashita R, Nishio M, Do RKG, Togashi K (2018) Convolutional neural networks: an overview and application in radiology. Insights Imag 9(4):611–629

5. Zhang X, Zou J, He K, Sun J (2015) Accelerating very deep convolutional networks for classification and detection, arXiv:1505.06798 [cs], May 2015
6. Vaswani A, Shazeer N, Parmar N, Uszkoreit J, Jones L, Gomez AN, Kaiser Ł, Polosukhin I (2017) Attention is all you need. In: Advances in neural information processing systems, pp 5998–6008
7. Dosovitskiy A, Beyer L, Kolesnikov A, Weissenborn D, Zhai X, Unterthiner T, Dehghani M et al (2020) An image is worth 16 × 16 words: transformers for image recognition at scale. arXiv preprint arXiv:2010.11929
8. Zuluaga-Gomez J, Al Masry Z, Benaggoune K, Meraghni S, Zerhouni N (2019) A CNN-based methodology for breast cancer diagnosis using thermal images. http://arxiv.org/abs/1910.13757
9. Silva LF, Saade DCM, Sequeiros GO, Silva AC, Paiva AC, Bravo RS, Conci A (2014) A new database for breast research with infrared image. J Med Imag Health Inform 4(1):92–100
10. Kakileti ST, Dalmia A, Manjunath G (2019) Exploring deep learning networks for tumor segmentation in infrared images. Quant Infr Thermogr J 17(3):153–168. https://doi.org/10.1080/17686733.2019.1619355
11. Torres-Galvan JC, Guevara E, Gonzalez FJ (2019) Comparison of deep learning architectures for pre-screening of breast cancer thermograms. In: Proceedings of Photonics North (PN), pp 2–3, May 2019. https://doi.org/10.1109/PN.2019.8819587
12. Fernández-ovies FJ, De Andrés EJ (2019) Detection of breast cancer using infrared thermography and deep neural networks. In: Bioinformatics and biomedical engineering. Springer, Berlin, Germany. https://doi.org/10.1007/978-3-030-17935-9
13. Mishra S, Prakash A, Roy SK, Sharan P, Mathur N (2020) Breast cancer detection using thermal images and deep learning. In: Proceedings of 7th international conference on computing for sustainable global development (INDIACom), pp 211–216, March 2020
14. Van Erkel AR, Pattynama PMT (1998) Receiver operating characteristic (ROC) analysis: basic principles and applications in radiology. Eur J Radiol 27:88–94
15. Sasaki Y (2007) The truth of the F-measure

An Improved Fourier Transformation Method for Single-Sample Ear Recognition

Ayush Raj Srivastava and Nitin Kumar

1 Introduction

Biometrics [1] are physical or behavioral characteristics that can uniquely identify a human being. Physical biometrics include—face, eye, retina, ear, fingerprint, palmprint, periocular, footprint, etc. Behavioral biometrics include voice matching, signature, handwriting, etc. There have been several applications [1] of biometrics in diverse areas such as ID cards, surveillance, authentication, security in banks, airports, corpse identification, etc. Ear [2] is a recent biometric which has drawn attention of the research community. This biometric possesses certain characteristics which distinguish it from other biometrics, e.g., less amount of information is required than face, where the person is standing in a profile manner to the camera, face recognition does not perform satisfactorily. Further, no user cooperation is required for ear recognition as required by other biometrics such as iris, fingerprint, etc.

Ear is one of those biometrices whose permanence attribute is very high. Unlike our face which changes considerably throughout our life, ear experiences very less changes. Further, it is fairly collectible and in the post-COVID scenario, it can be considered as a safer biometric since face and hands are covered with masks or gloves. It can be more acceptable if we do not bother a user for more number of samples. In real-world scenario, the problem of ear recognition becomes more complex when only a single training sample is available. Under these circumstances, One sample per person (OSPP) [3] architecture is used. This methodology has been highlighted in research community over all the problem domains such as face recognition [3, 4], ear recognition [5], and other biometrices. The reason of OSPP being popular is that the preparation of dataset; specifically the collection of samples from source is

A. R. Srivastava (✉) · N. Kumar
NIT Uttarakhand, 246174 Srinagar, Uttarkhand, India
e-mail: nitin@nituk.ac.in

© The Author(s), under exclusive license to Springer Nature Singapore Pte Ltd. 2023

H. Muthusamy et al. (eds.), *Robotics, Control and Computer Vision*, Lecture Notes in Electrical Engineering 1009, https://doi.org/10.1007/978-981-99-0236-1_14

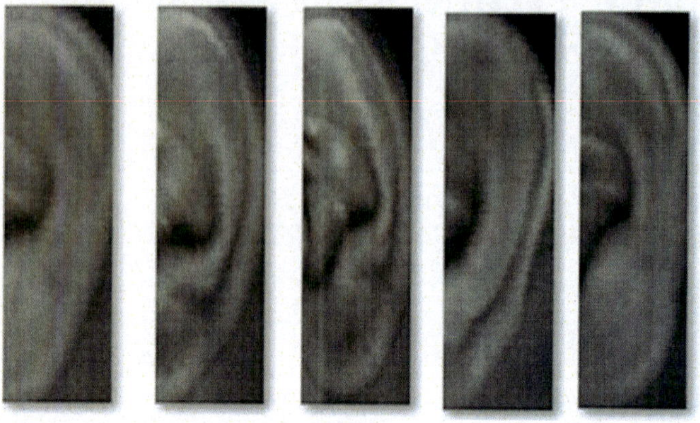

Fig. 1 Samples in IIT-Delhi dataset are tightly cropped

very easy. However, recognition becomes more complex due to the lack of samples. Hence, the model can not be trained in the best possible manner.

There are several methods suggested in literature by researchers for addressing OSPP for different biometric traits. Some of the popular methods include Principal Component Analysis (PCA), Kernel PCA, Wavelet transformation, Fourier transformation with frequency component masking, and wavelet transformation using subbands. Here, we propose an improved Fourier transform-based method for single-sample ear recognition. The biometrics image samples have been pre-processed using a morphological operation called opening. This is followed by the selection of high-frequency components using Fourier transformation and then PCA is used for feature extraction. Finally, SVM is used as a classifier. The performance of the proposed method is evaluated on the publicly available Indian Institute of Technology-Delhi (IIT-D) [6] ear dataset. Samples of dataset are shown in Fig. 1.

The rest of the paper is organized as follows: Sect. 2 presents the related work in single-sample ear recognition. Section 3 details the proposed improved Fourier transform-based method. Experimental setup and results are given in Sect. 4. Finally, the conclusion and future work are given in Sect. 5.

2 Related Work

PCA method was used for ear recognition by Zhang and Mu [9]. This method extracted local as well as global features. Linear Support Vector Machine (SVM) was used for classification. Later in 2009, Long and Chun [10] proposed using wavelet transformations for ear recognition. The proposed method was better than PCA and Linear Discriminant Analysis (LDA) [11] previously implemented. In 2011, Zhou et al. [12] used color Scale-Invariant Feature Transform (SIFT) method for representing the local features. In the same year, Wang and Yan [13] employed an ensemble of local binary pattern (LBP), direct LDA (linear discriminant analysis), and waterlet transformation methods for recognizing ears. The method was able to give accuracy up to 90% depending upon the feature dimension given as input. A robust method for ear recognition was introduced in 2012 by Yuan et al. [14]. They proposed an ensemble method of PCA, LDA, and random projection for feature extraction and sparse classifier for classification. The proposed was able to recognize partially occluded image samples. In 2014, Taertulakarn et al. [15] proposed ear recognition based on Gaussian curvature-based geometric invariance. The method was particularly robust against geometric transformations. In the same year, advanced form of wavelet transformation along with discrete cosine transformation was introduced by Ying et al. [16]. The wavelet used weighted distance which highlighted the contribution of low-frequency components in an image.

In 2016, Tian and Mu [17] used deep neural network for ear recognition. The proposed method also took advantage of CUDA cores for training the model. The final model was quite accurate against hair, pin, and glass occluded ear image. The same year, **One Sample Per Person** (OSPP) problem for ear biometric was tackled by Chen and Mu [18]. This method used an adaptive multi-keypoint descriptor sparse representation classifier. This method was occlusion-resistant and better than contemporary methods. The recognition time was little high in the band of 10–12 s. In 2017, Emersic et al. [8] introduced an extensive survey of methods of ear recognition. In this paper, different divisions were suggested for recognition approaches depending on the technique used for feature extraction, viz. holistic, geometric, local, and hybrid. Holistic approaches describe the ear with global properties. In this approach, ear sample is analyzed as a whole and local variations are not taken into consideration. Methods using geometrical characteristics of ear for feature representation are known as geometric approaches. Geometric characteristics of ear include location of specific ear parts, shape of the ear, etc. Local approaches describe local parts or local appearance of the ear and use these features for the purpose of recognition. Hybrid approaches involve those techniques which cannot be categorized into other categories or are an ensemble of different category methods. The paper also introduced a very diverse ear dataset called Annotated Web Ears (AWE) which has been used in this paper also.

In 2018, deep transfer learning method was proposed as a deep learning technique for ear biometric recognition by Ali et al. [19] over a pretrained CNN model called ALexNet. The methodology involved using a state-of-the-art training function called *Stochastic Gradient Descent with Momentum (SGDM)* and momentum of 0.9. Another deep learning-based method was suggested in 2019 by Petaitiemthong et al. [20]. In this method, a CNN architecture was employed for frontal-facing ear recognition. It was more acceptable due to the fact that the creation of face dataset simultaneously created the ear dataset. In the same year, Zarachoff et al. [21] proposed a variation of wavelet transformation and successive PCA for **single sample** ear recognition. In 2020, Omara et al. [22] introduced a variation of Support Vector Machine (SVM) for ear biometric recognition called "Learning distance Metric via DAG Support Vector Machine." In 2021, deep unsupervised active learning methodology was proposed by Khaldi et al. [23]. The labels were predicted by the model as it was unsupervised. Conditional Deep convolutional generative adversarial network (cDCGAN) was used to color the gray-scale image which further increased the accuracy of recognition.

Principal component analysis, or PCA [11], is a method used to reduce the dimensions of samples. It extracts those features which contain more variation in the intensity values and have higher contribution in image details. Reducing the number of variables of a dataset naturally comes at the expense of accuracy, but the trick in dimensionality reduction is to trade a little accuracy for simplicity. Because smaller datasets are easier to explore and visualize and make analyzing data much easier and faster for machine learning algorithms without extraneous variables to process.

PCA is a linear method which means that it can only be applied to datasets which are linearly separable. So, if we were to use it on non-linear datasets, higher chances are of getting inconsistent data. Kernel PCA [9] uses a kernel function to project dataset into a higher dimensional feature space, where the data is linearly separable. Hence, using the kernel, the original linear operations of PCA are performed in a reproducing kernel Hilbert space. Most frequently used kernels include cosine, linear, polynomial, radial basis function (rbf), sigmoid as well as pre-computed kernels. Depending upon the type of dataset on which these kernels are applied, different kernels may have different projection efficiency. Thus, the accuracy depends solely on the kernel used in the case of KPCA.

In the case of ear biometric, most of the data is contained in edges. In general case also, edge is the most important high-frequency information of a digital image. The traditional filters not only eliminate noise effectively but also make the image blurry. Blurring heavily deteriorates the edges. So, noise reduction becomes too costly in terms of information tradeoff. It is a top priority to retain the edge of the image when reducing the noise in an image. The wavelet analysis [10, 21] method is a time–frequency analysis method which selects the appropriate adaptive frequency band on the basis of the images' frequency component. Then the frequency band matches the spectrum which improves the time–frequency resolution. The wavelet transformation method has an obvious effect on the removal of noise in the signal. It also falls under the category of "local approaches". It preserves the locality of data while conversion from spatial/time to frequency domain. Hence, further operations can be applied in the frequency domain itself.

Fourier Transform [24] is a mathematical process that represents the image according to its frequency content. It is used for analyzing the signals. It involves the decomposition of the image components in the frequency domain in terms of infinite sinusoidal or cosinusoidal components. For a function of time, Fourier transform is a complex-valued function of frequency, whose magnitude gives the amount of that frequency present in the original function, and whose argument is the phase offset of the basic periodic wave in that frequency.

Unlike wavelet transformation which was a "local" approach, Fourier is a "holistic" approach. While converting from time/spatial domain to frequency domain, the locality of data is not preserved. Hence, data at each pixel in the resulting frequency map represents the components of the whole image in different proportions. Further operations in frequency domain become tricky, but the same "holistic" nature of this method increases its responsiveness towards other noise reduction techniques.

3 Proposed Work

Image pre-processing [24] using morphological operations [25] plays a vital role in improving the system performance. In morphology, two basic operations include dilation and erosion. Dilation operation is at the most basic level and XOR operation is performed on an image using a structuring element. It is used to fill holes as well as connect broken areas; subsequently, it widens the edges and increases the overall brightness of the images. Erosion, on the other hand, is the dual of dilation operator. It removes small anomalies as well as disconnects isthmus-like structures from images. Other advanced morphological operators are based on these two operators. One such operation is called *Opening*, which is successive dilation of the eroded image. The main aim of this operation is to remove small noise from the foreground. An illustration of these morphological operations is shown in Fig. 2.

We can see that erosion operation, although effectively removes the hair noise from the background, also increases the dimension of the ear periphery edge in foreground which is an important descriptor of the ear. Dilation removes that descriptor altogether as well as emphasizes the hair noise. Opening operation resembles the denoised ear to the maximum extent. Closing operation, although removes the hair occlusion effectively, also removes the periphery descriptor. Hence, in this proposed method, opening is preferred as a method of denoising the ear sample.

A schematic representation of the proposed method is shown in Fig. 3. After the pre-processing step, the Fourier transform is applied for finding low- and high-frequency components in the biometric image. Due to the fact that low-frequency components do not contribute much to the classification task, high-frequency components are selected using masking operation. The frequency components are arranged in descending order and the top 10% components are selected for image reconstruc-

Fig. 2 (left to right) Binary image, Images after erosion, dilation, opening, closing operations

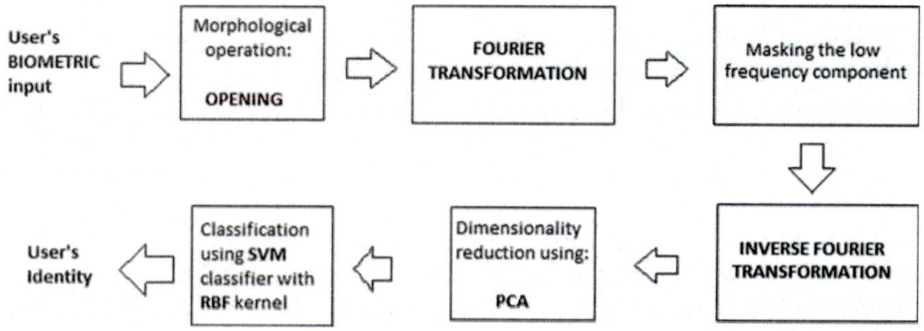

Fig. 3 Illustration of various steps involved in proposed approach

tion using Inverse Fourier transform (IFT) [24]. Subsequently, PCA is applied for feature extraction. Finally, support vector machine classifier is used for classification and Radial Basis Function (rbf) [26] kernel is used due to its property of projecting data into an infinite dimension.

Since data is finite, so infinite dimension won't be necessary. It guarantees the most optimum hyperplane since all data will be linearly separable in infinite dimensions. It contains two parameters: Regularization parameter (C) and Acceptance parameter (gamma). Regularization parameter indicates the complexity of decision boundary and a high value of this parameter will lead to overfitting since the boundary will be too complex to miss any point and a low value indicating that the boundary will be linear and model will underfit in training phase itself. Gamma is applicable to rbf kernel since it is based on the Gaussian function and has a classical inverted bell-shaped graph. Gamma indicates the significant region on that curve. A low value of gamma indicates that the model is too strict and may give low accuracy since it has very low tolerance towards deviation of samples. A high value of gamma will again lead to overfitting since it will accept any sample against any other sample. The proposed method improves the performance of the traditional Fourier transform-based method significantly. Experimental results presented in the next section also support this fact.

4 Experimental Results

In this section, we will compare the performance of the improved Fourier transformation method with other peer methods, viz. PCA, KPCA, and wavelet transformation using sub-bands in single-sample ear recognition scenario. The experiments are performed on the publicly available IIT-Delhi ear dataset. This dataset contains a total of 493 images corresponding to 125 identities with each image of size 50×180. One image per person is used for training and the remaining are used for testing. Each identity contains at least three images. The training is repeated three times by selecting one image of each identity in each iteration and forming the test set of the remaining images. The average classification accuracy of the three iterations is reported in this paper.

Each ear image is converted into a flattened feature vector of size 9000 (= 50 × 180). Thus, the size of the training data becomes 125 × 9000 whose covariance matrix will be of size 125 × 125. So, the maximum number of components after application of PCA is restricted to 125. Hence, the model is trained and tested on all possible number of principal components. The highest accuracy was obtained within the top 25 principal components in most cases. The performance in terms of average classification accuracy of proposed and compared methods on the basis of application of morphological operation is summarized in Table 1 and accuracy is plotted for all methods at all possible principal components in Fig. 4.

Now, we show the effect of kernel size of morphological operation on the classification accuracy as shown in Fig. 5. It can be observed that the kernel of size 6 × 14 performs optimally for the proposed method and result in a classification accuracy of 87.22%. It can also be observed that traditional PCA features are not

Table 1 Average classification accuracy of proposed and compared methods with and without morphological preprocessing

Method	Without opening		With opening		% Improvement
	Accuracy (%)	Components	Accuracy (%)	Components	
PCA [11]	71.59	6	76.05	21	6.23
KPCA [9]	71.03	8	78.26	102	10.18
Wavelet [21]	79.88	17	82.33	23	3.07
Proposed	74.15	18	**87.22**	22	**17.63**

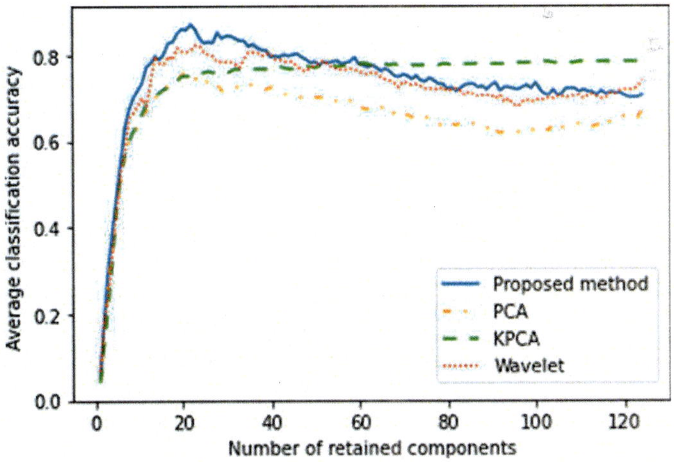

Fig. 4 Average classification accuracy of various methods against number of principal components

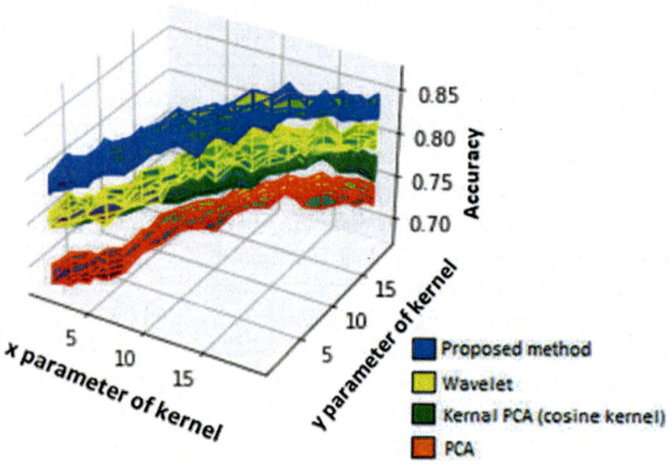

Fig. 5 Average classification accuracy of various methods against Kernel paremeters

Fig. 6 Average classification accuracy of various methods against classifier parameters

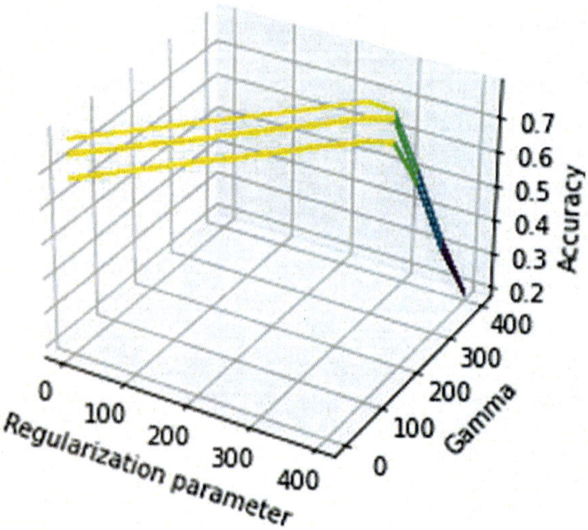

much suitable for single-sample ear recognition. Further, the effect of regularization and gamma parameters is shown in Fig. 6. It can be readily observed that the classification accuracy is not affected much over a large range of both these parameters. The classification accuracy decreases sharply when both these parameters take values of more than 250. The highest accuracy was obtained at parameters of classifier $C = 200$ and $gamma = 0.001$.

5 Conclusion and Future Work

Ear recognition has emerged as an attractive research area in the past few decades. This problem becomes more challenging when there is only one sample per person available for training. In this paper, we have proposed an improved method based on Fourier transformation for addressing single-sample ear recognition. Experimental results show that the proposed method performs better than the traditional Fourier transformation-based method. Further, it also performs better than several state-of-the-art methods. In future work, it can be explored how the deep learning-based methods can be exploited for single-sample ear recognition.

References

1. Jain A, Bolle R, Pankanti S (1996) Introduction to Biometrics. In: Jain AK, Bolle R, Pankanti S (eds) Biometrics. Springer, Boston, MA. https://doi.org/10.1007/0-306-47044-6_1
2. Yuan L, Mu Z, Xu Z (2005) Using ear biometrics for personal recognition. In: Li SZ, Sun Z, Tan T, Pankanti S, Chollet G, Zhang D (eds) Advances in biometric person authentication. In: IWBRS 2005. Lecture Notes in computer science, vol 3781. Springer, Berlin, Heidelberg. https://doi.org/10.1007/11569947_28
3. Kumar N, Garg V (2017) Single sample face recognition in the last decade: a survey. Int J Pattern Recogn Artif Intell. https://doi.org/10.1142/S0218001419560093
4. Zhao W, Chellappa R, Phillips P, Rosenfeld A (2003) Face recognition: a literature survey. ACM Comput Surv 35(4):399–458. https://doi.org/10.1145/954339.954342
5. Kumar N (2020) A novel three phase approach for single sample ear recognition. In: Boonyopakorn P, Meesad P, Sodsee S, Unger H (eds) Recent advances in information and communication technology 2019. Advances in intelligent systems and computing, vol 936. Springer, Cham. https://doi.org/10.1007/978-3-030-19861-9_8
6. Kumar A, Wu C (2012) Automated human identification using ear imaging. Pattern Recogn 41(5)
7. AMI Ear database. https://ctim.ulpgc.es/research_works/ami_ear_database/
8. Emeršič Z, Štruc V, Peer P (2017) Ear recognition: more than a survey. Neurocomputing 255:26–39. ISSN 0925-2312. https://doi.org/10.1016/j.neucom.2016.08.139. (https://www.sciencedirect.com/science/article/pii/S092523121730543X)
9. Zhang H, Mu Z (2008) Ear recognition method based on fusion features of global and local features. In: 2008 international conference on wavelet analysis and pattern recognition, pp 347–351. https://doi.org/10.1109/ICWAPR.2008.4635802
10. Long Z, Chun M (2009) Combining wavelet transform and orthogonal centroid algorithm for ear recognition. In: 2009 2nd IEEE international conference on computer science and information technology, pp 228–231. https://doi.org/10.1109/ICCSIT.2009.5234392
11. Kaçar Ü, Kirci M, Güneş E, İnan T (2015) A comparison of PCA, LDA and DCVA in ear biometrics classification using SVM. In: 2015 23nd signal processing and communications applications conference (SIU), pp 1260–1263. https://doi.org/10.1109/SIU.2015.7130067
12. Zhou J, Cadavid S, Mottaleb M (2011) Exploiting color SIFT features for 2D ear recognition. In: 2011 18th IEEE international conference on image processing, pp 553–556. https://doi.org/10.1109/ICIP.2011.6116405
13. Wang Z, Yan X (2011) Multi-scale feature extraction algorithm of ear image. In: 2011 international conference on electric information and control engineering, pp 528–531. https://doi.org/10.1109/ICEICE.2011.5777641

14. Yuan L, Li C, Mu Z (2012) Ear recognition under partial occlusion based on sparse representation. In: 2012 international conference on system science and engineering (ICSSE), pp 349–352. https://doi.org/10.1109/ICSSE.2012.6257205
15. Taertulakarn S, Tosranon P, Pintavirooj C (2014) Gaussian curvature-based geometric invariance for ear recognition. In: The 7th 2014 biomedical engineering international conference, pp 1–4. https://doi.org/10.1109/BMEiCON.2014.7017396
16. Ying T, Debin Z, Baihuan Z (2014) Ear recognition based on weighted wavelet transform and DCT. In: The 26th Chinese Control and decision conference (2014 CCDC), pp 4410–4414. https://doi.org/10.1109/CCDC.2014.6852957
17. Tian L, Mu Z (2016) Ear recognition based on deep convolutional network. In: 2016 9th international congress on image and signal processing, biomedical engineering and informatics (CISP-BMEI), pp 437–441. https://doi.org/10.1109/CISP-BMEI.2016.7852751
18. Chen L, Mu Z (2016) Partial data ear recognition from one sample per person. IEEE Trans Hum-Mach Syst 46(6):799–809. https://doi.org/10.1109/THMS.2016.2598763
19. Almisreb A, Jamil N, Din N (2018) Utilizing AlexNet deep transfer learning for ear recognition. In: 2018 fourth international conference on information retrieval and knowledge management (CAMP), pp 1–5. https://doi.org/10.1109/INFRKM.2018.8464769
20. Petaitiemthong N, Chuenpet P, Auephanwiriyakul S, Theera-Umpon N (2019) Person identification from ear images using convolutional neural networks. In: 2019 9th IEEE international conference on control system, computing and engineering (ICCSCE), pp 148–151. https://doi.org/10.1109/ICCSCE47578.2019.9068569
21. Zarachoff M, Sheikh-Akbari A, Monekosso D (2019) Single image ear recognition using wavelet-based multi-band PCA. In: 2019 27th European signal processing conference (EUSIPCO), pp 1–4. https://doi.org/10.23919/EUSIPCO.2019.8903090
22. Omara I, Ma G, Song E (2020) LDM-DAGSVM: learning distance metric via DAG support vector machine for ear recognition problem. In: 2020 IEEE international joint conference on biometrics (IJCB), pp 1–9. https://doi.org/10.1109/IJCB48548.2020.9304871
23. Khaldi Y, Benzaoui A, Ouahabi A, Jacques S, Ahmed A (2021) Ear recognition based on deep unsupervised active learning. IEEE Sens J 21(18):20704–20713. https://doi.org/10.1109/JSEN.2021.3100151
24. Gonzalez R, Woods R (2006) Digital image processing, 3rd edn. Prentice-Hall Inc., USA
25. Said M, Anuar K (2016) Jambek A, Sulaiman N (2016) A study of image processing using morphological opening and closing processes. Int J Control Theory Appl 9:15–21
26. Masood A, Siddiqui AM, Saleem M (2007) A radial basis function for registration of local features in images. In: Mery D, Rueda L (eds) Advances in image and video technology PSIVT, Lecture notes in computer science, vol 4872. Springer, Berlin, Heidelberg. https://doi.org/10.1007/978-3-540-77129-6_56

Driver Drowsiness Detection for Road Safety Using Deep Learning

Parul Saini, Krishan Kumar, Shamal Kashid, Alok Negi, and Ashray Saini

1 Introduction

Drowsiness is a state of lack of attention. It's a normal as well as transitory stage that happens as you're transitioning from becoming conscious toward being sleeping. Drowsiness can diminish a person's attention and raise the chance of an accident while they're doing things like driving a car, working a crane, or operating with heavy machinery like mine explosions. While driving, several indicators of driver drowsiness can be detected, such as inability to keep eyes open, frequent yawning, shifting the head forward, and so on. Various measures are used to determine the extent of driver drowsiness. Physiological, behavioural, and vehicle-based metrics are the three types of assessments [1].

Drowsy driving has resulted in several accidents and deaths. In a country like the United States, over 328,000 crashes happen each year. Each year, dollar 109 billion is spent on sleepy driving accidents [2]. To ensure that their vehicles are infallible, many automobile manufacturers employ various drowsy driver detecting technologies. Drowsy detection systems like driver alert and driver attention warning systems are incredibly effective and trustworthy, thanks to companies like Audi, BMW, and Bosch. There is, though, yet room to grow. There are a lot of different factors that may be utilised to identify tiredness in driver drowsiness detection systems. Behavioural data, physiological measurements, and vehicle-based data can all be used to detect criminal activity. Eye/face/head movement caught with a camera is considered behavioural data. Electrocardiogram (ECG) heart rate, electrooculogram (EOG), electroencephalogram (EEG), and others are examples of physiological measures [2].

P. Saini (✉) · K. Kumar · S. Kashid · A. Negi · A. Saini
Computer Science and Engineering, National Institute of Technology, Srinagar, Uttarakhand, India
e-mail: parulsaini.phd2020@nituk.ac.in

© The Author(s), under exclusive license to Springer Nature Singapore Pte Ltd. 2023

H. Muthusamy et al. (eds.), *Robotics, Control and Computer Vision*, Lecture Notes in Electrical Engineering 1009, https://doi.org/10.1007/978-981-99-0236-1_15

Steering wheel motion, vehicle speed, braking style, and lanes position deviation are all used to provide vehicle-based data. Questionnaires and electrophysical measurements can both be used to acquire data. However, getting meaningful feedback from a driver in a real-world driving situation is usually impossible or impracticable, and each of these methods has advantages and disadvantages. Physiological assessments are excessively intrusive, as they impair the driver's ability to drive safely. Hardware is required for vehicle-based measurements, which may be prohibitively expensive. Behavioural measurements, on the other hand, necessitate minimal technology, are very cost-effective, and do not impair the driver's ability to drive. Because of the benefits of behavioural data, we decided to use them as the foundation for the suggested detection system described in this study [2].

Behavioural measures are employed to detect driver tiredness in our suggested technique. Various face detection techniques [3] were employed in the Facial Detection phase to identify the face regions from the input photos. The problem of detecting human faces is simpler, but it is more challenging in computer vision. Face detection algorithms are divided into two categories: feature-based and image-based. Image-based algorithms for face detection have used statistical, neural network, and liner subspace methods. Different eye area detection techniques were employed in the second stage to detect and extract the eye region from facial photographs. After finding face regions, normalisation is performed in the preprocessing stage to reduce the impacts of illumination. Histogram equalisation can be used to adjust the contrast discrepancies between face images.

Extracted features was applied to the input eyes region images in the third stage. Appearance-based feature extraction and geometric-based feature extraction are the two basic methods for extracting features from photos. The geometric extraction approach extracts metrics relating to shape and position from the eyes and brows. In contrast, appearance-based feature extraction uses techniques like PCA [4], Discrete Cosine Transform (DCT) [5], and Linear Discriminant Analysis to extract skin appearance or face features (LDA). These approaches can be used to extract facial traits from the full face or specific parts of the face. Rise sleeping and non-sleeping images using the features extracted in the previous two steps. Deep Layered CNN was created to classify drowsy drivers [6].

2 Literature Review

This section describes Drowsiness Detection models [7, 8] and their limitations, along with some deep learning [9] processes that can automatically be fea tured directly from the raw data.

Babaeian et al. [10] introduced a unique technique for measuring driver drowsiness that uses biomedical signal analysis based on machine learning and is applied to heart rate variation (HRV), which is measured from an ECG signal. The wavelet transform (WT) and the short Fourier transform (SFT) are used in the procedure (STFT). It then uses the support vector machine (SVM) and k-nearest neighbour (KNN) methods to extract and select the desired features [10]. The applied technique achieves an accuracy of 80% or more as a result of this. The accuracy result for the SVM approach is 83.23% when using STFT and 87.5% when using WT methods in our research. The algorithm with the best accuracy resulted in a lower number of drowsiness-related accidents, as our findings demonstrate.

Jabbar et al. [11] proposed the model in which accuracy was improved by using facial landmarks detected by the camera and transmitted to a Convolutional Neural Network (CNN) to classify tiredness. With more than 88% for the category without glasses and more than 85% for the category night without glasses, study has demonstrated the ability to give a lightweight alternative to larger categorization models. In all areas, more than 83% accuracy was attained on average. Furthermore, the new proposed model has a significant reduction in model size, complexity, and storage when compared to the benchmark model, which has a maximum size of 75 KB. The suggested CNN-based model may be used to create a high-accuracy and simple-to-use real-time driver drowsiness detection system for embedded systems and Android devices.

Saifuddin et al. [12] proposed research used a cutting-edge cascade of regressors method, in which each regression refers to estimation of facial landmarks, to improve recognition under drastically variable illumination situations. To learn nonlinear data patterns, the proposed method uses a deep convolutional neural network (DCNN). In this case, the challenges of varying illumination, blurring, and reflections for robust pupil detection are overcome by using batch normalisation to stabilise distributions of internal activations during the training phase, reducing the impact of parameter initialization on the overall methodology. The accuracy rate of 98.97% was attained utilising a frame rate of 35 frames per second in the proposed research, which is greater than prior research results. Balam et al. [1] proposed unique deep learning architecture based on a convolutional neural network (CNN) for automatic drowsiness detection utilising a single-channel EEG input is proposed in this paper. Subject-wise, cross-subject- wise, and combined-subjects-wise validations have been used to improve the suggested method's generalisation performance. The entire project is based on pre-recorded sleep state EEG data from a benchmarked dataset. When compared to existing state-of-the-art drowsiness detection algorithms using single-channel EEG signals, the experimental results reveal a greater detection capability.

3 Dataset and Methodology

3.1 Dataset

The Deep Learning model developed here is trained on images obtained from open source driver drowsiness detection dataset. Open dataset is classified into two categories: Closed and open. 1234 images for training belonging to 2 classes. 218 images test belonging to 2 classes. These images are preprocessed to create frames for this study.

3.2 Proposed Model

The general architecture of driver sleepiness detection is shown in Fig. 1.

Step 1: Input Image and Data Preprocessing Images are fed for data preprocessing and resized into 224 X 224 with re-scaling to convert all the pixel values between 0 and 1. The idea of transfer learning and a pre-trained VGG-16 were utilised to extract features. The VGG16 deep convolutional neural network has 16 layers and was trained using the ImageNet dataset, which has 1000 classes and a vast number of images. It also used the ImageNet database to train. Despite being built for images with a size of 224 × 224 pixels, the network can also imply various sizes. Moderate features are learned using the ImageNet dataset's weight, and high-level features are extracted using three newly added fully connected layers.

Step 2: Data Augmentation It's also critical to have more data during the Deep Learning training phase so that the model can understand all of the complexities and differences in the images. Data augmentation is a standard way for increasing the training data points. VGG-16 model was used to build new images by conducting a series of augmentation operations on the images by using shear range (0.2), zoom range (0.2), and horizontal flip (True) as an augmentation parameter.

Step 3: VGG-16 model Training and Implementation The proposed work used the pretained VGG-16 model by freezing all the layers and then fully con nected layer are replaced with two new dense layers. The first dense layer uses the 128 hidden nodes with relu activation followed by dropout (0.5). The rectified linear activation function or ReLU is a linear function that, if the input is positive, outputs the input directly; else, it outputs zero. Because a model that utilises it is quicker to train and generally produces higher performance, it has become the default activation function

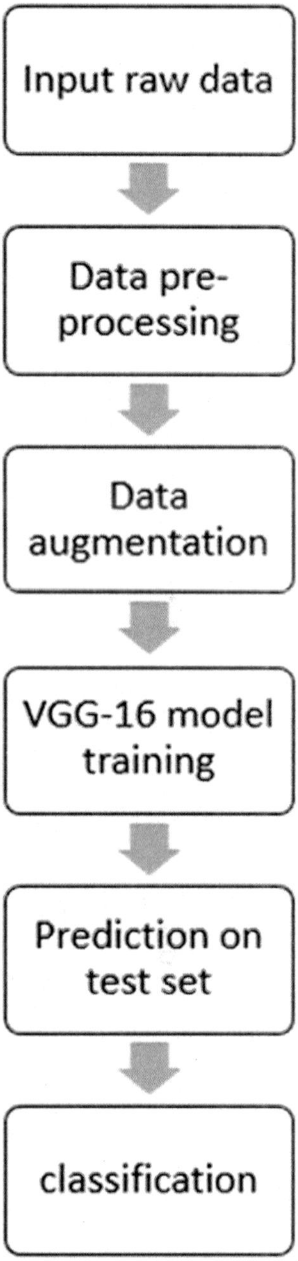

Fig. 1 General architecture of drowsiness detection

for many types of neural networks. The second dense layer is used for final output with 2 hidden nodes using softmax activation function. In neural network models that predict a multinomial probability distribution, the softmax function is utilised as the activation function in the output layer. Softmax is therefore utilised as the activation function for multiclass classification issues requiring class membership on more than two class labels.

Step 4: Transfer Learning To detect driver tiredness using hybrid features, a multi-layer based transfer learning strategy employing a convolutional neural network (CNN) was applied. A pre-train VGG-16 model, which is a sort of transfer learning approach, was employed to optimise feature.

4 Result and Discussion

The experiments were conducted on Google colab using python and model training runs for a total of 50 epochs with a batch size of 16. Image Data Generator is used for randomizing the training images for better performance of the model. Categorical cross entropy loss and accuracy are used as a metrics. A classifier's performance can be measured using a variety of indicators. Total accuracy, precision, recall and F1 Score measures are used in this paper and represented by Eqs. 1, 2, 3 and 4,

Accuracy is the number of correct predictions made as a ratio of all predictions made.

$$Acc = \frac{TP + TN}{FN + TP + TN + FP} \tag{1}$$

Precision analyzes the ability of the model to detect activeness when a subject is actually active.

$$Precision = TP/(TP + FP) \tag{2}$$

$$Recall = TP/(FN + TP) \tag{3}$$

F1 Score combines precision and Sensitivity results to balance the correct predictions rates of drowsy and active states.

$$F1\,Score = 2 \times (Precision * Recall)/(Precision + Recall) \tag{4}$$

The proposed work recorded 97.81% training accuracy with 0.07 loss and 96.79% accuracy with 0.08 loss score. The accuracy and loss curve are shown in Figs. 2 and 3. The precision, recall and F1 score are calculated as 97.22, 96.33 and 96.77% respectively. Confusion matrix are shown in Fig. 4. So, the eyes are certainly a crucial element in drowsiness classification in any setting, according to research and experimentation.

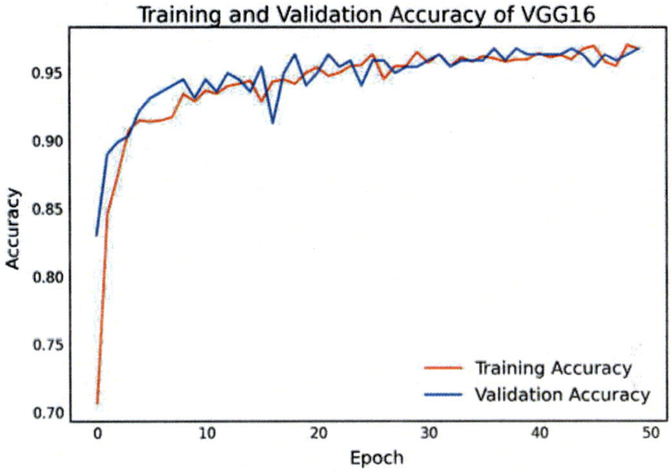

Fig. 2 Accuracy curve

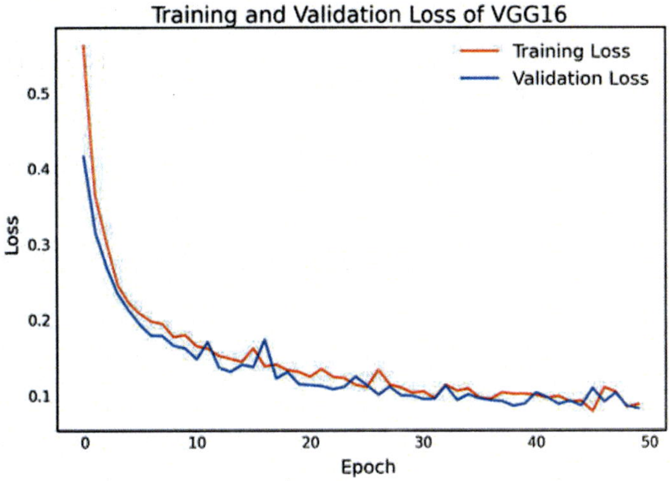

Fig. 3 Loss curve

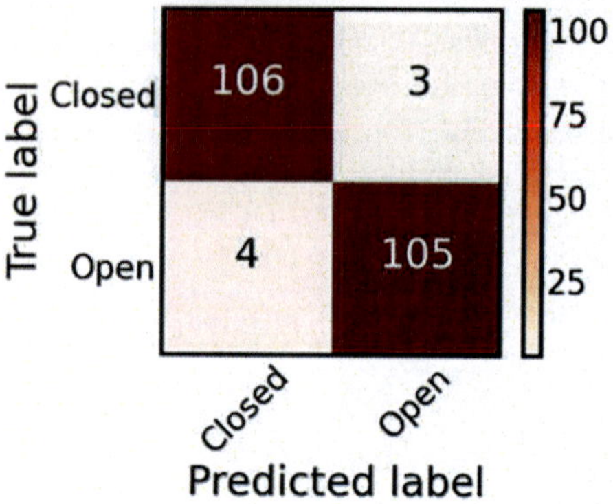

Fig. 4 Confusion matrix

5 Conclusion

Based on VGG-16 deep Learning, the research developed an enhanced drowsiness detection system. The major goal is to create a lightweight system that can be applied in VGG-16 and achieve excellent performance. The achievement in this case was the creation of a deep learning model that is minimal in size but precise. For all categories, the model described here has a total accuracy of 96.79%. This system may easily be linked into the next generation of car dashboards to support enhanced driver-assistance programs or even a mobile device to provide intervention when drivers are tired. This technology has drawbacks, such as obscuring facial features by wearing glasses [13, 14–17].

References

1. Balam VP, Sameer VU, Chinara S (2021) Automated classification system for drowsiness detection using convolutional neu ral network and electroencephalogram. IET Intell Transp Syst 15(4):514–524
2. Dua M, Singla R, Raj S, Jangra A (2021) Deep CNN models-based ensemble approach to driver drowsiness detection. Neural Comput Appl 33(8):3155–3168
3. Dang K, Sharma S (2017) Review and comparison of face detection algorithms. In: 2017 7th international conference on cloud computing, data scienceand engineering confluence. IEEE, pp 629–633
4. VenkataRamiReddy C, Kishore KK, Bhattacharyya D, Kim TH (2014) Multi-feature fusion based facial expression classification using DLBPand DCT. Int J Softw Eng Its Appl 8(9):55–68
5. Ramireddy CV, Kishore KK (2013)Facial expression classification using Kernel based PCA with fused DCT and GWT features. In: 2013 IEEE international conference on computational intelligence and computing research. IEEE, pp 1–6
6. Chirra VR, Reddy SR, Kolli VKK (2019) Deep CNN: a machine learning approach for driver drowsiness detection based on eye state. Rev d'Intelligence Artif 33(6):461–466
7. Altameem A, Kumar A, Poonia RC, Kumar S, Saudagar AKJ (2021) Early identification and detection of driver drowsiness by hybrid machine learning. IEEE Access 9:162805–162819

8. Esteves T, Pinto JR, Ferreira PM, Costa PA, Rodrigues LA, Antunes I, ... Rebelo A (2021) AUTOMOTIVE: a case study on AUTOmatic multiMOdal drowsiness detecTIon for smart VEhicles. IEEE Access 9:153678–153700
9. Negi A, Kumar K, Chauhan P, Rajput RS (2021) Deep neu ral architecture for face mask detection on simulated masked face dataset against COVID-19 pandemic. In: 2021 international conference on computing, communication, and intelligent systems (ICCCIS). IEEE, pp 595–600
10. Babaeian M, Mozumdar M (2019)Driver drowsiness detection algorithms using electrocardiogram data analysis. In: 2019 IEEE 9th annual computing and communication workshop and conference (CCWC). IEEE, pp 0001–0006
11. Jabbar R, Shinoy M, Kharbeche M, Al-Khalifa K, Krichen M, Barkaoui K (2020) Driver drowsiness detection model using convolutional neural networks techniques for android application. In: 2020 IEEE international conference on informatics, IoT, and enabling technologies (ICIoT). IEEE, pp237–242
12. Saifuddin AFM, Mahayuddin ZR (2020) Robust drowsiness detection for vehicle driver using deep convolutional neural network. Int J Adv Comput Sci Appl 11(10)
13. McDonald AD, Lee JD, Schwarz C, Brown TL (2018) A contextual and temporal algorithm for driver drowsiness detection. Accid Anal Prev 113:25–37
14. Zhao L, Wang Z, Wang X, Liu Q (2018) Driver drowsiness detection using facial dynamic fusion information and a DBN. IET Intel Transp Syst 12(2):127–133
15. Reddy B, Kim YH, Yun S, Seo C, Jang J (2017)Real-time driver drowsiness detection for embedded system using model compression of deep neural networks. In: Proceedings of the IEEE conference on computer vision and pattern recognition workshops, pp 121–128
16. Jabbar R, Al-Khalifa K, Kharbeche M, Alhajyaseen W, Jafari M, Jiang S (2018) Real-time driver drowsiness detection for android application using deep neural networks techniques. Procedia Comput Sci 130:400–407
17. Deng W, Ruoxue W (2019) Real-time driver-drowsiness detection system using facial features. IEEE Access 7:118727–118738

Performance Evaluation of Different Machine Learning Models in Crop Selection

Amit Bhola and Prabhat Kumar

1 Introduction

Agriculture is the world's primary source of food supply, and India is no exception. The pressure for food demand is increasing with growing population and reducing natural resources [1]. Hence, a more strategic approach with the use of modern technologies like artificial intelligence is need of the hour. Machine learning is a subsidiary of artificial intelligence, having two categories: supervised and unsupervised learning. Supervised learning algorithms perform classification or regression tasks, while unsupervised learning can cluster data based on similarity. ML techniques are being applied in various applications such as cybersecurity, agriculture, e-commerce, healthcare, and many more [2]. There are a variety of machine learning techniques that can assist in developing predictive models to solve real-world problems. ML is used in agriculture to solve various issues, including proper crop selection, weather forecasting, crop disease detection, agricultural production forecasting, and automated agricultural systems [3].

Traditional agricultural practices pose several challenges in terms of cost-effectiveness, and resource utilization including improper crop selection, declining crop yield, inappropriate usage of fertilizer and pesticides [4, 5]. Farmers and the agriculture community can benefit from machine learning technology to solve various issues by increasing crop yields and profits. Soil quality, climatic conditions, and water requirements play a vital role in crop selection for a specific piece of land [6]. In recent years, ML algorithms have been used in various aspects of agriculture like weather and yield prediction, disease detection, farmers risk assessment, and many more [7].

A. Bhola (✉) · P. Kumar
CSE Department, National Institute of Technology Patna, Bihar, India
e-mail: amitb.phd19.cs@nitp.ac.in

P. Kumar
e-mail: prabhat@nitp.ac.in

© The Author(s), under exclusive license to Springer Nature Singapore Pte Ltd. 2023
H. Muthusamy et al. (eds.), *Robotics, Control and Computer Vision*, Lecture Notes in Electrical Engineering 1009, https://doi.org/10.1007/978-981-99-0236-1_16

This paper implements six supervised machine learning models and analyze their performance for crop selection. The performance of these algorithms is evaluated in terms of accuracy parameter. This paper is organized as follows: Section I highlights the importance of ML in agriculture along with various agricultural issues. Section II discusses the related work in the field of crop selection. Various ML models used in this study are discussed in Section III. Section IV compares different crop prediction models with experimental data. Finally, Section V concludes the paper.

2 Related Work

Agriculture is the most important economic sector for any country, including India. Machine learning in agriculture aids scientists and researchers in predicting crop production, fertilizer, and pesticide use to boost crop yield and maximize resource utilization. Classification and prediction approaches using weather and soil data are analyzed for crop selection.

Paul et al. [8] provide a soil classification technique that divides soil into three categories based on its nutrient content: low, medium, and high. KNN classifies soil characteristics such as phosphorus, nitrogen, potassium, organic carbon, pH, and a few other micronutrients. The different soil categories help in determining the crop to a particular soil for optimal yield.

Kumar et al. [9] describe a crop selection approach for maximizing output yield. This research work also proposes crops plantation in a specific order annually to maximize the production. Crops are divided into two groups based on how long they take to grow, such as (1) crops available only at certain times of the year, (2) crops that can be grown throughout the year. Weather, soil, crop, and water density features are used for crop selection. This work also recommends crop sequencing depending on the crop sowing duration, time of plantation, and expected yield.

Tseng et al. [10] implement a crop selection approach that uses sensors to collect meteorological data such as humidity, temperature, etc. and soil data such as electrical conductivity, salinity, etc. 3D clustering is applied to examine crop growth used by farmers for a particular crop.

Pudumalar et al. [11] present an ensemble technique that uses Random forest, Naive Bayes, K-nearest neighbor, and CHAID to classify factors including soil colour, texture, depth, and drainage. This approach selects the crop for a given land using various input parameters.

Priya et al. [12] implement a Naive Bayes classification technique that uses weather parameters such as soil moisture, temperature, rainfall, and air pressure to determine the adaptability of crops such as rice, maize, cotton, and chilli. This approach also suggests the appropriate time for harvesting and sowing a specific crop.

Pratap et al. [13] implement a CART based system for fertilizer recommendation that uses ML model to determine the type and quantity of fertilizer to be used to maximize yield. This work tries to forecast the fertility of a particular soil sample in real-time by determining the soil nutrient content.

Chiche et al. [14] developed a neural network based crop yield prediction system. The proposed framework achieves a better prediction accuracy with 92.86% on 3281 instances collected from agricultural land dataset.

Kumar et al. [15] applied Logistic Regression, Support Vector Machine (SVM), and Decision Tree algorithms to predict the suitable crop based on agriculture parameters. These classification algorithms are compared and analyzed for crop prediction. The result shows that SVM performs better than other studied models.

Islam et al. [16] used Deep Neural Network (DNN) for agricultural crop selection and yield prediction. Various climatic and weather parameters are given as an input to the model for the crop prediction. The authors compared proposed DNN with SVM, Random forest, and Logistic regression. DNN outperforms other models in terms of accuracy.

From the literature review done on the existing work, it can be concluded that ML algorithms are being used in agriculture domain, but still there is a lot of scope in improving their performance in crop selection and yield prediction. Hence, this research work is conducted with a comparative study of supervised algorithms in crop selection. The following section discusses various machine learning models used in the agriculture domain.

3 Machine Learning Algorithms

Machine learning enables computers to make decisions based on their knowledge from data [8]. It applies to feature extraction, allowing the machines to extract essential properties from available data and information. ML applications includes fraud detection, disease detection, robot training using a set of rules, crop selection, yield prediction, etc. ML algorithms are broadly classified as supervised and unsupervised learning. This section discusses different supervised ML algorithms used in classification tasks and their performance in crop selection.

3.1 ML Algorithms for Crop Selection

This work implemented six different machine learning-based crop selecting algorithms. Different ML algorithms used are decision trees, random forests, support vector machines, naive Bayes, XGBoost, and k-NN to design and analyze crop-selecting models. The supervised machine learning algorithms are chosen for more accuracy in prediction tasks than unsupervised learning [17]. Various soil and weather parameters are used to implement these models. Soil parameters used are pH, nitrogen

(N), phosphorus (P), and potassium (K), and weather parameters used are temperature, humidity, and rainfall. Different machine learning models are discussed in the following subsection.

Decision Tree Classifier: A decision tree (DT) is a tree-structured classifier where internal nodes denote features, branches represent the decision rules, and each leaf node represents the outcome. The decisions or the tests are performed based on features of the given dataset. One of the DT techniques is classification or regression tree (CART). The tree begins with the root node, which contains all of the data, and splits the nodes using intelligent algorithms. It uses various impurity measures like the Gini Impurity Index, or Entropy to split the nodes. The Gini index and Entropy for a classification problem is defined in Eqs. (1) and (2) respectively, where n denotes total class and p_i is the probability of an object that is being classified to a particular class 'n'.

$$Gini = 1 - \sum_{i=1}^{n} (p_i)^2 \tag{1}$$

$$Entropy = \sum_{i=1}^{n} (-p_i) * log_2(p_i) \tag{2}$$

Naive Bayes: Naive Bayes is a classification technique based on Bayes' Theorem, assuming that all the features that predict the target value are independent of each other. It calculates the probability of each class and then picks the one with the highest probability. The Bayes' Theorem finds the likelihood of an event occurring using an already occurred event. The Bayes theorem is stated mathematically using Eq. (3), where 'X' and 'y' are different events, $p(X/y)$ is a conditional probability of event 'X' occurring given that 'y' is true, $p(y/X)$ is also a conditional probability of event 'y' occurring given X is true, $p(X)$ and $p(y)$ are the independent probabilities of X and y respectively.

$$p(y/X) = \frac{p(X/y)p(y)}{p(X)} \tag{3}$$

Support Vector Machines: Support Vector Machines are supervised machine learning approaches that are commonly used in multi-dimensional space categorization tasks. Different features of input data are plotted in an n-dimensional space, and the classifier model divides the input data into labels. The kernel functions of SVMs map the data from the bias functions to a potential higher-dimensional feature space. The support vectors are the subsets of the instances of the data. The hyperplane equation dividing the points (for classifying) is an approximation of a linear function in Eq. (4), where 'b' is bias of the hyperplane equation, 'w' is weight, and 'x' is input.

$$f(x) = w * x + b \tag{4}$$

Random forest (RF): A RF is a supervised machine learning algorithm that is constructed using decision tree algorithms. It can be used for classification or regression problems. RF is an ensemble method that operates on large number of individual decision trees and determines the outcome based on decision tree predictions. Each tree produces a class prediction, and the model's prediction is determined by the class with the most votes. The accuracy improves and problem of overfitting is prevented with the large number of trees. A Random Forest-based crop prediction model uses ensemble approaches to estimate the crop based on known soil and weather parameters.

Extreme Gradient Boosting (XGBoost): XGBoost is a gradient boosting-based decision-tree ensemble machine learning technique. It provides a parallel tree boosting to quickly solve many classification and prediction problems, including crop prediction. It is a supervised learning method work by optimizing loss functions and applying regularization techniques. The objective function (loss function and regularization) in Eq. (5) is required to be minimized, which contains two terms '*l*' is loss and 'Ω' is regularization, respectively. Here *l* is a differentiable convex loss function that measures the difference between the prediction '$\hat{y}_i$' and the target 'y_i'. 'Ω' penalizes the complexity of the model (i.e., the regression tree functions).

$$L^{(t)} = \sum_{i=1}^{n} l\left(y_i, \hat{y}_i^{(t-1)} + f_t(x_i)\right) + \Omega(f_t) \tag{5}$$

K-Nearest Neighbour (kNN): kNN is a supervised learning technique that can be used for regression or classification task. It predicts the correct class for classification task, by considering k nearest data points. In the case of regression, the value is the mean of the 'K' selected training points. kNN uses a crop prediction model that looks for similarities between important soil and weather properties to determine the best crop for a particular plot of land. The distance between a data point and its nearest neighbor can be calculated using any of the four distance method: Euclidean, Manhattan, Hamming, and Minkowski. Euclidean is commonly used distance function. Equation (6) shows the Euclidean distance formula, where x_i, y_i are points in Euclidean n-space.

$$d(y, x) = \sqrt{\sum_{i}^{n} (x_i - y_i)^2} \tag{6}$$

Artificial Neural Network (ANN): ANN is a computational network inspired by human brain, mimicking the way that biological neurons signal to one another. An ANN has weighted units called artificial neurons or nodes interconnected to each other forming a layered structure. The structure comprises of an input layer, one or more hidden layers, and an output layer. ANN uses training data to learn and upgrade their performance. The equation for the neural network is a linear combination of the independent variables and their respective weights and bias term for each neuron. Equation (7) shows the neural network formula, where W_0 is bias, $W_1, W_2 \ldots W_n$ are the weights, and $X_1, X_2 \ldots X_n$ are inputs. Here, each term represents neuron which is a combination of independent variables and their respective weights.

$$Z = W_0 + W_1X_1 + W_2X_2 + \cdots + W_nX_n \tag{7}$$

The discussed ML algorithms are designed to choose the optimum crop for a specific piece of land based on the soil and environmental properties of the land. These algorithms use soil attributes of a particular area and the required climatic conditions to recommend crops. The following section discuss the experimental setup, dataset description, results achieved and their discussion.

4 Experiment and Result Analysis

This section discusses the experimental setup to perform the analysis, dataset used, implementation specification, and discussion of results achieved.

4.1 Experimental Setup

The supervised machine learning-based crop selection models are implemented in Python programming. The implementation is carried out on a Windows platform having hardware configuration of Intel core i5 processor with 3.6 GHz quad-core × 64-based processor and 8 GB of RAM. The following subsection describes the dataset used in this study.

4.2 Dataset

The dataset considered in this study is collected from Kaggle [18]. The dataset includes soil properties like pH, phosphate (P), potassium (K), nitrogen (N), and environmental parameters that affect crop development like humidity, and precipitation. Table 1 presents the description of the features used in this study.

The data collected contains 2200 land samples and 22 different crops, with each crop containing 100 different land samples. The various crops included in the study

Table 1 Feature description

Feature(s)	Description	Unit
Nitrogen (N)	It is responsible for photosynthesis in the plant	kg/ha
Phosphorus (P)	It is crucial to the crop's development	kg/ha
Potassium (K)	It is required for reproduction of crops	kg/ha
pH level (pH)	It determines the availability of essential plant nutrients	pH value
Temperature	Temperature is a key factor in plant growth and development	degree Celsius
Humidity	Humidity is important for photosynthesis in plants	%
Rainfall	The primary source of water for agricultural production	mm

are maize, rice, banana, mango, grapes, watermelon, apple, orange, papaya, coconut, cotton, jute, coffee, muskmelon, lentil, black-gram, kidney beans, pigeon beans, mung beans, moth beans, and pomegranates. The following subsection analyzes the dataset used in this paper.

4.3 *Analyzing the Dataset*

This section analyses the soil and environmental data that affect the crop selection procedure among different crop data. Primary macronutrients play a vital role in increasing crop yield and quality. Nitrogen, phosphorus, and potassium (N, P, and K) are the three significant elements that must be present in large quantities for proper crop growth. Figure 1 shows the comparison of N, P, and K values required by various crops. The required amount of macronutrients for crop development is maximum in cotton, apple, and grapes, and minimum in lentils, blackgram, and orange, respectively.

Figure 2 shows the essential features for crop selection. It is inferred that rainfall and humidity are important features among all the weather parameters. Various soil macronutrients like N, P, and K have almost equal weightage for all the crops. Overall, rainfall has the highest importance, while pH is the least importance among all the used parameters.

The following sub-sections discuss the algorithm used in the study, followed with the results and discussion of the implemented machine learning based crop selection models.

4.4 *Crop Selection Procedure*

This section presents the algorithm used in the approach. Algorithm 1 explains the detailed steps involved in crop selection.

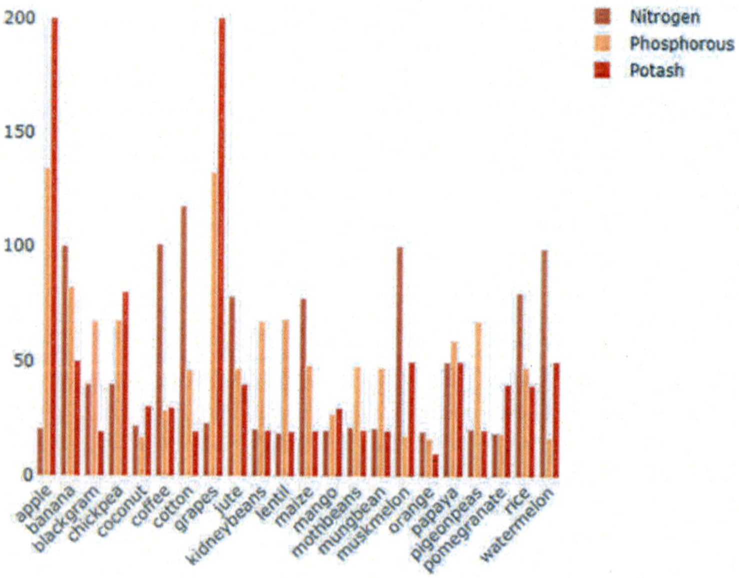

Fig. 1 N, P, K values required by different crops

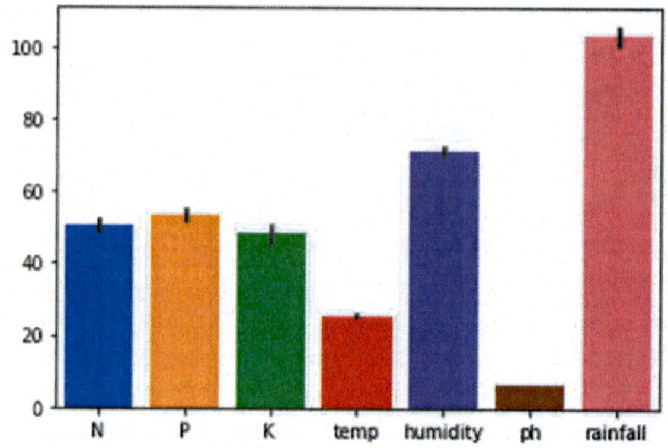

Fig. 2 Features importance

Algorithm 1: Crop Selection Procedure

Dataset: Soil data: N, P, K and pH
Weather data: temperature, humidity and rainfall
Crop data: 22 crops
Input: Soil and Weather data
Output: Crop

Step 1: Soil and Weather data is given as input.
Step 2: Data preprocessing steps are performed to fill any missing values, encoding categorical variables, etc.
Step 3: The preprocessed data is split into training and testing set in the ratio of 80:20.
Step 4: The machine learning model is applied to the training samples.
Step 5: The trained model is applied to the testing samples, to predict the most suitable crop for cultivation.
Step 6: Steps 1 to 5 are repeated for the discussed supervised algorithms.
Step 7: Finally, the performance of all the implemented algorithms are analyzed.

4.5 *Results and Discussion*

This section highlights the result obtained from ML techniques used on the crop data.

Machine Learning models can be evaluated using a variety of performance metrics like accuracy, precision, recall, Area under Curve (AUC), etc. This paper uses accuracy parameter to evaluate the models used in this study. These models are individually evaluated on the training and testing dataset as seen in Fig. 3. It shows the comparison of the training and testing accuracy of different ML models.

As seen in Fig. 3, the decision tree has the lowest training and testing accuracy of 88.18 and 90%, respectively. Random forest and XGBoost have the highest training accuracy of 100%, while XGBoost has the highest testing accuracy of 99.31%. As a result, in terms of testing accuracy, it can be concluded that Random Forest and XGBoost outperform all other supervised machine learning models.

The overall accuracy of all the crop prediction models is shown in Fig. 4. XGBoost has the highest accuracy in comparison to other models. Accuracy for Naive Bayes, SVM, Random Forest, and kNN are 99.09, 97.72, and 97.5%, respectively.

The Decision Tree is the worst performing model, with an accuracy of 90.0%. It can be concluded from the results achieved that Naive Bayes, Random Forest, and XGBoost perform better than other models for crop prediction, while XGBoost is the one which can be used for real applications, as it performed best in terms of overall accuracy. The following section concludes this paper, highlighting the research work done, results achieved and future scope.

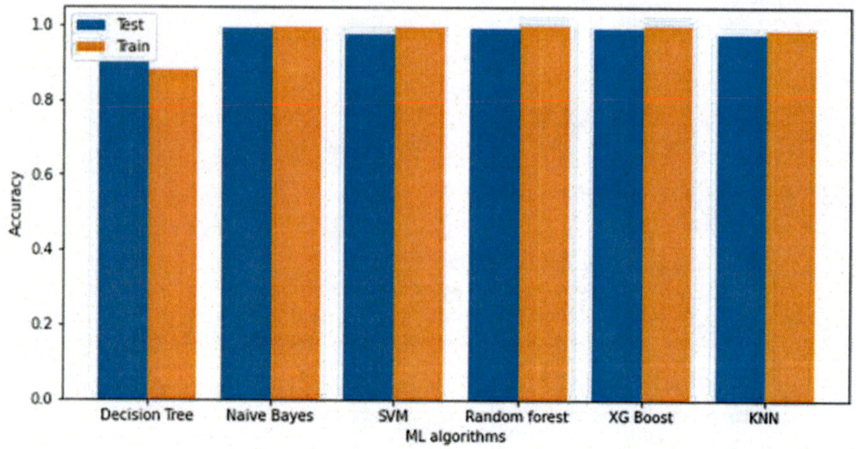

Fig. 3 Comparison of training and testing accuracies

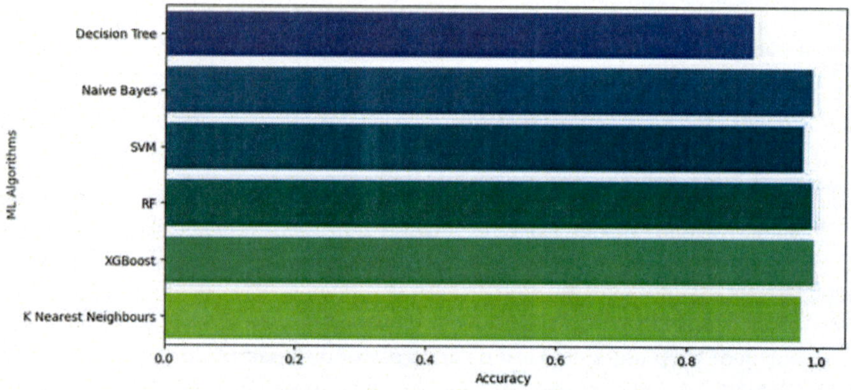

Fig. 4 Overall accuracy of different ML models

5 Conclusion

This paper compares six ML models to select crop based on soil and weather inputs. The models used are Decision Tree, Naive Bayes, Support Vector Machine, Random Forest, XGBoost, and K-Nearest Neighbor. The XGBoost supervised machine learning algorithm performed best with the testing accuracy of 99.31%, when compared with other used models. Crop selection models based on machine learning produces better results than traditional methods, as determined from the analysis done in this research work. Future work may include more number of parameters, such as water availability, irrigation facility, fertilizer requirement and market demand.

References

1. Gupta R, Sharma AK, Garg O, Modi K, Kasim S, Baharum Z, Mahdin H, Mostafa SA (2021) WB-CPI: weather based crop prediction in India using big data analytics. IEEE Access 9:137869–137885
2. Phadke M et al (2022) Designing an algorithm to support optimized crop selection by farmers. In: ICT analysis and applications. Springer, Singapore, pp 345–357
3. Kaur K (2016) Machine learning: applications in Indian agriculture. Int J Adv Res Comput Commun Eng 5(4):342–344
4. Jain K, Choudhary N (2022) Comparative analysis of machine learning techniques for predicting production capability of crop yield. Int J Syst Assur Eng Manag 1–11
5. Sinha A, Shrivastava G, Kumar P (2019) Architecting user-centric internet of things for smart agriculture. Sustain Comput: Inform Syst Sustain Comput: Inform Syst 23:88–102, 1 Sep 2019
6. Riaz F, Riaz M, Arif MS, Yasmeen T, Ashraf MA, Adil M, Ali S et al (2020) Alternative and non-conventional soil and crop management strategies for increasing water use efficiency. In: Environment, climate, plant and vegetation growth. Springer, Cham, pp 323–338
7. Suruliandi A, Mariammal G, Raja SP (2021) Crop prediction based on soil and environmental characteristics using feature selection techniques. Math Comput Model Dyn Syst 27(1):117–140
8. Paul M, Vishwakarma SK, Verma A (2015) Analysis of soil behaviour and prediction of crop yield using data mining approach. In 2015 international conference on computational intelligence and communication networks (CICN). IEEE, pp 766–771
9. Kumar R, Singh M, Kumar P, Singh J (2015) Crop selection method to maximize crop yield rate using machine learning technique. In: 2015 international conference on smart technologies and management for computing, communication, controls, energy and materials (ICSTM). IEEE, pp 138–145
10. Tseng FH, Cho HH, Wu HT (2019) Applying big data for intelligent agriculture-based crop selection analysis. IEEE Access 7:116965–116974
11. Pudumalar S, Ramanujam E, Rajashree RH, Kavya C, Kiruthika T, Nisha J (2017)Crop recommendation system for precision agriculture. In: 2016 eighth international conference on advanced computing (ICoAC). IEEE, pp 32–36
12. Priya R, Ramesh D, Khosla E (2018) Crop prediction on the region belts of India: a naive bayes mapreduceprecision agricultural model. In: 2018 international conference on advances in computing, communications and informatics (ICACCI). IEEE, pp 99–104
13. Pratap A, Sebastian R, Joseph N, Eapen RK, Thomas S (2019) Soil fertility analysis and fertilizer recommendation system. In: Proceedings of international conference on advancements in computing & management (ICACM)
14. Chiche A (2019) Hybrid decision support system framework for crop yield prediction and recommendation
15. Kumar A, Sarkar S, Pradhan C (2019)Recommendation system for crop identification and pest control technique in agriculture. In: 2019 international conference on communication and signal processing (ICCSP), IEEE, pp 0185–0189
16. Islam T, Chisty TA, Chakrabarty A (2018) A deep neural network approach for crop selection and yield prediction in Bangladesh. In: 2018 IEEE region 10 humanitarian technology conference (R10-HTC), pp 1–6, 6 Dec 2018
17. Jiang T, Gradus JL, Rosellini AJ (2020) Supervised machine learning: a brief primer. Behav Ther 51(5):675–687
18. https://www.kaggle.com/atharvaingle/crop-recommendation-dataset. Accessed 30 Nov 2021

Apriori Based Medicine Recommendation System

Indrashis Mitra, Souvik Karmakar, Kananbala Ray, and T. Kar

1 Introduction

COVID-19 carries an increased risk of serious consequences in some susceptible groups, such as the elderly, fragile, or those with several chronic illnesses. We can use such a categorization to put in place a method to combat medicine shortages. We strive to avoid shortages by using machine learning techniques to stock up on medicines that have been identified to be in high demand. Recently machine learning has been evolved from as a computational learning theory in artificial intelligence. It rose from an environment that was the integration of the interaction between available data, computing power, and statistical methodologies. Exponential growth of the available data compelled a spurt in computing power, which in turn stimulated the development of statistical methods to analyze large datasets.

Healthcare big data is a collection of patient, hospital, doctor, and medical treatment records that is so huge, complicated, scattered, and expanding at such a rapid rate that it is impossible to keep track of and analyze using typical data analytics methods [1]. To overcome these challenges, a big data analytics framework is used to apply machine learning algorithms to such a large quantity of data [2, 3]. Technology has also progressed significantly in the discovery and development of novel pharmaceuticals that have the potential to benefit patients with complex illnesses.

I. Mitra · S. Karmakar · K. Ray · T. Kar (✉)
KIIT Deemed to be University, Bhubaneswar, Odisha 750124, India
e-mail: tkarfet@kiit.ac.in

I. Mitra
e-mail: 1807274@kiit.ac.in

S. Karmakar
e-mail: 1807228@kiit.ac.in

K. Ray
e-mail: kbrayfet@kiit.ac.in

© The Author(s), under exclusive license to Springer Nature Singapore Pte Ltd. 2023

H. Muthusamy et al. (eds.), *Robotics, Control and Computer Vision*, Lecture Notes in Electrical Engineering 1009, https://doi.org/10.1007/978-981-99-0236-1_17

Given various amounts of accessible information about patients, a matching procedure for numerous different use cases has been developed [4]. Some large tech companies, such as IBM and Google, have developed machine learning tools to help patients find new therapy options. Precision medicine is an important concept in this discussion since it entails understanding mechanisms underlying complex disorders and developing new treatment options.

Although numerous semi-supervised strategies to give additional training data have been presented, automatically produced labels are typically too noisy to properly retrain models. As chronic diseases are long-lasting, it takes a lot of time to detect them [5]. Many machine learning and deep learning based models are proposed in the literature for disease detection and healthcare mangement [6–11]. The risk of severe complications from COVID-19 is higher for certain vulnerable populations, particularly people who are elderly, frail, or have multiple chronic conditions. Using such a classification we can implement a variety of measures for their betterment, such as a vaccine scheduler. Medicine shortages were a major contributor to the enormous number of fatalities. The current work suggests a solution to address this by a medicine recommendation system [3, 4, 12] using machine learning techniques to stockpile medications that have been identified as being in high demand, ensuring that there is no shortage and that we can provide them to people in need.

Big Data and the Cloud are two examples of new technologies that are helping to solve healthcare issues. Healthcare data is expanding at an exponential rate these days, necessitating an efficient, effective, and timely solution to cut mortality rates.

The importance of data collection, integration, processing, and reporting of underlying knowledge has been emphasized in the development of the concept of business intelligence and analysis, as well as how this knowledge can assist in making more appropriate business decisions and gaining a better understanding of market behaviors and trends. We have been able to unearth hidden information from data thanks to the massive expansion of data. Using current machine learning algorithms with minimal modifications, we may employ Big Data analysis for effective decision making in the healthcare industry. According to our findings, many academics are motivated to study machine learning algorithms in the health-care industry.

2 Proposed Model

The goal of this research is to use machine learning to help with drug supply. Using the Apriori algorithm's support metrics, the goal is to create a recommendation system for the medicine that a specific customer is most likely to buy, resulting in a win–win situation for both the customer and the shop owner: the customer gets the most appropriate medicine they want at all times and does not have to deal with the hassles of out-of-stock medicines; and the pharmacist learns the specific combination of medicines that is made available quickly. A lack of drug supply implies the medical black market is gone, which helps the economy thrive. The complete workflow of the proposed model is given in Fig. 1.

Fig. 1 Workflow of the model

Data Preprocessing

To support the laws and syntax that the specific ML model requires, the dataset must be preprocessed.

The following are the stages of preprocessing:

- Importing the desired libraries
- Importing datasets
- Dealing with missing data
- Encoding categorical data and encoding the dependent variable
- Feature scaling
- Splitting the dataset (training and test sets)

Apriori algorithm

- The Apriori algorithm [2, 13, 14] is an influential algorithm in determining frequent item sets for Boolean association rules.
- Apriori uses a "bottom up" approach, where frequent item sets are extended one item at a time (a step known as candidate generation, and groups of candidates are tested against the data).
- Apriori is designed to operate on datasets containing transactions. For example collection of items bought by customers.

Working of Apriori Model

The stages of the Apriori algorithm are given as follows. The flow chart of apriori model is depicted in Fig. 2.

1. Determine the itemsets' support in the transactional database and choose the lowest level of confidence and support.
2. Gather all of the dataset's support values that are greater than the minimum/selected support value.
3. Make a list of all the rules for subsets with a greater confidence value than the threshold or minimum confidence value.
4. Arrange the rules in order of decreasing lift.
5. The declining sequence of the lift will help us to better understand the relationship between the drugs.

Association Rule Learning

Association rule learning is a form of unsupervised machine learning approach that examines the reliance of one data item on another and maps appropriately to make

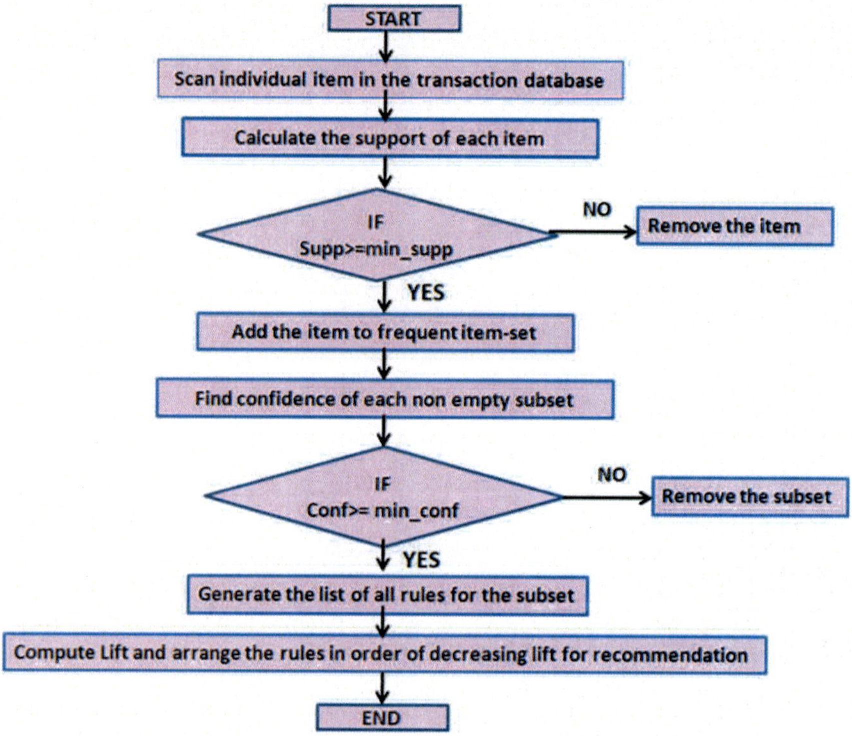

Fig. 2 Working of Apriori model

it more lucrative. It tries to uncover some interesting relationships or links between the dataset's variables. It uses a set of rules to find interesting relationships between variables in a database.

The discovery of frequent itemsets in a transactions database is a crucial aspect of association mining. It's used in a lot of data mining activities that aim to uncover interesting patterns in datasets, such association rules, episodes, classifiers, clustering, and correlation, and so on.

Model Description

In this project, we used the Apriori model to recommend the medicine combination that the customer is most likely to purchase. In 1994, Agrawal and Srikant introduced the Apriori technique [2], which uses recurring item sets to build association rules. It's designed to be used with transactional databases. These concepts can be used to determine how strongly or weakly particular items are connected.

The Apriori method uses a Hash tree and a breadth-first search algorithm to locate frequent items from a large dataset in an iterative fashion.

Association learning works on the if–then concept. The "If" element of association is called the Antecedent. The "Then" statement is called the Consequent. This type of relationship is called single cardinality. The metrics to find the association is given by the parameters namely **Support, Confidence and Lift.**

Support (Supp) is referred to as the frequency of X, or the number of times an item appears in a collection. It is the proportion of the transaction T that contains the itemset X as defined in (1).

$$\mathrm{Supp}(X) = \frac{\mathrm{Freq}(X)}{T} \tag{1}$$

Confidence (Conf) can be defined as the frequency with which a rule is correct which is reflected in its degree of confidence. It's the ratio of a transaction that contains X and Y to the number of records that include X and defined as given in (2).

$$\mathrm{Conf} = \frac{\mathrm{Freq}(X, Y)}{\mathrm{Freq}(X)} \tag{2}$$

Lift is the ratio of the observed support measure and expected support if X and Y are independent of each other as defined in (3).

$$\mathrm{Lift} = \frac{\mathrm{Supp}(X, Y)}{\mathrm{Supp}(X) \times \mathrm{Supp}(Y)} \tag{3}$$

It can have 3 values.

- Lift = 1: Antecedent and subsequent occurrence probabilities are independent of one another i.e. there is no association between the products.
- Lift > 1: Determines the degree to which the two items are interdependent i.e. the two products are more likely to be bought together.
- Lift < 1: It indicates that one object is a replacement for another, implying that one item causes harm to another i.e. the two products are unlikely to be bought together.

Higher the lift, more is the association between those elements.

3 Simulations and Result Analysis

The dataset used for simulation is a sample of medicine combinations that have been commonly bought by customers over the past 2 months. It is a random dataset that is made to illustrate the idea of medicine prediction and contains 7500 example records.

The dataset has been randomly generated thus ensuring the accuracy of the model in the context of its probability of getting lucky for a particular dataset. Since it is generated randomly, it verifies the model's correctness in terms of its likelihood of being fortunate for a certain dataset. The practical use case of this dataset is that it will

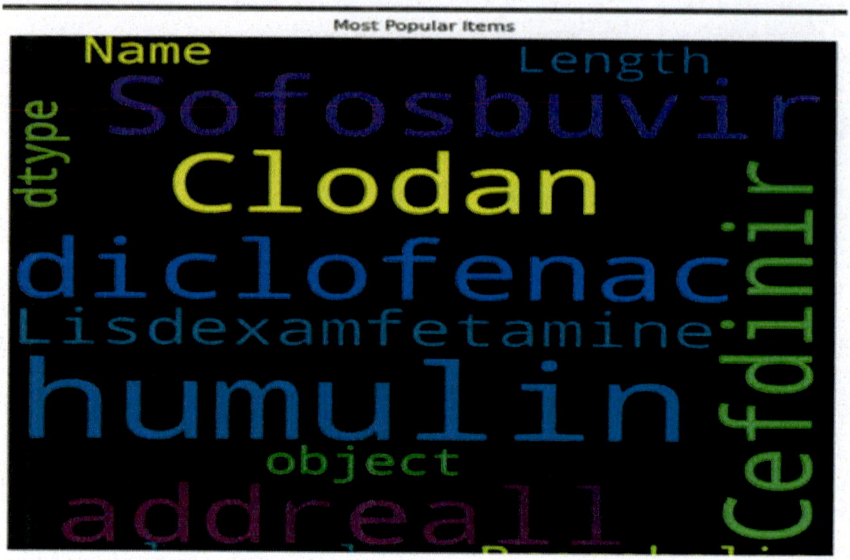

Fig. 3 Word cloud showing most popular items

be given by the chemist shop based on their previous sales. The Apriori algorithm will be executed on this for getting the preferred result.

The most commonly bought medicine items are shown in Fig. 3.

Figure 4 displays the most popular medicines as a frequency distribution. Figure 5 is a representation of the results obtained by using the algorithm to predict most common associations, presented as a descending order of their Lifts. Table 1 shows the labels for the different medicine combination. Figure 6 illustrates the association obtained for various medicine combinations, as recommended by the algorithm.

It is observed from Figs. 5 and 6 that the combination of the medicine Levothyroxin and Lisdexamfetamine denoted by (Le+Li) has the highest Lift which indicates that it is highly recommended. Similarly the combination of the medicine Sofosbuvir and Lupron denoted by (So+Lu) has the lowest Lift which indicates that the combination is least recommended.

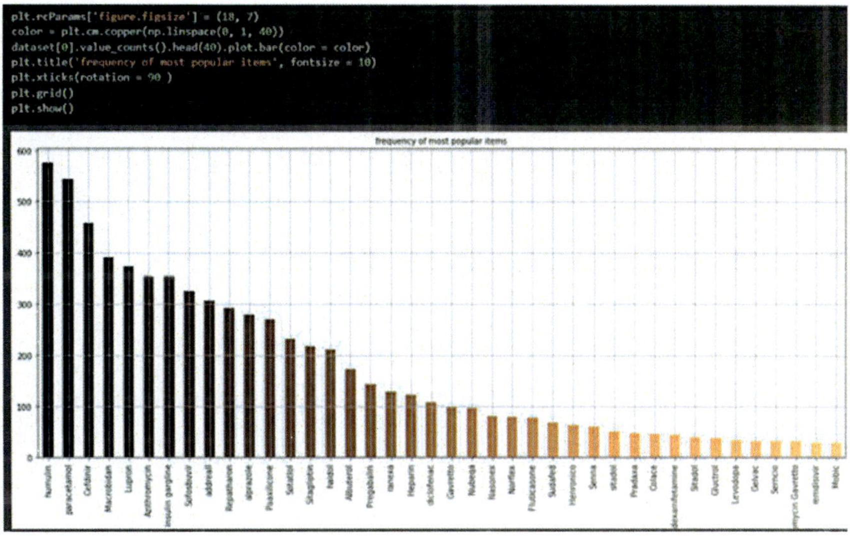

Fig. 4 Frequency distribution of most popular items

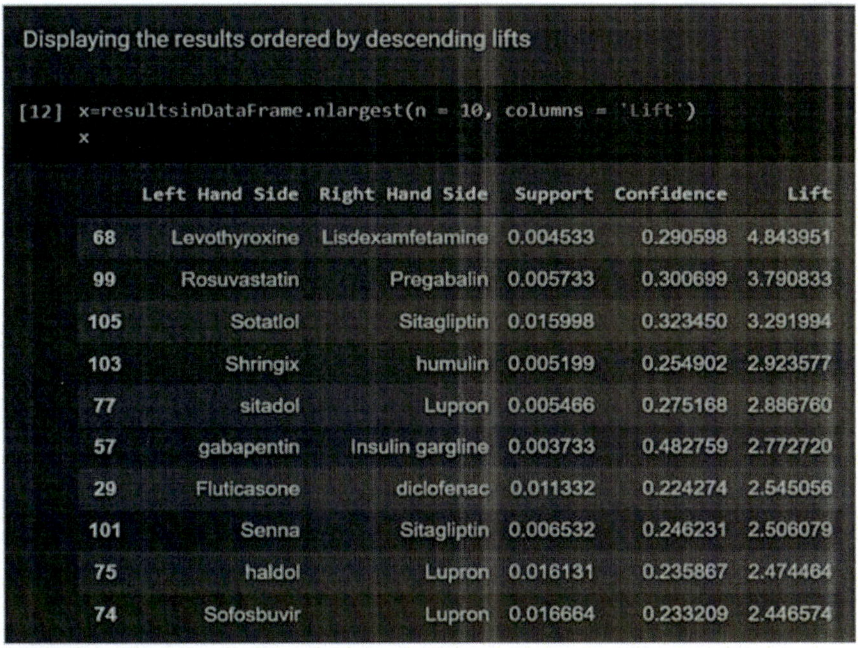

Displaying the results ordered by descending lifts

```
[12] x=resultsinDataFrame.nlargest(n = 10, columns = 'Lift')
     x
```

	Left Hand Side	Right Hand Side	Support	Confidence	Lift
68	Levothyroxine	Lisdexamfetamine	0.004533	0.290598	4.843951
99	Rosuvastatin	Pregabalin	0.005733	0.300699	3.790833
105	Sotatlol	Sitagliptin	0.015998	0.323450	3.291994
103	Shringix	humulin	0.005199	0.254902	2.923577
77	sitadol	Lupron	0.005466	0.275168	2.886760
57	gabapentin	Insulin gargline	0.003733	0.482759	2.772720
29	Fluticasone	diclofenac	0.011332	0.224274	2.545056
101	Senna	Sitagliptin	0.006532	0.246231	2.506079
75	haldol	Lupron	0.016131	0.235867	2.474464
74	Sofosbuvir	Lupron	0.016664	0.233209	2.446574

Fig. 5 Different medicine combination with their support, confidence and lift value

Table 1 Label for the medicine combination

Sl no.	Left hand side	Right hand side	Label
1	Levothyroxin	Lisdexamfetamine	Le + Li
2	Rosuvastatin	Pregabalin	Ro + Pr
3	Sotatlol	Sitagliptine	So + Si
4	Shringix	Humulin	So + hu
5	Sitadol	Lupron	Si + Lu
6	Gabapentin	Insulin gargling	Ga + In
7	Flucticasone	Diclofenac	Flu + dic
8	Senna	Sitagliptin	Se + Si
9	Haldol	Lupron	Ha + Lu
10	Sofosbuvir	Lupron	So + Lu

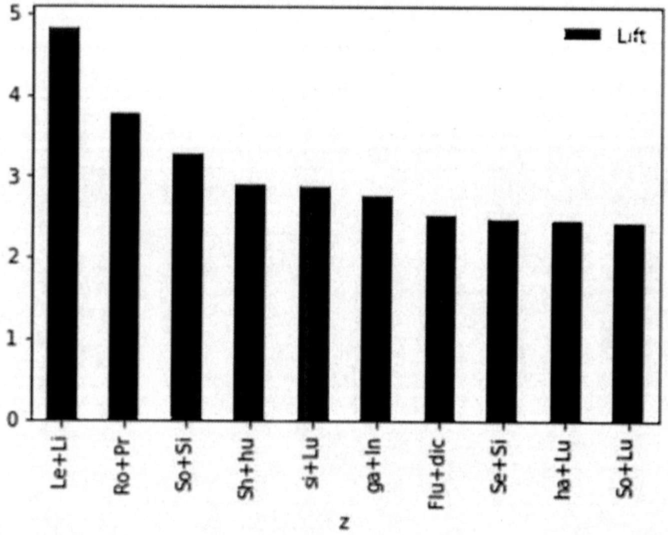

Fig. 6 Variation of the lift for the different medicine combination

Features of the Apriori Algorithm

- Uses large itemset property
- Easily parallelized
- Easy to implement

Disadvantages of Apriori Algorithm

- Assumes transaction database is memory resident
- Requires many database scans.

4 Conclusions

Patients and healthcare providers can use health recommender systems to help them make better health-related decisions. Shortages of key medicines will likely continue to be a problem. Our objective of, the medicine recommendation system will be helpful for the healthcare sector. People won't have to face the problem of unavailable medicines, since the stores will be stocked well in advance since they can know which medicines are most likely to be bought. Moreover, the economy will be helped since the medical black market will be eliminated as medicines are readily available so there will not be any shortage, thus no scope of dishonest people to dupe others by profiteering from selling medicines at exorbitant rates to the needy people. The future scope of this Apriori based machine learning recommendation model is that it will allow low infrastructural casualties in a healthcare center as it will always ensure that the best possible medicine or other health equipment are available at all times of the year. This will boost the lack of technical and managerial policies that are lacking today in different healthcare centers across India. This model can be further integrated with UI/UX apps which will allow a patient and his/her family to get a clear visual understanding of the current status of the different healthcare facilities that are available at a healthcare center in some developed areas without even travelling long distances in search of a preferable diagnostic center for the patient. This approach is expected to save many lives and thereby contribute to a better policy making for the common people.

References

1. Raghupathi W, Raghupathi V (2014) Big data analytics in healthcare: promise and potential. Health Inf Sci Syst 2:3. Published 2014 Feb 7. https://doi.org/10.1186/2047-2501-2-3
2. Al-Maolegi M, Arkok B (2014) An improved Apriori algorithm for association rules. Int J Nat Lang Comput. 3. https://doi.org/10.5121/ijnlc.2014.3103
3. Tran TNT, Felfernig A, Trattner C et al (2021) Recommender systems in the healthcare domain: state-of-the-art and research issues. J Intell Inf Syst 57:171–201
4. Han Q, Ji M, Martínez de Rituerto de Troya I, Gaur M, Zejnilovic L (2018) A hybrid recommender system for patient-doctor matchmaking in primary care. In: The 5th IEEE international conference on data science and advanced analytics (DSAA), pp 1–10
5. Kohli PS, Arora S (2018) Application of machine learning in disease prediction. In: 2018 4th international conference on computing communication and automation (ICCCA). IEEE, pp 1–4
6. Ferdous M, Debnath J, Chakraborty NR (2020) Machine learning algorithms in healthcare: a literature survey. In: 2020 11th international conference on computing, communication, and networking technologies (ICCCNT)
7. Ganiger S, Rajashekharaiah KMM (2018) Chronic diseases diagnosis using machine learning. In 2018 international conference on circuits and systems in digital enterprise technology (ICCSDET). IEEE, pp 1–6
8. Ramesh D, Suraj P, Saini L (2016) Big data analytics in healthcare: a survey approach. In: 2016 international conference on microelectronics, computing and communications (MicroCom). IEEE, pp 1–6

9. Ravì D et al (2017) Deep learning for health informatics. IEEE J Biomed Health Inform 21(1):4–21—Geron (2019) Hands-on machine learning with Scikit-Learn, Keras, and TensorFlow. O'Reilly Media, Inc, Canada
10. DeCaprio D, Gartner J, McCall C J, Burgess T, Garcia K, Kothari S, Sayed S (2020) Building a COVID-19 vulnerability index. J Med Artif Intell 3
11. Ahuja V, Nair LekshmiV (2021) Artificial intelligence and technology in COVID Era: a narrative review. J Anaesthesiol Clin Pharmacol 37:28. https://doi.org/10.4103/joacp.JOACP_558_20
12. Tran TNT, Atas M, Felfernig A, Le VM, Samer R, Stettinger M (2019) Towards social choice-based explanations in group recommender systems. In: Proceedings of the 27th ACM conference on user modeling, adaptation and personalization, UMAP'19. Association for Computing Machinery, New York, NY, USA, pp 13–21
13. Bagui S, Dhar PC (2019) Positive and negative association rule mining in Hadoop's MapReduce environment. J Big Data 6:75. https://doi.org/10.1186/s40537-019-0238-8
14. Zheng Y, Chen P, Chen B, Wei D, Wang M (2021) Application of apriori improvement algorithm in asthma case data mining. J Healthc Eng 2021:1–7. Article ID 9018408. https://doi.org/10.1155/2021/9018408

NPIS: Number Plate Identification System

Ashray Saini, Krishan Kumar, Alok Negi, Parul Saini, and Shamal Kashid

1 Introduction

Number plate recognition has been feasible vehicle monitoring in recent years. It may be used in a variety of public spaces for a variety of objectives such as traffic safety enforcement, automatic toll text collecting [1], car park system [2], and automated vehicle parking system [3]. The number plate identification systems use several methods to find vehicle number plates on automobiles and then extract vehicle numbers from the picture. This technology is also gaining popularity because it requires no other vehicle installation with a license plate. Although number plate detection algorithms have advanced significantly in recent years, it remains challenging to recognize license plates from photos with complicated backgrounds. Various scholars have offered different strategies for each phase, and each approach has advantages and disadvantages. The three primary steps for identifying license plates are as follows. That is the region of interest, extraction of plate numbers, and character recognition.

A. Saini (✉) · K. Kumar · A. Negi · P. Saini · S. Kashid
Computer Science and Engineering, National Institute of Technology Uttarakhand, Bhararisain, India
e-mail: mt20cse001@nituk.ac.in

K. Kumar
e-mail: kkberwal@nituk.ac.in

A. Negi
e-mail: aloknegi.phd2020@nituk.ac.in

P. Saini
e-mail: parulsaini.phd2020@nituk.ac.in

S. Kashid
e-mail: kashid.shamalphd2021@nituk.ac.in

© The Author(s), under exclusive license to Springer Nature Singapore Pte Ltd. 2023

H. Muthusamy et al. (eds.), *Robotics, Control and Computer Vision*, Lecture Notes in Electrical Engineering 1009, https://doi.org/10.1007/978-981-99-0236-1_18

Kim et al. [5] used a learning methodology to create a license plate recognition system. Inside the automobile detecting module, the camera collects an image. The result is then an image of the candidate region. Han et al. [6] proposed a system that tracks several targets and generates high-quality photos based on plate numbers. The author created a fine-tuned dual-camera computer with a fixed camera and a pan-tilt-zoom camera to track moving transportation in an open field. The CNN classifier has recognized the license plate consecutively for recognition. Because 64 cars entered this location, data was manually compiled from the science images, and 59 I.D.s were accurately recognized using this technology. Dhar et al. [7] developed an automated program to support it for identifying license plates. Prewitt operators performed the detection of the number plate to segment the edges. Morphological dilation was performed to accentuate the points. Eventually, deep CNN was used to accomplish the reconnaissance job.

As a result, technology needs to track vehicles used for illegal activities so that the criminal can be arrested and punished as quickly as feasible. Human vision is constrained by various elements such as speed, illumination, tiredness, and so on, therefore relying on human aptitude for such a task is not ideal. This technology is also gaining popularity because it does not require any other installation on vehicles that already have a number plate. Furthermore, the previous technique results in additional overhead due to our utilized learning parameters.

To manage the challenges noted above, we developed a vehicle number plate detection approach that can work in low-light and noisy environments to address the challenges above. Three categories divide the silent aspects of our work:

- The number plate identification problem was formulated as a complex and time-consuming machine learning problem. It attains a better computational complexity for vehicle plate number detection.
- Our method uses computer vision and deep learning to detect the vehicle number plate in low-light and noisy environments.
- This model relies on color and texture to detect the presence of multiple edges in images.

The outline of this paper is organized as follows. Related work of vehicle number plate detection is described in Sect. 2. The detection and recognition modules of our framework are described in Sect. 3. Experiments performed on test images and results obtained are summarized in Sect. 4. Finally, conclusions are drawn, and some comments, in general, are made in Sect. 5.

2 Literature Review

This section describes vehicle number plate identification models and their limitations, along with some deep learning processes featured directly from the raw data.

Prabuwono et al. [2] studied and designed a car park control system using Optical Character Recognition (OCR) devices which are presented in this work. The system is designed to work in a client–server scenario. The results reveal that the system can save log records, which will make it easier to track parking users, update user and parking credit databases, and monitor parking space availability.

Kim et al. [5] studied the construction of a license plate recognition system using a learning-based technique. Three modules made up the system. The car detection, license plate segmentation, and recognition modules are the three in question. The car detection module recognizes a car in a given image sequence collected from the camera with a simple color-based technique. The license plate in a detected car image is extracted utilizing Neural Networks (NNs) as filters to analyze the license plate's color and texture attributes. The recognition module then uses a Support Vector Machine (SVM)-based character recognizer to read the characters in the detected license plate.

Qadri et al. [14] proposed Automatic Number Plate Recognition (ANPR) is a methodology for image processing that utilizes a vehicle's plate number to recognize it. The developed system initially detects the car before taking a picture of it. The image segmentation is used to retrieve the vehicle number plate region. Character recognition is done using an optical character recognition approach. The gathered data is then compared to records in a database to determine specific information such as the vehicle's owner, registration location, and address.

Fahmy et al. [12] explained the place of each contained character is extracted using image processing procedures, and the Binary Associative Memories (BAM) neural network handles the character identification procedure. BAM is a neural network that may automatically read characters of a number plate. Even though BAM is a specific neural technique, it can rectify skewed input patterns.

3 Proposed Model

The general architecture of the number plate identification system is shown in Fig. 1. This section describes the proposed model of vehicle number plate detection steps in detail.

Step 1: Input image and Noise Reduction

During the picture capture, coding, transmission, and processing phases, noise is constantly present. Image noise is the random change of brightness or color information in collected photographs. In the first step, reduce the noise from the image to achieve better accuracy for our model by using noise-reducing filters. A common problem with a noise-reducing filter is that it can degrade image details or the edges present in the image.

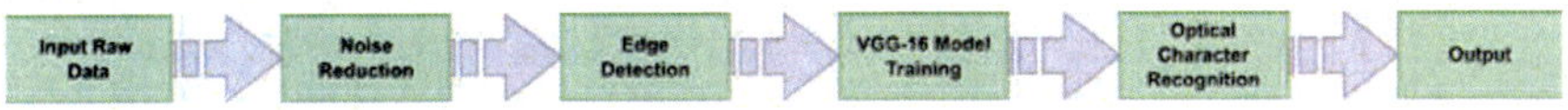

Fig. 1 Proposed model of vehicle number plate detection

So to eliminate the noise from the images while maintaining the features, the model uses a bilateral filter [4]. The bilateral filter is non-linear and edge-preserving in nature which employs the Gaussian filter, but it adds a multiplicative component based on pixel intensity difference. It guarantees that only pixel intensities identical to the center pixel are used when calculating the blurred intensity value. This filter is defined by Eq. (1), where the values of parameters of bilateral filters are as follows: diameter (Diameter of each pixel neighborhood) as 5, sigmaColor (Value in color space) and sigmaSpace (Value in coordinate space) both as 21.

$$BF[I]_p = \frac{1}{W_p} \sum_{x_i \in \Omega} I(x_i) f_r(||I(x_i) - I(x)||)\ g_s(||x_i - x||), \tag{1}$$

where $\boldsymbol{W_p}$ is a normalized term defined as

$$W_p = \sum_{x_i \in \Omega} f_r(||I(x_i) - I(x)||)\ g_s(||x_i - x||). \tag{2}$$

Step 2: Edge Detection

Edges are tiny fluctuations in the intensity of a picture. Edge detection is a critical mechanism for detecting and highlighting an object in an image and defining the borders between things and the background. The most common method for identifying significant discontinuities in intensity levels is edge detection. The edge representation of an image minimizes the amount of data to be processed while retaining important information about the forms of objects in the picture.

Gabor filter [8] has been used for edge detection and feature extraction. These filters include possessing optimal localization properties in both spatial and frequency fields and thus are well suited for texture segmentation issues. A Gabor filter can be described as a sinusoidal signal of a particular frequency and orientation, modulated by a Gaussian wave. The filter comprises a real and an imaginary component that represents orthogonal directions. The two parts can be combined to make a complex number or utilized separately. The Gabor filter is represented by Eqs. (3), (4) and (5). The values of parameter of Gabor filter are as follows: $\boldsymbol{\lambda}$ as 10, $\boldsymbol{\theta}$ as π, $\boldsymbol{\psi}$ as 5, $\boldsymbol{\sigma}$ as 1.9, and γ as 1.

$$\textbf{Complex: } g(x, y; \lambda, \theta, \psi, \sigma, \gamma) = \exp\left(-\frac{x'^2 + \gamma^2 y'^2}{2\sigma^2}\right) \exp\left(i\left(2\pi\frac{x'}{\lambda} + \psi\right)\right) \tag{3}$$

$$\textbf{Real: } g(x, y; \lambda, \theta, \psi, \sigma, \gamma) = \exp\left(-\frac{x'^2 + \gamma^2 y'^2}{2\sigma^2}\right) \cos\left(2\pi\frac{x'}{\lambda} + \psi\right) \tag{4}$$

$$\textbf{Imaginary: } g(x, y; \lambda, \theta, \psi, \sigma, \gamma) = \exp\left(-\frac{x'^2 + \gamma^2 y'^2}{2\sigma^2}\right) \sin\left(2\pi\frac{x'}{\lambda} + \psi\right) \quad (5)$$

$$\textbf{where } x' = x\cos\theta + y\sin\theta \textbf{ and } y' = -x\sin\theta + y\cos\theta$$

•

Step 3: VGG-16 Model Based on CNN

A typical CNN has several convolutional layers, pooling layers, and eventually fully linked layers in the final step. The convolution operation extracts high-level characteristics such as edges from the input picture. This output is transmitted to the next layer to identify more complex properties like corners and a combination of edges. As the network advances deeper, it detects increasingly difficult characteristics like things, faces, objects, and so on. The Pooling layer is in charge of lowering the spatial size of the convolved feature. Then the matrix is converted into a vector and sent into a fully linked layer, much like a neural network. Finally, it uses an activation function to classify or find particular points in pictures.

We have used transfer learning and customized VGG-16 architecture [9] to train the CNN model to recognize the number plate points. We have also augmented the data by using horizontal flip to True, vertical flip to True, zoom range as 0.2, and shear range as 0.2. All the layers use Rectified Linear Unit (ReLU) activation function except the last layer, which uses the linear activation function to predict the four points in the images of the vehicle number plate. The more detailed architecture that has been used to develop our model is given in Fig. 2.

Step 4: Optical Character Recognition

Optical Character Recognition (OCR) [10] systems convert a two-dimensional text picture, including machine-printed or handwritten text, from its image representation to machine-readable text. The initial phase is a connected component analysis, in which the component outlines are saved. Observing the layering of forms and the amount of child and grandchild outlines enables detecting and recognizing inverse text as straightforward as black-on-white writing. At this point, outlines are nested together to form Blobs.

Blobs are grouped into text lines, and the sequences and areas are evaluated to determine if the text is fixed pitch or proportional. Depending on the character spacing, text lines are divided into words in various ways. Character cells immediately cut selected pitch text. Balanced text is divided into words with definite and fuzzy spaces. In the first pass, an effort is made to recognize each word in turn. Each excellent term is used as training data by an adaptive classifier. The adaptive classifier is then allowed to detect text farther down the page more correctly. A second pass across the page is conducted because the adaptive classifier may have learned helpful information too late to contribute to the top of the page. Words that did not identify well enough are recognized again.

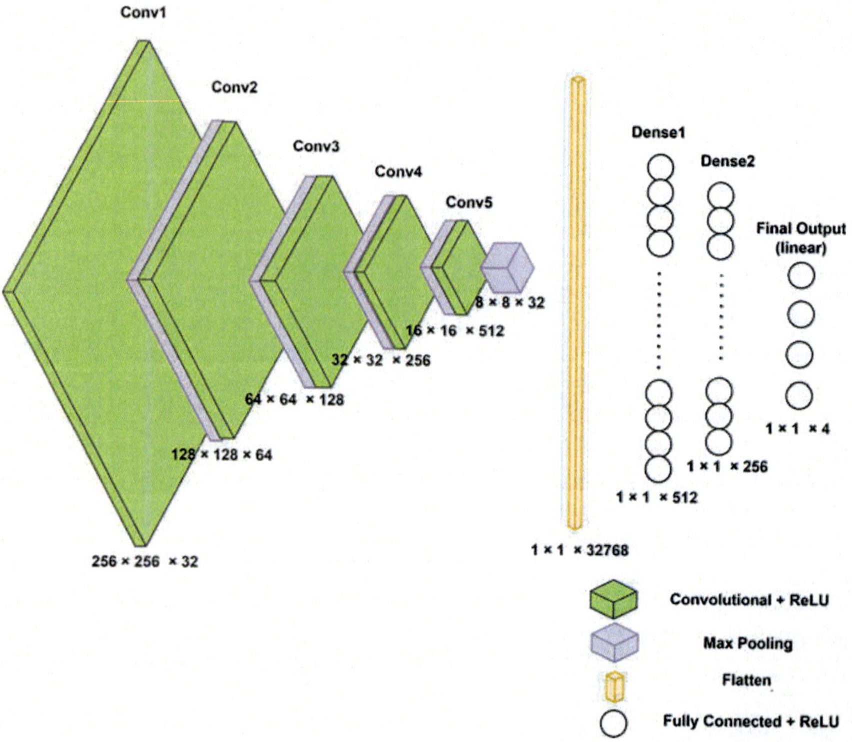

Fig. 2 Proposed convolutional neural network architecture

4 Experiments and Discussion

The NPIS model is built on a standard dual-core 2.6 GHz CPU on a six-core machine with an NVIDIA GeForce RTX 2060 GPU of 6GB. The experiment was carried out using a dataset of 664 images. The images were resized to $256 \times 256 \times 3$ pixels for training. The data is then normalized in the range [0, 1]. This data is then input into CNN architecture for training and testing purposes.

4.1 Quantitative Analysis

The proposed NPIS model correctly detected license numbers with great accuracy of 98.21% on the training dataset with 0.013 loss and 91.79% accuracy on the test dataset with 0.027 loss score. Our proposed model was trained up to 100 epochs, and the batch size was 11. The accuracy and loss curves of our model are shown in Fig. 3.

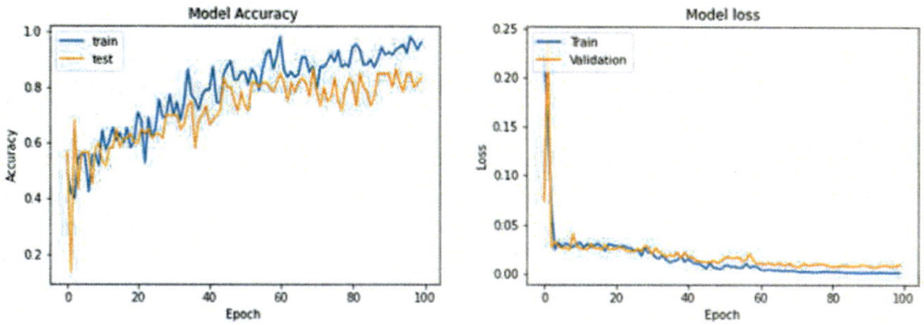

Fig. 3 Accuracy and loss of proposed model

Fig. 4 Output predicted by NPIS model

4.2 *Qualitative Analysis*

After completing the training of the model, it is used to predict the number plate. As shown in Fig. 4, our proposed model predicts the vehicle number plate inside the bounding box (which is shown in red color). The average time taken by our proposed model to predict the vehicle number plate in a single image is about 235 ms.

5 Conclusions

The proposed model is based on CNN architecture, NPIS (number plate identification system) system. Before processing, appropriate filters were applied to de-noise and sharpen low-quality photos resulting from high-speed vehicles. One of our strategy's primary characteristics is its scalability, which allows it to perform appropriately on various font styles and font sizes. The technology is so effective that it makes no difference whether the vehicle is stationary or moving at high speeds. The method given here may be applied in a cosmopolitan region, a rural location, an unpleasant background, poor light circumstances, a toll booth, any shielded parking lot, and so on. The primary drawback of this model is that it is not working on multiple vehicle number plates. The efficiency of larger datasets comprising a range of number plate styles from various countries will be improved in the future.

References

1. Chen Y-S, Cheng C-H (2010) A delphi based rough sets fusion model for extracting payment rules of vehicle license tax in the government sector. Exp Syst Appl 37(3):2161–2174
2. Prabuwono AS, Idris A (2008) A study of car park control system using optical character recognition. In: 2008 International conference on computer and electrical engineering. IEEE, pp 866–870
3. Albiol A, Sanchis L, Mossi JM (2011) Detection of parked vehicles using spatiotemporal maps. IEEE Trans Intell Transp Syst 12(4):1277–1291
4. Tomasi C, Manduchi R (1998) Bilateral filtering for gray and color images. In: Proceedings of the IEEE international conference on computer vision, pp 839–846
5. Kim KK, Kim KI, Kim JB, Kim HJ (2000) Learning-based approach for license plate recognition. In: Proceedings of the 2000 IEEE signal processing society workshop (Cat. No. 00TH8501). Neural Networks for Signal Processing X, vol 2. IEEE, pp 614–623
6. Han CC, Hsieh CT, Chen YN, Ho GF, Fan KC, Tsai CL (2007) License plate detection and recognition using a dual-camera module in a large space. In: 2007 41st annual IEEE international carnahan conference on security technology. IEEE, pp 307–312
7. Dhar P, Guha S, Biswas T, Abedin MZ (2018) A system design for license plate recognition by using edge detection and convolution neural network. In: 2018 International conference on computer, communication, chemical, material and electronic engineering (IC4ME2). IEEE, pp 1–4
8. Ji Y, Chang KH, Hung C-C (2004) Efficient edge detection and object segmentation using gabor filters. In: Proceedings of ACMSE-'04, pp 454–459, 2–3 April 2004

9. Simonyan K, Zisserman A (2015) Very deep convolutional networks for large-scale image recognition. In: ICLR
10. Verma R, Ali J (2012) A-survey of feature extraction and classification techniques in OCR systems. Int J Comput Appl Inform Technol 1(3):1–3
11. Lotufo RA, Morgan AD, Johnson AS (1990) Automatic number-plate recognition. In: IEE colloquium on image analysis for transport applications. IET, pp 1–6
12. Fahmy MM (1994) Automatic number-plate recognition: neural network approach. In: Proceedings of VNIS'94–1994 vehicle navigation and information systems conference. IEEE, pp 99–101
13. Kim KI, Jung K, Kim JH (2002) Color texture-based object detection: an application to license plate localization. In: International workshop on support vector machines. Springer, Berlin, Heidelberg, pp 293–309
14. Qadri MT, Asif M (2009) Automatic number plate recognition system for vehicle identification using optical character recognition. In: 2009 International conference on education technology and computer. IEEE, pp 335–338

Leveraging Advanced Convolutional Neural Networks and Transfer Learning for Vision-Based Human Activity Recognition

Prachi Chauhan, Hardwari Lal Mandoria, Alok Negi, Krishan Kumar, Amitava Choudhury, and Sanjay Dahiya

1 Introduction

Human activity recognition is essential in social contact and interpersonal relationships. It is hard to collect information on a specific individual, their personality, and psychological functioning. As a result of this study, numerous applications including security and surveillance have gained relevance in the vision community, particularly in crowded settings such as airports, retail malls, and social events, and require a multiple-action recognition system. The human ability to recognize the behaviors of another person is an important area of research in the computer vision and machine learning-based scientific fields.

Among the numerous categorization systems, two major questions arise: "What action?" (i.e., the difficulty with recognition) and "Where in the video?" (This is the localization issue.) A basic model for HAR in video frame patterns consists mostly of two steps. At the very first level, handcrafted features were retrieved from

P. Chauhan (✉) · H. L. Mandoria
Department of Information Technology, Govind Ballabh Pant University of Agriculture and Technology, Pantnagar 263153, Uttarakhand, India
e-mail: cprachi664@gmail.com

A. Negi · K. Kumar
Department of Computer Science and Engineering, National Institute of Technology, Srinagar (Garhwal) 246174, Uttarakhand, India
e-mail: aloknegi.phd2020@nituk.ac.in

K. Kumar
e-mail: kkberwal@nituk.ac

A. Choudhury
School of Computer Science, Pandit Deendayal Energy University, Gandhinagar, Gujrat, India

S. Dahiya
Ch. Devi Lal State Institute of Engineering and Technology, Panniwala Mota (Sirsa), Haryana, India

© The Author(s), under exclusive license to Springer Nature Singapore Pte Ltd. 2023

H. Muthusamy et al. (eds.), *Robotics, Control and Computer Vision*, Lecture Notes in Electrical Engineering 1009, https://doi.org/10.1007/978-981-99-0236-1_19

preprocessed data, and a classifier model was developed based on these features at the second level. The most prevalent HAR feature detectors include Directed Gradient Histograms, Optical Flow Histograms, Spatial-Temporal Interest Points, dense trajectories, and some others. Because the selection of characteristics varies in real time from problem to problem, extracting these features is indeed a time-consuming and challenging operation. To solve these issues, a deep learning model was presented and addressed to utilize the requirement for crafted features while reducing completeness.

Deep learning-based strategies [1, 2] have grown quite effective in recent years, outperforming conventional approaches to feature extraction to the extent of winning ImageNet contests. Because of its excellent accomplishment in multiple domains such as bio signal identification, gesture recognition, computer vision, bioinformatics, and so on, it might be completely utilized in human activity recognition. In the proposed study, transfer learning is utilized in conjunction with data augmentation, dropout, and batch normalization to train several advanced convolutional neural networks to categorize human activity photos into their appropriate classes.

This proposed work aims to recognize persons based on their position and motions using various advanced convolution neural networks. The research discussed in this paper makes two contributions to the field of human activity categorization. The first is activity detection and identification. The HAR system detects shapes or orientations based on implementation to task the system into executing a certain job, and activity detection is connected to the localization or position of a human at a given moment in a rigid image or succession of images, i.e., moving images. The quantitative comparative analysis of several advanced deep models is the second contribution.

2 Related Work

A lot of researchers have worked on HAR throughout the last few decades. For example, Liu et al. [3] presented a coupled hidden conditional random fields model for the UTKinect HAR dataset by taking the use of complementing properties on both RGB and depth modalities. The coupled hidden conditional random model expanded the standard hidden-state conditional random fields approach from one chain-structure sequential observation to multiple chain-structure sequential observations, synchronizing sequence information recorded in different modalities by merging RGB and depth sequential data. The authors established the graph structure for the interaction of several modalities and designed the associated potential functions for model formulation. The inference methods are then utilized to uncover the latent connection between depth and RGB data with the model temporal context within each modality.

Masum et al. [4] built an intelligent human activity recognition system employing Support Vector Machine (SVM), Random Forest (RF), Multilayer Perceptron (MLP), Naive Bayes (NB), and Deep CNN in the continuation of HAR research.

After that, sensors were employed for the data accumulation process such as the gyroscope, accelerometer, and magnetometer which removed the uniformity and null label instances of imbalanced classes.

For human motion in 3D space, Vemulapalli et al. [5] used translations and rotations to describe the 3D geometric connections between various body components in order to properly depict human motion in 3D space. Because 3D rigid body movements are always members of the special Euclidean group SE (3), the suggested skeleton representation is in the Lie group SE (3)*...*SE (3), which was a curved manifold. The authors transferred all of the curves to their Lie algebra, which was a vector space, and performed temporal modeling and classification in the Lie algebra, demonstrating that the suggested representation outperforms several existing skeleton representations on UTKinect action datasets.

On the other hand, for compact representation of postures from depth imagery, Xia et al. [6] developed a technique for HAR using histograms of 3D joint positions (HOJ3D) inside a modified spherical coordinate system for a concise depiction of orientations from depth imaging. The HOJ3D calculated from the action depth series is reprojected using LDA and then grouped into k posture visual words, which reflect the generic action poses. Discrete hidden Markov models are used to simulate the temporal variations of such visual words (HMMs). The authors also demonstrated considerable view invariance owing to the spherical coordinate system design and the robust 3D skeleton estimation from Kinect on a 3D action dataset consisting of 200 3D sequences of 10 indoor activities performed by 10 participants in different viewpoints.

In a similar vein, Phyo et al. [7] detected human everyday activities using human skeletal information, merging image processing and deep learning approaches. Because of the usage of Color Skl-MHI and RJI, the suggested system has a quite low computational cost. The processing time was calculated using the feature extraction times of Color Skl-MHI and RJI, as well as the classification time employing 15 frames per second of video data, as a result, the creation of an effective skeletal information-based HAR for usage as an embedded system. The studies were carried out with the use of two well-known public datasets Color Skl-MHI and RJI of human everyday activities.

In terms of 3D space-time, Zhao et al. [8] suggested a fusion-based action recognition system made up of three components: a 3D space-time CNN, a human skeletal manifold depiction, and classifier fusion. The strong correlation among human activity was considered throughout the time domain, followed by the deep mobility map series as input to another stream of the 3D space-time CNN. Furthermore, the related 3D skeleton sequence data was assigned as the recognition framework's third input. For the additional fusion step, the computational cost was in the tens of milliseconds range. As a result, the proposed approach might be used in parallel. In the past few years, we have seen significant development in HAR for RGB videos using handcrafted features.

Liu et al. [9] proposed a simple and effective HAR technique based on depth sequence skeletal joint information. To begin, the authors computed three feature vectors that collect angle and position data between joints. The resulting vectors were then utilized as inputs to three independent support vector machine (SVM) classifiers. Finally, action recognition was carried out by combining the SVM classification findings. Because the retrieved vectors primarily featured angle and normalization relative position based on joint coordinates, the attributes are perspective-invariant. By employing interpolation to standardize action videos of varying temporal durations to a constant size, the extracted features have the same dimension for different videos while retaining the main movement patterns, making the suggested technique time-invariant. The experimental findings showed that the suggested technique outperformed state-of-the-art methods on the UTKinect-Action3D dataset while being more efficient and easier.

3 Proposed Work

The goal of the proposed study is to develop and implement a unique paradigm that uses advanced convolutional models (CNN, VGG-16, VGG-19, ResNet50, ResNet101, ResNet152, and YOLOv5) to classify human behavior into ten categories, making it a multiclass classification problem in machine learning terms.

- Firstly, UTKinect dataset is divided into training and testing sets, and data augmentation is performed to get a clear view of an image sample from different angles.
- Initially, a base CNN is implemented and then pertained ImageNet is used to fine-tune the VGG-16, VGG-19, ResNet50, ResNet101, and ResNet152 architecture. At last, YOLOv5 model is implemented to leverage the power of deep learning.
- For advanced CNN models, a fully connected layer is designed by exploring the use of dropout and normalization techniques. Two new Dense layers with dropout and batch normalization are added to the top and a dense layer with a softmax activation function is added to predict the final image.
- For YOLOv5, Darknet 52 works as a backbone that is used as a feature extractor, which gives us a feature map representation of the input. Neck is the subset of the backbone which enhances the feature of discrimination so YOLOv5 uses PAN as a Neck. If the prediction made is composed of one stage, then it is called Dense Prediction.
- Finally, a comparison study of different advanced deep CNN models and YOLOv5 are performed for the best score.

4 Result and Analysis

Input images in the UTKinect-Action dataset are of various sizes and resolutions, so they were reduced to 256 x 256 x 3 to reduce file size, and 1610 of the total 1896 images are in training, while the remaining 286 are in validation. To avoid

Table 1 Experiment results for best logloss score and accuracy

Model	Training loss	Validation loss	Training accuracy	Validation accuracy
Deep CNN with Augmentation + Dense Layers + Dropout + BatchNormalization	0.0698	0.0938	97.45	96.94
VGG-16 with Augmentation + Dense Layers + Dropout + BatchNormalization	0.2165	0.1708	92.02	93.09
VGG-19 withAugmentation + Dense Layers + Dropout + BatchNormalization	0.2361	0.2003	91.43	92.52
ResNet50 withAugmentation + Dense Layers + Dropout + BatchNormalization	0.2217	0.1764	91.66	92.52
ResNet101 withAugmentation + Dense Layers + Dropout + BatchNormalization	0.2022	0.1883	92.42	93.66
ResNet152 withAugmentation + Dense Layers + Dropout+ BatchNormalization	0.1922	0.1771	92.75	92.70

overfitting, the proposed model uses the transfer learning technique along with data augmentation, dropout, and batch normalization. Fully connected layers are excluded from each model and pre-trained weights are used. The Accuracy and Loss curve with data augmentation, dense layers, dropout, and batch normalization have recorded the models per epoch for 50 epochs. Table 1 displays all the experiments performed along with their results. In a Convolution neural network, the input layer reads the image so there is no parameter. There are $(n \times m \times l + 1) \times k$ total parameters in the convolution layer which takes l and k feature maps as input and output using $n \times m$ filter size. The pooling layer has no parameters because it is used to reduce the dimension. The fully connected layer has the $(n + 1) \times m$ total parameters.

For Deep CNN, total parameters are 3,209,322 out of which 3,207,274 are trainable parameters and 2,048 are non-trainable parameters. As shown in Fig. 5, the training accuracy is 97.45 % near the end of 48 epochs and validation accuracy is also about 96.94 % near the end of the 44 epochs in the diagram. Similarly, the best training loss is close to 0.0698 and the validation loss is around 0.0938 as shown in Fig. 1.

For VGG-16, total parameters are 49,338,186 out of which 34,619,402 are trainable parameters and 14,718,784 are non-trainable parameters. As shown in Fig. 7, the training accuracy is 92.02 % near the end of 49 epochs and validation accuracy is also about 93.09 % near the end of the 31 epochs in the diagram. Similarly, the best training loss is close to 0.2165 and the validation loss is around 0.1708 as shown in Fig. 2.

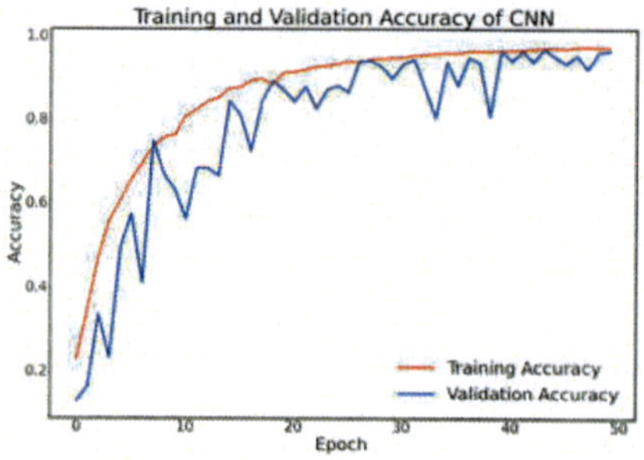

(a) Accuracy Curve

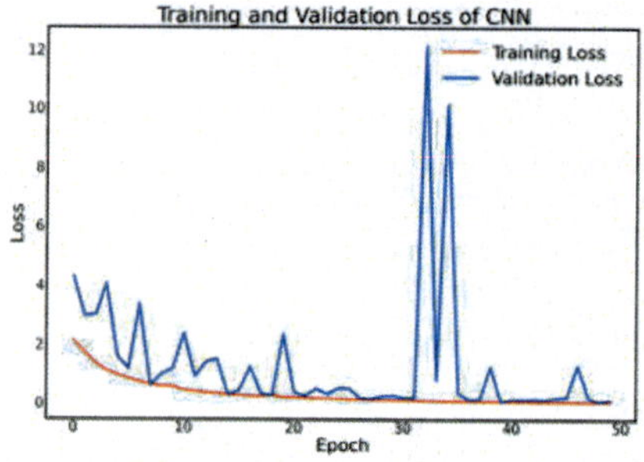

(b) Loss Curve

Fig. 1 Accuracy and Loss curve of Deep CNN

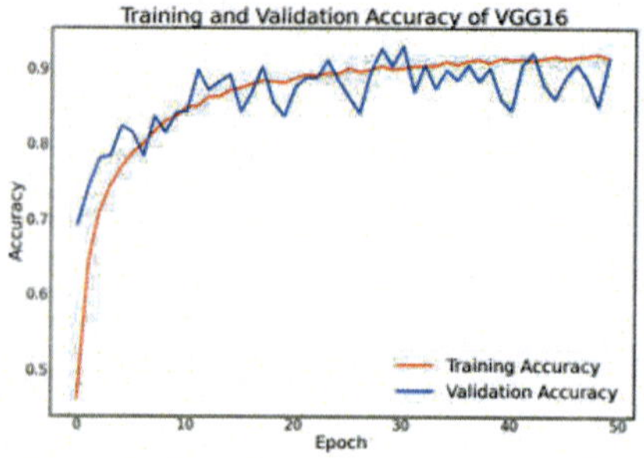

(a) Accuracy Curve

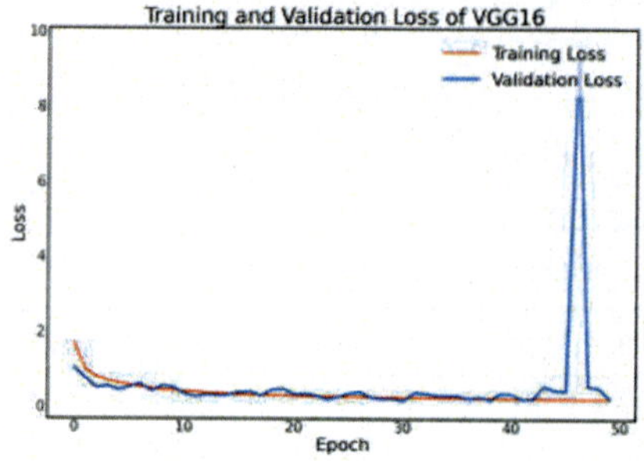

(b) Loss Curve

Fig. 2 Accuracy and Loss curve of VGG-16

VGG-19 have total 37,073,994 parameters out of which 17,047,562 are trainable parameters and 20,026,432 are non-trainable parameters. As shown in Fig. 9 the training accuracy is 91.43 % near the end of 45 epochs and validation accuracy is also about 92.52 % near the end of the 44 epochs in the diagram. Similarly, the best training loss is close to 0.2361 and validation loss is around 0.2003 as shown in Fig. 3.

ResNet50 has total parameters 90,968,970 out of which 67,379,210 are trainable parameters and 23,589,760 are non-trainable parameters. As shown in Fig. 11 the training accuracy is 91.66 % near the end of 48 epochs and validation accuracy is also about 92.52 % near the end of the 42 epochs in the diagram. Similarly, the best training loss is close to 0.2217 and the validation loss is around 0.1764 as shown in Fig. 4.

For ResNet101, total parameters are 110,039,434 out of which 67,379,210 are trainable parameters and 42,660,224 are non-trainable parameters. As shown in Fig. 13, the training accuracy is 92.42 % near the end of 49 epochs and validation accuracy is also about 93.66 % near the end of the 44 epochs in the diagram. Similarly, the best training loss is close to 0.2022 and the validation loss is around 0.1883 as shown in Fig. 5.

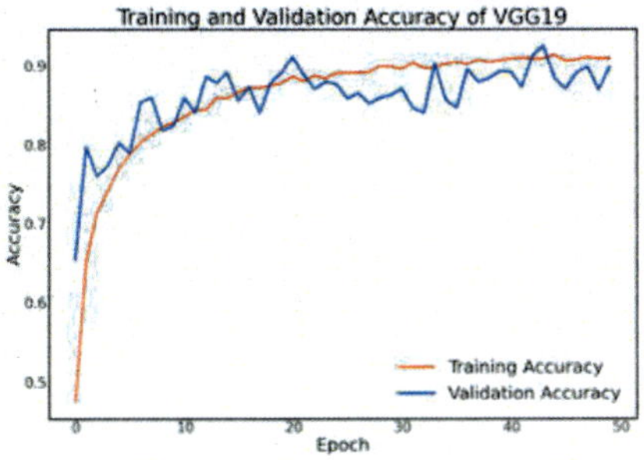

(a) Accuracy Curve

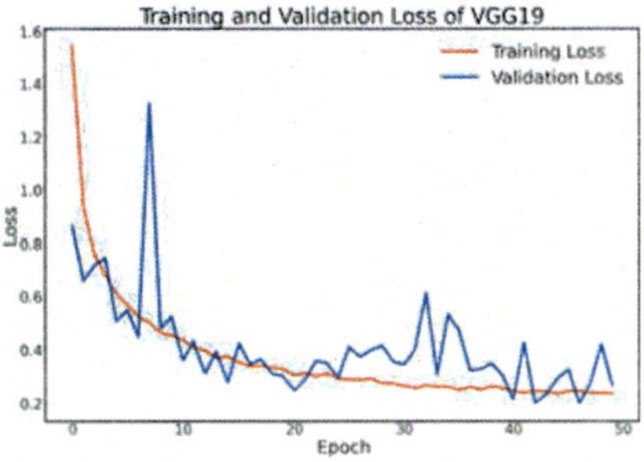

(b) Loss Curve

Fig. 3 Accuracy and Loss curve of VGG-19

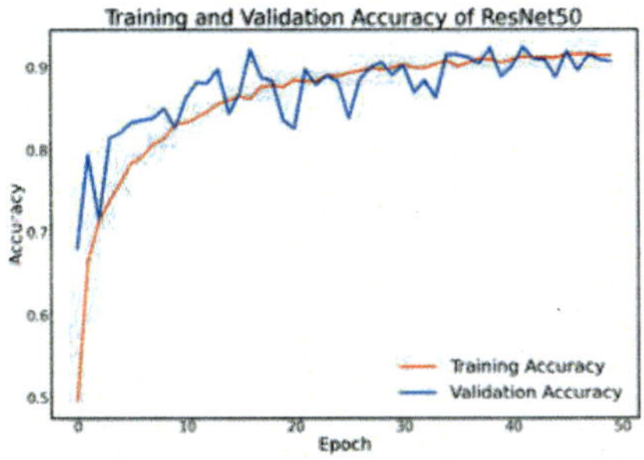

(a) Accuracy Curve

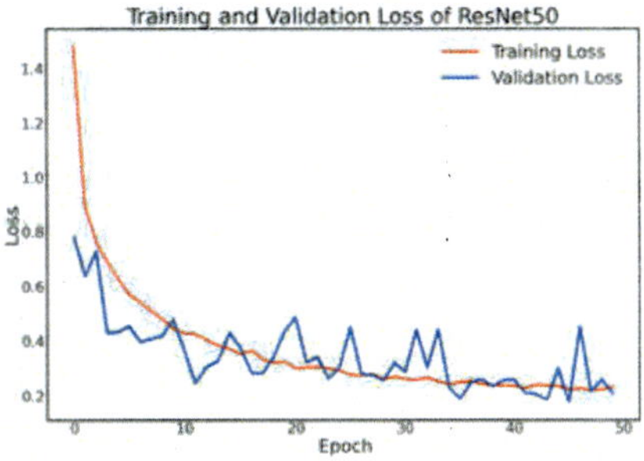

(b) Loss Curve

Fig. 4 Accuracy and Loss curve of ResNet50

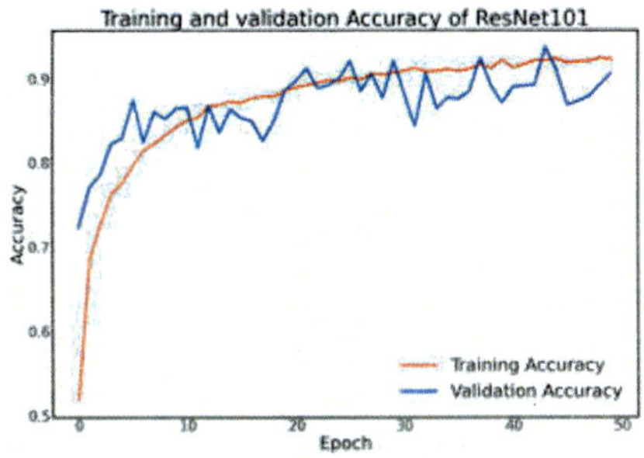

(a) Accuracy Curve

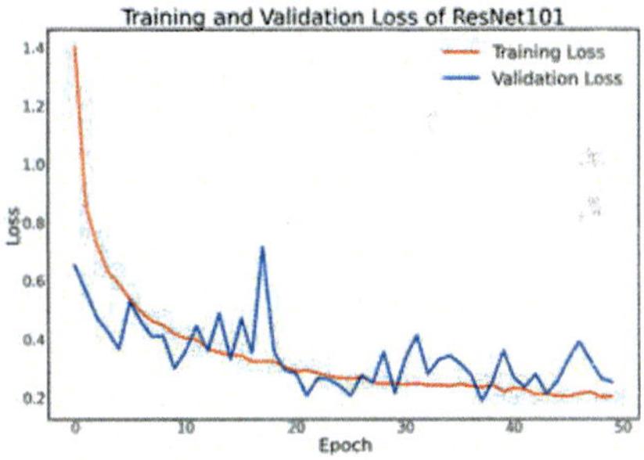

(b) Loss Curve

Fig. 5 Accuracy and Loss curve of ResNet101

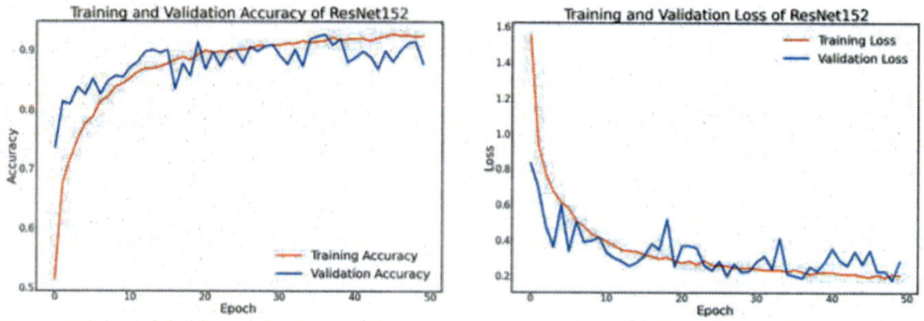

(a) Accuracy Curve (b) Loss Curve

Fig. 6 Accuracy and Loss curve of ResNet152

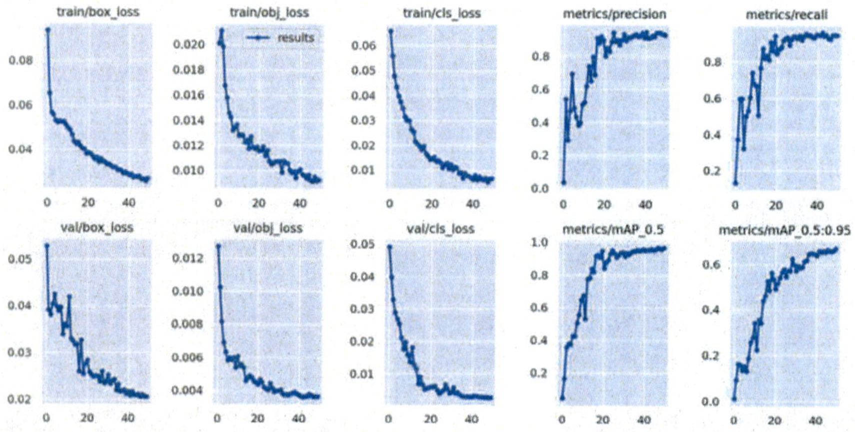

Fig. 7 YOLOv5 results

For ResNet152, total parameters are 193,657,738 out of which 135,282,698 are trainable parameters and 58,375,040 are non-trainable parameters. As shown in Fig. 15, the training accuracy is 92.75 % near the end of 46 epochs and validation accuracy is also about 92.70 % near the end of the 37 epochs in the diagram. Similarly, the best training loss is close to 0.1922 and the validation loss is around 0.1771 as shown in Fig. 6.

At last, the YOLOv5 model is trained for 50 epochs using the batch size 8 in 0.558 h. Only 802 images and 130 images are used for training and validation for YOLOv5. The model uses 213 layers, 7037095 parameters, and 0 gradients. Precision, recall, and mean average precision are recorded at 92.9 %, 94.5 %, and 96.6 %, respectively. Mean average precision computes the average precision value for recall value over 0 to 1. Figure 7 shows the results of this experiment.

The activity detection task is challenging to complete since the human stance in the image changes depending on whether the person is sitting, standing, walking, or sleeping. The rotation can occur both within and outside of the plane. Therefore, as a solution to the mentioned problem, we introduced the concept of YOLOv5 as one stage detector. The YOLOv5 firstly classifies and then localizes every movement in different activities.

5 Comparison with the State-of-the-Art Models on UTKinect Action Dataset

We evaluated the mAP and classification accuracy of our proposed system to that of other systems, as given in Table 1, and found that some methods had better as well as closer accuracy, starting from [3] in which the authors recorded 92 % accuracy on the same dataset. In another work [6], the author recorded 90.92 % accuracy. In another work [5] 97.08 % accuracy was calculated.

6 Conclusion

In our research, we examined the effectiveness of numerous well-known classifiers and preprocessing approaches used in human activity recognition. The findings also included the selection of the best preprocessing approaches from among promptly applicable methods. On the UTKinect-Action dataset, we also calculated the optimal classifiers for the specified preprocessing methodology. The Deep CNN classifier and YOLOv5 performed better in detecting and localizing human activity, with 96.9 % and 96.6 % accuracy, respectively. We expect that the research findings will help other researchers in this field to choose classifiers and data preparation approaches. This work also highlights the benefit of recognizing human actions and indicates a viable route for completing recognition tasks via depth information. Traditional RGB data may also be merged with depth data to generate additional data and algorithms with higher recognition rates and resilience.

References

1. Negi A, Kumar K, Chauhan P, Rajput R (2021) Deep neural architecture for face mask detection on simulated masked face dataset against covid-19 pandemic. In: 2021 International Conference on Computing, Communication, and Intelligent Systems (ICCCIS). IEEE, pp 595–600
2. Negi A, Chauhan P, Kumar K, Rajput RS (2020) Face mask detection classifier and model pruning with Keras-Surgeon. In: 2020 5th IEEE International Conference on Recent Advances and Innovations in Engineering (ICRAIE). IEEE, pp 1–6
3. Liu AA, Nie WZ, Su YT, Ma L, Hao T, Yang ZX (2015) Coupled hidden conditional random fields for RGB-D human action recognition. Signal Process 112:74–82

4. Masum AKM, Hossain ME, Humayra A, Islam S, Barua A, Alam GR (2019) A statistical and deep learning approach for human activity recognition. In: 2019 3rd International Conference on Trends in Electronics and Informatics (ICOEI). IEEE, pp 1332–1337
5. Vemulapalli R, Arrate F, Chellappa R (2014) Human action recognition by representing 3d skeletons as points in a lie group. In: Proceedings of the IEEE conference on computer vision and pattern recognition. pp 588–595
6. Xia L, Chen CC, Aggarwal JK (2012) View invariant human action recognition using histograms of 3d joints. In: 2012 IEEE computer society conference on computer vision and pattern recognition workshops. IEEE, pp 20–27
7. Phyo CN, Zin TT, Tin P (2019) Deep learning for recognizing human activities using motions of skeletal joints. IEEE Trans Consum Electron 65(2):243–252
8. Zhao C, Chen M, Zhao J, Wang Q, Shen Y (2019) 3d behavior recognition based on multi-modal deep space-time learning. Appl Sci 9(4):716
9. Liu Z, Feng X, Tian Y (2015) An effective view and time-invariant action recognition method based on depth videos. In: 2015 Visual Communications and Image Processing (VCIP). IEEE, pp 1–4
10. Verma KK, Singh BM, Mandoria HL, Chauhan P (2020) Two-stage human activity recognition using 2D-ConvNet. Int J Interact Multimed Artif Intell 6(2)

Control Techniques and Their Applications

Real Power Loss Reduction by Chaotic Based Riodinidae Optimization Algorithm

Lenin Kanagasabai

1 Introduction

Loss lessening is a precarious assignment in power systems since its plays foremost role in better operation. Conversely, in this matter the aforementioned owns an indisputable influence on upholding the solidity and protected power course. Commonly, this problem is smeared to optimum controlling of the bases in links targeting at underrating losses and taming the power silhouette. Loss lessening is a momentous commission in network. Loss is primarily self-possessed and instigated by flow of power. Supplementary loss not solitary upsurges production cost, nevertheless lessening the power factor of the organism. Consequently, the loss is unique and and is a key function. Munificent conformist approaches [1–6] previously employed and Evolutionary techniques [7–16] are smeared. Meta-heuristic procedures fluctuate from approaches and methodically moving towards nearby conceivable optimum location throughout the calculation procedure, sidestepping premature convergence to indigenous optima [17]. In addition, these approaches frequently agonize on or after the succeeding inadequacies. Primarily, a considerable calculation encumbrance is obligatory owing to monotonous power course computation, and creation of actual phase of application is perplexing. Furthermore, procedure's enactment is strappingly reliant on the structure prototype's accurateness. In this paper Chaotic based Riodinidae (CRO) optimization algorithm is used to condense the loss. In Riodinidae the optimization examination process owns twofold possessions of Riodinidae. Tinkerbell chaotic map engendering standards are implemented. Riodinidae algorithm has been integrated with the Firefly algorithm's examination. In IEEE 118 and 300 bus systems, Chaotic based Riodinidae (CRO) optimization

L. Kanagasabai (✉)
Prasad V.Potluri, Siddhartha Institute of Technology, Kanuru, Vijayawada, Andhra Pradesh 520007, India
e-mail: gklenin@gmail.com

© The Author(s), under exclusive license to Springer Nature Singapore Pte Ltd. 2023
H. Muthusamy et al. (eds.), *Robotics, Control and Computer Vision*, Lecture Notes in Electrical Engineering 1009, https://doi.org/10.1007/978-981-99-0236-1_20

algorithm legitimacy has been weighed. Appraisal of loss is done with standard procedures. Projected Chaotic based Riodinidae (CRO) optimization algorithm abridged the loss adeptly.

2 Problem Formulation

Loss reduction scientifically delineated as,

$$F = P_L = \sum_{k \in Nbr} g_k\left(V_i^2 + V_j^2 - 2V_i V_j \cos\theta_{ij}\right) \tag{1}$$

$$F = P_L + \omega_v \times VDV \tag{2}$$

$$VDV = \sum_{i=1}^{Npq} |V_i - 1| \tag{3}$$

Parity and Disparity constraints are,

$$P_G = P_D + P_L \tag{4}$$

$$P_{gslack}^{min} \le P_{gslack} \le P_{gslack}^{max} \tag{5}$$

$$Q_{gi}^{min} \le Q_{gi} \le Q_{gi}^{max},\ i \in N_g \tag{6}$$

$$V_i^{min} \le V_i \le V_i^{max},\ i \in N \tag{7}$$

$$T_i^{min} \le T_i \le T_i^{max},\ i \in N_T \tag{8}$$

$$Q_c^{min} \le Q_c \le Q_C^{max},\ i \in N_C \tag{9}$$

3 Chaotic Based Riodinidae Optimization Algorithm

Riodinidae optimization procedure examination process owns double possessions of Riodinidae. Levy flights and over feminine Riodinidae progenies' are created.

$$R_{i,k}^{t+1} = R1_{a1,k}^{t} \tag{10}$$

Computation of ratio "crt" is done by,

$$crt = \text{a}.p \tag{11}$$

where a is arbitary

Freshly created Riodinidae is calculated as,

$$R_{i,k}^{t+1} = R_{a2,k}^{t} \tag{12}$$

Riodinidae exploration process is quantified as,

$$R_{j,k}^{t+1} = R_{b,k}^{t} \tag{13}$$

Freshly created Riodinidae is premeditated as,

$$R_{i,k}^{t+1} = R_{a3,k}^{t} \tag{14}$$

Adaptability rate defined as

$$R_{j,k}^{t+1} = R_{j,k}^{t} + \gamma \times (AR_k - 0.50) \tag{15}$$

In the proposed Chaotic based Riodinidae (CRO) optimization procedure, the exploration process is boosted by applying the Firefly procedure's exploration equivalence. Figure 1 shows the Schematic diagram of Chaotic based Riodinidae (CRO) optimization algorithm.

$$R_i^{t+1} = R_i^t + \beta_o r^{-\gamma r_{i,j}^2}\left(R_j^t - R_i^t\right) + \gamma(a - 0.50) \tag{16}$$

Tinkerbell chaotic map [18] engendering standards are implemented.

$$e_{t+1} = e_t^2 - f_t^2 + a \cdot e_t + b \cdot f_t \tag{17}$$

$$f_{t+1} = 2e_t f_t + c \cdot e_t + d \cdot f_t \tag{18}$$

where a, b, c and d are non-zero parameters

The functional value by linear scaling in Tinkerbell chaotic map is demarcated as,

$$e_{t+1}^{*} = e_{t+1} - min(e)/max(e) - minimum(e) \tag{19}$$

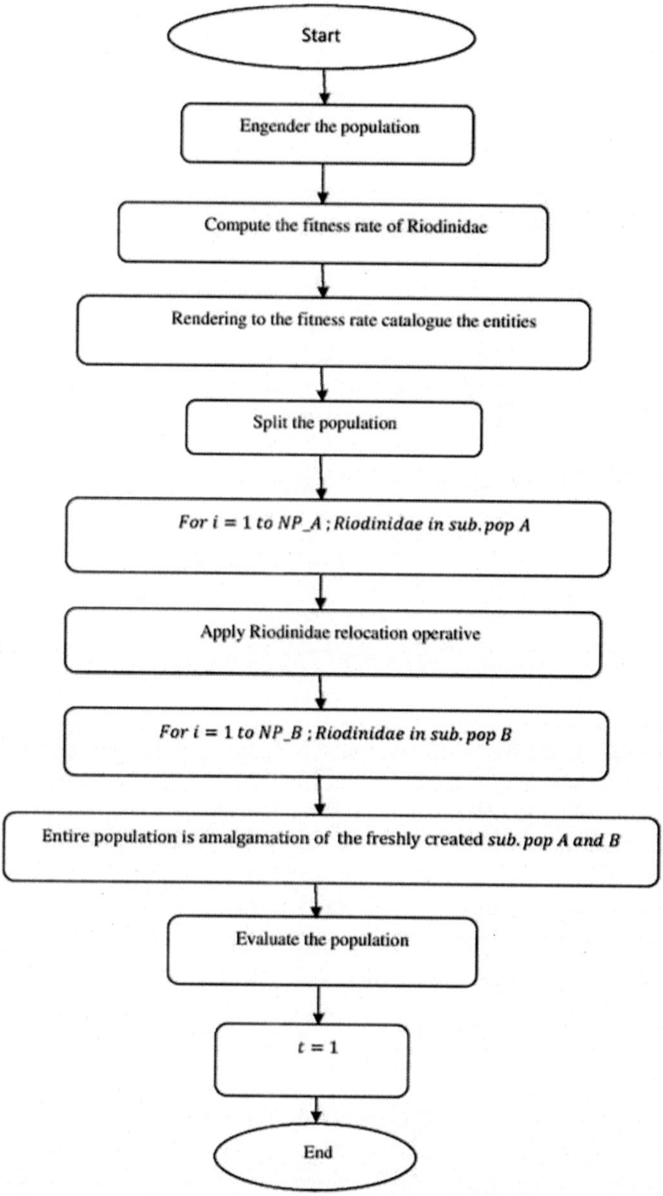

Fig. 1 Schematic diagram of Chaotic based Riodinidae (CRO) optimization algorithm

a. Start
b. Engender the population
c. Compute the fitness rate of Riodinidae
d. *while t < max.gen do*
e. Rendering to the fitness rate catalogue the entities
f. Split the population
g. *For i = 1 to* NP_A*; Riodinidae in sub pop A*
h. Apply Riodinidae relocation operative
i. Create fresh entities
j. End for
k. *For i = 1 to NP_B; Riodinidae in sub pop B*
l. *if t < max. gen: 0.50, then*
m. *Engender sub.pop by Riodinidae regulative op.*
n. *otherwise*
o. *Engender new pop. In sub. pop B by Riodinidae regulative op*
p. *Apply Tinkerbell chaotic map*
q. $e_{t+1} = e_t^2 - f_t^2 + a \cdot e_t + b \cdot f_t$
r. $f_{t+1} = 2e_t f_t + c \cdot e_t + d \cdot f_t$
s. End if
t. End for
u. Entire population is amalgamation of the freshly created *sub. Pop A and B*
v. Rendering to the freshly rationalised locations, appraise the populace
w. t = 1
x. End while
y. choose the exceptional unit form complete populace
z. End.

4 Simulation Results

Validity of the Chaotic based Riodinidae (CRO) optimization algorithm is verified in IEEE 118 and 300 bus systems [19]. Tables 1 and 2 give the comparison results. Figures 2 and 3 show the loss comparison with other described algorithms.

Table 1 Valuation of loss (IEEE 118 Bus system)

Parameter	True loss (MW)
Base value [20]	132.8
ImPSO [20]	117.19
BaPSO [21]	119.34
BaEPSO [22]	131.99
BaCLPSO [22]	130.96
CRO	112.19

Table 2 Loss valuation (IEEE 300 Bus system)

Parameter	True loss (MW)
AdGA [23]	646.299800
FaEA [23]	650.602700
BaCSO [24]	635.894200
CRO	625.020208

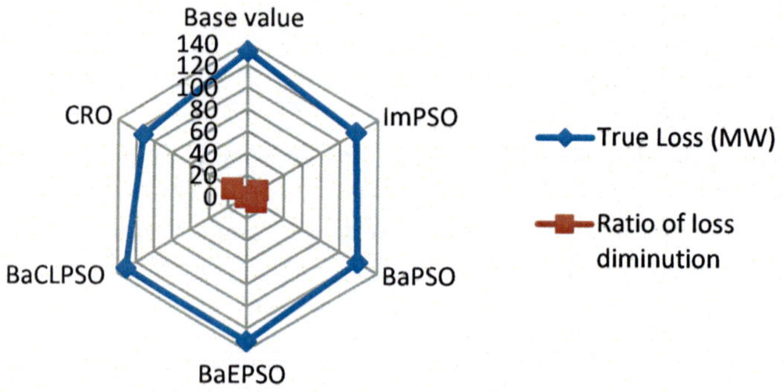

Fig. 2 Loss assessment (IEEE 118 bus system)

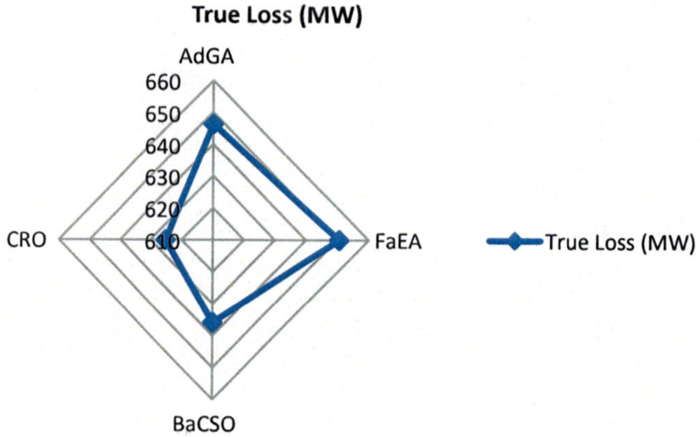

Fig. 3 Loss appraisal (IEEE 300 bus system)

Table 3 shows the convergence characteristics of Chaotic based Riodinidae (CRO) optimization algorithm.

Table 3 Convergence characteristics

CRO	Loss in MW	Time (S)	No. of iter
IEEE 118	112.19	38.79	29
IEEE 300	625.020208	68.62	36

5 Conclusion

In this paper Chaotic based Riodinidae (CRO) optimization algorithm competently abridged the loss. Loss lessening is a momentous commission in the network. Loss is primarily self-possessed and instigated by flow of power. Supplementary loss not solitary. upsurges production cost, nevertheless lessens the power factor of the organism. Consequently, the loss is unique and a key function. In Riodinidae the optimization examination process owns twofold possessions of Riodinidae. Riodinidae algorithm has been integrated with the Firefly algorithm's examination. Tinkerbell chaotic map engendering standards are implemented. In IEEE 118 and 300 bus systems, Chaotic based Riodinidae (CRO) optimization algorithms legitimacy has been weighed. Appraisal of loss is done with standard procedures. Projected Chaotic based Riodinidae (CRO) optimization algorithm abridged the loss adeptly.

References

1. Lee K (1984) Fuel-cost minimisation for both real and reactive-power dispatches. Proc Gener, Transm Distrib Conf 131(3):85–93
2. Deeb N (1998) An efficient technique for reactive power dispatch using a revised linear programming approach. Electr Power Syst Res 15(2):121–134
3. Bjelogrlic M (1990) Application of Newton's optimal power flow in voltage/reactive power control. IEEE Trans Power System 5(4):1447–1454
4. Granville S (1994) Optimal reactive dispatch through interior point methods. IEEE Trans Power Syst 9(1):136–146
5. Grudinin N (1998) Reactive power optimization using successive quadratic programming method. IEEE Trans Power Syst 13(4):1219–1225
6. Sinsuphan N (2013) Optimal power flow solution using the improved harmony search method. Appl Soft Comput 13(5):2364–2374
7. Valipour K (2017) Using a new modified harmony search algorithm to solve multi-objective reactive power dispatch in deterministic and stochastic models. AI Data Min 5(1):89–100
8. Naidji (2020) Stochastic multi-objective optimal reactive power dispatch considering load and renewable energy sources uncertainties: a case study of the Adrar isolated power system. Int Trans Electr Energy Syst 6(30):1–12
9. Farid (2021) A novel power management strategies in PV-wind-based grid connected hybrid renewable energy system using proportional distribution algorithm. Int Trans Electr Energy Syst 31(7):1–20
10. Sheila (2021) A novel ameliorated Harris hawk optimizer for solving complex engineering optimization problems. Int J Intell Syst 36(12):7641–7681
11. Prashant (2021) Design and stability analysis of a control system for a grid-independent direct current micro grid with hybrid energy storage system. Comput & Electr Eng 93(1):1–15
12. Chen. : Optimal reactive power dispatch by improved GSA-based algorithm with the novel strategies to handle constraints. Appl Soft Computing, 50(1), 58–70 (2017).

13. Mei (2017) Optimal reactive power dispatch solution by loss minimization using moth flame optimization technique. Appl Soft Comput 59(1):210–222
14. Uney (2019) New metaheuristic algorithms for reactive power optimization. Tehnički Vjesnik 26(1):1427–1433
15. Abaci K (2017) Optimal reactive-power dispatch using differential search algorithm. Electr. Engineering 99(1):213–225
16. Huang (2012) Combined differential evolution algorithm and ant system for optimal reactive power dispatch. Energy Procedia 14(1):1238–1243
17. Kanatip R, Keerati C (2021) Probabilistic optimal power flow considering load and solar power uncertainties using particle swarm optimization. GMSARN Int J 15:37–43
18. Inoue (2000) Application of chaos degree to some dynamical systems. Chaos, Solut Fractals 11 (1):1377–1385
19. Salimi (2015) Stochastic fractal search: a powerful metaheuristic algorithm. Knowl-Based Syst 75(1):1–18
20. IEEE (1993) The IEEE-test systems. http://www.ee.washington.edu/trsearch/pstca/
21. Dai C (2009) Seeker optimization algorithm for optimal reactive power dispatch". IEEE T Power System 24(3):1218–1231
22. Reddy (2014) Faster evolutionary algorithm based optimal power flow using incremental variables. Electr Power Energy Syst 54(1):198–210
23. Reddy S (2017) Optimal reactive power scheduling using cuckoo search algorithm. Int J Electr Comput Engineering 7(5):2349–2356
24. Hussain AN (2018) Modified particle swarm optimization for solution of reactive power dispatch. Res J Appl Sci, Eng Technol 15(8):316–327

5G Enabled IoT Based Automatic Industrial Plant Monitoring System

Kshitij Shinghal, Amit Saxena, Amit Sharma, and Rajul Misra

1 Introduction

In modern day industrial plants electrical machines i.e., motors, generator, transformers etc. are the prime elements. No industry can run without the use of electrical machines to drive the system. If an electrical machine fails, it may result in several consequences such as break in continuity of production time, failure of system or even complete shutdown of the plant and in some cases may even pose threat of injury or even human life. Thereby failure of an electrical machine may result in lots of revenue, production, product quality and risk to safety of workers. The Fig. 1 depicts how the electrical machines and automation has become a key element of modern day Industries.

Therefore, condition monitoring of parameters of electrical machines like vibration, temperature, current, voltages etc. becomes important in order to timely identify defect development of a fault in machine. Condition monitoring plays a vital rook in predictive maintenance. With the help of proper condition monitoring necessary maintenance can be scheduled ensuring complete health of the machines.

This will prevent the consequential damages to the machine and further implications. Figure 2 shows a typical industrial setup deployed for condition monitoring of industrial machines.

K. Shinghal · A. Saxena (✉)
Department of Electronics and Communication Engineering, Moradabad Institute of Technology, Moradabad, U.P, India
e-mail: amitssaksena@gmail.com

A. Sharma
Department of Electronics and Communication Engineering, Teerthanker Mahaveer University, Moradabad, U.P, India

R. Misra
Department of Electrical Engineering, Moradabad Institute of Technology, Moradabad, U.P, India

© The Author(s), under exclusive license to Springer Nature Singapore Pte Ltd. 2023

H. Muthusamy et al. (eds.), *Robotics, Control and Computer Vision*, Lecture Notes in Electrical Engineering 1009, https://doi.org/10.1007/978-981-99-0236-1_21

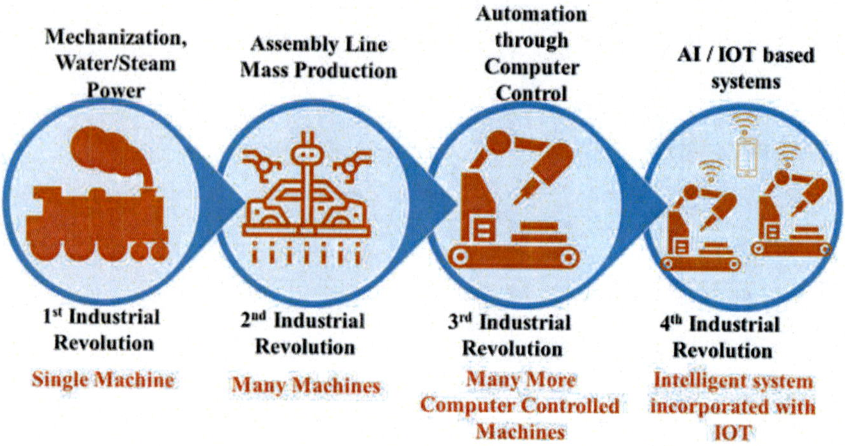

Fig. 1 Various stages of industrial revolution

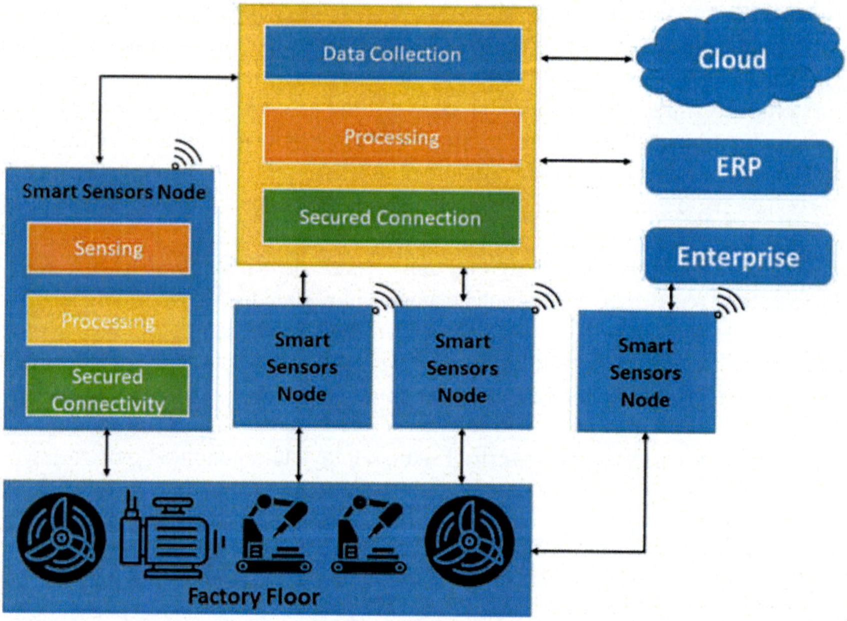

Fig. 2 Industrial setup for condition monitoring of industrial machines

The proposed system uses 5G technologies based IoT System for the purpose of communication with end servers. Figure 3 depicts the various generations of mobile communication.

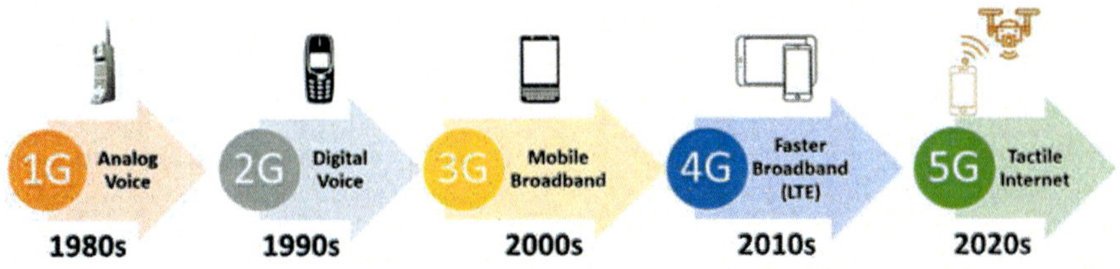

Fig. 3 Different generation of mobile communication technology

5G wireless technologies are growing at a rapid speed and will find numerous applications in the coming years. It has several advantages over existing 4G wireless technologies such as faster network speed, low delay in data rate i.e. huge increase in responsiveness and a smooth experience (must for real time applications). Figure 4 depicts the advantage of using 5G wireless technologies for IoT based industrial plant monitoring systems.

Rest of the paper is organized as follows: literature review, problem identification and gap in existing technology is carried out in Sect. 2, the proposed 5G enabled monitoring system is presented in Sect. 3 followed by the experimental setup & methodology in Sect. 4. The results are discussed in Sect. 5 and finally the conclusion and future work of the proposed work are given in Sect. 6.

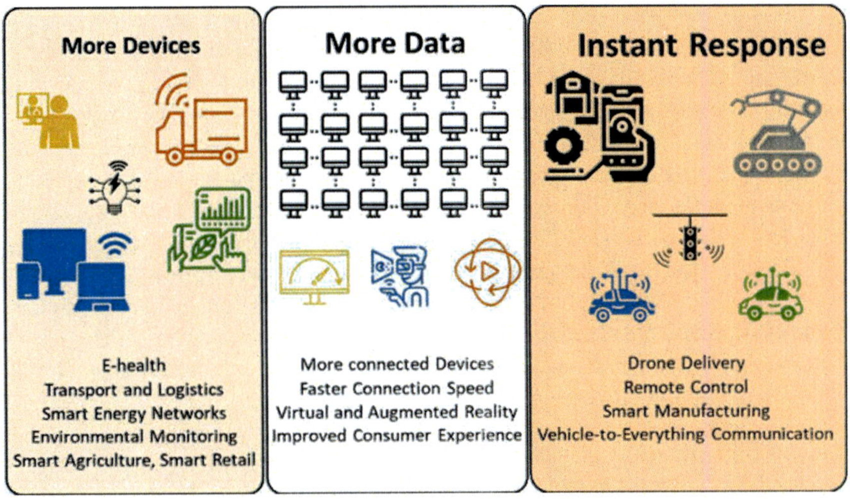

Fig. 4 Advantages of using 5G wireless technologies for IoT based industrial plant monitoring systems

2 Literature Review

Karemore et al. in their paper titled a review of IoT based smart industrial system for controlling and monitoring presented proposed a framework required in industries for controlling, monitoring, security and safety of various exercises. The monitoring frame incorporates detectors like fire detector, fume detector, ultrasonic detector, humidity and temperature detector, current and voltage with Wi-Fi module for control operations. With the advantages of unusual practices, reasonable conditioning will be actuated. [1]. Gore et al. in their paper titled Bluetooth based sensor monitoring in industrial IoT plants presented that typical industrial IoT use cases involve acquiring data from detector bias in factory and communicating the same to the internet for original or remote monitoring and control. They described how Bluetooth low energy (BLE) technology can be used to connect detector bumps to Internet-grounded services and operations using gateway in an artificial factory [2]. A. Vakaloudis et al. in their paper titled A framework for rapid integration of IoT Systems with industrial environments proposed a comprehensive start to finish perspective extending from sensor devices to interfacing with the end user where all product and equipment components of the framework are being thought of and addressed [3]. Zhao et al. in their paper titled design of an industrial IoT based monitoring system for power substations gave a reasonable application that was executed and tried in a real power substation. The framework joins the highlights of an IoT stage with the necessities of high-speed real-time applications while utilizing a solitary high-resolution time source as the reference for both steady-state and transient conditions [4]. Picot et al. in their paper titled Industry 4.0 LabVIEW Based Industrial Condition Monitoring System for Industrial IoT System presented a platform to host varied operations, the industry standard fieldbus protocol Modbus TCP was used in confluence with the LabVIEW development ambient, where a bespoke graphical UI was created to give control and a visual depiction of the information gathered. In addition, one of the bases went about as the yield for outfit shows, which in turn corresponded the alert status of the UI [5]. Khan et al. in their paper titled IoT Based Health Monitoring System for Electrical Motors presented internet of things (IoT) predicated system is designed for the electrical motor. The electrical motor health is covered by measuring the parameters similar as vibration, current and temperature. It can be measured through the detectors, like accelerometer, current detector and thermocouple. To avoid the limitation of the internet, the signals of these detectors were also transferred to the receiver through global system for mobile (GSM) because it can also work in the areas where the internet isn't available [6]. Gore et al. in their paper titled IoT based equipment identification and location for maintenance in large deployment industrial plants presented condition monitoring system that employs fusion of detectors and uses acquired data in health evaluation algorithms to distinguish faults. In a standard factory deployment, each machine similar as motor would be convoyed by a respective health monitoring unit. Condition monitoring operation system integrated with regulator in the factory control room, gathers condition monitoring data from the varied sub-systems and generates automated cautions upon failure discovery [7]. Lyu et al. in their paper titled 5G Enabled Codesign of Energy-Efficient Transmission and Estimation for Industrial IoT Systems introduced a transmission assessment codesign structure to set out the establishment for ensuring the endorsed assessment exactness with restricted correspondence assets. The proposed approach is then optimized by planning a compelled minimization issue, which is blended integer nonlinear programming and addressed effectively with a block coordinate descent based deterioration technique. At last, simulation results show that the proposed approach has superiorities in improving both the assessment precision and the energy productivity [8].

From the literature review it is evident that monitoring applications are having higher real time responsiveness requirements. In critical machines such as assembly lines, conveyer belts where the product is continuously being supplied, immediate response on detecting a fault is required to save the product from damage, the use of reliable monitoring infrastructure based on superior qualities of 5G network is needed where there is risk of human life in case of failure.

3 5G Wireless Technology Enabled Monitoring System

The advanced monitoring system requires large number of connected things, machines, sensors actuators and controllers that are part of the IoT enabled system [9–11]. The use of 5G wireless technologies will require investment in terms of deploying antennas, switches, repeaters etc. Further to reduce latency mobile edge computing will be implemented [12, 13]. The system will consist of wireless routers that are strategically placed in vicinity of the wireless infrastructure of the monitoring system to reduce the latency and increase the response time. Figure 5 depicts the block diagram of such 5G enabled industrial plat monitoring system.

Figure 5 depicts typical motors used in any industrial plant. The condition of these electrical machines is monitored as per the requirement by deploying local sensors. In the presented case three sensors are placed for condition monitoring i.e. temperature sensor, vibration sensor and current sensor as shown in Fig. 5. These sensors monitor the status of machines and feed the data to the IoT gateway. The IoT gateway consists of Odyssey–X86J4105800 which processes the data and sends it through 5G wireless technology for high end processing, monitoring and control dashboard. The high end processing, processes the signal and issues commands to the control & actuator subunit for operating the relays, valves and actuators.

4 Experimental Setup and Methodology

The proposed system comprises of Odyssey–X86J4105800. It has inbuilt Intel Celeron J4105, Quad core processor at 1.5–2.5 GHz and a dual band frequency 2.5/5 GHz Wi-Fi/Bluetooth and 5G LTE gateway for IoT applications. It also has

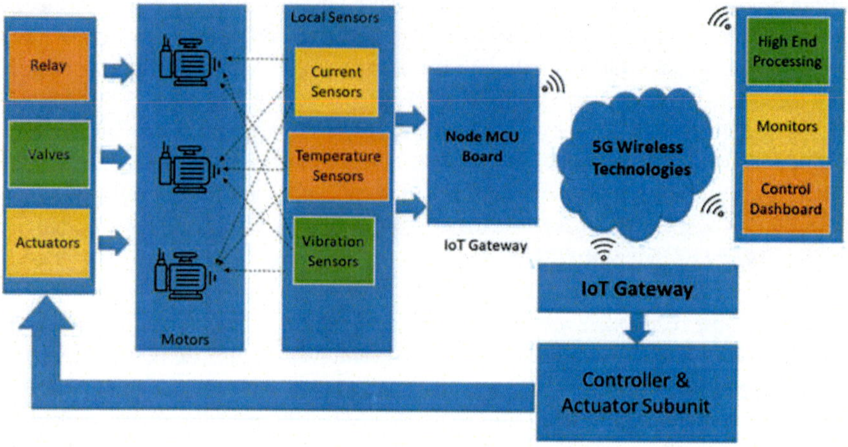

Fig. 5 Block diagram of typical 5G enabled industrial plant monitoring system

an Arduino coprocessor on board to connect with local sensors required for monitoring of electrical machines and also with controller actuators subunit for controlling relays, switches and valves etc. All experiments were conducted in laboratory with the same local area network within the radius of 6-m. Figure 6 shows the laboratory experimental setup for conducting the experiments.

Table 1 Outlines hardware configuration of IoT node and 5G wireless gate-way. The local sensors were installed to monitor stator current and temperature.

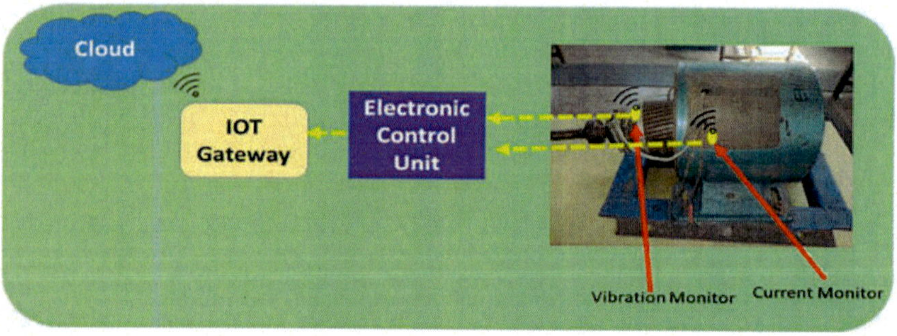

Fig. 6 Laboratory experimental setup for conditioning monitoring of Induction Motor (IM)

Table 1 Hardware configuration of IoT node

Node	RAM (GB)	Storage (GB)	CPU
Odyssey-X86J4105800 wifi 802.11 a/b/g/n/ac	LPDDR4 8 GB	16 GB SSD	Intel celeron J4105 microchip ATSAMD21G18 32-Bit ARM Cortex Mot (coprocessor)

Experimental studies of eighty-four (84) cases have been carried out, out of which eight (08) cases have been reported here for rotor fault detection. The specifications of Induction Motors (IM) under observation are tabulated in Table 2. The current patterns for all the cases are shown in Figs. 7, 8, 9, 10, 11, 12, 13, 14. The rating of the IM is varying from 180 – 661 KW.

Table 2 Specification of IMs

Case studies	Output (KW)	Rotor bars	Pole pairs	PPF (Hz)	Load (%)	USB (dB)	LSB (dB)	R_{HI}
1	480	88	4	0.963	85.7	−61.9	−65.8	0.0245
2	350	58	3	1.060	98.20	−80.3	−76	0.0437
3	360	58	3	1.444	105.20	−83.7	−73.4	0.0547
4	180	36	2	0.619	52.80	−57.7	−60	0.3223
5	550	66	3	0.712	79.30	−62.1	−49	0.8683
6	550	66	3	0.619	68.50	−52.8	−59.9	0.9019
7	661	58	3	0.481	65.00	−45.2	−56	1.4489
8	661	58	2	0.573	68.5	−38.4	−41.8	1.9635

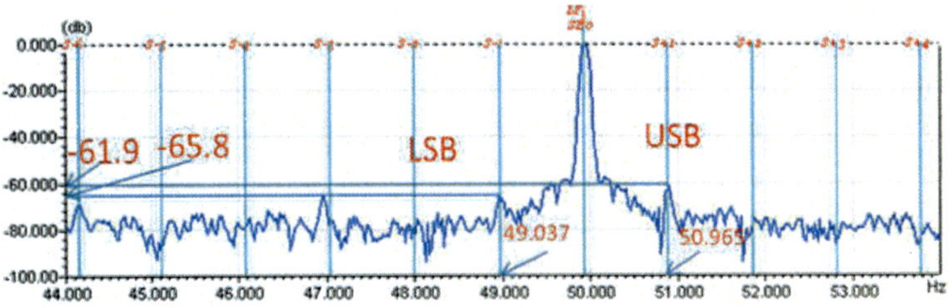

Fig. 7 Current signatures for case 1

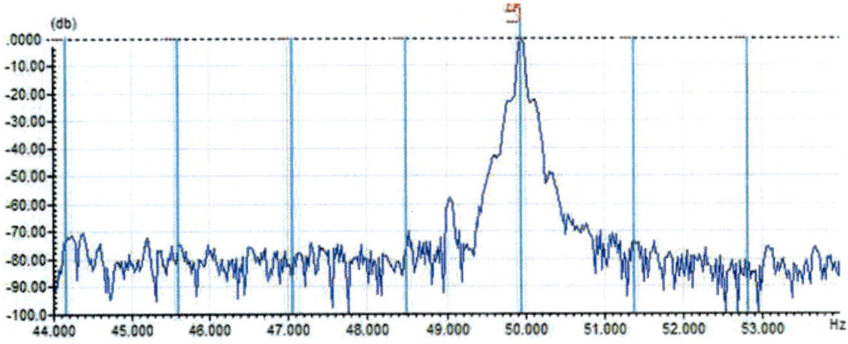

Fig. 8 Current signatures for case 2

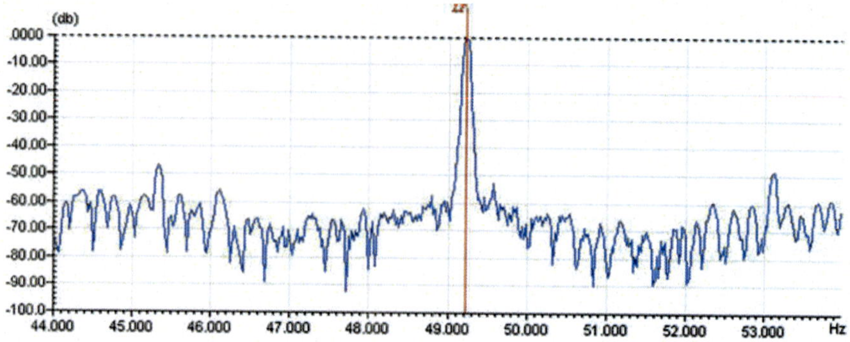

Fig. 9 Current signatures for case 3

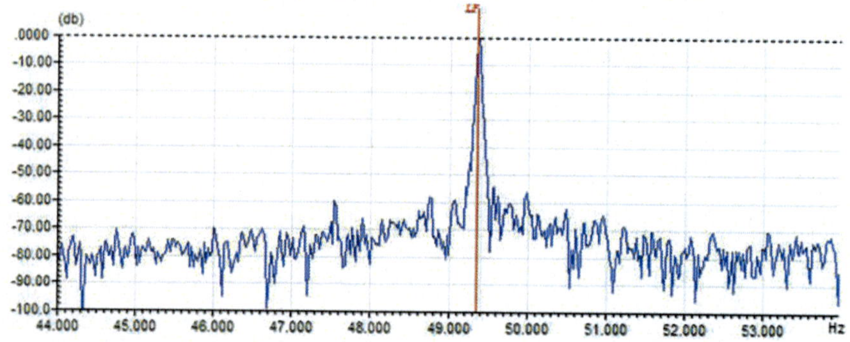

Fig. 10 Current signatures for case 4

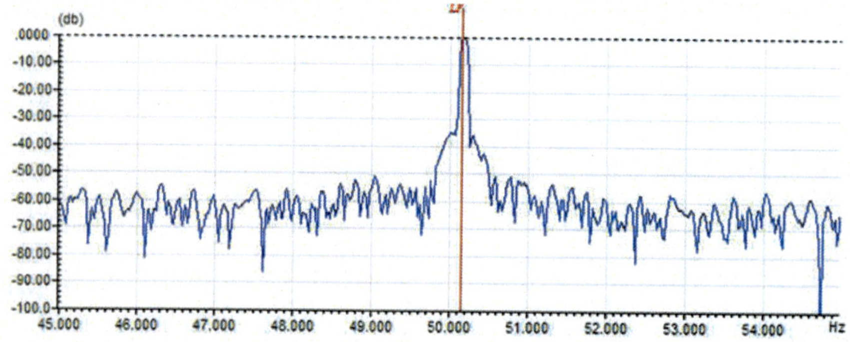

Fig. 11 Current signatures for case 5

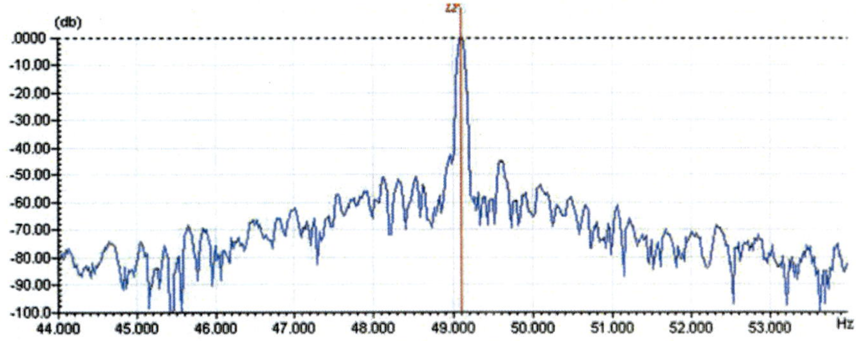

Fig. 12 Current signatures for case 6

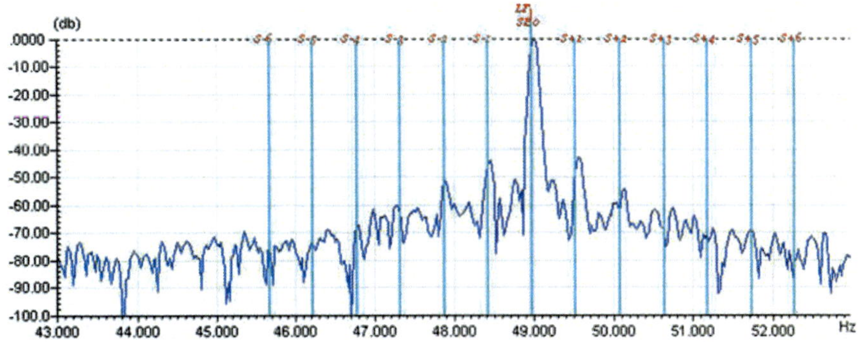

Fig. 13 Current signatures for case 7

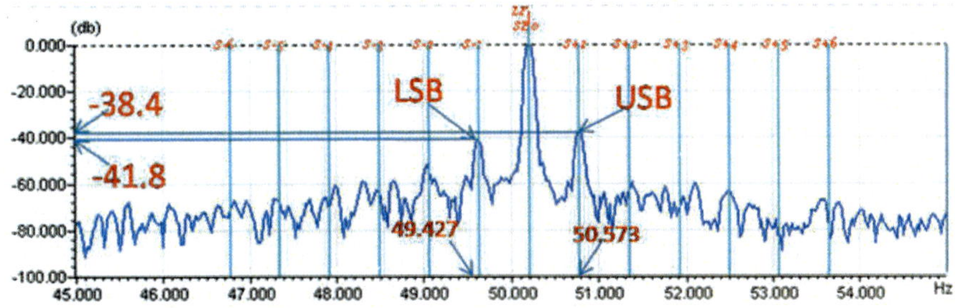

Fig. 14 Current signatures for case 8

5 Results and Discussion

The proposed setup was evaluated using a prototype system deployed in the laboratory to study the behavior of the 5G based IoT enabled plant monitoring system. Latency in terms of actuator & control subunit response time and reliability was evaluated. Earlier

Table 3 Response time

Experiment	Latency between sensor unit & high end processing (msec)	Latency between high end processing & controller and actuator subunit (msec)	Latency between controller and actuator subunit & plant (msec)	Total latency of the system (msec)	Total latency of the system as per [14, 15] (msec)
Case 1	76	62	25	163	252
Case 2	65	50	15	130	241
Case 3	62	54	20	136	245
Case 4	59	52	20	131	239
Case 5	64	48	15	127	238
Case 6	60	50	16	126	239
Case 7	60	51	17	128	237
Case 8	72	59	29	160	248

it was considered as highly reliable and its low latency is achievable only through wired connections. The use of 5G based wireless technologies enabled developing wireless condition monitory systems using IoT. This helped manufacturers in increased productivity with increased safety & reliability of complete systems Table 3.

It can be observed from the results shown in Fig. 15 that latency for case 1 and case 8 is maximum i.e. 163 and 160 ms, respectively. It can be seen that even the worst case latency in case of 5G IoT network is approximate 15% better than a standard 4G network latency [14, 15]. Further from Table 4 it is observed that the resource utilization is more in case 1 and case 8 and the storage device required is a solid state storage type which is costlier than conventional storage devices but are faster and are more reliable.

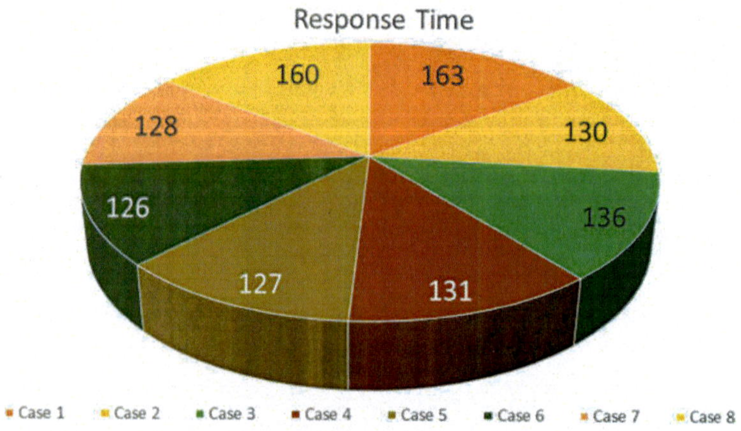

Fig. 15 Total response time for various cases

Table 4 Resource utilization and reliability

	SSD* Usage GB	[%]	Storage GB	[%]	CPU (%)
Case 1	0.431	43	8.598/29	30	8
Case 2	0.329	33	7.42/29	26	2
Case 3	0.452	45	8.391/29	29	5
Case 4	0.434	43	6.542/29	23	3
Case 5	0.396	40	7.562/29	26	2
Case 6	0.424	42	8.147/29	28	3
Case 7	0.381	38	5.98/29	20	2
Case 8	0.452	45	12.123/29	42	9

* Solid state storage device

6 Conclusion and Future Work

The purpose of this paper is to develop a 5G technology based automatic industrial plant monitoring system. The automatic plant monitoring system will monitor the various parameters of the motor using sensors and transmit it using 5G technologies. The system will take to its leverage inherent properties of 5G and IoT for its benefit and to overcome the limitations posed by the 4G/LTE technologies.

The analysis results showed that the proposed method was able to monitor the condition of the motors. It was also able to utilize 5G abled IoT technologies ensuring reduced date delay and increased reliability in terms of quality of service by 15%.

The limitations of 5G network is that it is not available in remote areas, and still establishing of towers and network coverage is required. Further developing and establishing network requires huge expenses for infrastructure setup.

As day by day use of artificial intelligence (AI), machine learning (ML), Industrial Internet of Things (IIoT) is growing, the proposed system will be adopted for its greater speed of transmission, lower latency and therefore less downtime for industries. In near future with development of 5G technologies the proposed system can be implemented in all Small and medium-sized enterprises (SME) and Micro, Small & Medium Enterprises (MSME).

Acknowledgements The authors are thankful to Prof. Rohit Garg, Director MIT & the Management of MITGI for constant motivation and support.

References

1. Karemore P, Jagtap PP (2020) A review of IoT based smart industrial system for con-trolling and monitoring. In: 2020 Fourth International Conference on Computing Methodologies and Communication (ICCMC). Erode, India, pp 67–69. https://doi.org/10.1109/ICCMC48092.2020.ICCMC-00012

2. Gore RN, Kour H, Gandhi M, Tandur D, Varghese A (2019) Bluetooth based Sensor Monitoring in Industrial IoT Plants. In: 2019 International Conference on Data Science and Communication (IconDSC). Bangalore, India, pp 1–6. https://doi.org/10.1109/IconDSC.2019.8816906
3. Vakaloudis A, O'Leary C (2019) A framework for rapid integration of IoT Systems with industrial environments. In: 2019 IEEE 5th World Forum on Internet of Things (WF-IoT). Limerick, Ireland, pp 601-605. https://doi.org/10.1109/WF-IoT.2019.8767224
4. Zhao L, Matsuo, Zhou Y, Lee W (2019) Design of an Industrial IoT-Based Monitoring System for Power Substations. In: 2019 IEEE/IAS 55th Industrial and Commercial Power Systems Technical Conference (I&CPS). Calgary, AB, Canada, pp 1-6. https://doi.org/10.1109/ICPS.2019.8733348
5. Picot HW, Ateeq M, Abdullah B, Cullen J (2019) Industry 4.0 LabVIEW Based Industrial Condition Monitoring System for Industrial IoT System. In: 2019 12th International Conference on Developments in eSystems Engineering (DeSE). Kazan, Russia, pp 1020–1025. https://doi.org/10.1109/DeSE.2019.00189
6. Khan N, Rafiq F, Abedin F, Khan FU (2019) IoT based health monitoring system for electrical motors. In: 2019 15th International Conference on Emerging Technologies (ICET). Peshawar, Pakistan, pp 1–6. https://doi.org/10.1109/ICET48972.2019.8994398
7. Gore RN, Kour H, Gandhi M (2018) IoT based equipment identification and location for maintenance in large deployment industrial plants. In: 2018 10th International Conference on Communication Systems & Networks (COMSNETS). Bengaluru, pp 461–463. https://doi.org/10.1109/COMSNETS.2018.8328244
8. Lyu L, Chen C, Zhu S, Guan X (2018) 5G enabled codesign of energy-efficient trans-mission and estimation for industrial IoT systems. IEEE Trans Industr Inf 14(6):2690–2704. https://doi.org/10.1109/TII.2018.2799685
9. Acharya V, Hegde VV, Anjan K, Kumar M (2017) IoT (Internet of Things) based efficiency monitoring system for bio-gas plants. In: 2017 2nd International Conference on Computational Systems and Information Technology for Sustainable Solution (CSITSS). Banga-lore, pp 1–5. https://doi.org/10.1109/CSITSS.2017.8447567
10. Shyamala D, Swathi D, Prasanna JL, Ajitha A (2017) IoT platform for condition monitoring of industrial motors. In: 2017 2nd International Conference on Communication and Electronics Systems (ICCES). Coimbatore, pp. 260–265. https://doi.org/10.1109/CESYS.2017.8321278
11. Zhang F, Liu M, Zhou Z, Shen W (2016) An IoT-based online monitoring system for continuous steel casting. IEEE Internet Things J 3(6):1355–1363. https://doi.org/10.1109/JIOT.2016.2600630
12. Rahman A, Hossain MRT, Siddiquee MS (2021) IoT based bidirectional speed control and monitoring of single phase induction motors. In: Vasant P, Zelinka I, Weber GW (eds) Intelligent computing and optimization. ICO 2020. Advances in intelligent systems and computing, vol 1324. Springer, Cham. https://doi.org/10.1007/978-3-030-68154-8_88
13. Kannan R, Solai Manohar S, Senthil Kumaran M (2019) IoT-based condition monitoring and fault detection for induction motor. In: Krishna C, Dutta M, Kumar R (eds) Proceedings of 2nd international conference on communication, computing and networking. Lecture notes in networks and systems, vol 46. Springer, Singapore. https://doi.org/10.1007/978-981-13-1217-5_21
14. D. K. M. Dr. V. Khanaa, "4G Technology", International Journal of Engineering and Computer Science, vol. 2, no. 02, Feb. 2013.
15. Gopal BG (2015) A comparative study on 4G and 5G technology for wireless applications. IOSR J Electron Commun Eng (IOSR-JECE), vol.10, issue 6, Dec. 2015

Criterion to Determine the Stability of Systems with Finite Wordlength and Delays Using Bessel-Legendre Inequalities

Rishi Nigam and Siva Kumar Tadepalli

1 Introduction

During the design of controllers for robots many hardware are employed that are based on fixed point representation of data. Usually, the fixed point hardware have limited wordlength known as finite wordlength. Further, many of the mobile robot systems are controlled using wired control or wireless control as in the case of drones. There may arise propogation delays during the control of such mobile robots. The presence of delays and the finite wordlength nature of the hardware employed may lead to instabilities in the system. This paper is concerned with the instabilities that arise in discrete systems during their digital implementation and due to the time-varying delays present in the system. Due to limited wordlength being employed, overflow arises in the digital implementation of discrete systems. To overcome the overflow saturation finite wordlength nonlinearity is widely employed [2–4, 10–12].

The delays are another source of instability in a system. Various summation inequalities such as Jensen, Reciprocally Convex and Wirtinger have been employed to deal with the sum terms that arise in the forward difference of Lyapunov functions [1, 6, 9].

The system considered in this paper represents a class of systems under the influence of finite wordlength nonlinearities and time-varying delays. Such systems have been studied for example in [2, 10, 14, 15]. In [2], a delay-dependent stability criterion was proposed for discrete systems with saturation nonlinearities, time-varying delays and uncertainties. Free-weighting matrix method was employed to obtain the criterion. The delay-partitioning method was employed in [14] to obtain less conservative results as compared to [2]. Further improvement in conservativeness was reported in a criterion presented in [15]. The nonlinear characterization was similar to [2, 14], the improvement was due to Wirtinger-based inequality employed to deal

R. Nigam · S. K. Tadepalli (✉)
National Institute of Technology Uttarakhand, Srinagar, Uttarakhand 246174, India
e-mail: sktadepalli@nituk.ac.in

© The Author(s), under exclusive license to Springer Nature Singapore Pte Ltd. 2023
H. Muthusamy et al. (eds.), *Robotics, Control and Computer Vision*, Lecture Notes in Electrical Engineering 1009, https://doi.org/10.1007/978-981-99-0236-1_22

with the sum terms in the forward difference of the Lyapunov function. In [10], the problem was extended for the case of two-dimensional discrete systems represented by the Fornasini-Marchesini Second Local State Space (FMSLSSS) model.

Through better nonlinear characterization and employing better summation inequalities, there is still further scope to obtain less conservative results. This is the motivation behind the work presented in this paper. Following is the contribution of this paper:

(1) A new criterion is presented by employing Bessel-Legendre summation inequalities.
(2) The criterion is compared with the previously reported criterion.
(3) Numerical example is proposed to highlight the significance of the work.

Section 2 describes the system and specifies the lemmas employed to obtain the main results of the paper; Sect. 3 presents the main results of the paper; a Numerical Example is provided in Sect. 4 and comparisons are made with previous works available in the literature.

2 Problem Formulation

The discrete time system with saturation nonlinearities is considered as follows:

$$\varpi(\iota+1) = \mathcal{F}(\mathfrak{y}(\iota)) \tag{1a}$$

$$\mathfrak{y}(\iota) = \mathcal{A}\varpi(k) + \mathcal{A}_d\varpi(\iota-\tau(\iota)) \tag{1b}$$

$$\varpi(\iota) = \boldsymbol{\Phi}(\iota), \ \ \iota = -\tau_2, \ldots, 0, \tag{1c}$$

where $\varpi(\iota) \in \mathbb{R}^n$ is vector representing the system state; $\mathcal{A}, \mathcal{A}_d \in \mathbb{R}^{n\times n}$ are system matrices; $\boldsymbol{\Phi}(\iota) \in \mathbb{R}^n$ represents initial condition at time ι; $\mathcal{F}(\cdot)$ is the saturation nonlinear function and $\tau(\iota)$ is a time-varying delay satisfying $\tau(\iota) \in [\tau_2, \tau_1]$.

The following Lemmas have been used for obtaining the main results of the paper.

Lemma 1 ([5]) *For $\mathcal{N} > \mathbf{0}$, integers c, d, r satisfying $c \le d-1$, the following holds for a vector $\varpi : [c, d-1] \cap \mathbb{Z} \longrightarrow \mathbb{R}^n$:*

$$\sum_{i=c}^{d-1} \eta^T(i)\mathcal{N}\eta(i) \ge \frac{1}{d-c}\mathbb{G}_r^T(c, d-1)\omega_r(\mathcal{N})\mathbb{G}_r(c, d-1), \tag{2}$$

where

$$\mathbb{G}_r(c, d-1) = col\{G_0(c, d-1), \ldots, G_r(c, d-1)\}, \tag{3}$$

$$\omega_r(\mathcal{N}) = diag\{\mathcal{N}, 3\mathcal{N}, \ldots, (2r+1)\mathcal{N}\} \tag{4}$$

$$G_r(a, b-1) = \varpi(b) - \sum_{l=0}^{\iota} \frac{\alpha_l^{\iota} l!}{(b-a+1)^{\bar{l}}} J_{l-1}(a, b) \tag{5}$$

where

$$J_l(a,b) = \varpi(a), \quad if \;\; r = -1$$
$$= \sum_{i_{r+1}=a}^{b} \cdots \sum_{i_2=i_3}^{b} \sum_{i_1=i_2}^{b} \varpi(i) if \;\; r \geq 0 \tag{6}$$

$a_l^{\iota} = (-1)^{\iota+l}\binom{\iota}{l}\binom{\iota+l}{l}$, $\binom{\iota}{l}$ *represent the Binomial Coefficients given by* $\frac{\iota!}{(\iota-l)!l!}$.

Remark 1 Lemma 1 is based on the discrete Legendre Polynomial employed to Bessel inequality. The inequality presented in Lemma 1 is known as the Bessel-Legendre Inequality.

Lemma 2 ([5, 7]) *For a positive definite matrix* $\mathcal{R} \in \mathbb{R}^n$ *and matrices* $\mathcal{S}_1$, $\mathcal{S}_2 \in \mathbb{R}^{2n \times n}$, *the following inequality holds:*

$$\begin{bmatrix} \frac{1}{\alpha}\mathcal{R} & \mathbf{0} \\ \mathbf{0} & \frac{1}{1-\alpha}\mathcal{R} \end{bmatrix} \geq \boldsymbol{He}(\mathcal{S}_1[\boldsymbol{I} \quad \mathbf{0}] + \mathcal{S}_2[\mathbf{0} \quad \boldsymbol{I}]) - \alpha\mathcal{S}_1\mathcal{R}^{-1}\mathcal{S}_1^T$$
$$- (1-\alpha)\mathcal{S}_2\mathcal{R}^{-1}\mathcal{S}_2^T \quad \forall \alpha \in (0, 1). \tag{7}$$

3 Main Results

Theorem 1 *For a time-varying delay* $\tau(\iota)$ *and nonnegative integer* r, *the system described by (1) is asymptotically stable if there exist positive definite matrices* $\mathcal{P}_r \in \mathbb{R}^{(r+2)n}$, $\mathcal{Q}_1$, $\mathcal{Q}_2$, $\mathcal{R}_1$, $\mathcal{R}_2 \in \mathbb{R}^n$, *matrices* $\mathcal{S}_1$, $\mathcal{S}_2 \in \mathbb{R}^{4n \times 2n}$ *and matrices* $\mathcal{M}$, $\mathcal{N}$ *and* $\mathcal{Q}$ *such that*

$$\begin{bmatrix} \Psi(\tau_1) & \mathcal{E}_2 \\ * & -\omega_1(\mathcal{R}_2) \end{bmatrix} < 0, \quad \begin{bmatrix} \Psi(\tau_2) & \mathcal{E}_1 \\ * & -\omega_1(\mathcal{R}_2) \end{bmatrix} < 0 \tag{8}$$

$$\mathfrak{d}_i \geq \phi_i, \;\; i = 1, 2, \ldots n \tag{9}$$

where

$$\mathfrak{d}_i = \min\{(\mathfrak{n}_{ii} + k_s\mathfrak{m}_{ii}), (\mathfrak{n}_{ii} + \mathfrak{m}_{ii})\}, \;\; i = 1, 2, \ldots, n \tag{10}$$

$$\phi_i = \sum_{s=1,s\neq i}^{n} \max\{|\mathfrak{n}_{si} + k_s \mathfrak{m}_{si}|, |\mathfrak{n}_{si} + \mathfrak{m}_{si}|\} + \sum_{s=1}^{n} |\mathfrak{q}_{si}|, \quad i = 1, 2, \ldots, n \tag{11}$$

k_s denotes $\sum_{m=1}^{2n} |a_{sm}|$ where a_{sm} is the element of $\mathcal{A} = [\mathcal{A}\mathcal{A}_d]$.

Proof We employ the Lyapunov Functional [5] shown below to obtain the stability criterion:

$$\mathcal{V}(\iota) = \mathcal{V}_1(\iota) + \mathcal{V}_2(\iota) + \mathcal{V}_3(\iota) \tag{12a}$$

$$\mathcal{V}_1(\iota) = \boldsymbol{\Lambda}_r^T(\iota)\mathcal{P}_r\boldsymbol{\Lambda}_r(\iota) \tag{12b}$$

$$\mathcal{V}_2(\iota) = \sum_{i=\iota-\tau_1}^{\iota-1} x^T(i)\mathcal{Q}_1 x(i) + \sum_{i=\iota=\tau_2} x^T(i)\mathcal{R}_1 x(i) \tag{12c}$$

$$\mathcal{V}_3(\iota) = \tau_1 \sum_{j=-\tau_1}^{-1} \sum_{i=\iota+j}^{\iota-1} \eta^T(i)\mathcal{Q}_2\eta(i) + \tau_{12} \sum_{j=-\tau_2}^{-\tau_1-1} \sum_{i=\iota+j}^{\iota-1} \eta^T(i)\mathcal{R}_2\eta(i) \tag{12d}$$

where $\eta(\iota) = \varpi(\iota+1) - \varpi(\iota)$

$$\boldsymbol{\Lambda}_r(\iota) = col\left[\hat{\Lambda}_0, \hat{\Lambda}_1, \ldots, \hat{\Lambda}_r\right] \tag{13}$$

where

$$\hat{\Lambda}_0 = col\left[\varpi(\iota), \sum_{i=\iota-\tau_2}^{\iota-\tau_1-1} \varpi(i)\right] \tag{14a}$$

$$\hat{\Lambda}_r = J_{r-1}(\iota-\tau_1, \iota) - \binom{\tau_1 + r - 1}{r-1}\varpi(\iota) \quad \forall r \geq 1 \tag{14b}$$

where

$$J_r(a, b) = \varpi(a) \quad \text{if} \quad r = -1 \tag{15}$$

$$J_r(a, b) = \sum_{i_{r+1}=a}^{b} \cdots \sum_{i_2=i_3}^{b} \sum_{i_1=i_2}^{b} \varpi(i_1) \quad \text{if} \quad r \geq 0. \tag{16}$$

Finding the forward difference of the Lyapunov Functional gives

$$\Delta\mathcal{V}(\iota) = \Delta\mathcal{V}_1(\iota) + \Delta\mathcal{V}_2(\iota) + \Delta\mathcal{V}_3(\iota) \tag{17}$$

$$\Delta\mathcal{V}_1(\iota) = \boldsymbol{\Lambda}_r^T(\iota+1)\mathcal{P}_r\boldsymbol{\Lambda}_r(\iota+1) - \boldsymbol{\Lambda}_r^T(\iota)\mathcal{P}_r\boldsymbol{\Lambda}_r(\iota). \tag{18}$$

Let

$$\boldsymbol{\Lambda}_r(\iota) = \chi_r^T(\tau(\iota))\boldsymbol{\xi}_r(\iota), \quad \Delta\boldsymbol{\Lambda}_r(\iota) = U_r^T\boldsymbol{\xi}_r(\iota) \tag{19}$$

where

$$\mathcal{U}_r = \left[\bar{\mathcal{U}}_0, \bar{\mathcal{U}}_1, \bar{\mathcal{U}}_2, \ldots, \bar{\mathcal{U}}_r\right] \tag{20}$$

$$\bar{\mathcal{U}}_0 = [\mathfrak{e}_{r+7} - \mathfrak{e}_1, \mathfrak{e}_2 - \mathfrak{e}_4] \tag{21}$$

$$\bar{\mathcal{U}}_1 = \mathfrak{e}_1 - \mathfrak{e}_2 \tag{22}$$

$$\bar{\mathcal{U}}_r = \binom{\tau_1 + r - 1}{r-1}(\mathfrak{e}_1 - \mathfrak{e}_{r+5}) \quad \text{for} \quad r \geq 2 \tag{23}$$

$$\chi_r(\tau(\iota))) = \left[\bar{\chi}_0(\tau(\iota))), \bar{\chi}_1(\tau(\iota))), \bar{\chi}_2(\tau(\iota)))\cdots, \bar{\chi}_r(\tau(\iota)))\right] \tag{24}$$

$$\bar{\chi}_0(\tau(\iota))) = [\mathfrak{e}_1, \ (\tau(\iota)) - \tau_1 + 1)\mathfrak{e}_5 - \mathfrak{e}_2 + (\tau_2 - \tau(\iota)) + 1)\mathfrak{e}_6 - \mathfrak{e}_3] \tag{25}$$

$$\bar{\chi}_r(\tau(\iota))) = \binom{\tau_1 + r}{r}\mathfrak{e}_{r+6} - \binom{\tau_1 + r - 1}{r-1}\mathfrak{e}_1 \quad \text{for} \quad r \geq 1 \tag{26}$$

$$\begin{aligned}\boldsymbol{\xi}_r^T(\iota) = \big[&\varpi^T(\iota) \quad \varpi^T(\iota-\tau_1) \quad \varpi^T(\iota-\tau(\iota))) \quad \varpi^T(\iota-\tau_2) \\ &\Gamma_{2,1}^T(\iota) \quad \Gamma_{3,1}^T(\iota) \quad \mathcal{F}^T(\mathfrak{y}(\iota))\big], \quad \text{if r=0}\end{aligned} \tag{27}$$

$$\begin{aligned}\boldsymbol{\xi}_r^T(\iota) = \big[&\varpi^T(\iota) \quad \varpi^T(\iota-\tau_1) \quad \varpi^T(\iota-\tau(\iota))) \quad \varpi^T(\iota-\tau_2) \\ &\Gamma_{2,1}^T(\iota) \quad \Gamma_{3,1}^T(\iota) \quad \Gamma_{1,1}^T(\iota) \\ &\Gamma_{1,2}^T(\iota) \quad \cdots \quad \Gamma_{1,r}^T(\iota) \quad \mathcal{F}^T(\mathfrak{y}(\iota))\big], \quad \text{if} \quad r \geq 1\end{aligned} \tag{28}$$

where

$$\Gamma_{1,r}(\iota) = \frac{r!}{(\tau_1+1)^{\bar{r}}} J_{r-1}(\iota - \tau_1, \iota) \tag{29}$$

$$\Gamma_{2,1}(\iota) = \frac{r!}{(\tau(\iota)) - \tau_1 + 1)^{\bar{r}}} J_{r-1}(\iota - \tau(\iota)), \iota - \tau_1) \tag{30}$$

$$\Gamma_{3,1}(\iota) = \frac{r!}{(\tau_2 - \tau(\iota)) + 1)^{\bar{r}}} J_{r-1}(\iota - \tau_2, k - \tau(\iota))). \tag{31}$$

For the case when $r = 5$ we have

$$\Gamma_{2,1} = \frac{1}{(\tau(\iota) - \tau_1 + 1)} \sum_{j_1=\iota-\tau(\iota)}^{\iota-\tau_1} \varpi(j_1) \tag{32}$$

$$\Gamma_{3,1} = \frac{1}{(\tau_2 - \tau(\iota) + 1)} \sum_{j_1=\iota-\tau_2}^{\iota-\tau(\iota)} \varpi(j_1) \tag{33}$$

$$\Gamma_{1,1} = \frac{1}{\tau_1 + 1} \sum_{j_1=\iota-\tau_1}^{\iota} \varpi(j_1) \tag{34}$$

$$\Gamma_{1,2} = \frac{2}{(\tau_1 + 1)(\tau_1 + 2)} \sum_{j_2=\iota-\tau_1}^{\iota} \sum_{j_1=j_2}^{\iota} \varpi(j_1) \tag{35}$$

$$\Gamma_{1,3} = \frac{6}{(\tau_1 + 1)(\tau_1 + 2)(\tau_1 + 3)} \sum_{j_3=\iota-\tau_1}^{\iota} \sum_{j_2=j_3}^{\iota} \sum_{j_1=j_2}^{\iota} \varpi(j_1) \tag{36}$$

$$\Gamma_{1,4} = \frac{24}{(\tau_1 + 1)(\tau_1 + 2)(\tau_1 + 3)(\tau_1 + 4)} \times \sum_{j_4=\iota-\tau_1}^{\iota} \sum_{j_3=j_4}^{\iota} \sum_{j_2=j_3}^{\iota} \sum_{j_1=j_2}^{\iota} \varpi(j_1) \tag{37}$$

$$\Gamma_{1,5} = \frac{120}{(\tau_1 + 1)(\tau_1 + 2)(\tau_1 + 3)(\tau_1 + 4)(\tau_1 + 5)} \times \sum_{j_5=\iota-\tau_1}^{\iota} \sum_{j_4=j_5}^{\iota} \sum_{j_3=j_4}^{\iota} \sum_{j_2=j_3}^{\iota} \sum_{j_1=j_2}^{\iota} \varpi(j_1). \tag{38}$$

This yields

$$\begin{aligned} \Delta\mathcal{V}_1(\iota) &= \boldsymbol{\xi}_r^T(\iota)[\boldsymbol{\mathcal{U}}_r + \chi_r(\tau(\iota)))]P_r[\boldsymbol{\mathcal{U}}_r^T + \chi_r^T(\tau(\iota)))]\boldsymbol{\xi}_r(\iota) \\ &\quad - \boldsymbol{\xi}_r^T(\iota)[\chi_r(\tau(\iota)))]P_r[\chi_r^T(\tau(\iota)))]\boldsymbol{\xi}_r(\iota) \end{aligned} \tag{39}$$

$$\Delta\mathcal{V}_1(\iota) = \Delta\boldsymbol{\Lambda}_r^T(\iota)\boldsymbol{\mathcal{P}}_r\Delta\boldsymbol{\Lambda}_r(\iota) + He\left\{\boldsymbol{\Lambda}_r^T(\iota)\boldsymbol{\mathcal{P}}_r\Delta\boldsymbol{\Lambda}_r(\iota)\right\} \tag{40}$$

$$= \boldsymbol{\xi}_r^T(\iota)\left[\boldsymbol{\mathcal{U}}_r\boldsymbol{\mathcal{P}}_r\boldsymbol{\mathcal{U}}_r^T + He\left\{\chi_r(\tau(\iota)))\boldsymbol{\mathcal{P}}_r\boldsymbol{\mathcal{U}}_r^T\right\}\right]\boldsymbol{\xi}_r(\iota). \tag{41}$$

Also,

$$\begin{aligned} \Delta\mathcal{V}_2(\iota) &= \varpi^T(\iota)\mathcal{Q}_1\varpi(\iota) - \varpi^T(\iota-\tau_1)\mathcal{Q}_1\varpi(\iota-\tau_1) \\ &\quad + \varpi^T(\iota-\tau_1)\mathcal{R}_1\varpi(\iota-\tau_1) - \varpi^T(\iota-\tau_2)\mathcal{R}_1\varpi(\iota-\tau_2) \end{aligned} \tag{42}$$

$$\Delta\mathcal{V}_2(\iota) = \boldsymbol{\xi}_r^T(\iota)\boldsymbol{\Omega}_1\boldsymbol{\xi}_r(\iota) \tag{43}$$

$$\boldsymbol{\Omega}_1 = \mathbf{e}_1\mathcal{Q}_1\mathbf{e}_1^T - \mathbf{e}_2\mathcal{Q}_1\mathbf{e}_2^T + \mathbf{e}_2\mathcal{R}_1\mathbf{e}_2^T - \mathbf{e}_4\mathcal{R}_1\mathbf{e}_4^T \tag{44}$$

$$\Delta\mathcal{V}_3(\iota) = \boldsymbol{\xi}_r^T(\iota)\boldsymbol{\Omega}_2\boldsymbol{\xi}_r(\iota) - \tau_1 \sum_{i=\iota-\tau_1}^{\iota-1} \Delta\boldsymbol{\varpi}^T(i)\mathcal{Q}_2\Delta\boldsymbol{\varpi}(i)$$
$$- \tau_{12} \sum_{i=\iota-\tau_2}^{\iota-\tau(\iota))-1} \boldsymbol{\varpi}^T(i)\mathcal{R}_2\Delta\boldsymbol{\varpi}(i)$$
$$- \tau_{12} \sum_{i=\iota-\tau(\iota))}^{\iota-\tau_1-1} \boldsymbol{\varpi}^T(i)\mathcal{R}_2\Delta\boldsymbol{\varpi}(i) \tag{45}$$
$$\boldsymbol{\Omega}_2 = d_1^2(\mathbf{e}_{r+7} - \mathbf{e}_1)\mathcal{Q}_2(\mathbf{e}_{r+7} - \mathbf{e}_1)^T + d_{12}^2(\mathbf{e}_{r+7} - \mathbf{e}_1)\mathcal{R}_2(\mathbf{e}_{r+7} - \mathbf{e}_1)^T. \tag{46}$$

Using Lemma 1 [5]

$$- \tau_1 \sum_{i=\iota-\tau_1}^{\iota-1} \Delta\boldsymbol{\varpi}^T(i)\mathcal{Q}_2\Delta\boldsymbol{\varpi}(i) \leq -\boldsymbol{\xi}_r^T(\iota)\boldsymbol{\Upsilon}_r(\omega_r(\mathcal{Q}_2))\boldsymbol{\Upsilon}_r^T\xi_r(\iota) \tag{47}$$
$$\boldsymbol{\Upsilon}_r = \left[\bar{\Upsilon}_0, \bar{\Upsilon}_1, \ldots, \bar{\Upsilon}_r\right] \tag{48}$$
$$\bar{\Upsilon}_0 = \mathbf{e}_1 - \mathbf{e}_2 \tag{49}$$
$$\bar{\Upsilon}_r = \mathbf{e}_1 - a_0^r\mathbf{e}_2 - \sum_{i=1}^{r} a_i^r\,\mathbf{e}_{i+6}, \quad \text{for} \quad r \geq 1 \tag{50}$$

also using $r = 1$ will lead to

$$- \tau_{12} \sum_{i=\iota-\tau_2}^{\iota-\tau(\iota))-1} \Delta\boldsymbol{\varpi}^T(i)\mathcal{R}_2\Delta\boldsymbol{\varpi}(i) - \tau_{12} \sum_{i=\iota-\tau(\iota))}^{\iota-\tau_1-1} \Delta\boldsymbol{\varpi}^T(i)\mathcal{R}_2\Delta\boldsymbol{\varpi}(i)$$
$$\leq \frac{-\tau_{12}}{\tau_2 - \tau(\iota))}\left[Y_1^T(\iota - \tau_2, \iota - \tau(\iota)) - 1)\omega_1(\mathcal{R}_2) \times Y_1(\iota - \tau_2, \iota - \tau(\iota)) - 1)\right]$$
$$- \frac{\tau_{12}}{\tau(\iota)) - \tau_1}\left[Y_1^T(\iota - \tau(\iota)), \iota - \tau_1 - 1)\omega_1(\mathcal{R}_2)\times Y_1(\iota - \tau(\iota)), \iota - \tau_1 - 1)\right] \tag{51}$$
$$\leq \frac{-\tau_{12}}{\tau(\iota)) - \tau_1}\left[\boldsymbol{\xi}_r^T(\iota)\varsigma_2\omega_1(\mathcal{R}_2)\varsigma_2^T\xi_r(\iota)\right] \frac{-\tau_{12}}{\tau_2 - \tau(\iota))}\left[\boldsymbol{\xi}_r^T(\iota)\varsigma_1\omega_1(\mathcal{R}_2)\varsigma_1^T\xi_r(\iota)\right] \tag{52}$$
$$\leq -\boldsymbol{\xi}_r^T(\iota)\left\{[\varsigma_1 \quad \varsigma_2]\begin{bmatrix} \frac{1}{\alpha}\omega_1(\mathcal{R}_2) & \mathbf{0} \\ \mathbf{0} & \frac{1}{1-\alpha}\omega_1(\mathcal{R}_2) \end{bmatrix}[\varsigma_1 \quad \varsigma_2]^T\right\}\xi_r(\iota). \tag{53}$$

Using Lemma 2

$$
\begin{aligned}
&-\xi_r^T(\iota)\left\{[\varsigma_1 \quad \varsigma_2]\begin{bmatrix}\frac{1}{\alpha}\omega_1(\mathcal{R}_2) & \mathbf{0}\\ \mathbf{0} & \frac{1}{1-\alpha}\omega_1(\mathcal{R}_2)\end{bmatrix}[\varsigma_1 \quad \varsigma_2]^T\right\}\xi_r(\iota)\\
&\leq -\xi_r^T(\iota)\{[\varsigma_1 \quad \varsigma_2][He(\mathcal{S}_1[\boldsymbol{I}_n \quad \mathbf{0}_n]+\mathcal{S}_2[\mathbf{0}_n \quad \boldsymbol{I}_n])\\
&-\alpha\mathcal{S}_1\omega_1(\mathcal{R}_2)^{-1}\mathcal{S}_1^T-(1-\alpha)\mathcal{S}_2\omega_1(\mathcal{R}_2)^{-1}\mathcal{S}_2^T][\varsigma_1 \quad \varsigma_2]^T\}\xi_r(\iota).
\end{aligned}
\tag{54}
$$

We further obtain

$$
\leq -\xi_r^T(\iota)\left\{[\varsigma_1 \quad \varsigma_2]\begin{bmatrix}\frac{1}{\alpha}\omega_1(\mathcal{R}_2) & \mathbf{0}\\ \mathbf{0} & \frac{1}{1-\alpha}\omega_1(\mathcal{R}_2)\end{bmatrix}[\varsigma_1 \quad \varsigma_2]^T\right\}\xi_r(\iota) \tag{55}
$$

$$
\begin{aligned}
&\leq \xi_r^T(\iota)\boldsymbol{\Omega}_3\xi_r(\iota)+\xi_r^T(\iota)\{\alpha\mathcal{E}_1\omega_1(\mathcal{R}_2)^{-1}\mathcal{E}_1^T\\
&\quad+(1-\alpha)\mathcal{E}_2\omega_1(\mathcal{R}_2)^{-1}\mathcal{E}_2^T\}\xi_r(\iota)
\end{aligned}
\tag{56}
$$

where

$$\boldsymbol{\Omega}_3 = -He(\mathcal{E}_1\varsigma_1^T+\mathcal{E}_2\varsigma_2^T) \tag{57}$$

$$\mathcal{E}_1 = [\varsigma_1 \quad \varsigma_2]\mathcal{S}_1 \tag{58}$$

$$\mathcal{E}_2 = [\varsigma_1 \quad \varsigma_2]\mathcal{S}_2 \tag{59}$$

$$
\begin{aligned}
\varsigma_1 &= [\mathfrak{e}_2-\mathfrak{e}_3, \quad \mathfrak{e}_2+\mathfrak{e}_3-2\mathfrak{e}_5]\\
\varsigma_2 &= [\mathfrak{e}_3-\mathfrak{e}_4, \quad \mathfrak{e}_3+\mathfrak{e}_4-2\mathfrak{e}_6].
\end{aligned}
\tag{60}
$$

Combining (41), (43), (45), (47) and (55) yields

$$
\begin{aligned}
\Delta\mathcal{V}(\iota) \leq\ & \xi_r^T(\iota)\left[\mathcal{U}_r\mathcal{P}_r\mathcal{U}_r^T+He\{\chi_r(\tau(\iota)))\mathcal{P}_r\mathcal{U}_r^T\}\right.\\
&\left.+\boldsymbol{\Omega}_1+\boldsymbol{\Omega}_2+\boldsymbol{\Omega}_3-\boldsymbol{\Upsilon}_r(\omega_r(\mathcal{Q}_2))\boldsymbol{\Upsilon}_r^T\right]\xi_r(\iota)\\
&+\xi_r^T(\iota)\{\alpha\mathcal{E}_1\omega_1(\mathcal{R}_2)^{-1}\mathcal{E}_1^T\\
&\quad+(1-\alpha)\mathcal{E}_2\omega_1(\mathcal{R}_2)^{-1}\mathcal{E}_2^T\}\xi_r(\iota).
\end{aligned}
\tag{61}
$$

Next, employing the nonlinear characterization yields

$$
\begin{aligned}
\Delta\mathcal{V}(\iota) \leq\ & \xi_r^T(\iota)\{\Psi(\tau(\iota)))+\alpha\mathcal{E}_1\omega_1(\mathcal{R}_2)^{-1}\mathcal{E}_1^T\\
&+(1-\alpha)\mathcal{E}_2\omega_1(\mathcal{R}_2)^{-1}\mathcal{E}_2^T\}\xi_r(\iota)-\varrho
\end{aligned}
\tag{62}
$$

$$\alpha = \frac{\tau(\iota))-\tau_1}{\tau_{12}} \tag{63}$$

$$\begin{aligned}\boldsymbol{\Psi}(\tau(\iota))) =& \mathcal{U}_r\mathcal{P}_r\mathcal{U}_r^T + He\{\chi_r(\tau(\iota)))\mathcal{P}_rU_r^T\} + \boldsymbol{\Omega}_1 \\ &+ \boldsymbol{\Omega}_2 + \boldsymbol{\Omega}_3 - \boldsymbol{\Upsilon}_r(\omega_r(\mathcal{Q}_2))\boldsymbol{\Upsilon}_r^T \\ &+ (\alpha\mathcal{E}_1\omega_1(\mathcal{R}_2)^{-1}\mathcal{E}_1^T + (1-\alpha)\mathcal{E}_2\omega_1(\mathcal{R}_2)^{-1}\mathcal{E}_2^T) \\ &+ \mathfrak{e}_1(\mathcal{A}^T\mathcal{M}\mathcal{A} + \mathcal{Q}\mathcal{A} + \mathcal{A}^T\mathcal{Q}^T + \mathcal{A}^T\mathcal{M}^T\mathcal{A})\mathfrak{e}_1^T \\ &+ \mathfrak{e}_3(\mathcal{A}_d^T\mathcal{M}\mathcal{A}_d + \mathcal{A}_d^T\mathcal{M}^T\mathcal{A}_d)\mathfrak{e}_3^T + \mathfrak{e}_{r+7}(-\mathcal{N} - \mathcal{N}^T)\mathfrak{e}_{r+7}^T \\ &+ He\{\mathfrak{e}_1(\mathcal{A}^T\mathcal{M}\mathcal{A}_d + \mathcal{Q}\mathcal{A}_d + \mathcal{A}^T\mathcal{M}^T\mathcal{A}_d)\mathfrak{e}_3^T \\ &+\mathfrak{e}_1(\mathcal{A}^T\mathcal{N}^T - \mathcal{A}^T\mathcal{M} - \mathcal{Q})\mathfrak{e}_{12}^T + \mathfrak{e}_3(\mathcal{A}_d^T\mathcal{N}^T - \mathcal{A}_d^T\mathcal{M})\mathfrak{e}_{12}^T\}\end{aligned} \quad (64)$$

$$\Delta\mathcal{V}(\iota) \leq \boldsymbol{\xi}_r^T(\iota)\boldsymbol{\Xi}(\tau(\iota)))\boldsymbol{\xi}_r(\iota) - \varrho. \quad (65)$$

Since ϱ is a positive quantity [12], then $\Delta\mathcal{V}(\iota) \leq \mathbf{0}$ if and only if $\boldsymbol{\Xi}(\tau(\iota))) < \mathbf{0}$. Here, $\boldsymbol{\Xi}(\tau(\iota))) < \mathbf{0}$ implies $\boldsymbol{\Xi}(\tau_1) < \mathbf{0}$ and $\boldsymbol{\Xi}(\tau_2) < \mathbf{0}$. Using the Schurs complement would yield the LMIs

$$\begin{bmatrix} \boldsymbol{\Psi}(\tau_1) & \mathcal{E}_2 \\ * & -\omega_1(\mathcal{R}_2) \end{bmatrix} < 0, \quad \begin{bmatrix} \boldsymbol{\Psi}(\tau_2) & \mathcal{E}_1 \\ * & -\omega_1(\mathcal{R}_2) \end{bmatrix} < 0. \quad (66)$$

This completes the proof of the main result.

4 Numerical Example

In this section, we consider a numerical example to highlight the usefulness of the presented criterion.

Example 1 Consider the system (1) with the following system parameters:

$$\mathcal{A} = \begin{bmatrix} 0.68 & -0.45 \\ 0.45 & 0.65 \end{bmatrix}, \quad \mathcal{A}_d = \begin{bmatrix} -0.1 & -0.2 \\ -0.2 & -0.1 \end{bmatrix}. \quad (67)$$

We find the upper delay bound τ_2 for a given lower delay bound τ_1 to compare the conservativeness of the presented criterion with the previously reported criterion. Here we use SeDuMi solver [13] and YALMIP parser [8] along with MATLAB to obtain the results.

Table 1 presents the upper delay bound for various lower delay bounds. 'X' denotes the inability of the criterion to determine the stability of the system under consideration. It can be observed that Theorem 2 [15] and Theorem 3.1 [2] are unable to test the stability of the system. Theorem 2.1 is used with different r, this yields different results. By increasing r the conservativeness decreases and the criterion is able to determine the stability of the system.

Table 1 τ_2 for corresponding τ_1

Methods/τ_1	4	6	7	9
Theorem 2 [15]	X	X	X	X
Theorem 3.1 [2]	X	X	X	X
Theorem 2.1 ($r = 0$)	X	X	X	X
Theorem 2.1 ($r = 3$)	5	7	8	X
Theorem 2.1 ($r = 5$)	5	8	8	10

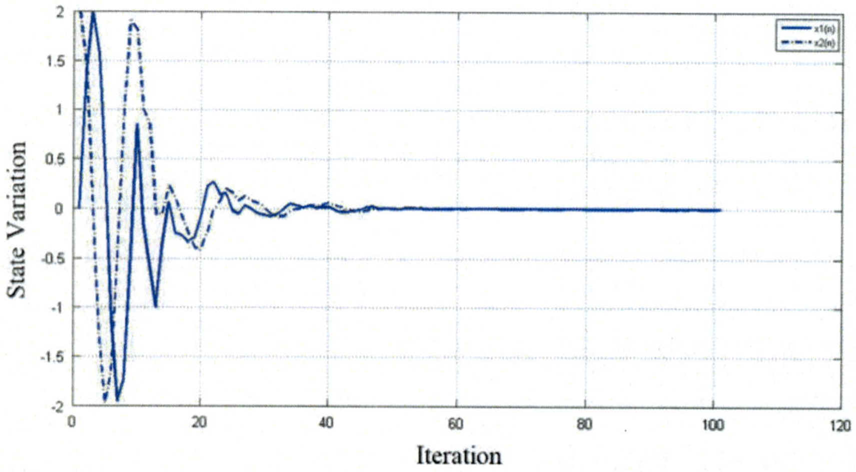

Fig. 1 State Trajectory of the system considered in Example 1

Therefore, the presented criterion is less conservative as compared to previously reported criterion Theorem 2 [15] and Theorem 3.1 [2].

The trajectory of the system under consideration is shown in Fig. 1. From Fig. 1, it can be stated that the conclusions arrived at regarding the stability of the system using the proposed Theorem 2.1 are correct.

Remark 2 In the proposed Theorem 2.1, employing $r = 0$ is equivalent to employing Jensen inequality, and $r = 3$ represents employing auxiliary-function-based summation inequalities. It may be noted that increasing the value of r may yield less conservative results as shown in Table 1.

5 Conclusion and Future Scope

This paper presented a stability criterion for discrete systems with saturation finite wordlength nonlinearity and time-varying delays. The criterion is based on the Bessel-Legendre Summation Inequalities. With the help of a numerical example, it was shown that the criterion is less conservative as compared to a previously reported criterion. The criterion presented can be extended to the stability analysis of 2-D systems and discrete fuzzy systems with uncertainties.

References

1. Hien LV, Trinh H (2016) New finite-sum inequalities with applications to stability of discrete time-delay systems. Automatica 71:197–201
2. Kandanvli VKR, Kar H (2013) Delay-dependent stability criterion for discrete-time uncertain state-delayed systems employing saturation nonlinearities. Arab J Sci Eng 38(10):2911–2920
3. Kokil P, Jogi S, Ahn CK, Kar H (2020) An improved local stability criterion for digital filters with interference and overflow nonlinearity. IEEE Trans Circuits Syst II Express Briefs 67(3):595–599. https://doi.org/10.1109/TCSII.2019.2918788
4. Kokil P, Jogi S, Ahn CK, Kar H (2021) Stability of digital filters with state-delay and external interference. Circuits Syst Signal Process. https://doi.org/10.1007/s00034-021-01650-8
5. Lee SY, Park J, Park P (2018) Bessel summation inequalities for stability analysis of discrete-time systems with time-varying delays. Int J Robust Nonlinear Control 29(2):473–491
6. Liu J, Zhang J (2012) Note on stability of discrete-time time-varying delay systems. IET Control Theory Appl 6(2):335–339
7. Liu K, Seuret A, Xia Y (2017) Stability analysis of systems with time-varying delays via the second-order Bessel-Legendre inequality. Automatica 76:138–142
8. Lofberg J (2004) Yalmip: a toolbox for modeling and optimization in MATLAB. In: Proceedings of computer aided control systems design conference, Taipei, Taiwan, pp 284–289
9. Nam PT, Pathirana PN, Trinh H (2015) Discrete Wirtinger-based inequality and its application. J Franklin Inst 352:1893–1905
10. Pandey S, Tadepalli SK (2021) Improved criterion for stability of 2-D discrete systems involving saturation nonlinearities and variable delays. ICIC Express Lett 15(3):273–283
11. Rani P, Kumar MK, Kar H (2019) Hankel norm performance of interfered fixed-point state-space digital filters with quantization/overflow nonlinearities. Circuits Syst Signal Process 38:3762–3777
12. Shen T, Yuan Z, Wang X (2012) Stability analysis for digital filters with multiple saturation nonlinearities. Automatica 48(10):2717–2720
13. Sturm J (1999) Using SeDuMi 1.02, a MATLAB toolbox for optimization over symmetric cones. Optim Methods Softw 11–12:625–653, version 1.05. http://fewcal.kub.nl/sturm
14. Tadepalli SK, Kandanvli VKR (2016) Improved stability results for uncertain discrete-time state-delayed systems in the presence of nonlinearities. Trans Inst Meas Control 38(1):33–43
15. Tadepalli SK, Kandanvli VKR, Vishwakarma A (2018) Criteria for stability of uncertain discrete-time systems with time-varying delays and finite wordlength nonlinearities. Trans Inst Meas Control 40(9):2868–2880

Adaptive Control for Stabilization of Ball and Beam System Using H_∞ Control

Sudhir Raj

1 Introduction

Control of the ball and beam system is an interesting problem in the control theory. The proposed non-linear controller is applied for the stabilization of underactuated system. H_∞-based adaptive control can be applied for the control of underactuated systems. The problem considered is the ball and beam system and the objective is to develop the controller for stabilization of underactuated systems.

The proposed method [1] combines the state feedback controller with observer-based control for the stabilization of ball beam system. State-dependent saturation controller [2] is used for the stabilization of ball and beam system. Energy shaping [3]-based inverse Lyapunov controller is applied for the control of ball beam system. Convex optimization-based optimal control [4] is carried out for the stabilization of ball beam system. Three different controllers [5] are applied for the control of ball beam system. Experimental results are presented to validate the controllers. Static and dynamic-based sliding mode controller [6] is applied for the control of ball beam system which avoids chattering in the system. The proposed ant colony optimization [7] is proposed for the control of ball beam system. The interpolating sliding mode observer [8]-based control is carried out for the stabilization of ball beam system. Adaptive control [9] based on recurrent neural network is applied for the stabilization of ball and beam system which gives better performance as compared to Linear quadratic regulator. The passivity-based controller [10] is applied for the control of ball beam system. The input output linearization [11]-based controller is carried out for the stabilization of ball beam system. The algebraic Riccati equation approach [12] is applied to H_∞-based state feedback controller. Adaptive sliding mode control [13] is proposed for non-linear underactuated systems.

S. Raj (✉)
SRM University, Amaravati, Andhra Pradesh, India
e-mail: sudhir.r@srmap.edu.in

© The Author(s), under exclusive license to Springer Nature Singapore Pte Ltd. 2023

H. Muthusamy et al. (eds.), *Robotics, Control and Computer Vision*, Lecture Notes in Electrical Engineering 1009, https://doi.org/10.1007/978-981-99-0236-1_23

H_∞-based adaptive control was not reported in the earlier work for stabilization of ball and beam system. The objective of this controller is to bring the states of the system to the origin after a perturbation was started. The control objective is to find a decentralized control that will bring an arbitrary initial state to the equilibrium point of the system. The main contribution of this work is to develop a robust controller for the stabilization of underactuated systems.

2 Dynamic Model of Ball and Beam System

The control of ball beam becomes difficult due to its non-linear dynamics. Figure 1 shows the diagram of ball and beam system. The ball rolls on the beam and the rotation of the beam is controlled by the motor. H_∞-based adaptive control is applied so that the position of the ball can be controlled.

The state space model of the ball and beam system is given by the equation number (1). The disturbances for ball and beam system is taken as w_1.

$$\dot{x}_1 = A_1 x_1 + B_{21} u_1 + B_{11} w_1 \tag{1}$$

2.1 *Analysis of the Force Balance of the Ball*

The equation of the ball beam for force balance can be written as

$$\begin{aligned} \sum F_b &= M_{ball} g sin\theta - Fr \\ &= M_{ball}\ddot{x} + b_1\dot{x} \end{aligned} \tag{2}$$

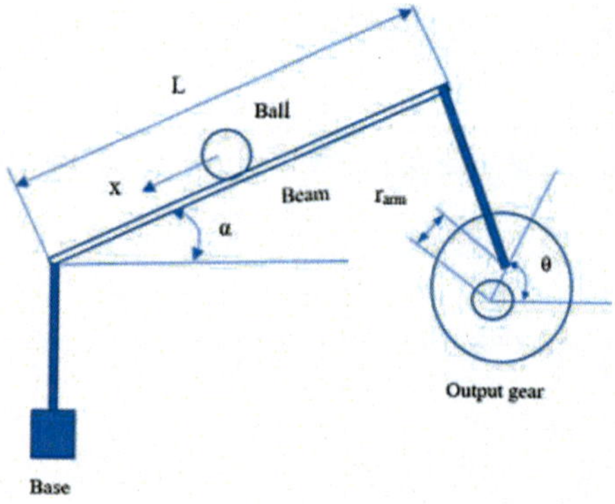

Fig. 1 Ball and beam system

x is taken as the vertical distance between the center of ball and center of the shaft. b_1 is taken as the friction constant. θ gives the beam's tilt angle from the horizontal position. F_r gives the value of the force which is applied externally. The position of the ball is given by equation number (3). The rotational angle and radius of ball are taken as α and a_1, respectively.

$$x = \alpha a_1 \tag{3}$$

The equation of torque balance of the ball can be taken as

$$\begin{aligned}\sum \tau_b &= F_r a_1 \\ &= J_{ball}\ddot{\alpha}\end{aligned} \tag{4}$$

J_{ball} gives the moment of inertia of the ball.

$$J_{ball} = \frac{2}{5} M_{ball} R_b^2 \tag{5}$$

Equation (6) can be derived from Eqs. (2)–(5).

$$\left[1 + \frac{2}{5}\left(\frac{R_b}{a_1}\right)^2\right]\ddot{x} + \frac{b_1 \dot{x}}{M_{ball}} = gsin\theta \tag{6}$$

2.2 *Torque Balance of the Motor and Beam*

The torque balance of the motor is given by equation number (7).

$$T_{motor} = KI - J_{motor}\ddot{\theta} - b\dot{\theta} \tag{7}$$

K is the electromotive force constant as taken in equation number (7). I gives the current which flows in the motor. The term b is taken as the damping constant of the rotational system. The torque by the ball and beam is given by the equation numbers (8) and (9), respectively.

$$T_{ball} = -x M_{ball} g cos\theta \tag{8}$$

$$T_{beam} = T_{motor} + T_{ball} \tag{9}$$

The moment of inertia of the beam is expressed as follows:

$$J_{bm} = J_{beam} + J_{motor} \tag{10}$$

$$J_{beam} = \frac{1}{12} M_{beam} J_{beam}^2 \tag{11}$$

Equation (12) can be obtained from Eqs. (7)–(11).

$$\ddot{\theta} = \frac{KI - xM_{ball}gcos\theta - b\dot{\theta}}{J_{bm}} \tag{12}$$

2.3 DC Motor Equation

Equation (13) is found by combining Newton's law with the Kirchhoff's law since the DC motor is armature-controlled.

$$L\frac{dI}{dt} + RI = V - K_e\dot{\theta} \tag{13}$$

L is taken as the armature induction. K_e is the motor constant and R is taken as the armature resistance. Equation (14) can be derived by rearranging equation number (14).

$$\dot{I} = \frac{V - RI - K_e\dot{\theta}}{L} \tag{14}$$

2.4 State Space Model

The state space equation is given by equation number (16).

$$\dot{X} = \begin{bmatrix} \dot{x} \\ \ddot{x} \\ \dot{\theta} \\ \ddot{\theta} \\ \dot{I} \end{bmatrix} \tag{15}$$

$$\dot{X} = AX + BV \tag{16}$$

$$A = \begin{bmatrix} 0 & 1 & 0 & 0 & 0 \\ 0 & \frac{b_1}{\left[1+\frac{2}{5}\left(\frac{R_b}{a}\right)^2\right]M_{ball}} & \frac{g}{1+\frac{2}{5}\left(\frac{R_b}{a_1}\right)^2} & 0 & 0 \\ 0 & 0 & 0 & 1 & 0 \\ -\frac{M_{ball}g}{J_{bm}} & 0 & 0 & -\frac{b}{J_{bm}} & \frac{K}{J_{bm}} \\ 0 & 0 & 0 & -\frac{K_e}{L} & -\frac{R}{L} \end{bmatrix}$$

$$B = \begin{bmatrix} 0 \\ 0 \\ 0 \\ 0 \\ \frac{1}{L} \end{bmatrix}$$

$$y = \begin{bmatrix} 1\,0\,0\,0\,0 \\ 0\,0\,1\,0\,0 \end{bmatrix} X \tag{17}$$

The equation number (18) is derived from equation number (14) since the armature resistance is very small.

$$V = RI + K_e \dot{\theta} \tag{18}$$

Equation number (19) is obtained using Eqs. (12) and (18).

$$\ddot{\theta} = \frac{\frac{K}{R}V - \left(\frac{KK_e}{R} + b\right)\dot{\theta} - XM_{ball}gcos\theta}{J_{bm}} \tag{19}$$

The equation number (20) is derived from equation numbers (19) and (6) assuming that friction constant b_1 is zero.

$$\begin{bmatrix} \dot{x} \\ \ddot{x} \\ \dot{\theta} \\ \ddot{\theta} \end{bmatrix} = \begin{bmatrix} 0 & 1 & 0 & 0 \\ 0 & 0 & \frac{g}{1+\frac{2}{5}\left(\frac{R_b}{a_1}\right)^2} & 0 \\ 0 & 0 & 0 & 1 \\ -\frac{M_{ball}}{J_{bm}} & 0 & 0 & -\frac{\left(\frac{KK_e}{R}+b\right)}{J_{bm}} \end{bmatrix} \begin{bmatrix} x \\ \dot{x} \\ \theta \\ \dot{\theta} \end{bmatrix} + \begin{bmatrix} 0 \\ 0 \\ 0 \\ 0 \\ \frac{K}{RJ_{bm}} \end{bmatrix} V \tag{20}$$

$$y = \begin{bmatrix} 1\,0\,0\,0 \\ 0\,0\,1\,0 \end{bmatrix} X \tag{21}$$

The system parameters are given in Table 1. The system parameters are substituted in system equations.

$$\begin{bmatrix} \dot{x} \\ \ddot{x} \\ \dot{\theta} \\ \ddot{\theta} \end{bmatrix} = \begin{bmatrix} 0 & 1 & 0 & 0 \\ 0 & 0 & 3.7731 & 0 \\ 0 & 0 & 0 & 1 \\ -5.170 & 0 & 0 & -105.1 \end{bmatrix} \begin{bmatrix} x \\ \dot{x} \\ \theta \\ \dot{\theta} \end{bmatrix} + \begin{bmatrix} 0 \\ 0 \\ 0 \\ 0 \\ 16.85 \end{bmatrix} V$$

Table 1 Parameter of the Ball and beam

Symbol	Value
M_{ball}	0.0327 Kg
J_{bm}	0.062 Kgm^2
R_b	0.01 m
a_1	0.005 m
b	1.5279 Nm/ (rad/s)
K	4.91 Nm/A
K_e	4.77 V/ (rad/s)
R	4.7Ω
L	0.0016 H
L_{beam}	0.7 m
M_{beam}	0.381 Kg
J_{motor}	0.043 Kgm^2

3 H_∞-Based Integral Sliding Mode Control

The control of an underactuated system is carried out using H_∞-based adaptive control.

$$PA_1 + A_1^T P - P\left(B_{21}B_{21}^T - \gamma^{-2}B_{11}B_{11}^T\right)P + C_1^T C_1 = 0 \tag{22}$$

P is a positive definite solution of the Algebraic Riccati Equation and the system becomes stable.

$$u_0 = -B_{21}^T P x_1 \tag{23}$$

3.1 H_∞-Based Integral Sliding Mode Control

Equation number (24) gives the sliding surface for ball beam system. G is same as B_{21}^+, and B_{21}^+ is taken as the Pseudoinverse of matrix B_{21}.

$$s(x,t) = G[x_1(t) - x_1(t_0) - \int_{t_0}^{t}(A_1 - B_{21}B_{21}^T P)x_1(t)dt]] \tag{24}$$

Differentiation of the sliding surface s is done with respect to time.

$$\begin{aligned}\dot{s} &= G\left[\dot{x}_1(t) - \left(A_1 - B_{21}B_{21}^T P\right)x_1\right] \\ &= G\left[A_1x_1 + B_{21}u_1 + B_{11}w_1\right] \\ &- G\left[\left(A_1 - B_{21}B_{21}^T P\right)x_1\right] \\ &= G\left[B_{21}u_1 + B_{11}w_1 + B_{21}B_{21}^T P x_1\right]\end{aligned} \tag{25}$$

The terms u_{eq} and u_{sw} are taken as the equivalent and switching control, respectively.

$$u_1 = u_{eq} + u_{sw} \tag{26}$$

The derivative of the sliding surface is taken as zero to find the equivalent control.

$$G\left[B_{21}u_1 + B_{11}w_1 + +B_{21}B_{21}^T P x_1\right] = 0$$

$$u_{eq} = -B_{21}^T P x_1 - (GB_{21})^{-1} GB_{11}w_1 \tag{27}$$

The switching control law is found using the Lyapunov theorem. The Lyapunov function is taken as equation number (28).

$$V = \frac{s^T s}{2} \tag{28}$$

Equation number (29) gives the condition for the convergence of the sliding mode. The term η is a positive constant.

$$\dot{V} = s\dot{s} \leq -\eta|s| \tag{29}$$

The computation of $\dot{V}$ is given by equation number (30).

$$\begin{aligned}
\dot{V} &= s\dot{s} \\
&= sG\left[B_{21}u_1 + B_{11}w_1 + B_{21}B_{21}^T P x_1\right] \\
&= sG\left[B_{21}\left(u_{eq} + u_{sw}\right) + B_{11}w_1 + B_{21}B_{21}^T P x_1\right] \\
&= sG\left[B_{21}u_{eq} + B_{21}u_{sw} + B_{11}w_1 + B_{21}B_{21}^T P x_1\right] \\
&= s\left[-GB_{21}B_{21}^T P x_1 - GB_{21}(GB_{21})^{-1} GB_{11}w_1\right] \\
&+ s\left[GB_{21}u_{sw} + GB_{11}w_1 + GB_{21}B_{21}^T P x_1\right] \\
&= s\left[GB_{21}u_{sw}\right]
\end{aligned} \tag{30}$$

The switching control law for stabilization of ball and beam system can be found by the equation number (31).

$$\begin{aligned}
\dot{V} = s\left[GB_{21}u_{sw}\right] &\leq -\eta|s| \\
u_{sw} &\leq -\frac{\eta sign(s)}{GB_{21}} \\
u_{sw} &= -u_0 sign(s)
\end{aligned} \tag{31}$$

The sliding mode control input is given by equation number (32) for the stabilization of ball and beam system.

$$u_1 = -B_{21}^T P x_1 - u_0 sign(s) \tag{32}$$

ΔA_1 is taken as the uncertainties of the system matrix A. Equation number (33) gives the state space equation of ball and beam system.

$$\dot{x}_1 = (A_1 + \Delta A_1) x_1 + B_{21} u_1 + B_{11} w_1 \tag{33}$$

Equation number (34) gives the derivative of the sliding surface.

$$\begin{aligned}
\dot{s} &= G\left[\dot{x}_1(t) - \left(A_1 - B_{21} B_{21}^T P\right) x_1\right] \\
&= G\left[(A_1 + \Delta A_1) x_1 + B_{21} u_1 + B_{11} w_1\right] \\
&- G\left[\left(A_1 - B_{21} B_{21}^T P\right) x_1\right] \\
&= G\left[A_1 x_1 + \Delta A_1 x_1 + B_{21} u_1 + B_{11} w_1\right] \\
&+ G\left[-A_1 x_1 + B_{21} B_{21}^T P x_1\right] \\
&= G\left[\Delta A_1 x_1 + B_{21} u_1 + B_{11} w_1 + B_{21} B_{21}^T P x_1\right] \\
&= G\left[\Delta A_1 x_1 - B_{21} B_{21}^T P x_1 - B_{21} u_0 sign(s)\right] \\
&+ G\left[B_{11} w_1 + B_{21} B_{21}^T P x_1\right] \\
&= G\left[\Delta A_1 x_1 - B_{21} u_0 sign(s) + B_{11} w_1\right]
\end{aligned} \tag{34}$$

The reaching condition for ball and beam system can be taken as equation number (35).

$$\begin{aligned}
s\dot{s} &< 0 \\
sG\left[\Delta A_1 x_1 - B_{21} u_0 sign(s) + B_{11} w_1\right] &< 0 \\
s\left[B_{21}^+ \Delta A x_1 + B_{21}^+ B_{11} w - u_0 sign\,(s)\right] &< 0 \\
\left(\| B_{21}^+ \Delta A x_1 \|\right) \| s \| & \\
+ \left(\| B_{21}^+ B_{11} \| \gamma - u_0\right) \| s \| &< 0
\end{aligned} \tag{35}$$

The condition for u_o can be obtained as equation number (36).

$$u_0 > \| B_{21}^+ \Delta A x_1 \| + \| B_{21}^+ B_{11} \| \gamma \tag{36}$$

The term u_0 is greater than $\| B_{21}^+ \Delta A x_1 \| + \| B_{21}^+ \Delta A_{12} x_2 \| + \| B_{21}^+ B_{11} \| \gamma$ by a constant υ .

$$
\begin{aligned}
V &= \frac{s^T s}{2} \\
\dot{V} &= s\dot{s} \\
&= \left(\| B_{21}^{+} \Delta A x_1 \|\right) \| s \| + \left(\| B_{21}^{+} B_{11} \| \gamma - u_0\right) \| s \| \\
&= \left(\| B_{21}^{+} \Delta A x_1 \|\right) \| s \| + \left(\| B_{21}^{+} B_{11} \| \gamma\right) \| s \| \\
&- \left(\| B_{21}^{+} \Delta A x_1 \|\right) \| s \| - \left(\| B_{21}^{+} B_{11} \| \gamma + v\right) \| s \| \\
&= -v \| s \| < 0
\end{aligned} \tag{37}
$$

The negative sign of the derivative of the Lyapunov function V ensures the stabilization of the ball and beam system.

3.2 H_∞-Based Adaptive Integral Sliding Mode Control

H_∞-based adaptive control is proposed. Lyapunov theory is used to verify the proposed controller. The modified control law can be written as

$$
u = u_{eq} + u_{sm} \tag{38}
$$

The term u_{eq} is the same as in Eq. (27) used for the nominal system. The adaptive term u_{sw} is modified as equation number (39).

$$
u_{sm} = -\hat{\Gamma} sgn\,(s) \tag{39}
$$

The adaptive law is taken as equation number (40).

$$
\dot{\hat{\Gamma}} = \frac{| s |}{\alpha} \tag{40}
$$

The term $\dot{\hat{\Gamma}}$ is an adjustable gain constant. The term α is the adaptation gain and $\alpha > 0$. The adaptation speed of $\hat{\Gamma}$ can be tuned by α. The adaptation error is defined as equation number (41).

$$
\tilde{\Gamma} = \hat{\Gamma} - \Gamma_d \tag{41}
$$

Equation number (42) gives the Lyapunov function for the modified controller.

$$
V = \frac{s^T s}{2} + \frac{1}{2}\alpha\tilde{\Gamma}^2 \tag{42}
$$

The derivative of the sliding surface can be taken as equation number (43).

$$\begin{aligned}
\dot{V} &= s\dot{s} + \alpha\tilde{\Gamma}\dot{\tilde{\Gamma}} \\
\dot{s} &= G\left[A_1x_1 + \Delta A_1x_1 + B_{21}u_1 + B_{11}w_1\right] \\
&\quad + G\left[-A_1x_1 + B_{21}B_{21}^TPx_1\right] \\
&= G\left[A_1x_1 + \Delta A_1x_1 + B_{21}\left(u_{eq} + u_{sm}\right) + B_{11}w_1\right] \\
&\quad + G\left[-A_1x_1 + B_{21}B_{21}^TPx_1\right] \\
&= G\left[A_1x_1 + \Delta A_1x_1\right] \\
&\quad + G\left[B_{21}\left(-B_{21}^TPx_1 - (GB_{21})^{-1}GB_{11}w_1 - \hat{\Gamma}sgn(s)\right)\right] \\
&\quad + G\left[B_{11}w_1 - A_1x_1 + B_{21}B_{21}^TPx_1\right] \\
&= G\left[A_1x_1 + \Delta A_1x_1 - B_{21}B_{21}^TPx_1\right] \\
&\quad + G\left[-B_{21}(GB_{21})^{-1}GB_{11}w_1 - B_{21}\hat{\Gamma}sgn(s)\right] \\
&\quad + G\left[B_{11}w_1 - A_1x_1 + GB_{21}B_{21}^TPx_1\right] \\
&= G\left[\Delta A_1x_1 - B_{21}\hat{\Gamma}sgn(s)\right] \\
s\dot{s} + \alpha\tilde{\Gamma}\dot{\tilde{\Gamma}} &= sG\left[\Delta A_1x_1 - B_{21}\hat{\Gamma}sgn(s)\right] \\
&\quad + \alpha\left(\hat{\Gamma} - \Gamma_d\right)\dot{\hat{\Gamma}} \\
&= sG\left[\Delta A_1x_1 - B_{21}\hat{\Gamma}sgn(s)\right] + s\left(\hat{\Gamma} - \Gamma_d\right)sgn(s) \\
&= sG\Delta A_1x_1 - s\Gamma_d sgn(s) \\
&= s\left(G\Delta A_1x_1 - \Gamma_d sgn(s)\right) \\
&= s\left(G\Delta A_1x_1\right) - \Gamma_d \mid s \mid < 0
\end{aligned} \tag{43}$$

The term Γ_d satisfies the inequality as given in equation number (44).

$$\Gamma_d > \mid G\Delta A_1x_1 \mid \tag{44}$$

The convergence of s and $\tilde{\Gamma}$ is proved using Lyapunov theorem.

4 Simulation Results

Simulation of ball and beam system was carried out in MATLAB. The different parameter values are taken from Table 1 for the simulation of the ball and beam system. Two initial states are considered for the ball and beam system as:

$$X_0 = [1.2, 0, 0, 0]^T$$
$$X_1 = [0.09, 0, 0.0873, 0]^T$$

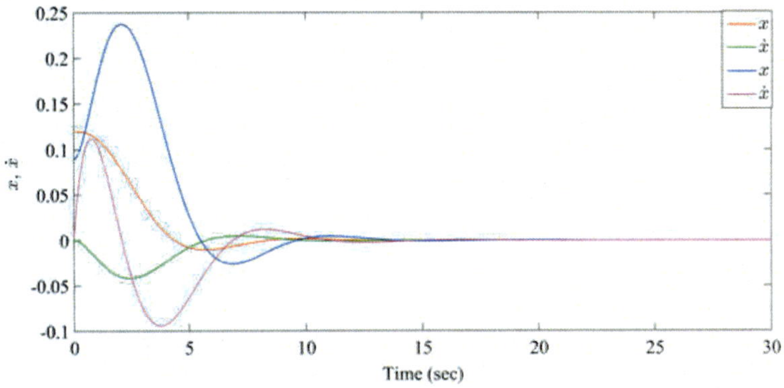

Fig. 2 Plot of x and $\dot{x}$ versus time in H_∞-based adaptive control

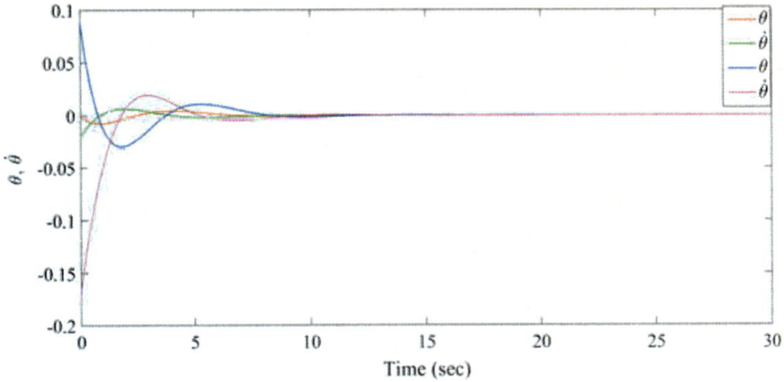

Fig. 3 Plot of θ and $\dot{\theta}$ versus time in H_∞-based adaptive control

The proposed controller is applied for the stabilization for ball and beam system. Simulation results for the ball and beam system, controlled by the proposed controller, are shown in Figs. 2, 3, 4, and 5. Figures 2 and 3 show the trajectories of the ball and beam system using the proposed controller. The corresponding control input is shown in Fig. 4. Figure 5 shows the variation of sliding surfaces s_1 and s_2 for the ball and beam system using H_∞-based adaptive control.

5 Conclusion

H_∞-based adaptive control was applied for the stabilization of underactuated non-linear systems. The effectiveness of the proposed controller is shown considering various initial conditions for stabilization of ball and beam system. The proposed controller can be applied to many other non-linear underactuated control problems.

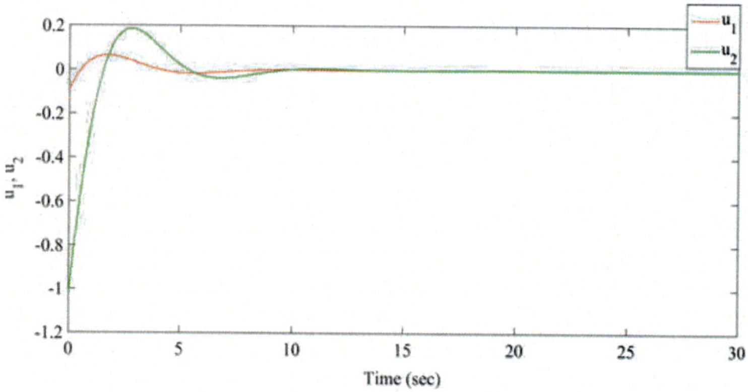

Fig. 4 Plot of u_1 and u_2 versus time in H_∞-based adaptive control

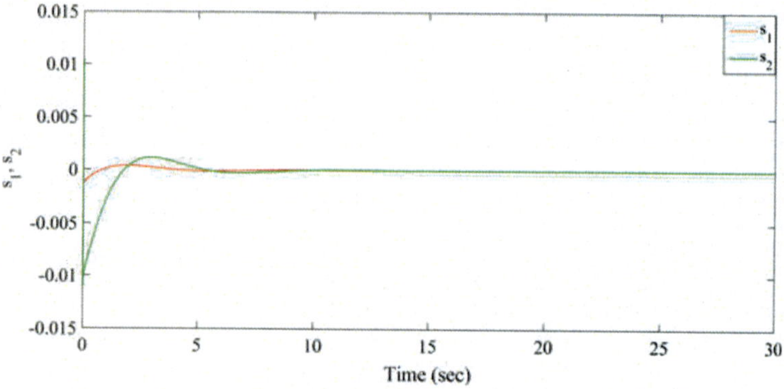

Fig. 5 Plot of s_1 and s_2 versus time in H_∞-based adaptive control

References

1. Rapp P, Sawodny O, Tarin C (2013) Stabilization of the ball and beam system by dynamic output feedback using incremental measurements. In: European control conference. Zurich, Switzerland
2. Ye H, Gui W, Yang C (2011) Novel stabilization designs for the ball-and-beam system. In: Proceedings of the 18th world congress, Italy
3. Aguilar-Ibanez C, Suarez Castanon MS, de Jesus Rubio J (2012) Stabilization of the ball on the beam system by means of the inverse Lyapunov approach. Math Prob Eng
4. Lian J, Zhao J (2019) Stabilisation of ball and beam module using relatively optimal control. Int J Mech Eng Robot Res 8(2):265–272
5. Keshmiri M, Jahromi AF, Mohebbi A, Amoozgar MH, Xie W-F (2012) Modelling and control of ball and beam system using model based and non model based control approaches. Int J Smart Sens Intell Syst 5
6. Naif B (2010) Almutairi and Mohamed Zribi: on the sliding mode control of a ball on a beam system. Nonlinear Dyn 59:221–238
7. Changa YH, Chang C-W, Tao C-W, Lin H-W, Taurd J-H (2012) Fuzzy sliding mode control for ball and beam system with fuzzy ant colony optimization. Experts Syst Appl 39:3624–3633

8. Hammadih ML, Al Hosani K, Boiko I (2016) Interpolating sliding mode observer for a ball and beam system. Int J Control 39:3624–3633
9. Tack HH, Choo YG, Kim CG, Jung MW (1999) The stabilization control of a ball-beam using self-recurrent neural networks. In: International conference on knowledge-based intelligent information engineering systems, Australia
10. Muralidharana V, Anantharamanb S, Mahindrakara AD (2010) Asymptotic stabilisation of the ball and beam system: design of energy-based control law and experimental results. Int J Control 83(6):1193–1198
11. Hauser J, Sastry S, Kokotovid P (1992) Nonlinear control via approximate input-output linearization: the ball and beam example. IEEE Trans Autom Control 31(3)
12. Yan X-G, Edwards C, Spurgeonl SK (2004) Strengthened H infinity control via state feedback: a majorization approach using algebraic Riccati inequalities. IEEE Trans Autom Control 49:824–827
13. Huang Y-J, Kuo T-C, Chang S-H (2008) Adaptive sliding mode control for nonlinear systems with uncertain parameters. IEEE Trans Syst Man Cybern 38(2):534–539

Optimal Robust Controller Design for a Reduced Model AVR System Using CDM and FOPI$^{\lambda}$D$^{\mu}$

Manjusha Silas and Surekha Bhusnur

1 Introduction

Most engineering applications provide an appropriate solution of dynamic system by converting them into accurate model with fixed physical parameters. However, in reality these parameters show some uncertainty in physical phenomena due to less accuracy in system modeling. Randomness in system parameters influences the characteristic behavior of the deterministic model. Hence, numerous probabilistic methods have been developed to incorporate this parametric uncertainty in mathematical models. Fractional calculus (FC) and coefficient diagram method (CDM) are such techniques to handle these parametric uncertainties. The concept of fractional differential-integro operator in FC has immense impact to enhance the controller output. CDM is one of the effective controller design methods that can output a system response with zero overshoot and that can settle fast within the tolerance band. Hence, in order to adjust the parameters of classical proportional–integral derivative controller the fractional calculus and coefficient diagram method are blended. The first impetus of CDM was contributed by Prof. Shunji Manabe in the year 1998. CDM is a controller design and analysis method based on algebraic approach combining the classical control design as well as modern control design methodologies [1, 2]. The characteristic polynomial of plant and controller structure is introduced in this method to avoid cancellation of poles and zeros. The main features of CDM are its lucidity and reliable parameter selection rules used in design. The concept applied in CDM constructs a target characteristic equation which fulfilled the required performance of the response. This method also includes a coefficient diagram which is a semi log diagram and is used to contrast system behavior like stability, system response speed and robustness. Many process control systems are better approximated in a non-integer order (fractional order) form as compared to the integer order

M. Silas (✉) · S. Bhusnur
Bhilai Institute of Technology, Durg, Chhattisgarh, India
e-mail: manjushamasih@gmail.com

© The Author(s), under exclusive license to Springer Nature Singapore Pte Ltd. 2023
H. Muthusamy et al. (eds.), *Robotics, Control and Computer Vision*, Lecture Notes in Electrical Engineering 1009, https://doi.org/10.1007/978-981-99-0236-1_24

form which describes its accurate features. The differential-integro operator in FC is the extended form of the full order calculus in which the integrative and differentiative orders are non-integers, commonly represented respectively as λ and μ [3]. Since the last two decades several researchers have used fractional calculus for implementing, modeling and analysis in control applications for improving system quality performances [4–7].

The significance of an AVR in power system is to restrict the voltage output of an alternator to a constant value, within a specified limit under various load conditions. By the use of AVR system, the excitation field regulates the generated emf as well as the reactive power flow [8]. Various techniques have been implemented from past few decades to design a classical PID [9, 10] and FOPID [11–13] controller for an AVR system.

Even though the ZN controller tuning gives better performance, still researchers work on new evolution techniques for the enhancement of the system performance because at the same time not all design specifications are fulfilled. At the beginning CDM involves simultaneous approach to fix the controller polynomial's type and degree and characteristic polynomials of closed loop response. In this research domain, to improve AVR system behavior, tuning of FOC is incorporated with merits of CDM control strategy.

2 Automatic Voltage Regulator

In an alternator, the rotor and the remaining part of the system is interlocked through electromechanical coupling and the assembly just behaves like a R-L-C system which oscillates around the steady state. Turbine output fluctuates in an oscillatory manner due to the occurrence of sudden transitions in loads and variation in parameters of transmission line. The most crucial measure to strengthen the power system stability is synchronous generator excitation control. Ignoring the saturation effect and other non-linearities, the mathematical modeling of the system is presented in Fig. 1. AVR system parameter range chosen for simulation is as follows (Tables 1 and 2).

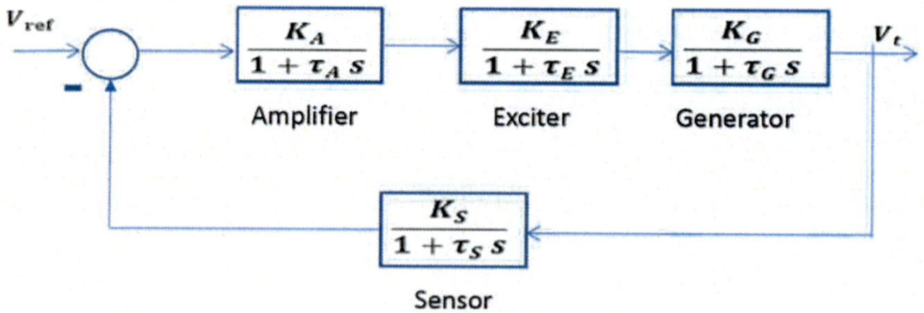

Fig. 1 Structure of AVR

Table 1 Parameter range

Component parameter	Amplifier	Exciter	Generator	Sensor
Gain	[10 400]	[1 400]	[0.7 1.0]	[1.0 2.0]
Time constant	[0.02 0.1]	[0.4 1.0]	[1.0 2.0]	[0.001 0.06]

Table 2 AVR parameters

	Gain	Time constant
Amplifier	10.0	0.10
Exciter	1.0	0.40
Generator	1.0	1.0
Sensor	1.0	0.01

The AVR parameter values considered here are mention below.

Hence, the AVR closed loop system devoid of a controller can be denoted as:

$$G_{AVR} = \frac{V_{ter}(s)}{V_{ref}(s)} = \frac{0.1s + 10}{0.0004s^4 + 0.045s^3 + 0.555s^2 + 1.51s + 11} \quad (1)$$

The closed loop response of AVR is stable but its nature is oscillatory. So further this higher degree function is changed to a lower degree using model reduction technique for easy design of the controller, system analysis and representation. Figures 2 and 3 depict the step response and the bode plot of the reduced order AVR system. (Fig. 3).

The reduced order system shows a similar response as the original AVR system and hence it can be used for modeling. Transfer function of the reduced second order AVR is as follows:

$$G_{AVR} = \frac{18.41}{s^2 + 1.147s + 20.25} \quad (2)$$

3 Overview of Coefficient Diagram Method

Classical theory and modern control theory are combined in the CDM method, which enables an efficient algebraic design and analysis of the controller [14, 15]. CDM is an effective technique for control system design, controller parameter adjustment and to observe the effect of parameter variations. Stability indices Υ_i, stability limit Υ_i^* and equivalent time constant τ are significant parameters in designing of CDM [16]. They, respectively, depict the transient behavior and the stability of the system in the time domain. Further, the robustness during parameter variations can be observed. By adapting the Lipatov's stability conditions, Manabe modified the range of stability indices. The new form is called as the Manabe Standard Form [17]. CDM design procedure is abridged as:

Initially, a mathematical model of a plant is described in polynomial form and next step is concerned with the assumption of suitable controller order and configuration in polynomial format. The desired design specifications are translated into the characteristic equation and the controller coefficients are deduced by solving the Diophantine equation. Finally, a coefficient diagram is drawn, to visualize and make inferences about the stability and robustness. Two prominent factors, equivalent time constant τ and stability indices Υ_i are chosen to compute the coefficient of CDM controller polynomials.

The standard CDM control structure is presented in Fig. 4. In plant transfer function $N_p(s)$ and $D_p(s)$ are numerator and denominator polynomials, $A_c(s)$ and $B_c(s)$ polynomials of the CDM controller to fix a desired transient response and pre-filter F(s) takes care of the steady state gain. The symbols u, d, r and y are controller signal, external disturbance signal, reference input and system output respectively.

From Fig. 4, the closed response of the system is derived as

$$y = \frac{N_p(s)F(s)}{P(s)} r + \frac{A_c(s)N_p(s)}{P(s)} d \quad (3)$$

where, the closed-loop characteristic polynomial P(s) is the Hurwitz polynomial with positive real coefficients and is given by

$$\begin{aligned} P(s) &= A_c(s)D_p(s) + B_c(s)N_p(s) \\ &= a_n s^n + a_{n-1}s^{n-1} + \text{------} + a_1 s + a_0 \end{aligned} \quad (4)$$

$$= \sum_{i=0}^{n} a_i s^i \quad (5)$$

The plant of the un-tuned design system is expressed as:

$$G(s) = \frac{N_P(s)}{D_P(s)} = \frac{a_m s^m + a_{m-1}s^{m-1} + \_\_\_\_\_ + a_1 s + a_0}{b_n s^n + b_{n-1}s^{n-1} + \_\_\_\_\_ + b_1 s + b_0} \quad (6)$$

In Eq. (6), $N_P(s)$ and $D_P(s)$ are independent of each other and their degree are related by the condition $m \leq n$.

Controller polynomials $A_c(s)$ and $B_c(s)$ are chosen as:

$$A_c(s) = \sum_{i=0}^{p} l_i s^i \text{ and } B_c(s) = \sum_{i=0}^{q} k_i s^i$$

CDM controller polynomials with coefficients l_i and k_i must satisfy the condition $p \geq q$ for practical implementation.

Design parameters of CDM, are defined as

$$\tau = \frac{a_1}{a_0} \tag{7}$$

$$\gamma_i = \frac{a_i^2}{a_{i+1}a_{i-1}}, \; i = 1, 2, \_\_\_\_\_\_\_ (n-1) \tag{8}$$

$$\gamma_i^* = \frac{1}{\gamma_{i+1}} + \frac{1}{\gamma_{i-1}}, \; i = 1, 2, \_\_\_\_\_\_ (n-1), \; \gamma_n = \gamma_0 = \infty \tag{9}$$

The system stability is determined by stability indices and stability limits; the equivalent time constant determines the speed of the time domain response. The required settling time $t_{s,}$ is resolved before the design procedure is started. The relation between the user defined settling time (t_s) and equivalent time constant (τ) is expressed as

$$\tau = \frac{t_s}{(2.5 \simeq 3)}$$

There is conflict amidst τ and the control signal magnitude. Control signal diminishes and the system becomes slow when τ increases. When the response becomes faster due to small τ, the control signal grows in size. Accordingly, the value of τ should be chosen in view of the aforesaid conflict.

4 Tuning of PID Controller Using CDM

PID controllers are one of the prominent amongst controllers designed for various industrial applications and also it is the most popular practical controllers implemented. In the above context CDM-PID controller design is proposed.

CDM-PID controller design for AVR system covers the following steps:

i. Higher order AVR is approximated in second order using model reduction technique is as follows:

$$G_p(s) = \frac{N_p(s)}{D_p(s)} = \frac{18.41}{s^2 + 1.147s + 20.25}$$

ii. The CDM-PID controller polynomials are chosen as

$$
\begin{aligned}
F(s) &= P(s)/N_p(s)_{s=0} = P(0)/N(0) = 1/K = k_0 \\
G_c(s) &= B_c(s)/A_c(s) = \left(k_2 s^2 + k_1 s + k_0\right)/l_1 s
\end{aligned}
$$

where k_2, k_1, k_0, l_1 are coefficients.

iii. The target characteristic polynomial is as follows

$$
\begin{aligned}
P_{target}(s) &= a_0\left[\left\{\sum_{i=2}^{n}\left(\prod_{i=1}^{i-1}\frac{1}{\gamma_{i-j}^{i}}\right)(\tau s)^i\right\} + \tau s + 1\right] \\
&= a_0\left[\frac{\tau^3}{\gamma_1^2\gamma_2}s^3 + \frac{\tau^2}{\gamma_1}s^2 + \tau s + 1\right]
\end{aligned} \tag{10}
$$

where γ_1 and γ_2 are stability indices and τ is the equivalent time constant

iv. Characteristic polynomial is formulated as

$$
P(s) = A_c(s)D_p(s) + B_c(s)N_p(s) \tag{11}
$$

v. By comparing the corresponding terms of (10) and (11), $l_1 = 0.01728$, $k_2 = 0.006745$, $k_1 = 0.01358$, $k_0 = 0.054$ are obtained. As per Manabe's rules τ is chosen as, $\tau = \left(\frac{t_s}{2.5}\right)$, where t_s denotes the desired settling time.

vi. By matching the coefficients of the CDM-PID controller and the traditional controller, parameters of CDM-PID are deduced as follows

$$
\begin{aligned}
C(s) &= K_c + \frac{K_c}{T_i s} + K_c T_D s \\
&\frac{K_1}{l_1} + \frac{K_0}{l_1 s} + \frac{K_2 s}{l_1} \\
K_c &= \frac{K_1}{l_1},\ T_i = \frac{K_1}{K_0},\ \text{and } T_D = \frac{K_2}{K_1}
\end{aligned} \tag{12}
$$

vii. By putting the value of l_1, k_2, k_1 and k_0 in (12), CDM-PID parameters are computed as follows:

$$
K_c = 0.7861,\ T_i = 0.2515,\ T_D = 0.4967
$$

5 Description of Fractional Calculus

This concept originated in 1695, when two scientists L' Hospital and Leibnitz communicated through letter about the concept with respect to half-order derivative, non-integer order. This mathematical concept is mapped and represent as integration and differentiation in term of non-integer order as ${}_aD_t^{\alpha}$, here operating limits are α and t.

5.1 *Preliminaries*

In the continuous domain, Integro-differential operator is presented as:

$$ {}_aD_t^{\alpha} = \begin{Bmatrix} \frac{d^{\alpha}}{dt^{\alpha}} & \alpha > 0 \\ 1 & \alpha = 0 \\ \int_a^t (dt)^{-\alpha} & \alpha < 0 \end{Bmatrix} \tag{13} $$

where, α is the differintegral operator which is either a real or a complex number. The mostly used explanation of fractional order differentiation and integration is described in the literature [18].

(i) The Grunwald–Letnikov (GL) Explanation:

$$ {}_aD_t^{\alpha}f(t) = \lim_{h\to 0}\frac{1}{h^{\alpha}}\sum_{j=0}^{\left[\frac{t-a}{h}\right]}(-1)^j\binom{\alpha}{j}f(t-jh) \tag{14} $$

where, $w_j^{\alpha} = (-1)^j\binom{\alpha}{j}$ presents the coefficients of the polynomial $(1-z)^{\alpha}$. Alternatively, recursively they can be derived from $w_0^{\alpha} = 1 w_j^{\alpha} = \left(1 - \frac{\alpha+1}{j}\right)w_{j-1}^{\alpha}$ j = 1, 2,....

(ii) The Riemann–Liouville (RL) Explanation:

$$ {}_aD_t^{-\alpha} f(t) = \frac{1}{\Gamma(\alpha)}\int_a^t (t-\tau)^{\alpha-1}f(\tau)d\tau \tag{15} $$

Here, 'a' presents initial time instance vary between 0 < α < 1. RL explanation is prominently used in FC and in fractional order differentiation if its order satisfied (n–1 < α ≤ n) and it is given as:

$$
{}_aD_t^{\alpha} f(t) = \frac{1}{\Gamma(n-\alpha)} \frac{d^n}{dt^n} \int_a^t \frac{f(\tau)}{(t-\tau)^{\alpha-n+1}} d\tau \tag{16}
$$

5.2 *Fractional Order PID Controller*

FOCs are extended kind of classical PID controllers. FOPID is used for enhancing flexibility, stability and robustness of the system. Despite the existence of uncertainties, the aim of using non-integer models is to get robust performance. In FOCs besides the nominal three parameters, two additional parameters, add to further complexity as well as flexibility in tuning the control parameters. There are abundant analytical methods and numerical techniques in [19–23] that have been trialed for optimum tuning of five parameters of FOCs. Therefore, it has five parameters that make the FOCs flexible and less sensitive towards change in parameter. Various toolboxes like NINTEGER [24], CRONE [25], FOMCON [26] aid in design of the fractional order system in which many optimization techniques have been provided within the toolbox itself. The standard mathematical output response of FOCs is presented a

$$
\begin{aligned}
C_{FOPID}(s) &= K_p + \frac{K_i}{s^{\lambda}} + K_d.s^{\mu}, 0 < (\lambda, \mu) < 2 \\
C_{FOPID}(s) &= K_p\left(1 + \frac{1}{T_i s^{\lambda}} + T_D s^{\mu}\right)
\end{aligned} \tag{17}
$$

All the conventional PID controllers can be obtained by the FOPID controller because it is a particular case of the fractional controller and its converging region in the two-dimensional plane is given as (Fig 5).

Firstly, parameters such as K_p, K_i, K_d, λ, and μ of control variables were optimized and then the fractional term of the controller was into the integer term. There are several approximation techniques which convert fractional term into integer order [27].

5.3 *Oustaloup's Approximation Algorithm*

There are many methods available for realization of FOFT into integer order in continuous domain [28–30]. In a given specified frequency band $[w_b, w_h]$, Oustaloup's recursive method is an ubiquitous approach to approximate the fractional term into an integer order.The generalized non-integer order representation of the differentiator s^{α} can be presented as:

$$G(s) = (C_0)^{\alpha} \prod_{k=-N}^{N} \frac{s + w_k'}{s + w_k} \tag{18}$$

where, $w_k' = w_b \left(\frac{w_b}{w_\mu}\right)^{\frac{k+N+\frac{1}{2}+\frac{\alpha}{2}}{2N+1}}$ and $w_k = w_b \left(\frac{w_b}{w_\mu}\right)^{\frac{k+N+\frac{1}{2}-\frac{\alpha}{2}}{2N+1}}$ are the rank k zeros and poles respectively and (2N + 1) is their total number.

6 Simulation Results

Closed loop response of unity feedback without the controller for AVR is shown in Fig. 2. Although, the Z-N method gives an enhanced response, yet research work is on to cast around to magnify the quality, performance and robustness of the controller. Further, many researchers have designed and implemented the fractional order $PI^{\lambda}D^{\mu}$ Controller for improvement in the performance of AVR [31–33]. The unit response of an AVR with the $FOPI^{\lambda}D^{\mu}$ controller is revealed in Fig. 6.

Further system performance is improved by employing CDM-PID with FOCs to develop a new CDM-$FOPI^{\lambda}D^{\mu}$ control technique for tuning. CDM-$FOPI^{\lambda}D^{\mu}$. Controller is established by entailing the CDM-PID controller parameters (K_p = 0.7861, K_i = 3.125, K_d = 0.3903) and its transfer function is given as:

$$C_{FOPID}(s) = K_p + \frac{K_i}{s^{\lambda}} + K_d.s^{\mu} \tag{19}$$

Nelder-Mead optimization is used to compute an optimum value of fractional integral and differentiation order in time domain using the FOMCON toolbox based on ITAE

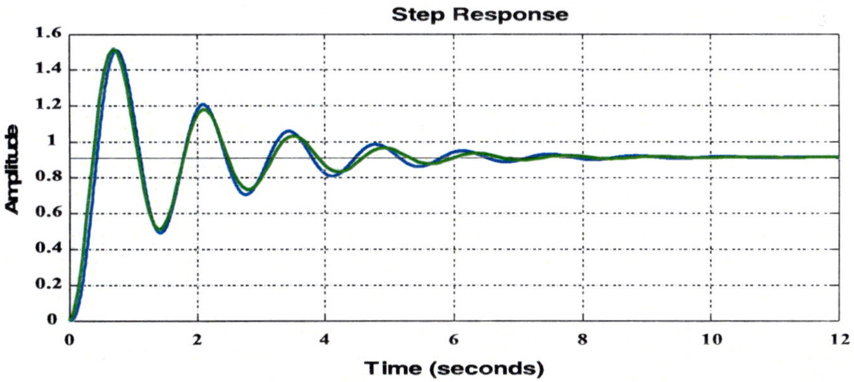

Fig. 2 Comparison of original and reduced order step response of AVR system

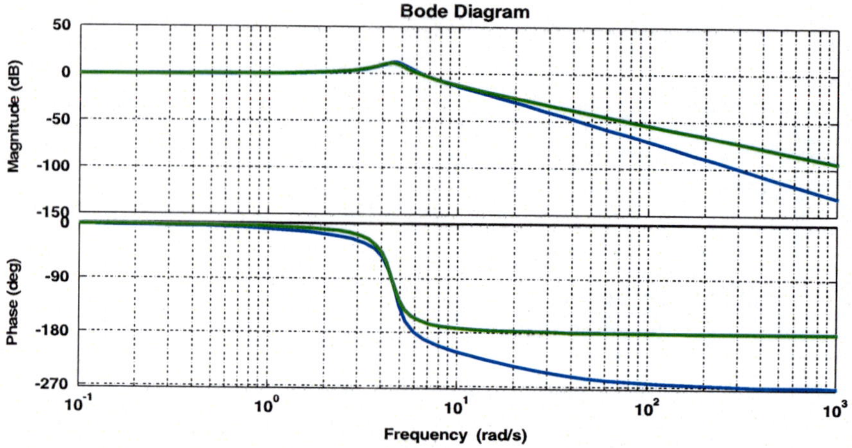

Fig. 3 Bode-plot of original and reduced order AVR

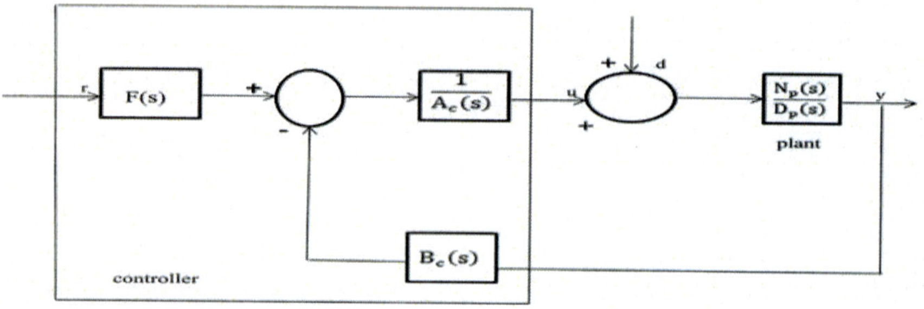

Fig. 4 Closed-loop structure of CDM

Fig. 5 Coverage of FOPID controller

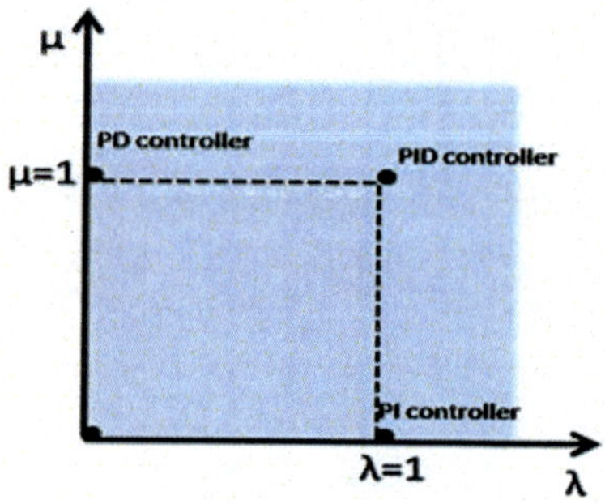

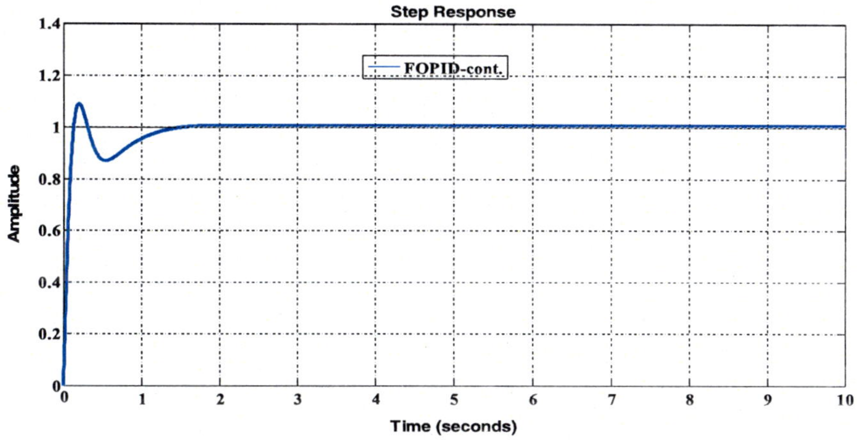

Fig. 6 AVR response with FOPID controller

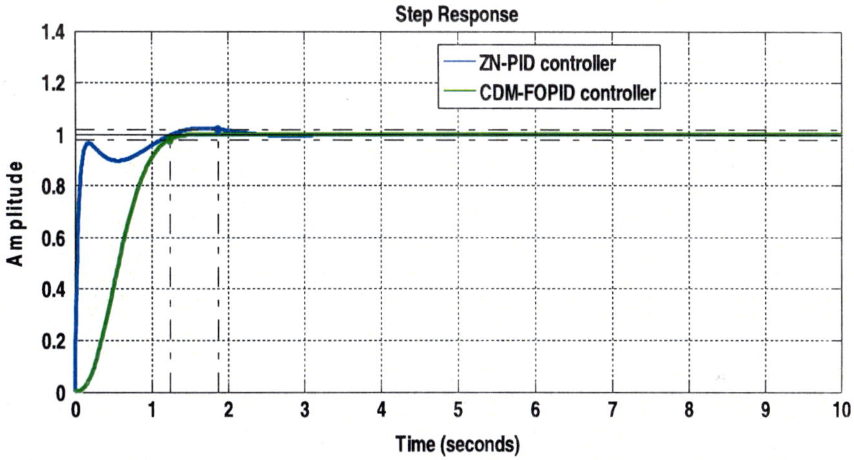

Fig. 7 AVR step response with CDM-FOPI$^{\lambda}$D$^{\mu}$ controller

criteria as $\lambda = 0.9997$ and $\mu = 0.9744$ respectively. Hence, CDM-FOPI$^{\lambda}$D$^{\mu}$ transfer function is formulated as given below and response is shown in Fig. 7 and Table 3

$$C_{CDM-FOPID}(s) = 0.7861 + \frac{3.125}{s^{0.9997}} + 0.3903.s^{0.9744} \tag{20}$$

Table 3 Comparison of performance characteristics

Parameters	Without controller	IOPID-ZN	FOPID-NM	CDM-FOPID
Kp	–	1.155	0.06	0.7861
Ki	–	2.25	18.519	3.125
Kd	–	0.1422	1.136	0.3903
Λ	–	1	0.995	0.9989
μ	–	1	0.861	0.9745
Settling time	6.9865 s	3.512 s	1.22 s	1.2446 s
Peak time	0.7522 s	0.6126 s	0.208 s	1.6278 s
Rise time	0.261 s	0.2204 s	0.0992	0.7286 s
Max overshoot	65.7%	58.87%	8.85%	0.0%
Peak amplitude	1.51	1.5887	1.14	0.9982

7 Analysis of Robustness of CDM-FOPI$^{\lambda}$D$^{\mu}$ Controller

Robustness of CDM-FOPI$^{\lambda}$D$^{\mu}$ controller is found by allowing for parametric uncertainties in the AVR system.

7.1 Effect of Amplifier Parametric Uncertainty

Considering the change in parameters of the amplifier from $K_A = 10$, $\tau_A = 0.1$to $K_A = 12$, $\tau_A = 0.005$, the terminal voltage response to step input with the CDM-FOPI$^{\lambda}$D$^{\mu}$ controller is revealed in Fig. 8.

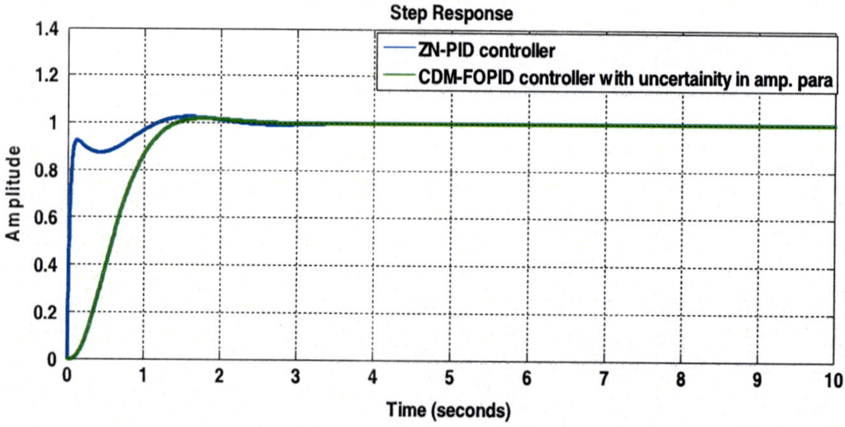

Fig. 8 AVR step response with parameter uncertainty in amplifier

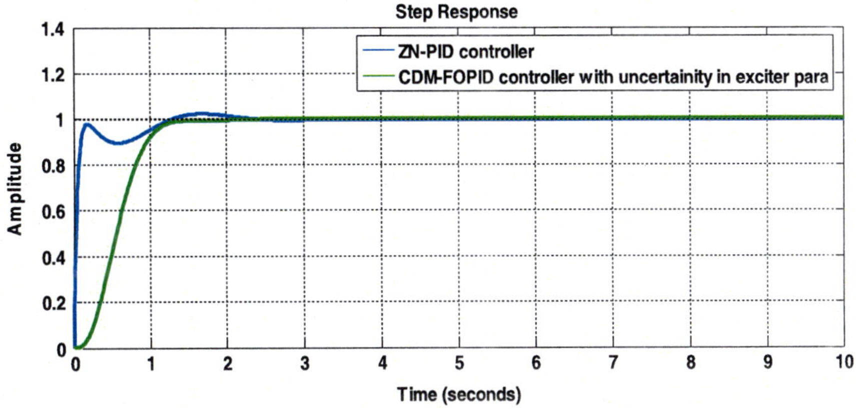

Fig. 9 AVR response with uncertainty in amplifier parameter

7.2 *Effect of Exciter Parametric Uncertainty*

Considering a change in exciter Parameter from $K_E = 1-1.2$ and $\tau_E = 0.4-0.5$. The terminal voltage to step input with the CDM-FOPI$^{\lambda}$D$^{\mu}$ controller is shown in Fig. 9.

7.3 *Effect of Generator Parametric Uncertainty*

Considering a change in the generator parameter from $K_G = 1$, $\tau_G = 1$to $K_G = 0.8$, $\tau_G = 1.4$. The terminal voltage response to step input with the CDM-FOPI$^{\lambda}$D$^{\mu}$ controller is shown in Fig. 10.

The responses show that the behavior of CDM-FOPI$^{\lambda}$D$^{\mu}$ is robust in the aura of perturbations in AVR parameters.

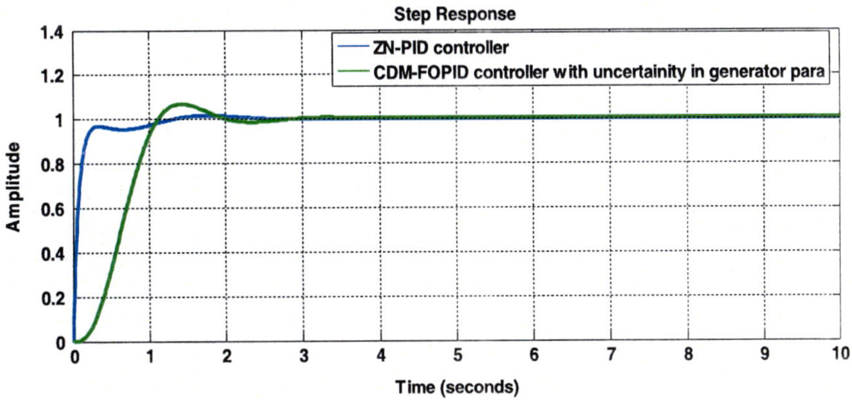

Fig. 10 AVR step response with uncertainty in generator parameter

8 Conclusion and Future Directions

According to this work a new CDM-FOPI$^{\lambda}$D$^{\mu}$ controller was designed for AVR system by blending features of CDM and fractional calculus to optimize the controller parameters. The response of the AVR with the proposed controller gives better result as compared to prevailing techniques for PID and FOPID controllers. Simulation results show effectiveness of CDM-FOPI$^{\lambda}$D$^{\mu}$ controller as contrasted to the conventional technique. The standard performance specifications are fully achieved by the CDM-FOPI$^{\lambda}$D$^{\mu}$ controller. The variation in step response in the presence of uncertainty is trivial, which confirms the robustness.

Incorporating the proposed method relative stability analysis can be investigated by comparing with other methods using Kharitonov theorem, Edge theorem etc. Although fractional order controller design is computationally complex, it provides greater flexibility and control over system performance.

References

1. Manabe,S.:Coefficient diagram method.In:IFAC Proceedings Volumes, 31(21), pp.211–222,(1998).
2. Bhusnur S (2020) An optimal robust controller design for automatic voltage regulator system using coefficient diagram method. J Inst Eng (India), Ser (B) 101(5):443–450
3. Podlubny I (1999) Fractional-order systems and PIλDμ controllers. IEEE Trans Autom Control 44(1):208–214
4. Monje CA, Vinagre BM, Feliu V, Chen Y (2008) Tuning and auto-tuning of fractional order controllers for industry applications. Control Eng Pract 16(7):798–812
5. Padula F, Visioli A (2010) Tuning rules for optimal PID and fractional-order PID controllers. J Process Control 21(7):69–81
6. Shah P, Agashe S (2016) Review of fractional PID controller. Mechatronics 38:29–41
7. Silas M, Bhusnur S (2021) Augmenting DC buck converter dynamic response using an optimally designed fractional order PI controller. Design Eng: 4836–4849
8. Saadat H (1999) Power system analysis. McGraw-Hill, New-York
9. Gaing ZL (2004) A particle swarm optimization approach for optimum design of PID controller in AVR system. IEEE Trans Energy Convers 19(2):384–391
10. Amer ML, Hassan HH, Youssef HM (2008) Modified evolutionary particle swarm optimization for AVR-PID tuning. In: Communications and information technology, systems and signals. pp 164–173
11. Pan I, Das S (2012) Chaotic multi-objective optimization based design of fractional order PIλDμ controller in AVR system. Int J Electr Power Energy Syst 43(1):393–407
12. Verma SK, Yadav S, Nagar SK (2017) Optimization of fractional order PID controller using grey wolf optimizer. J Control Autom Electr Syst 28(3): 318–322
13. Zamani M, Karimi-Ghartemani M, Sadat N, Parniani M (2009) Design of a fractional order PID controller for an AVR using particle swarm optimization. Control Eng Pract 17(12): 1380–1387
14. Manabe, S (2002) Brief tutorial and survey of coefficient diagram method. In: 4th Asian control conference. pp 25–27
15. Kim YC, Manabe S (2001) Introduction to coefficient diagram method. In: IFAC Proceedings vol 34, no 13. pp 147–152
16. Bhusnur S (2015) Effect of stability indices on robustness and system response in coefficient diagram method. Int J Res Eng Technology 4(10):282–287

17. Manabe S (1999) Sufficient condition for stability and instability by Lipatov and its application to the coefficient diagram method. In: 9-th Workshop on Astrodynamics and Flight Mechanics, Sagamihara, ISAS, pp 440–449
18. Monje CA, Chen Y,Vinagre BM, Xue D, Feliu-Batlle V (2010) Fractional-order systems and controls fundamentals and applications. Springer Science & Business Media
19. Chen Y, Petras I, Xue D (2009) Fractional order control-a tutorial. In: American control conference, 2009. ACC'09. IEEE, pp 1397–411
20. Valerio D, Costa JS.da (2010) A review of tuning methods for fractional PIDs. In: 4th IFAC Workshop on fractional differentiation and its applications, FDA, vol 10
21. Yeroglu C, Tan N (2011) Note on fractional-order proportional–integral–differential controller design. IET Control Theory Appl 5(17):1978–1989
22. Xue D, Zhao C, Chen YQ (2006) Fractional order PID control of a DC-motor with elastic shaft: a case study. In: American control conference. pp 3182–3187
23. Monje,C.A. et al.: Proposals for fractional P I λD μ tuning. In: Proceedings of The First IFAC Symposium on Fractional Differentiation and its Applications (FDA04)., vol. 38, pp. 369–381,(2004).
24. Valério D, Costa J.Sá da (2004) Ninteger, a non-integer control toolbox for MatLab. In: Proc First IFAC Work Fract Differ Appl Bordeaux. pp 208–213
25. Oustaloup A, Melchior P, Lanusse P, Cois O, Dancla F (2000) The CRONE toolbox for Matlab. In: CACSD. Conference Proceedings. IEEE International symposium on Computer-Aided Control System Design (Cat.No.00TH8537). pp 190–195
26. Tepljakov A, Petlenkov E, Belikov J (2011) FOMCON: Fractional-order modeling and control toolbox for MATLAB. In: Mixed Design of Integrated Circuits and Systems (MIXDES), 2011 Proceedings of the 18th International Conference IEEE. pp 684–689
27. Vinagre BM, Podlubny I, Hernandez A, Feliu V (2000) Some approximations of fractional order operators used in control theroy and applications. Fract Calc Appl Anal 3(3):231–248
28. Maione G (2008) Continued fractions approximation of the impulse response of fractional-order dynamic systems. IET Control Theory Appl 2(7):564–572
29. Xue,D., Zhao,C.,Chen,Y.Q.:A modified approximation method of fractional order system.In: Proc. 2006 IEEE Int. Conf.Mechatron. Autom., pp. 1043–1048 ,Jun(2006).
30. Khanra,M., Pal,J.,Biswasl,K.:Rational approximation and analog realization of fractional order transfer function with multiple fractional powered terms. Asian J. Control, vol. 15, no. 4, (2013).
31. Verma SK, Nagar SK (2018) Design and optimization of fractional order PIλDμ controller using grey wolf optimizer for automatic voltage regulator system. Recent Advances in Electrical & Electronics Engineering (Formerly Recent Patents on Electrical & Electronics Engineering), vol. 11, no. 2. pp. 217–226
32. Tang Y, Cui M, Hua C, Li L, Yang YY (2012) Optimum design of fractional order PIλDμ controller for AVR system using chaotic ant swarm. Expert Syst Appl 39(8):6887–6896
33. Majid Zamani NS, Karimi-Ghartemani M (2007) Fopid controller design for robust performance using practicle swarm Optimization. Fract Calc Appl Anal An Int J Theory Appl 10(2):169–187

Neural Network Based DSTATCOM Control for Power Quality Enhancement

Islavatu Srikanth and Pradeep Kumar

1 Introduction

The primary goal of the power distribution network is to provide hormonic-free electricity to end users and utilities. Both reactive power compensation and harmonics are controlled by DSTATCOM in both balanced and unbalanced load circumstances [1]. The use of solid-state controllers, unplanned expansion of distribution network lead to the PQ problems in AC distribution network. High reactive power burden, harmonic currents, load imbalance, and excessive neutral current are some of the issues. As per the IEEE-519 standard, the power quality is regulated at point of common coupling (PCC) [2]. In the distribution system, a group of devices namely the custom power devices like DSTATCOM, connected across the load can suppress the current based problems [14]. A dynamic voltage restorer (DVR) linked in series with the load can suppress voltage problems. Unified power quality conditioner having both DVR and DSTATCOM connected in a grid is used to solve both voltage and current based power quality problems. At the distribution level, nonlinear load currents cause nonlinearity in supply currents, which DSTATCOM can reduce at the PCC. By employing a suitable control technique, DSTATCOM-performance was improved in terms of computation time, dependability, and simplicity [2, 3].

Under varying load conditions, the PI controller will not perform in an optimum manner because its structure is fixed and simple. Hence, the advent controllers like, fuzzy logic, artificial network and genetic algorithms are developed [4]. In the past, the 3-ϕ 3-Wire arrangement was used by PI controllers. Many controllers, including instantaneous reactive power theory (IRPT), synchronous reference theory (SRFT), power balance theory, synchronous detection (SD) theorem, and NN controllers, can

I. Srikanth (✉) · P. Kumar
EEED, NIT, Ravangla, Sikkim, India
e-mail: phee210004@nitsikkim.ac.in

P. Kumar
e-mail: pradeep@nitsikkim.ac.in

© The Author(s), under exclusive license to Springer Nature Singapore Pte Ltd. 2023
H. Muthusamy et al. (eds.), *Robotics, Control and Computer Vision*, Lecture Notes in Electrical Engineering 1009, https://doi.org/10.1007/978-981-99-0236-1_25

subsequently be used to generate DSTATCOM reference currents. A SRFT based controller is widely used to generate reference currents generation for the three phase system. SRF controller deals with dc quantities, and so easy to implement [5]. A simple and realistic control technique has been presented to correct for current-based power quality issues [6]. To produce switching pulses for VSC, with the help of neural network we can track the reference source currents and compared with sensed currents [9–13]. This article presents ADALINE algorithm with the least mean square method (LMS), which was applied to calculate the reference current components for a DSTATCOM type compensator [7, 8, 10]. The main contribution of this work is to develop the ADALINE LMS control based DSTATCOM system with nonlinear load and to achieve reduced harmonic content in the source current.

2 DSTATCOM Topology

The proposed system configuration is a 3-ϕ 3-Wire system connected directly to a nonlinear load. The nonlinear load is used as the uncontrolled rectifier with Resistive-Inductive load. A 3-leg voltage source converter (VSI) is linked through the PCC over the interfacing inductor (L_r). The voltage source converter includes one DC capacitor and IGBT switches. IGBT is a high-speed switching device and it does not require the commutation. Gate pulses to IGBT of VSI are produced by using the neural network algorithm. ADALINE based LMS control algorithm was used as the control strategy for DSTATCOM as shown in Fig. 1 [12].

3 NN Based Control Strategy

Estimation of reference supply currents with use of unit vectors through Adaline NN-based control technique is discussed here. In each phase, the fundamental active load current component is extracted i.e., reference source current. The neural network LMS-Adaline based extraction algorithm uses the PCC voltages and load current. Weights are obtained from each phase in this technique, i.e., W_p, W_q. Figures 2 and 3 demonstrate the control algorithm for computing active and reactive weight components. Using the LMS algorithm, the weights are derived from the load currents and unit vectors, and the loss dc component is added to provide reference currents for each phase.

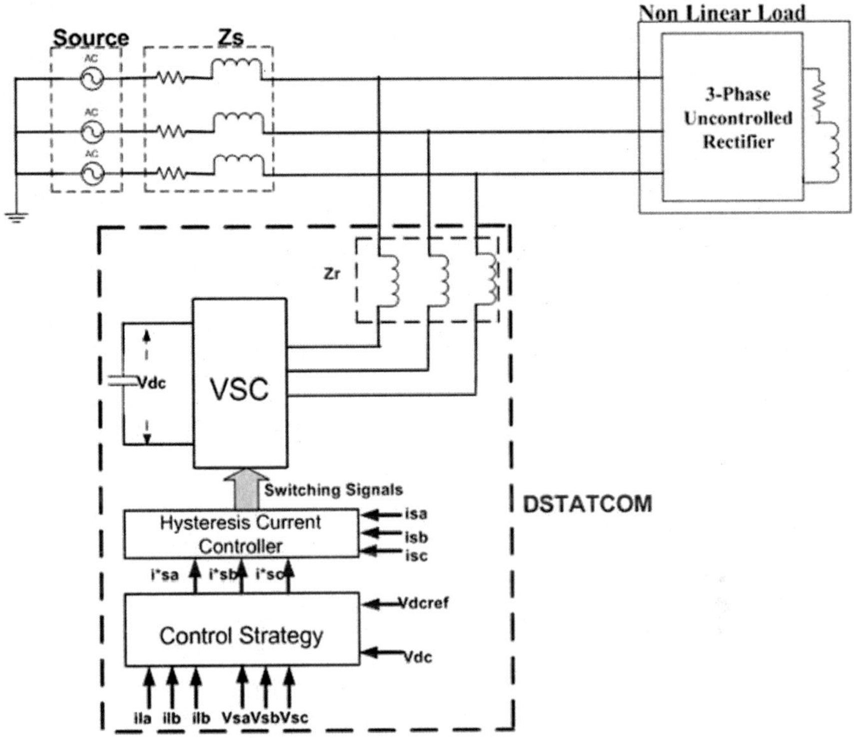

Fig. 1 3-ϕ 3-Wire distribution STATCOM

4 Calculation of Active Component Currents

The sensed 3-ϕ PCC voltages are filtered and the amplitude is given by

$$\mathrm{v}_t = \left\{(2/3)(\mathrm{v}_{sa}^2 + \mathrm{v}_{sb}^2 + \mathrm{v}_{sc}^2)\right\}^{1/2} \tag{1}$$

The in-phase unit vectors as

$$u*_a = \frac{\mathrm{v_{sa}}}{\mathrm{v_t}},\ u*_b = \frac{\mathrm{v_{sb}}}{\mathrm{v_t}},\ u*_c = \frac{\mathrm{v_{sc}}}{\mathrm{v_t}} \tag{2}$$

At the ith sample interval, the error signal produced is

$$V_{DCe}(i) = \mathrm{V}_{DCeref}(i) = \mathrm{V}_{DCeref}(i) - \mathrm{V}_{DC}(i) \tag{3}$$

The PI controller output at the ith sampling interval is

$$\omega_L(i) = \omega_L(i-1) + k_{pd}\{\mathrm{V}_{DCe}(i) - \mathrm{V}_{DCe}(i-1)\} + k_{id}\mathrm{V}_{DCe}(i) \tag{4}$$

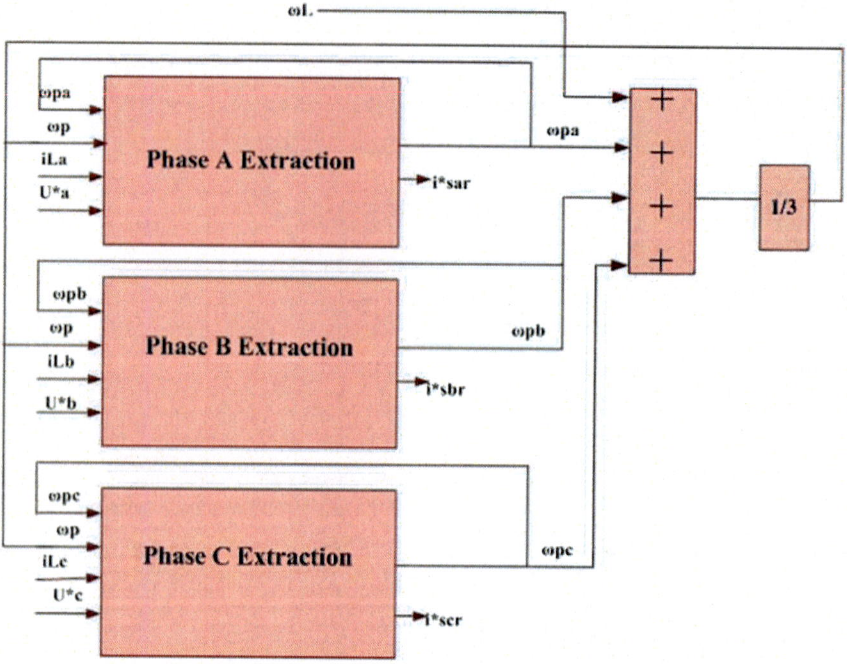

Fig. 2 Extraction of real components in a 3-ϕ system using adaline

where $\omega_L(n)$ is the active components of supply currents and k_{pd} are proportional and k_{id} integral gain constants.

The active component of the supply currents' mean weight is

$$\omega_L(i) = \{\omega_L(i) + \omega_{pa}(i) + \omega_{pb}(i) + \omega_{pc}(i)\}/3 \tag{5}$$

The extraction of weights of the basic d-axis components of the load currents may be done using the least mean square (LMS) technique, and weights can be trained using the Adaline neural network algorithm. The weights of 3- ϕ load currents' d-axis components are assessed as follows:

$$\omega_{pa}(i) = [\omega_{pa}(i-1) + \eta\{i_{La}(i) - \omega_{pa}(i-1)^* u_{pa}(i)\}^* u_a^*(i)] \tag{6}$$

$$\omega_{pb}(i) = [\omega_{pb}(i-1) + \eta\{i_{Lb}(i) - \omega_{pb}(i-1)^* u_{pb}(i)\}^* u_b^*(i)] \tag{7}$$

$$\omega_{pc}(i) = [\omega_{pc}(i-1) + \eta\{i_{Lc}(i) - \omega_{pc}(i-1)^* u_c^*(i)\}^* u_c^*(i)] \tag{8}$$

where η is the convergence factor and the value of η diverges from 0.01 to 1. The 3-ϕ active components of load currents of the weights were extracted using Adaline in

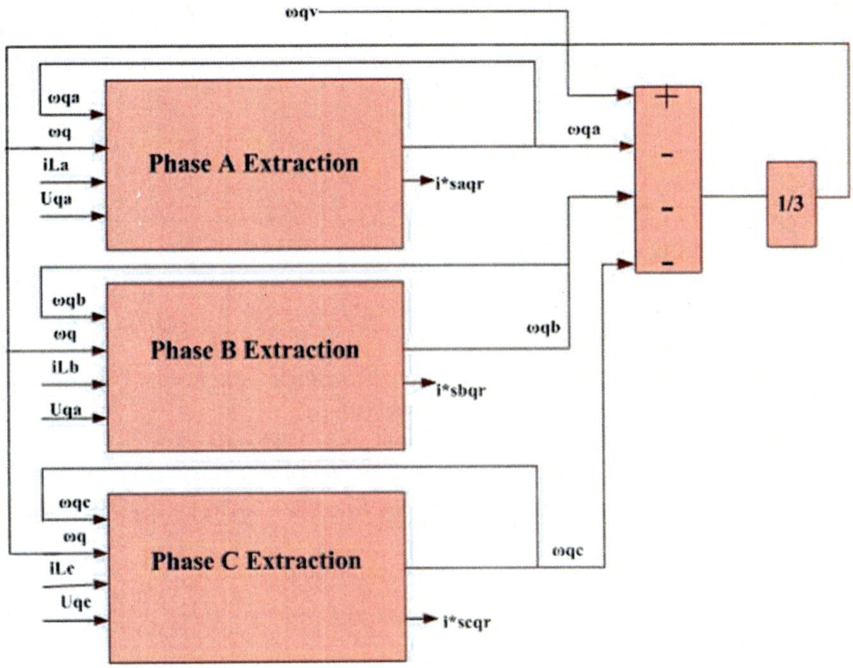

Fig. 3 Extraction of reactive components in a 3-ϕ system using adaline

each phase. The fundamental 3-ϕ reference active components of the supply currents are computed as

$$i^*_{sapr} = \omega_p u*_a,\ i^*_{sbp}r = \omega_p u*_b,\ i^*_{scpr} = \omega_p u*_c \tag{9}$$

4.1 Calculation of Reactive Power Components

The unit vectors of quadrature are obtained using phase unit vectors as

$$u_{qa} = \frac{(-u*_b + u*_c)}{\sqrt{3}},\ u_{qb} = \left(\frac{\sqrt{3}}{2}\right) * (u*_a) + \frac{(u*_b - u*_c)}{\left(2\sqrt{3}\right)},$$
$$u_{qc} = \left(-\frac{\sqrt{3}}{2}\right) * u*_a + \frac{(u*_b - u*_c)}{\left(2\sqrt{3}\right)} \tag{10}$$

The measured PCC voltages and the PCC voltage reference value are sent into the AC PI Controller as the terminal voltage. At the ith sample instant, the AC voltage error is

$$V_{te}(i) = V_{tr}(i) - V_t(i) \tag{11}$$

The output of PCC voltage from the AC Voltage PI Controller at the ith sampling instant.

$$\omega_{qv}(i) = \omega_{qv}(i-1) + k_{pa}\{\mathrm{V}_{te}(i) - \mathrm{V}_{te}(i-1)\} + k_{ia}\mathrm{V}_{te}(i) \tag{12}$$

where $\omega_{qv}(i)$ is the d-axis component of the supply currents and k_{pa} is the proportional gain, k_{ia} are the integral gain constants.

The 3-ϕ weights of the reactive components of the load currents are computed as

$$\omega_{qa}(i) = [\omega_{qa}(i-1) + \eta\{i_{La}(i) - \omega_{qa}(i-1) * u_{qa}(i)\} * u_{qa}(i)] \tag{13}$$

$$\omega_{qb}(i) = [\omega_{qb}(i-1) + \eta\{i_{Lb}(i) - \omega_{qb}(i-1) * u_{qb}(i)\} * u_{qb}(i)] \tag{14}$$

$$\omega_{qc}(i) = [\omega_{qc}(i-1) + \eta\{i_{Lc}(i) - \omega_{qc}(i-1) * u_{qc}(i)\} * u_{qc}(i)] \tag{15}$$

The reactive component average weight of the supply currents is given as

$$\omega_q(i) = [\omega_{qv}(i) - \{\omega_{qa}(i) + \omega_{qb}(i) + \omega_{qc}(i)\}]/3 \tag{16}$$

The reactive components of the source currents in the 3-φ system is given as

$$i^*_{saqr} = \omega_q u_{qa},\ i^*_{sbqr} = \omega_q u_{qb},\ i^*_{scqr} = \omega_q u_{qc} \tag{17}$$

The sum of the active and reactive power components is used to calculate the total reference supply currents.

$$i^*_{sa} = i^*_{sapr} + i^*_{saqr},\ i^*_{sb} = i^*_{sbpr} + i^*_{sbqr},\ i^*_{sc} = i^*_{scpr} + i^*_{scqr} \tag{18}$$

The sensed feedback currents are related with the assessed reference supply to generate the error signal. The output error signal is given to the IGBT of the VSC through the hysteresis current controller.

5 Simulink Based Outcomes

The characteristics of the 3-ϕ system, when the DSTATCOM is in the operating mode and not in the operating mode are discussed. The simulation results were validated through the MATLAB/Simulink software.

Case.1: Performance of the 3-ϕ system not connected to the DSTATCOM.

Because of the nonlinear load, i.e., an unregulated rectifier with *R-L* load, the supply current waveform of a 3-ϕ system is non-sinusoidal. The DSTATCOM injected current is also zero, and the DC-link voltage constant is to be $V_{dcref} = 700$ V. The load current (i_{Labc}) exhibits a non-sinusoidal waveform due to the connected 3phase uncontrolled rectifier as shown in Fig. 4.

Case.2: Performance of the 3-ϕ system connected to DSTATCOM.

The 3-ϕ supply currents are sinusoidal in nature, as seen by the waveform in Fig. 5. DSATACOM injects currents (i_{DST}) in to PCC and the DC link Voltage (v_{DC}) is constant throughout the simulation period.

From Figs. 6 and 7, it has been observed that the THD percentage of the supply current without DSTATCOM is 26.66%, and the THD percentage with DSTATCOM

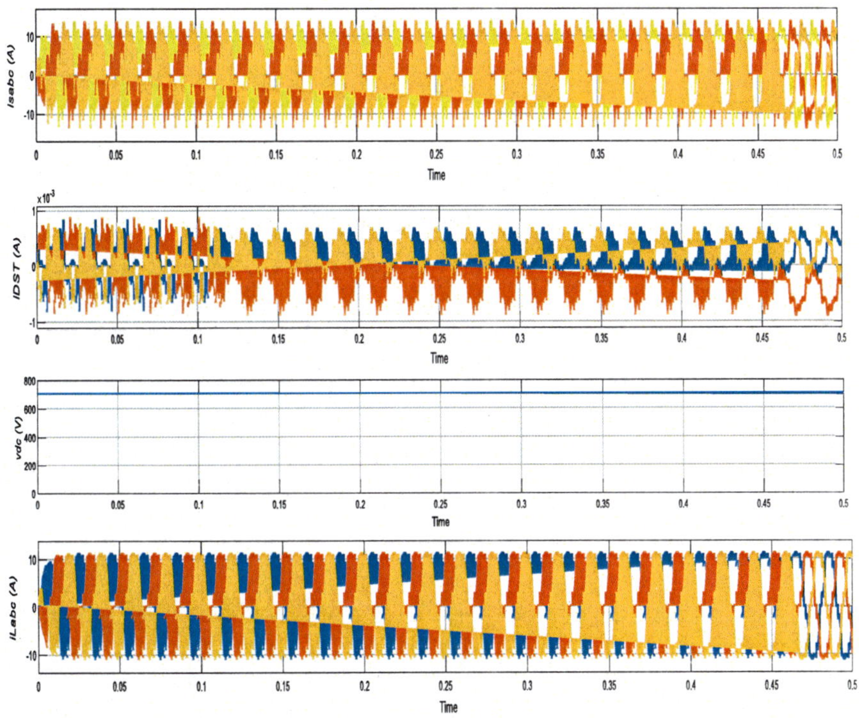

Fig. 4 Simulation wave forms without DSTATCOM under non linear load

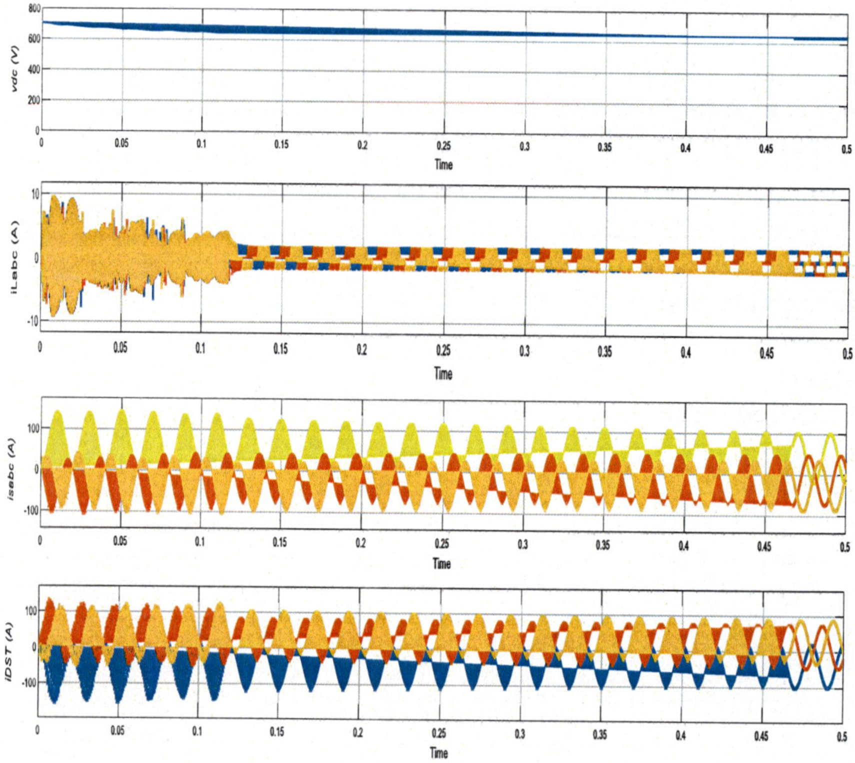

Fig. 5 Simulation wave forms with DSTATCOM under non-linear load

is 1.20%. The results reveal that the neural network control method performs well when it comes to removing harmonic distortion. According to the IEEE-519 standard, it should be less than 5%, which is attained by the neural network control.

Fig. 6 Percentage THD of supply current without DSTATCOM

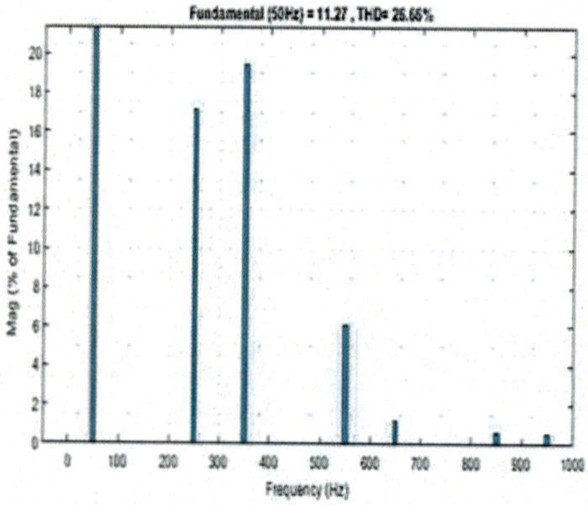

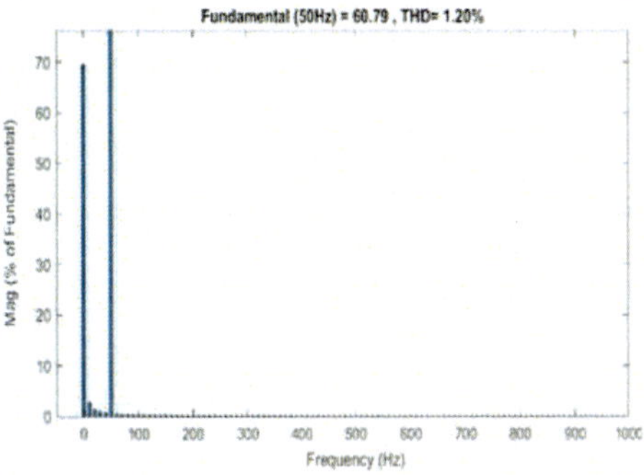

Fig. 7 Percentage THD of supply current with DSTATCOM

6 Conclusion

This paper mainly elaborates the ADALINE neural network-based LMS algorithm for DSTATCOM. The DC-link voltage is kept constant throughout the simulation, making the system more stable and without harmonics. The addition of DSTATCOM using the neural network control algorithm compensates the harmonics in the supply currents. Its performance improved under a nonlinear load condition. Also, the simulation results indicate that the source current limitation complies with the IEEE-519 THD standard.

Appendix

System Parameters for Simulation Studies:

Grid Parameters: Source Voltage: 415 V, 50 Hz, Source Inductance: 15 mH, *Load Parameters*: 3-phase rectifier with RL load $R_L = 50\ \Omega$, $L_L = 150$mH,

VSC Parameter: $V_{dcref} = 700$ V, $C_{dc} = 13000\ \mu F$, $L_c = 4$mH.

References

1. Singh B, Jayaprakash P, Kumar S, Kothari DP (2011) Implementation of neural-network-controlled three-leg VSC and a transformer as three-phase four-wire DSTATCOM. IEEE Transct. on Indust. Applns 47(4):1892–1901. https://doi.org/10.1109/TIA.2011.2153811
2. Ahmad MT, Kumar N, Singh B (2017) Generalized neural network-based control algorithm for DSTATCOM in distribution systems. IET Pow Electr 10. pp 1529–1538. https://doi.org/10.1049/iet-pel.2016.0680
3. Jayachandran J, Sachithanandam RM (2016) ANN based controller for three phase four leg shunt active filter for power quality improvement. Ain Shams Eng J 7(1). pp 275–292
4. Mittal C, Srivastava S (2020) Comparison of ANN and ANFIS controller based hysteresis current control scheme of DSTATCOM for fault analysis to improve power quality. International

Conference on Electronics and Sustainable Communication Systems (ICESC) 2020:149–156. https://doi.org/10.1109/ICESC48915.2020.9155619
5. Balasubramanian M, Selvam P, Gopinath S, Anna baby, Sreehari S, Jenopaul P (2021) Novel LMS-neural network based DSTATCOM for improving power quality. Ann Rom Soc Cell Biol:13524–13535. Retrieved from https://www.annalsofrscb.ro/index.php/journal/article/view/4368
6. Mangaraj M, Panda AK, Penthia T (2015) Neural network control technique-based sensor less DSTATCOM for the power conditioning. In: 2015 Annual IEEE India Conference (INDICON), pp 1–6. https://doi.org/10.1109/INDICON.2015.7443184
7. Mangaraj M, KumarPanda A (2018) DSTATCOM deploying CGBP based icos ϕ neural network technique for power conditioning. ASE J 9(4). pp 1535-1546. https://doi.org/10.1016/j.asej.2016.11.009
8. Jayachandran J, Murali Sachithanandam R (2015) Neural network-based control algorithm for DSTATCOM under nonideal source voltage and varying load conditions. Canadi Journ Elec Comp Engg 38 (4):307–317. https://doi.org/10.1109/CJECE.2015.2464109
9. Ahmad M, Kirmani S (2021) Simulation and analysis of a grid integrated distribution system based on LMS algorithm for hybrid types of loads. Int J Syst Assur Eng Manag. https://doi.org/10.1007/s13198-021-01392-5
10. Kumar A, Kumar P (2021) Power quality improvement for grid-connected PV system based on distribution static compensator with fuzzy logic controller and UVT/ADALINE-based least mean Square controller. J Mod Power Syst Clean Energy 9(6):1289–1299. https://doi.org/10.35833/MPCE.2021.000285
11. Singh B, Arya SR (2014) Back-propagation control algorithm for power quality improvement using DSTATCOM. IEEE Trac on Ind Elecx 61(3):1204–1212. https://doi.org/10.1109/TIE.2013.2258303
12. Mangaraj M, Panda AK, Penthia T (2016) Investigating the performance of DSTATCOM using ADALINE based LMS algorithm. In: 2016 IEEE 6th International Conference on Power System (ICPS). pp 1-5. https://doi.org/10.1109/ICPES.2016.7584062
13. Jyothi KRS, Kumar PV, Kumar J (2021) A review of different configurations and control techniques for DSTATCOM in the distribution system. In: J.E3S Web of conferences; Les Ulis, vol. 309. https://doi.org/10.1051/e3sconf/202130901119

An Extensive Critique on FACTS Controllers and Its Utilization in Micro Grid and Smart Grid Power Systems

D. Sarathkumar, Albert Alexander Stonier, and M. Srinivasan

1 Introduction

FACTS devices are extensively used for the effective power utilization, demand management, stabilization of voltage, improvement of power quality, mitigation of harmonic and power factor improvement [1, 2]. The additional benefits of these controllers include compensation of reactive power, control of power flow, voltage regulation, enhancement of steady state and transient stability, minimization of power losses, and conditioning of power systems [3, 4]. Emerging trends in nonconventional and distributed energy sources stimulated FACTS devices to play a critical role to maintain the effective energy usage, improvement of reliability and security of the power grid [1].

The advantages of this controller are utilized in standalone microgrids for the purpose of effective usage of distributed power sources to deliver power intended for the remote locations [2]. With the help of power electronic converters, performance of the system is collectively improved. The expected outcomes are in enhancement of quality of power in the point of common coupling.

The utilities, domestic, industrial and commercial customers face a very big challenge to mitigate the various power quality indices existing in the system [3]. Several FACTS controllers and its control methodologies can support to overcome the power quality issues. To utilize the power sources in a much effective and secure manner,

D. Sarathkumar (✉) · M. Srinivasan
Electrical and Electronics Engineering, Kongu Engineering College, Erode, Tamilnadu 638 060, India
e-mail: sarath.eee@kongu.ac.in

A. A. Stonier
School of Electrical Engineering, Vellore Institute of Technology, Vellore, Tamilnadu 632 014, India

© The Author(s), under exclusive license to Springer Nature Singapore Pte Ltd. 2023

H. Muthusamy et al. (eds.), *Robotics, Control and Computer Vision*, Lecture Notes in Electrical Engineering 1009, https://doi.org/10.1007/978-981-99-0236-1_26

FACTS devices begun its debut incorporate in the power system during 1970s. Fundamental operation of these components depends on various control methodologies to control reactive as well as real flow of power [4].

The recent research concentrates on the architectures and control strategies of power electronic converters to enhance the overall efficiency of controllers in power electrical networks and also subsequently improve the security of the power system [5, 6]. Currently, FACTS controllers and smart control approaches became a most dominant device in power generation through distributed power sources such as solar photo voltaic, wind farm as well as fuel cell [6]. More number of researchers concentrated on maximum power extraction from renewable energy sources. The effective usage of these controllers for micro-grid and smart grid integrated with the non-conventional system paved a new avenue for the overall performance improvement [7, 8]. The major objective of the article is to survey the advantages of FACTS controllers for micro-grid and smart grid which was integrated with renewable energy sources.

The paper comprises of six sections. In Sect. 2, the basic concept of power quality in power system networks was explained. The overview of transmission side FACTS controller and its role was presented in Sect. 3. Section 4 deals with the distribution side FACTS controllers and its task was elaborated. Section 5 postulates the role of FACTS controllers in the micro grid and smart grid environments. In Sect. 6, the conclusion and future focus of FACTS controllers in the micro grid and smart grid was explained.

2 Basic Concepts of Flexible Alternating Current Transmission System and Power Quality

Power quality issues results voltage or current distortions in the electrical systems or deviations of frequency causing the faults or abnormal operations of consumer components. Moreover, the electrical energy is provided to the customers to be safe, secure and also continuous with pure sine waveform with constant frequency and magnitude which need to be ensured at all levels.

Commonly, power quality issues leads to maximization of power losses, maloperation of apparatus which are interconnected with adjacent power networks too. The more utilization of power electronic devices results in the minimization of current and harmonics and also maximizes the reactive and real power [9]. Nowadays, the improvement of power quality is a very difficult phenomenon and creates the serious tasks in various levels of electrical networks. Hence, this problem creates more effects in the electrical networks. So, power quality problems are getting increased attention and awareness amongst customers and power companies [10]. Sustaining quality of power in the permissible range was the major challenging task. Major issues in poor power quality are clearly explained in the paper. [11].

Table 1 Power quality occurrence and its effects

Problem	Causes	Effects
Harmonics	Electromagnetic intcrfcrcncc from appliances, machines, radio and TV broadcasts	Continuous distortion of normal voltage, Random data errors
Voltage sags/swells	Major equipment startup or shutdown. Sort circuitsf faults), Undersized electrical wiring, Temporary voltage rise or drop	Memory loss. Data errors, Dim or bright lights. Shrinking display screens. Equipment shutdown
Interruption	Switching Operator, Attempting to isolate electrical problem and maintain power to power distribution area	Equipment trips off. Programming is lost, Disk drive crashes
Flicker	Arc furnace. Voltage fluctuations on utility transmission and distribution systems	Visual irritation, introduction of many harmonic components in the supply power and their associated equipment
Transients	Lightning, Turning major equipment on or off, Utility switching	Tripping, Processing errors, data loss. Burned circuit boards

Table 1 explains the continuous effects, origin and description of power quality indices and its occurrence in an electrical network. It is noted that the occurrence of swells in voltages have the largest level which is approximately 35% and the minimum level of occurrence is transients in voltage which is nearly 8%. More usage of critical loads, create the harmonics and non-sinusoidal voltages of around 20% and 18% consequently. From the 30 years of Scopus database, 3264 papers was published in FACTSs controllers from the year 1987–2017.

3 Facts Controllers

FACTS controllers with the combination of power electronic circuits and high speed operation control methods are used in recent micro grids comprising of alternating current to direct current distributed power sources. It depends on the following fundamental strategies:

(1) Reactance was connected at PCC
(2) Supplying the alternating current systems in any one combination with the power network junctions
(3) Injecting total power and reactive current in point of real energy flow operation.

The operation tools depend upon current, power, phase angle or real current flow operation, applying PID tuning, optimum regulation, analytical optimization operational methodologies, heuristic-optimization control execution index.

The converter strategies are categorized as:

(a) Current supply-rectifier interface
(b) DC-power supply fed converters
(c) Dynamic capacitors or inductors
(d) Passive filter strategies

Output current and voltage which have interferences also induce harmonics and based on its dynamic behaviour for converter topologies, extra filters were generally needed. In recent years, the rising requirement of power in several countries, power utilities are in the position to construct additional power lines, power towers and raising the rating in power lines are highly improved.

Building the additional power lines also needs a large capital cost and choosing the optimal results to minimize the costs for power utilities is a great argument. The primary intention of FACTS controllers is to enhance the stable transmission rating of power lines and to regulate the energy flow in the planned transmission paths [12]. FACTS controllers are also applied to enhance the quality of power. The several categories of FACTS controllers are: SVC, TCSC, STATCOM, SSSC, UPFC, IPFC etc., are available based on the controlling methods, connection and technology improvement. Figure 1 shows the FACTS controllers on transmission and distribution environment.

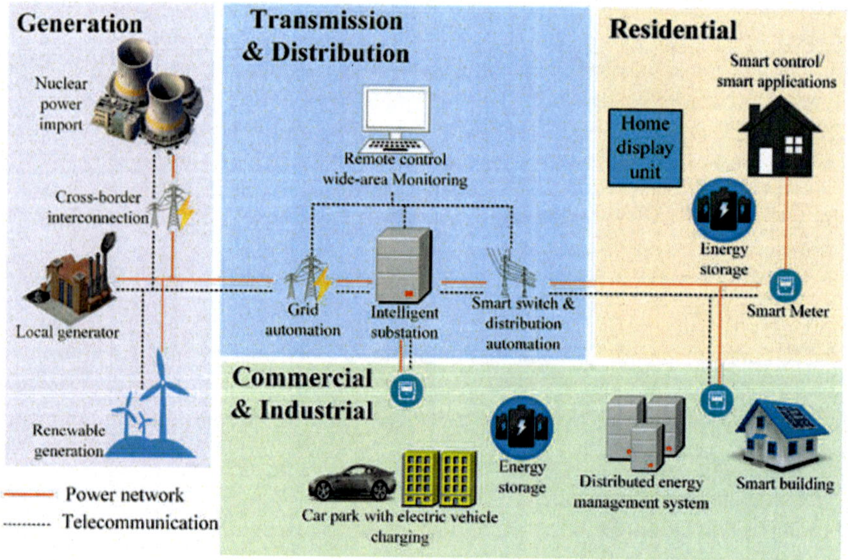

Fig. 1 FACTS controllers in transmission and distribution environment

3.1 Static VAR Compensator (SVC)

This device was implemented in the late 1970s which is the initial inventions of FACTS controllers. The SVC was interconnected in a parallel connection in the point of common coupling to inject or absorb the reactive power that is competent of interchanging the inductive and capacitive power to regulate the particular parameters in an existing electrical system [13]. In the year 1974, General Electric Company implemented the initial SVC. Around 500 SVCs in reactive power ratings ranging as 50–500 MVAR is installed by power companies till now.

SVCs are used to enhance the rotor angle stability through dynamically controlling the voltage in various places as well as transient stability in supporting to enhance the dynamic of power oscillation. The availability, effectiveness and speed response of SVCs is enabled to give superior action related to the control of transient and steady state parameters. Moreover, this device is used for improving alternator rotor angle stability, swinging of damping power oscillations and minimization of power losses through controlling of reactive power [14]. The SVC can be functioned in two modes namely, VAR regulator and voltage regulation mode. The steady state behaviour of SVC in the voltage regulation state was given in Fig. 2.

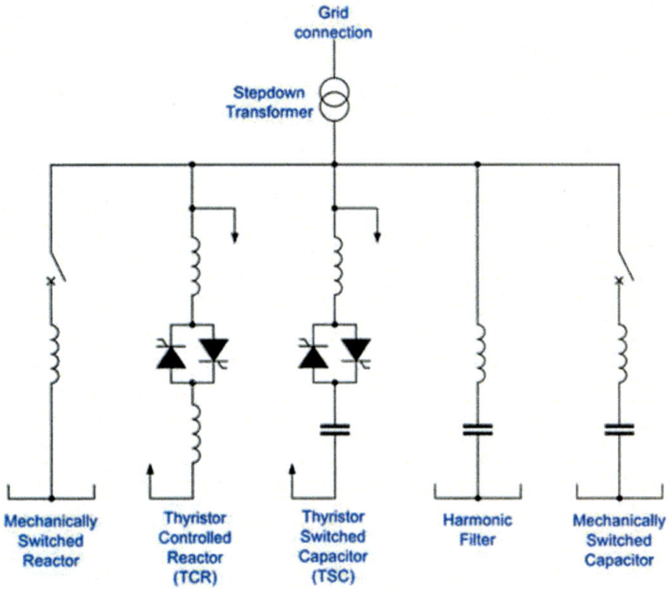

Fig. 2 SVC for voltage regulation

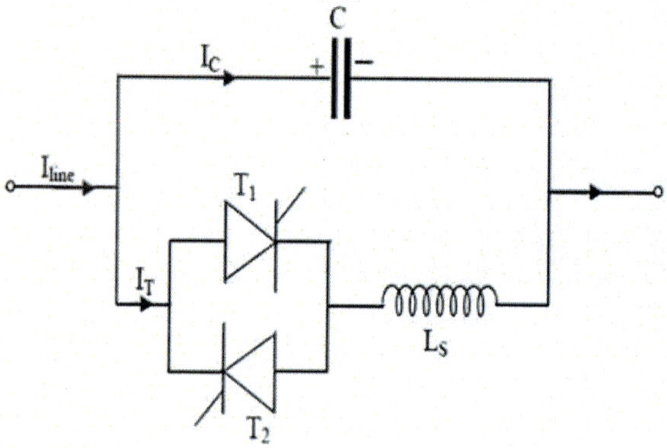

Fig. 3 TCSC for power quality problems mitigation

3.2 *Thyristor Controlled Series Compensator (TCSC)*

This device which is a combination in series form of capacitors parallel with the silicon-controlled reactor gives a variable series capacitive reactance in flexible manner [15]. TCSC plays important role in the functioning and regulation of electrical systems like power flow improvement, short circuit current limiting, improving the dynamic and transient stability.

The important features of TCSC components is enhancing the real power flow, damping of power oscillations, and control of line power flow [5, 6]. The starting TCSC was first implemented in Arizona power substation in the year late 1994 functioning in 220 kV and is utilized to enhance the transfer of power flow capacity. After implementing this capability, the power network was increased by approximately 30%. Figure 3 depicts the TCSC for power quality problems mitigation.

3.3 *Static Synchronous Compensator (STATCOM)*

This device was combined through static var compensator and commonly depending on gate turn-off thyristor based SCRs. This device was capable to reactive power supplying or absorbing in the receiving end side. It also functions with real power flow it should integrate from a power supply or energy storage systems with proper ranging.

The initial STACTOM was implemented in Japan during the year 1994 in the Inumaya power substation. It was capacity of ± 60 MVAR and supports voltage stability improvement. The intention of this controller implementation is to support variable reactive power compensation.

STATCOM does not require more capacitive and inductive components to support capacitive and inductive reactive power in large power transmission networks as needed in SVCs [16]. The primary advantage of STATCOM is the requirement of a minimum area and large output reactive power in minimum grid networks.

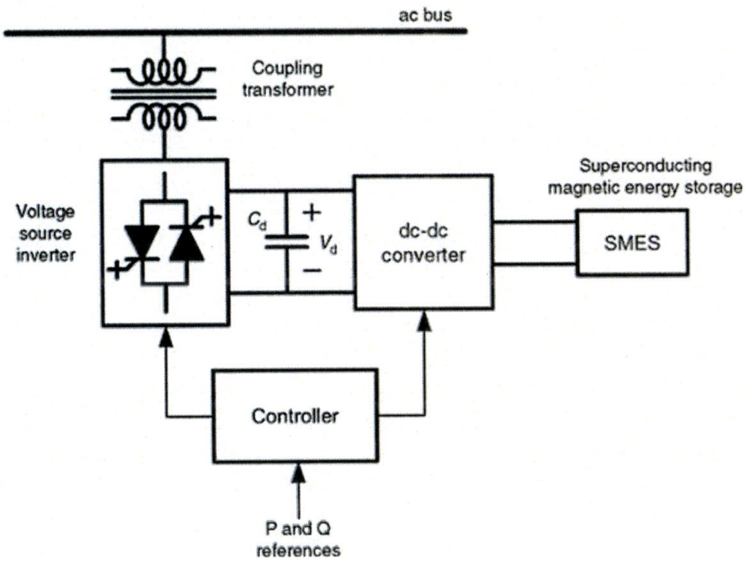

Fig. 4 STATCOM for VAR regulation and voltage control

STATCOM provides a current source while it is not depending in grid supply voltage. Also, it provides better variable stability in the exact location and STATCOM gives best damping behaviour than SVC. It also transiently interchanges the real power of the networks. Commonly, a STATCOM is functioning in two modes such as VAR regulation and voltage control mode. Figure 3 depicts the STATCOM for VAR regulation and voltage control Fig. 4.

3.4 *Static Synchronous Series Compensator (SSSC)*

SSSC is series-combination of voltage source converter based FACTS device. It supplies the power in a regulated amplitude and power angle in the system frequency and has the capacity to regulate energy flow and also enhances rotor angle stability margin with damping of oscillations [17]. The control system and illustration of SSSC is given in Fig. 5.

3.5 *Distributed FACTS Controllers (D-FACTS Controllers)*

In increasing applications of renewable energy sources and distributed generation of power distribution systems, the strategies of contribution in the power networks and regulation of electrical system was changed [5, 6]. In paper [18] the authors

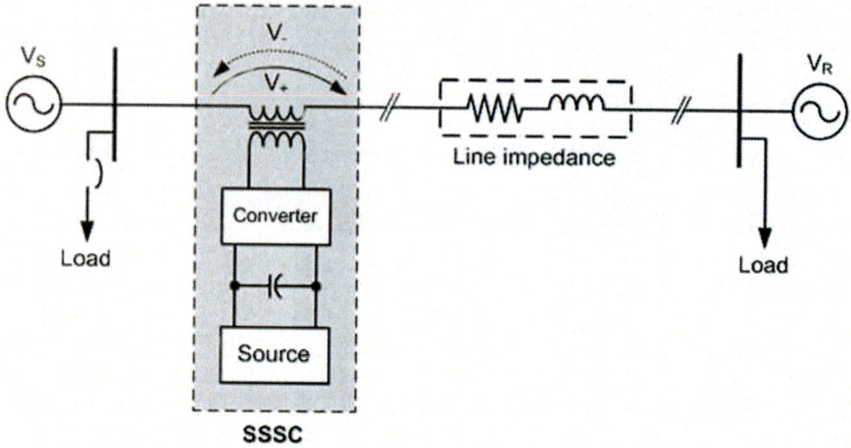

Fig. 5 SSSC for power quality problems mitigation

presented a novel method of distribution controllers and proposed a best solution to mitigate the major issues in the previous generation of FACTS controllers while it provides economically efficient control of power flow.

Recently, distribution side FACTS controllers are used to design and target various controlling methods of power flow issues. Distribution FACTS controllers are used for variable regulation of the efficient system reactance. It is from the electrical system context, this controller provides several additional features as its less cost and also minimum size compared with transmission side controllers. It provides best solution in large scale arrangement [7, 8]. The best essential distribution side controllers used for micro grids and smart grids was given in paper [9]. Increasing the minimum power FACTS controllers, i.e. Distribution FACTS controllers, supports a best characteristics and minimum price tool for improving micro and smart grids reliability, security and controllability and also improves the source usage and customer power quality by reducing the environmental pollution and reducing the total cost [9, 10].

However, to mitigate several issues in micro and smart grids, different power transferring and regulation components were implemented to support and regulate the different levels of power systems. The essential solution to major power quality issues was indicated. In paper [6], the authors have given smart grid architecture along with several categories of distribution FACTS controllers (Table 2).

Table 2 A short survey in control attributes of various FACTS controllers

S.no	Control Attributes	Facts controllers				
		SVC	TCSC	STATCOM	SSSC	D-FACTS Controllers
1	Power flow control		✔		✔	✔
2	Voltage profile improvement	✔		✔		✔
3	Line commutated	✔	✔			
4	Forced commutated			✔	✔	✔
5	Voltage source converter			✔	✔	✔
6	Current source converter	✔	✔	✔	✔	
7	Transient and dynamic converter	✔	✔		✔	✔
8	Damping oscillation	✔	✔	✔	✔	✔
9	Fault current limiting		✔			✔
10	Voltage stability	✔	✔	✔	✔	✔

4 Facts Controllers to Enhance the Power Quality Problems in Micro Grid and Smart Grid

Developing smart grids along with distributed generation and renewable energy sources needs the help of FACTS controllers and power electronic circuit's stabilization, combined with super behaviour operation methodologies [7, 8]. Advanced FACTS controller is developed to assure decoupled alternating current to direct current integration, enhanced power security, compensation of reactive power, improvement of voltage and power factor and minimization of loss [9, 10]. It also improved the reliability in distribution side micro grids networks, stand-alone alternating current to direct current distribution generation strategies through non-conventional energy systems. FACTS controllers involve along with the voltage source converters, the passive filters [6–10].

Advanced electrical networks along with additional demand advance metering infrastructure and distributed generation integration comprise solar photovoltaic, wind energies need the newly designed advanced-soft computing tools, operational methodologies and improved power electronic circuits infrastructure to assure reliability, safety, efficiency without involving the short circuit currents and transient over-voltages [19]. Enhanced power usage and efficient power regulation is the main interconnections line regulations control the rating of extra additional or substitute generation [4]. Clean and non-conventional energy production was able to deliver 30–35% of total energy in the year 2040 in various sources. The advance implementation of FACTS controller strategies is aimed for generation and transmission system components [20] of smart grid.

5 Conclusion

This article examined a detailed survey and application of FACTS controllers' integration with renewable energy sources for minimizing the power quality problems in micro grid as well as in smart grid technology. The presently available FACTS controllers is subjected to various modifications in the design depending on optimization of control methods by applying the smart grid control methods which also serves several functions like control of power flow, enhancement of stability, and compensation of reactive power.

The article also surveyed various FACTS control solutions, while the regulation methods for better usage of linear, nonlinear and critical loads, power quality problems in the smart grid and micro-grid environments are also presented. The overview of this survey is intended for effective power utilization; minimize the losses, stabilization of voltage, and enhancement in power quality, and minimizing harmonics in the PCC of transmission. Another issue of grid integration issues in the weak alternating power utility networks was examined. Future of these controllers are exciting and welcomed by more and optimal usage distributed energy sources in domestic building, office buildings, commercial based buildings, industries and create the awareness in hybrid power systems and power grid to E-vehicles, energy storage technologies, better lighting schemes and use of energy efficient motors.

References

1. Darabian M, Jalilvand A (2017) A power control strategy to improve power system stability in the presence of wind farms using FACTS devices and predictive control. Int J Electr Power Energy Syst 85(2):50–66
2. Subasri CK, Charles Raja S, Venkatesh P (2015) Power quality improvement in a wind farm connected to grid using FACTS device. Power Electron Renew Energy Syst. 326(4):1203–1212
3. Liao H, Milanović JV (2017) On capability of different FACTS devices to mitigate a range of power quality phenomena. IET Gener Transm Distrib 11(5):2002–2012
4. Yan R, Marais B, Saha TK (2014) Impacts of residential photovoltaic power fluctuation on on-load tap changer operation and a solution using DSTATCOM. Electr Power Syst Res. 111:185–193
5. Hemeida MG, Rezk H, Hamada MM (2017) A comprehensive comparison of STATCOM versus SVC-based fuzzy controller for stability improvement of wind farm connected to multi-machine power system. Electr Eng 99: 1–17
6. Bhaskar MA, Sarathkumar D, Anand M (2014) Transient stability enhancement by using fuel cell as STATCOM. In: 2014 International conference on electronics and communication systems (ICECS). pp 1–5
7. Sarathkumar D, Srinivasan M, Stonier AA, Samikannu R (2021) A research survey on microgrid faults and protection approaches. In: IOP Conference series: Materials science and engineering, vol 1055. pp 012128
8. Sarathkumar D, Srinivasan M, Stonier AA, Samikannu R, Dasari NR, Raj RA (2021) A technical review on classification of various faults in smart grid systems. In: IOP conference series: Materials science and engineering, Vol 1055. pp 012152

9. Sarathkumar D, Srinivasan M, Stonier AA, Samikannu R, Dasari NR, Raj RA (2021) A technical review on self-healing control strategy for smart grid power system. In: IOP conference series: Materials science and engineering, vol 1055. pp 012153
10. Sarathkumar D, Srinivasan M, Stonier AA, Samikannu R, Vijay Anand D (2021) Design of intelligent controller for hybrid PV/wind energy based smart grid for energy management applications. In: IOP Conference series: Materials science and engineering, vol 1055. pp 012129
11. Vanaja DS, Stonier AA, Mani G, Murugesan S (2021) Investigation and validation of solar photovoltaic-fed modular multilevel inverter for marine water-pumping applications. Electr Eng. https://doi.org/10.1007/s00202-021-01370-x
12. Stonier AA, Lehman B (2018) An intelligent-based fault-tolerant system for solar-fed cascaded multilevel inverters. IEEE Trans Energy Convers
13. Alexander A, Thathan M (2014) Modelling and analysis of modular multilevel converter for solar photovoltaic applications to improve power quality. IET Renew Power Gener
14. Albert Alexander S, Manigandan T (2014) Power quality improvement in solar photovoltaic system to reduce harmonic distortions using intelligent techniques. J Renew Sustain Energy
15. Albert Alexander S, Manigandan T (2014) Digital control strategy for solar photovoltaic fed inverter to improve power quality. J Renew Sustain Energy
16. Sarathkumar D, Srinivasan M, Stonier AA, Kumar S, Vanaja DS (2021) A brief review on optimization techniques for smart grid operation and control. In: 2021 International conference on advancements in electrical, electronics, communication, computing and automation (ICAECA). pp 1–5. https://doi.org/10.1109/ICAECA52838.2021.9675618
17. Sarathkumar D, Srinivasan M, Stonier AA, Kumar S, Vanaja DS (2021) A review on renewable energy based self-healing approaches for smart grid. In: 2021 International conference on advancements in electrical, electronics, communication, computing and automation (ICAECA). pp 1–6. https://doi.org/10.1109/ICAECA52838.2021.9675495
18. Stonier A, Yazhini M, Vanaja DS, Srinivasan M, Sarathkumar D (2021) Multi level inverter and its applications—An extensive survey. In: 2021 International conference on advancements in electrical, electronics, communication, computing and automation (ICAECA). pp 1–6. https://doi.org/10.1109/ICAECA52838.2021.9675535
19. Sarathkumar D, Kavithamani V, Velmurugan S, Santhakumar C, Srinivasan M, Samikannu R (2021) Power system stability enhancement in two machine system by using fuel cell as STATCOM (static synchronous compensator). Mater Today: Proc 45, Part 2:2130–2138. ISSN 2214–7853. https://doi.org/10.1016/j.matpr.2020.09.730. 9
20. Sarathkumar D, Venkateswaran K, Vijayalaxmi A (2020) Design and implementation of solar powered hydroponics systems for agriculture plant cultivation. Int J Adv Sci Technol (IJAST) 29(05):3266–3271

Arctangent Framework Based Least Mean Square/Fourth Algorithm for System Identification

Soumili Saha, Ansuman Patnaik, and Sarita Nanda

1 Introduction

One of the major challenges in the study of adaptive filters is the selection of a suitable cost function [1, 2]. The efficiency of adaptive filters is primarily determined by the design technique of the filter and the cost function (CF) used. Mean Square Error (MSE) is preferably a widely used cost function for Gaussian signals or noise distribution because of its low computational tractability, simplicity, optimal performance and convexity. Some of the adaptation algorithms developed utilizing this criterion are least mean square (LMS), normalized LMS (NLMS) and variable step-size LMS (VSS-LMS) [1, 2]. In practical scenarios, MSE based algorithms can sometimes deviate and degrade its performance where noise is non-Gaussian or impulsive [2, 3].

The cost function used for noise or signal with a light-tailed impulsive distribution should be higher-order moment of the error measurement. The family of least mean fourth (LMF) algorithm [2] uses this property. However, the instability issue hampers its performance. This results in the development of a least mean square/fourth (LMS/F) algorithm combining the strengths of both LMF and LMS algorithms [4] where the LMS/F algorithm's behavioral impact in the Gaussian noise environment was studied and the algorithm's behavior in the presence of non-Gaussian noise environment was compared in [5]. However, with the constant presence of such impulsive noise, the algorithm's performance was not satisfactory. Later in [6], a reweighted zero-attracting modified variable step-size continuous mixed p− norm algorithm was developed to exploit sparsity in a system against impulsive noise.

Arctangent, being one of the saturation properties of non-linearity error that can enhance the behavior of the adaptive algorithms. A novel cost function framework

S. Saha (✉) · A. Patnaik · S. Nanda
School of Electronics Engineering, KIIT Deemed to be University, Bhubaneswar, India
e-mail: soumilisaha08@gmail.com

© The Author(s), under exclusive license to Springer Nature Singapore Pte Ltd. 2023

H. Muthusamy et al. (eds.), *Robotics, Control and Computer Vision*, Lecture Notes in Electrical Engineering 1009, https://doi.org/10.1007/978-981-99-0236-1_27

called the arctangent framework was proposed exploiting the property of the arctangent function. Proposed algorithms such as arctangent sign algorithm (ATSA), arctangent least mean square (ATLMS), arctangent least mean fourth (ATLMF), and arctangent generalized maximum correntropy criterion algorithm are all based on the arctangent framework [7]. Since the LMS/F algorithm outperforms the standard LMS and LMF algorithms while maintaining their flexibility and stability [4], an arctangent Least Mean Square/Fourth (ATLMS/F) algorithm is presented, and its response is evaluated using various simulations in MATLAB in a noisy environment system identification model. Sect. 2 reviews the arctangent framework based cost function. Sect. 3 explains the proposed algorithm. Sect. 4 discusses the simulation and observations while sect. 5 states the conclusion.

2 Arctangent Framework Based Cost Function

Consider a system identification problem whose block diagram is provided in Fig. 1

Let $X(n)$ be the tapped delay input given to the physical system of length M, where the weight vector of the filter is defined as $\phi(n) = [\phi_1, \phi_2, \phi_3, \ldots., \phi_M]^T$. The output of the system identification system is corrupted with the additive noise $\upsilon(n)$ uncorrelated to the input signal. The noise in the system is a mix of impulsive and Gaussian noise. The desired output signal of the system $d(n)$ is determined as

$$d(n) = \phi^T(n)X(n) + \upsilon(n) \tag{1}$$

where $X(n) = [x(n), x(n-1), \ldots.., x(n-M+1)]^T$ represents the input signal vector and $\hat{\phi}(n)$ is the filter coefficient. Defining the weight coefficients of the adaptive filter as $\hat{\phi}(n) = [\hat{\phi}_1, \hat{\phi}_2, \hat{\phi}_3, \ldots., \hat{\phi}_M]^T$ and transmitting $X(n)$ through the adaptive

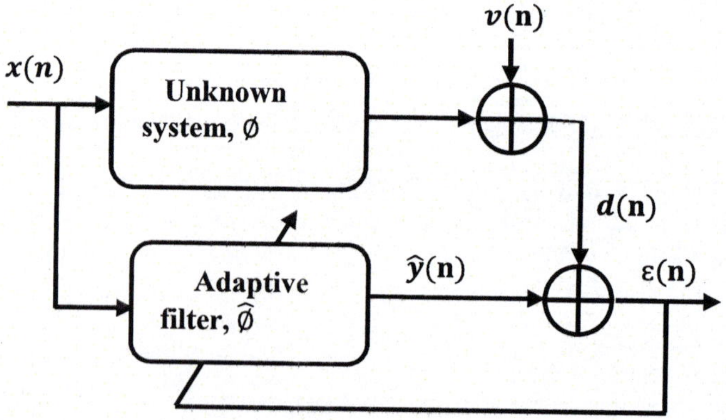

Fig. 1 Block diagram of adaptive system identification

filter provides the output signal $\hat{y}(n)$ and error $\varepsilon(n)$ as follows

$$\hat{y}(n) = \hat{\varphi}^T(n)X(n) \tag{2}$$

$$\varepsilon(n) = d(n) - \hat{y}(n) \tag{3}$$

The values of the weight coefficients of the adaptive system can be optimized by either reducing or maximizing the CF. It has been recognized that the saturation attributes of non-linearity error provide resilience against random impulsive disturbances [3, 8]. Based on saturation properties of the arctangent function, an arctangent framework dependent cost function was introduced as [7]

$$\psi(n) = \tan^{-1}[\alpha\xi(n)] \tag{4}$$

where controlling constant $\alpha > 0$ controls steepness of the arctangent cost function. Gradient of Eq. (4) is denoted as

$$\nabla_\varphi \psi(n) = \frac{\partial\psi(n)}{\partial\varphi(n)} = \frac{\alpha\nabla_\varphi\xi(n)}{1 + [\alpha\xi(n)]^2} \tag{5}$$

In addition to the gradient of the conventional CF defined by $\nabla_\varphi \varsigma(n)$, an extra parameter $1 + [\alpha\varsigma(n)]^2$ is included which is essential for reducing the higher steady-state misalignment. Hence, for extensive errors, the gradient of the arctangent cost function results in robustness and is more bounded in comparison to the gradient of the existing cost function. The weight update for the arctangent algorithm can be stated based on the gradient descent approach as follows [2]

$$\varphi(n+1) = \varphi(n) - \beta\frac{\partial\psi(n)}{\partial\varphi(n)} \tag{6}$$

where β represents step-size of the weight upgradation. Combining Eqs. (4) and (6), the updated weight vector is

$$\varphi(n+1) = \varphi(n) - \beta'\frac{\nabla_\varphi\xi(n)}{1 + [\alpha\xi(n)]^2} \tag{7}$$

where $\beta' = \beta\alpha$ is defined as the cumulative step-size. In the next section, based on the arctangent framework cost function, an LMS/F algorithm is derived.

3 Arctangent LMS/F Algorithm (ATLMS/F)

The cost function constructed for LMS/F algorithm is given as [4]

$$\xi(n) = \frac{1}{2}\varepsilon^2(n) - \frac{1}{2}\lambda \ln\{\varepsilon^2(n) + \lambda\} \quad (8)$$

Integrating LMS/F algorithm's CF, $\xi(n)$ with the conventional arctangent framework provided in (7), the updated arctangent LMS/F (ATLMS/F) algorithm's weight vector is defined as

$$\varphi(n+1) = \varphi(n) + \mu \frac{\varepsilon^3(n).X(n)}{(\varepsilon^2(n)+\lambda)\left[1 + [\alpha\{\frac{1}{2}\varepsilon^2(n) - \frac{1}{2}\lambda \ln\{\varepsilon^2(n)+\lambda\}\}]^2\right]} \quad (9)$$

From (9) it is observed that an extra term $1 + \left[\alpha\left\{\frac{1}{2}\varepsilon^2(n) - \frac{1}{2}\lambda \ln\{\varepsilon^2(n) + \lambda\}\right\}\right]^2$ in the weight update equation of ATLMS/F algorithm compared to the conventional LMS/F algorithm counteracts any change in the weight updation under the influence of impulsive noise making the ATLMS/F algorithm stable in comparison to the typical LMS/F algorithm.

4 Simulation and Results

The performance of the presented algorithm for system identification in an impulsive noise environment is analyzed. The input signal considered are normally distributed sequences with zero mean and unit variance. The system noise v(n) is a combination of white Gaussian noise signal with a 20 dB signal to noise ratio and distributed Bernoulli-Gaussian (BG) impulsive noise. The BG noise is derived as $\varrho(\mathrm{n}) = \mathrm{K_m(n)}\mathcal{B}_\mathrm{i}(\mathrm{n})$, where Bernoulli process is denoted by $\mathrm{K_m(n)}$ and Gaussian random procedure by $\mathcal{B}_\mathrm{i}(\mathrm{n})$ having zero mean and variance $\sigma_\mathrm{a}^2 = 10^4/12$, $\mathrm{K_m(n)}$ is elaborated in terms of probability as $\mathrm{P(K_m(n) = 1) = P_i}$ and $\mathrm{P(K_m(n) = 0) = 1 - P_i}$ with $\mathrm{P_i} = 0.01$ [2]. The criteria used to determine the performance of the proposed algorithm is normalized mean square deviation is given as

$$NMSD(n) = 10\log_{10}\frac{\|\varphi - \hat{\varphi}\|_2^2}{\|\varphi\|_2^2} \quad (10)$$

where $||.||_2$ is the l2 norm. The calculated NMSD is for n = 20,000 iterations taking the average of 100 independent trials for analyzing the outcomes. The performance of the suggested algorithm is compared to that of the LMS/F algorithm. The step-size parameter used for the LMS/F algorithm is $\beta = 0.002$ whereas the cumulative step-size used for the ATLMS/F algorithm is $\beta' = 0.01$ where $\beta = 0.1$ and $\alpha = 0.1$ for both the experiments based on system identification.

A system identification case is considered where the impulse response is constructed synthetically using the method given in [9]. The approach begins by defining a vector U

$$U_{Mx1} = \left[O_{M_p x1} 1 e^{-\frac{1}{\tau}} e^{-\frac{2}{\tau}} .. e^{-\frac{(M_u - 1)}{\tau}}\right]^T \quad (11)$$

where M_p is the length of the bulk delay and $M_u = M - M_p$ represents the length of the decaying window that can be regulated by τ. The synthetic impulse is represented as

$$h(n) = \begin{bmatrix} O_{M_p x M_p} & O_{M_p x M_u} \\ O_{M_u x M_p} & B_{M_u x M_p} \end{bmatrix} u + P \quad (12)$$

where $B_{M_u x M_p} = \text{diag}(b)$, P and b represents zero mean white Gaussian noise vectors of length M and M_u respectively. The simulation parameters used for the generation of impulse response shown in Fig. 1 are $M = 128$, $M_p = 30$ and $\tau = 2$.

The impulse response of the echo path generated for the first experiment of length 128 is provided in Fig. 2 whereas Fig. 3 shows the NMSD behavior of the proposed algorithm in comparison to the standard algorithm.

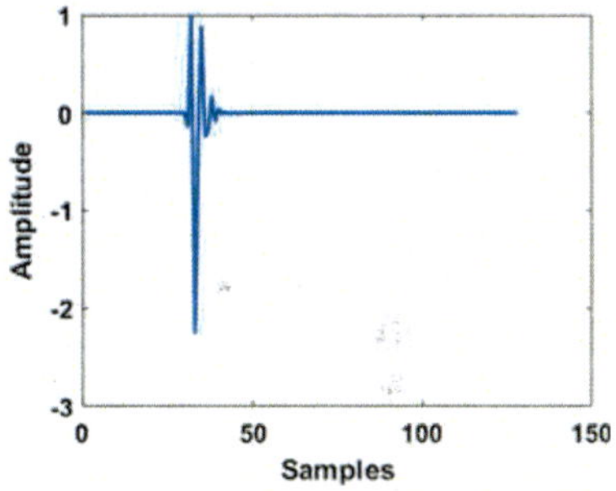

Fig. 2 Impulse response of the system

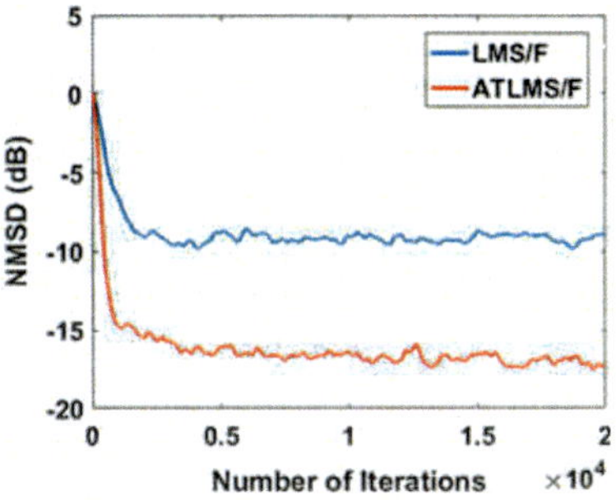

Fig. 3 NMSD comparison of the proposed algorithm

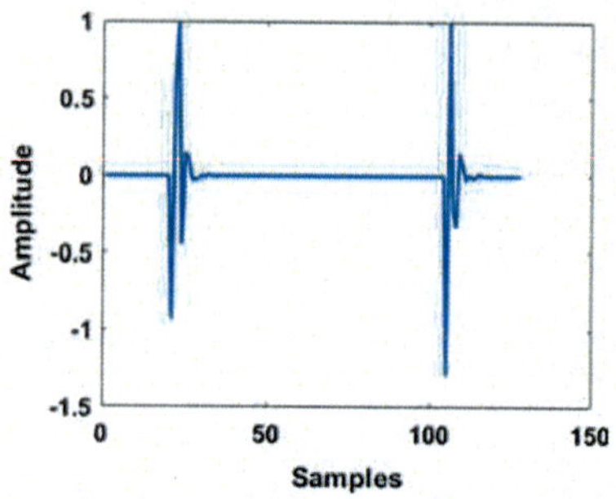

Fig. 4 Concatenating impulse response of the system

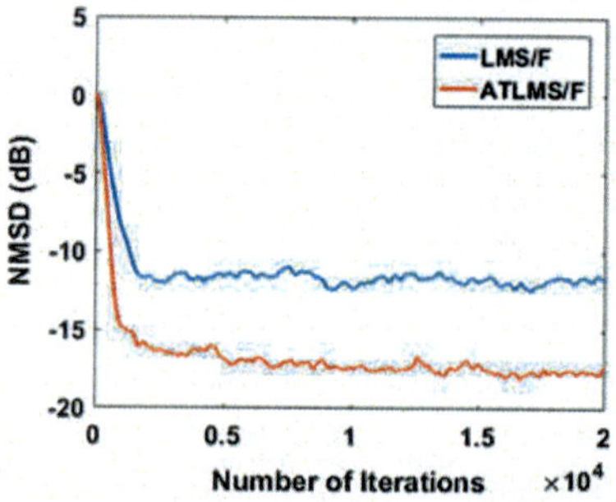

Fig. 5 NMSD comparison of the proposed algorithm

In comparison to the LMS/F algorithm, the ATLMS/F algorithm gives a reduced steady-state NMSD as shown in Fig. 3. The suggested algorithm achieves a lower steady-state NMSD value of approximately −17.53 dB, compared to around −9.8 dB for the LMS/F algorithm.

In the second, experiment a concatenated impulse response of length 128 is provided in Fig. 4 whereas Fig. 5 shows the NMSD variation of the proposed algorithm in comparison to the standard algorithm.

In comparison to the LMS/F algorithm, the ATLMS/F algorithm gives a reduced steady-state NMSD as shown in Fig. 5. The proposed algorithm produces a lower steady-state NMSD value of approximately −17.96 dB, compared to around −12.46 dB for the LMS/F algorithm.

5 Conclusion

A novel arctangent least mean square/fourth algorithm was proposed in this work. It was developed by embedding the standard LMS/F algorithm cost function into the arctangent framework. The ATLMS/F algorithm's performance was compared with the standard LMS/F algorithm for system identification cases under impulsive noise effect. The simulation results provided better steady-state values compared to the standard algorithm.

References

1. Diniz PS (2020) Introduction to adaptive filtering. adaptive filtering. Springer, Cham, pp 1–8
2. Wang S, Wang W, Xiong K, Iu HH, Chi KT (2019) Logarithmic hyperbolic cosine adaptive filter and its performance analysis. IEEE Trans Syst, Man, Cybern: Syst
3. Chen B, Xing L, Zhao H, Zheng N, Prı JC (2016) Generalized correntropy for robust adaptive filtering. IEEE Trans Signal Process 64(13):3376–3387
4. Gui G, Peng W, Adachi F (2014) Adaptive system identification using robust LMS/F algorithm. Int J Commun Syst 27(11):2956–2963
5. Patnaik A, Nanda S (2020) The variable step-size LMS/F algorithm using nonparametric method for adaptive system identification. Int J Adapt Control Signal Process 34(12):1799–1811
6. Patnaik A, Nanda S (2021) Reweighted zero-attracting modified variable step-size continuous mixed p-norm algorithm for identification of sparse system against impulsive noise. In: Proceedings of international conference on communication, circuits, and ystems: IC3S 2020, vol 728. Springer Nature, p 509
7. Kumar K, Pandey R, Bora SS, George NV (2021) A robust family of algorithms for adaptive filtering based on the arctangent framework. Express Briefs, IEEE Transactions on Circuits and Systems II
8. Das RL, Narwaria M (2017) Lorentzian based adaptive filters for impulsive noise environments. IEEE Trans Circuits Syst I Regul Pap 64(6):1529–1539
9. Khong AW, Naylor PA (2006) October. Efficient use of sparse adaptive filters. In:2006 Fortieth asilomar conference on signals, systems and computers. IEEE, pp 1375–1379

Robotics and Autonomous Vehicles

Stabilization of Ball Balancing Robots Using Hierarchical Sliding Mode Control with State-Dependent Switching Gain

Sudhir Raj

1 Introduction

Trajectory planning and control of an underactuated system becomes difficult due to less control inputs than the degree of freedom. The uncertainty is caused by the simplified model of the ball bot. Linear controllers are not robust to the uncertainties or disturbances of the ball bot system. Ball bots can be used as a carrier robots and humans can sit on the seat of the car-like structure which is fixed on a single spherical wheel. The ball bot can go in confined spaces. The structure of the ball bot is tall and thin like the human which makes it suitable for use in the workplace. Linear controller such as Linear Quadratic Regulator is not robust to the uncertainties or disturbances for the ball bot robot. Therefore, nonlinear controllers are required for the stabilization of the ball bot robot. The proposed controller Is nonlinear which makes it suitable for the control of underactuated systems such as ball bots.

Ball bot robot [1] is an example of an underactuated system. The objective of the proposed hierarchical sliding mode control is to keep the body in its vertical position in the presence of disturbances. Simulation and experimental results are used to verify the effectiveness of the proposed controller. Trajectory tracking and balancing [2] of a ball bot is achieved using virtual angle-based sliding mode control. Simulation results show that the proposed controller is effective in trajectory tracking and balancing of the ball bot. Kalman filter [3] is used for the estimation of the states.

Experimental results show that the proposed algorithm gives better results than the extended Kalman filter. Extended Kalman filter [4] is used in estimating the states of the ball bot using sensor information. Experimental results are presented to validate the algorithm for the ball bot. The proposed algorithm [5] is verified using experimental results. Trajectory tracking of ball bot [6] is achieved using a Feedback controller. Simulation and experimental results show the efficacy of the

S. Raj (✉)
SRM University, Amaravati, Andhra Pradesh, India
e-mail: sudhir.r@srmap.edu.in

© The Author(s), under exclusive license to Springer Nature Singapore Pte Ltd. 2023

H. Muthusamy et al. (eds.), *Robotics, Control and Computer Vision*, Lecture Notes in Electrical Engineering 1009, https://doi.org/10.1007/978-981-99-0236-1_28

proposed controller. Extended Kalman filter [7] -based state estimation is carried out for the ball bot to maintain its upright position. The proposed robot [8] consists of three omnidirectional wheels with stepping motors. The observer is designed for the stabilization of the ball beam system. The proposed sliding mode control [9] gives better performance as compared to other linear controllers for the stabilization and tracking of the ball bot. Neural network-based control [10] for trajectory tracking and balancing of a ball balancing robot is carried out considering uncertainties.

The vertical position is achieved using the proposed controller, and it requires less time to stabilize the ball bot system. The control input of the state-dependent switching gain is less as compared to the Hierarchical sliding mode controller. The objective of this work is to stabilize the ball bot in less time as compared to the previous controllers as reported in the literature review. The comparison between the two controllers is carried out to show the effectiveness of the proposed controller.

2 Dynamic Model of Ball Bot System

The ball bot is an underactuated system with four degrees of freedom and two control inputs. There are three omni wheel motors in the ball bot. It is assumed that no slip is occurring between the ball and the floor and between the ball and the wheels. The equation of the ball bot is derived using the Euler-Lagrange formulation. The motion of the ball bot is derived in the x-z and y-z planes. Figure 1 shows the ball bot in the x-z plane. The Lagrangian L is calculated as the difference between the kinetic and potential energy of the ball bot:

$$L = T - V \tag{1}$$

$$= T_{kx} + T_{wx} + T_{ax} - (V_{kx} + V_{wx} + V_{ax}) \tag{2}$$

$$\begin{aligned} &= \frac{1}{2}\left(m_k + \frac{I_k}{r_k^2}\right)\dot{y}_k^2 + \frac{3I_w cos^2\alpha}{4r_w^2}\left(\dot{y}_k + r_k\dot{\theta}_x\right)^2 \\ &+ \frac{1}{2}I_x\dot{\theta}_x^2 + \frac{1}{2}m_a\left(\dot{y}_k - I\dot{\theta}_x cos\theta_x\right)^2 \\ &+ \frac{1}{2}m_a l^2\dot{\theta}_x^2 sin^2\theta_x - m_a g l cos\theta_x \end{aligned} \tag{3}$$

Therefore, the Lagrangian dynamics for the ball bot can be calculated as equation number (4):

$$\frac{d}{dt}\left(\frac{\partial L_x}{\partial \dot{q}_x}\right) - \frac{\partial L_x}{\partial q_x} = \frac{1}{r_w}\left[\frac{1}{r_k}\right]\tau_x - D\left(\dot{q}_x\right) \tag{4}$$

The equations of the ball bot in the y-z plane can be taken as equation numbers (5) and (6):

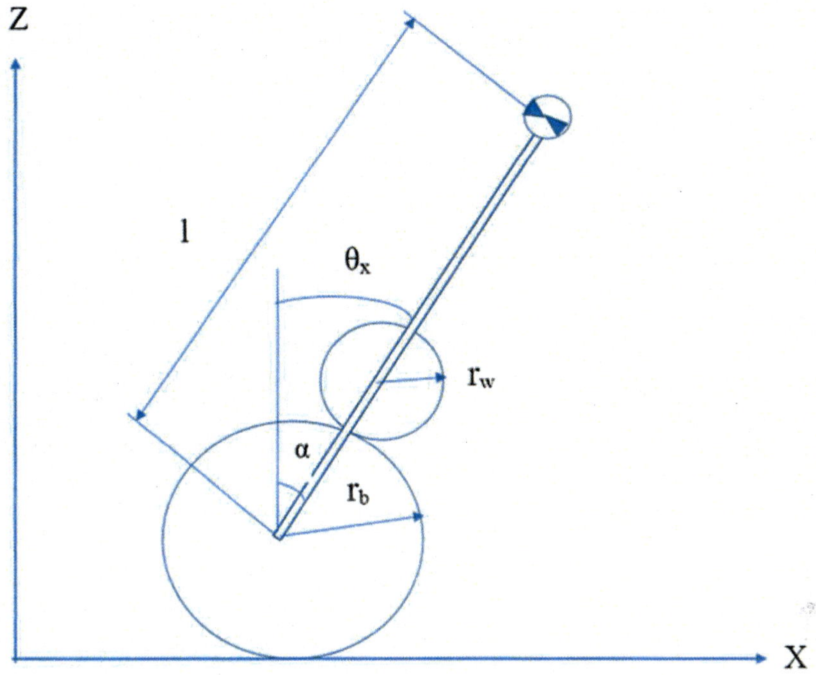

Fig. 1 Ball bot system in x-z plane

$$\ddot{y}_k a_1 + (a_4 - a_3 cos\theta_x)\,\ddot{\theta}_x + a_3\dot{\theta}_x^2 sin\theta_x + b_y\dot{y}_k = r_w^{-1}\tau_x \tag{5}$$

$$(a_4 - a_3 cos\theta_x)\,\ddot{y}_k + \ddot{\theta}_x a_2 + b_{rx}\dot{\theta}_x - a_5 sin\theta_x = r_k r_w^{-1}\tau_x \tag{6}$$

The constants defined in Eqs. (5) and (6) are taken as

$$a_1 = m_k + \frac{I_{kx}}{r_k^2} + m_a + \frac{3I_w cos^2\alpha}{2r_w^2}$$

$$a_2 = m_a l^2 + \frac{3I_w r_k^2 cos^2\alpha}{2r_w^2} + I_x$$

$$a_3 = m_a l$$

$$a_4 = \frac{3I_w cos^2\alpha}{2r_w^2} r_k$$

$$a_5 = m_a g l$$

The system equations of the ball bot in the y-z plane are taken as equation numbers (7) and (8):

$$\ddot{y}_k = F_{x1}(q_x, \dot{q}_x) + G_{x1}(q_x)\tau_x \quad (7)$$

$$\ddot{\theta}_x = F_{x2}(q_x, \dot{q}_x) + G_{x2}(q_x)\tau_x \quad (8)$$

$$F_{x1}(q_x, \dot{q}_x) = A_x^{-1}[(a_3 cos\theta_x - a_4)\left(a_5 sin\theta_x - b_{rx}\dot{\theta}_x\right) - a_2\left(a_3\dot{\theta}_x^2 sin\theta_x + b_y\dot{y}_k\right)]$$

$$G_{x1}(q_x) = A_x^{-1} r_w^{-1}(a_2 + a_3 r_k cos\theta_x - a_4 r_k)$$

$$F_{x2}(q_x, \dot{q}_x) = A_x^{-1}[(a_4 - a_3 cos\theta_x)\left(a_3\dot{\theta}_x^2 sin\theta_x + b_y\dot{y}_k\right) + a_1\left(a_5 sin\theta_x - b_{rx}\dot{\theta}_x\right)]$$

$$G_{x2}(q_x) = A_x^{-1} r_w^{-1}(a_3 cos\theta_x - a_4 + a_1 r_k)$$

$$A_x = a_1 a_2 - (a_4 - a_3 cos\theta_x)^2$$

The mathematical equations describe the ball segway system dynamics in the x-z plane as follows:

$$\ddot{x}_x b_1 + \left(b_4 cos\theta_y - b_3\right)\ddot{\theta}_y - b_4\dot{\theta}_y^2 sin\theta_y + b_x\acute{x}_k = -r_w^{-1}\tau_y \quad (9)$$

$$\left(b_4 cos\theta_y - b_3\right)\ddot{x}_k + \ddot{\theta}_y b_2 - b_5 sin\theta_y + b_{ry}\dot{\theta}_y = r_k r_w^{-1}\tau_y \quad (10)$$

where

$$b_1 = m_k + \frac{I_k}{r_k^2} + m_a + \frac{3I_w cos^2\alpha}{2r_w^2}$$

$$b_2 = m_a l^2 + \frac{3I_w r_k^2 cos^2\alpha}{2r_w^2} + I_y$$

$$b_3 = \frac{3I_w cos^2\alpha}{2r_w^2} r_k$$

$$b_4 = m_a l$$

$$b_5 = m_a g l$$

The system equations are

$$\ddot{x}_k = F_{y1}\left(q_y, \dot{q}_y\right) + G_{y1}\left(q_y\right)\tau_y \quad (11)$$

$$\ddot{\theta}_y = F_{y2}\left(q_y, \dot{q}_y\right) + G_{y2}\left(q_y\right)\tau_y \quad (12)$$

where

$$F_{y1}\left(q_y,\dot{q}_y\right)=A_y^{-1}[b_2\left(b_4sin\theta_y\dot{\theta}_y^2-b_x\dot{x}_k\right)$$
$$+\left(b_3-b_4cos\theta_y\right)\left(b_5sin\theta_y-b_{ry}\dot{\theta}_y\right)]$$
$$G_{y1}\left(q_y\right)=A_y^{-1}r_w^{-1}\left(b_2-b_3r_k+b_4r_kcos\theta_y\right)$$
$$F_{y2}\left(q_y,\dot{q}_y\right)=A_y^{-1}[\left(b_3-b_4cos\theta_y\right)\left(b_4sin\theta_y\dot{\theta}_y^2-b_x\dot{x}_k\right)$$
$$+b_1\left(b_5sin\theta_y-b_{ry}\dot{\theta}_y\right)]$$
$$G_{y2}\left(q_y\right)=A_y^{-1}r_w^{-1}\left(r_kb_1-b_3+b_4cos\theta_y\right)$$
$$A_y=b_1a_2-\left(b_4cos\theta_y-b_3\right)^2$$

The sliding mode surfaces for the y-z plane are given by equation numbers (13) and (14):

$$s_{x1}=c_{x1}e_{x1}+\dot{e}_{x1} \tag{13}$$
$$s_{x2}=c_{x2}e_{x2}+\dot{e}_{x2} \tag{14}$$

where c_{x1} and c_{x2} are constants, and e_{x1} and e_{x2} are taken as tracking errors:

$$e_{x1}=y_k-y_{kd} \tag{15}$$
$$e_{x2}=\theta_x-\theta_{xd} \tag{16}$$

Equation (13) can be written as equation number (17):

$$s_{x1}=c_{x1}\left(y_k-y_{kd}\right)+\dot{y}_k \tag{17}$$
$$s_{x2}=c_{x2}\theta_x+\dot{\theta}_x \tag{18}$$

$\dot{s}_{x1}$ and $\dot{s}_{x2}$ are equated to zero for finding the equivalent control of subsystems:

$$\tau_{xeq1}=-G_{x1}^{-1}\left(q_x\right)\left[c_{x1}\dot{y}_k+F_{x1}\left(q_x,\dot{q}_x\right)\right] \tag{19}$$
$$\tau_{xeq2}=-G_{x2}^{-1}\left(q_x\right)\left[c_{x2}\dot{\theta}_x+F_{x2}\left(q_x,\dot{q}_x\right)\right] \tag{20}$$

The hierarchical sliding mode control can be taken as $S_{x1}=s_{x1}$. Equation number (21) gives the sliding mode control law for the first layer. The Lyapunov function is taken as equation number (22):

$$\tau_{x1}=\tau_{xeq1}+\tau_{xsw1} \tag{21}$$
$$V_{x1}\left(t\right)=0.5S_{x1}^2 \tag{22}$$

The τ_{xsw1} is the switching control of the first layer of Sliding mode control. $V_{x1}\left(t\right)$ is differentiated with respect to time t:

$$\dot{V}_{x1}\left(t\right)=S_{x1}\dot{S}_{x1} \tag{23}$$

$$\dot{S}_{x1} = k_{x1} S_{x1} - \eta_{x1} sign \left(S_{x1} \right) \tag{24}$$

where k_{x1} and η_{x1} are positive constants:

$$\tau_{x1} = \tau_{xeq1} + G_{x1}^{-1} \left(q_x \right) \dot{S}_{x1} \tag{25}$$

The sliding mode control for the second layer can be taken as S_1 and s_2, respectively:

$$S_{x2} = \alpha_x S_{x1} + s_{x2} \tag{26}$$

where α_x is the sliding mode parameter. The sliding mode control law for the second layer can be taken as equation number (27):

$$\tau_{x2} = \tau_{x1} + \tau_{xeq2} + \tau_{xsw2} \tag{27}$$

The Lyapunov function can be taken as equation number (28):

$$V_{x2} \left(t \right) = 0.5 S_{x2}^2 \tag{28}$$

where τ_{xsw2} is the switching control of the second layer of sliding mode control. $V_{x2} \left(t \right)$ is differentiated with respect to time t:

$$\dot{V}_{x2} \left(t \right) = S_{x2} \dot{S}_{x2} \tag{29}$$

The control law can be taken as equation number (30):

$$\dot{S}_{x2} = -\eta_{x2} sign \left(S_{x2} \right) \tag{30}$$

where k_{x2} and η_{x2} are positive constants.

The control law for the ball bot in the y-z and x-z planes can be taken as equation numbers (31) and (32), respectively:

$$\tau_{x2} = \frac{\alpha_x G_{x1} \left(q_x \right) \tau_{xeq1} + G_{x2} \left(q_x \right) \tau_{xeq2} + \dot{S}_{x2}}{\alpha_x G_{x1} \left(q_x \right) + G_{x2} \left(q_x \right)} \tag{31}$$

$$\tau_{y2} = \frac{\alpha_y G_{y1} \left(q_y \right) \tau_{yeq1} + G_{y2} \left(q_x \right) \tau_{yeq2} + \dot{S}_{y2}}{\alpha_y G_{y1} \left(q_y \right) + G_{y2} \left(q_y \right)} \tag{32}$$

3 State-Dependent Switching Gain-Based Controller

The Lyapunov function is defined as equation number (33):

$$V_{x1}(t) = 0.5S_{x1}^2 \tag{33}$$

$$\dot{V} = S_{x1}\dot{S}_{x1} \tag{34}$$

$\dot{S}_{x1}$ is defined as equation number (35) to make $\dot{V}$ negative definite:

$$\dot{S}_{x1} = -\eta_{x1}.sat(S_{x1}) \tag{35}$$

The state-dependent switching gain is selected as

$$\eta_{x1} = \beta\left(S_{x1}^2 + \gamma\right)$$

β and γ are taken as positive constants. The switching gain η_{x1} is a function of the state variable. Integrating both sides of the equation from 0 to t,

$$\int_0^t \dot{V}dx = -\int_0^t \eta_{x1}S_{x1}sat\left(S_{x1}\right)dx$$

$$V(t) - V(0) = -\int_0^t \eta_{x1}S_{x1}sat\left(S_{x1}\right)dx$$

$$V(0) = V(t) + \int_0^t \eta_{x1}S_{x1}sat\left(S_{x1}\right)dx$$

$$V(0) \geqslant \int_0^t \eta_{x1}S_{x1}sat\left(S_{x1}\right)dx$$

The steady state form of the above equation is given by

$$\lim_{t\to\infty}\int_0^t \eta_{x1}S_{x1}sat\left(S_{x1}\right)dx \le V(0) < \infty$$

According to Barbalat lemma

$$\lim_{t\to\infty}\eta_{x1}S_{x1}sat\left(S_{x1}\right) = 0 \tag{36}$$

It follows from Eq. (36) that $\lim_{t\to\infty} S_{x1} = 0$. As a consequence of this, the second-level sliding surface is asymptotically stable.

4 Simulation Results

Hierarchical sliding mode controller (HSMC) and state-dependent switching gain-based sliding mode controller (SDSG) are applied to the ball bot system and Simulation was carried out in MATLAB. The decoupled dynamics are given by Eqs. (7), (8), (11) and (12) with the proposed control algorithm of Eqs. (31) and (32) being numerically simulated based on Matlab/Simulink real-time environment. Simulation of ball bot [1] is done using the following parameters: $m_a = 116\,\text{kg}$, $I_x = 16.25\,\text{kgm}^2$, $I_y = 15.85\ \text{kgm}^2$, $r_w = 0.1\ \text{m}$, $l = 0.23\,\text{m}$, $I_w = 0.26\,\text{kgm}^2$, $r_k = 0.19\,\text{m}$, $m_k = 11.4\,\text{kg}$, $I_k = 0.165\,\text{kgm}^2$, $b_x = b_y = 5\,\text{Ns/m}$, $b_{rx} = b_{ry} = 3.68\ Nms/rad$ and the zenith angle $\alpha = 56°$. The control parameters used for the simulation of the ball bot are taken as $c_{x1} = 0.01$, $c_{x2} = 35$, $\alpha_x = 0.05$, $\eta_{x2} = 0.1$, $k_{x2} = 10$, $c_{y1} = 0.01$, $c_{y2} = 17$, $\alpha_y = 0.05$, $\eta_{y2} = 0.1$, $k_{y2} = 10$. The control parameters are selected to increase the speed of the system response in the reaching phase for the ball bot system. The ratio of wheel rotation to ball rotation is determined by the zenith angle, which is an important parameter of the ball bot system.

The response time of the proposed controller is fast as compared to the Hierarchical sliding mode control. The proposed controller stabilizes the ball bot system much faster than conventional Hierarchical sliding mode control. Simulation of the ball bot is carried out in the x-z plane, and it is shown in Figs. 2, 3, 4, 5 and 6. The initial conditions of the ball bot in the x-z plane $\left(x, \dot{x}, \theta_y, \dot{\theta_y}\right)$ are taken as −25, 0, 6.5°, 0.

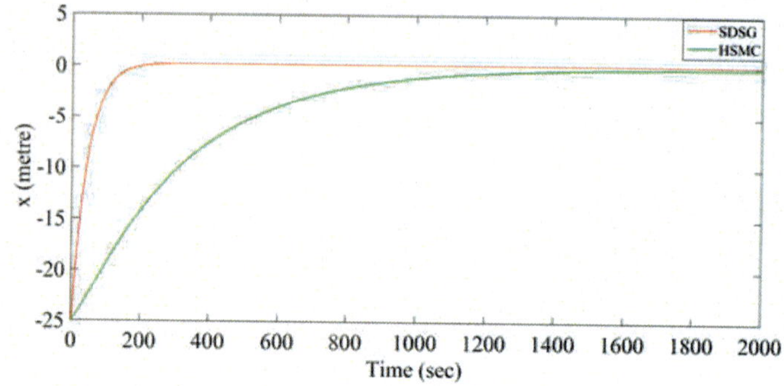

Fig. 2 Plot of x versus time

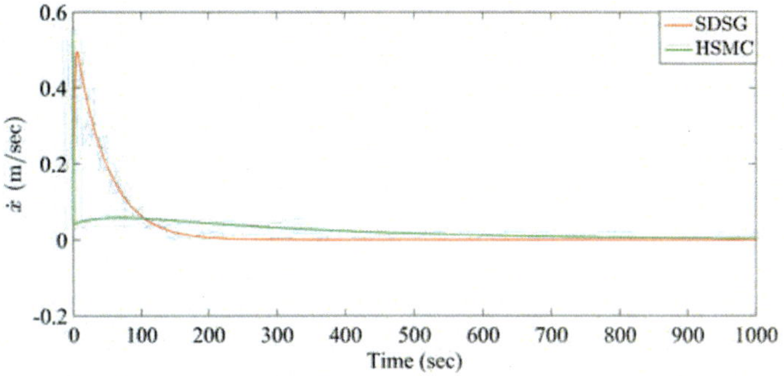

Fig. 3 Plot of $\dot{x}$ versus time

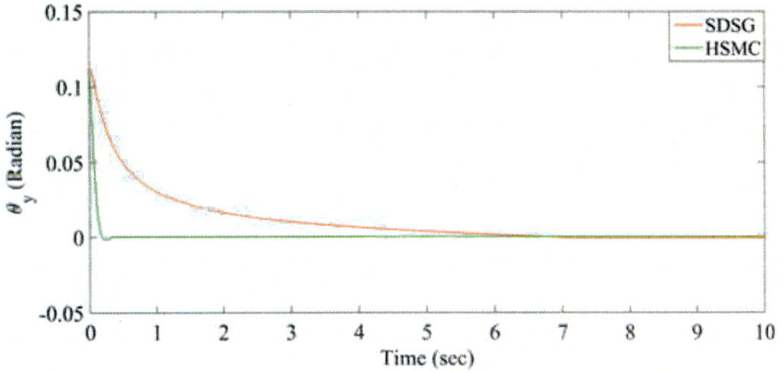

Fig. 4 Plot of θ_y versus time in the x-z plane

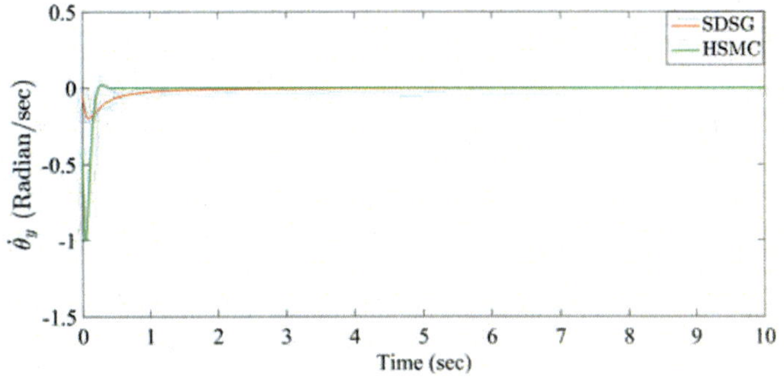

Fig. 5 Plot of $\dot{\theta}_y$ versus time in the x-z plane

Simulation results for the ball bot in the y-z plane are shown in Figs. 7, 8, 9, 10 and 11, respectively. The initial conditions of the ball bot in the y-z plane $\left(y, \dot{y}, \theta_x, \dot{\theta_x}\right)$ are taken as $-25, 0, 6.5°, 0$.

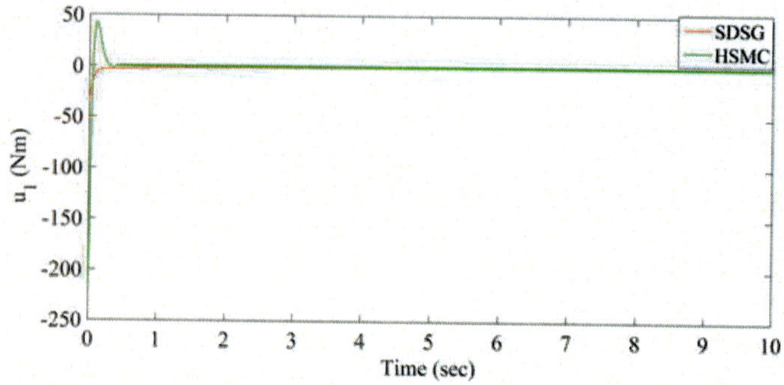

Fig. 6 Plot of u_1 versus time

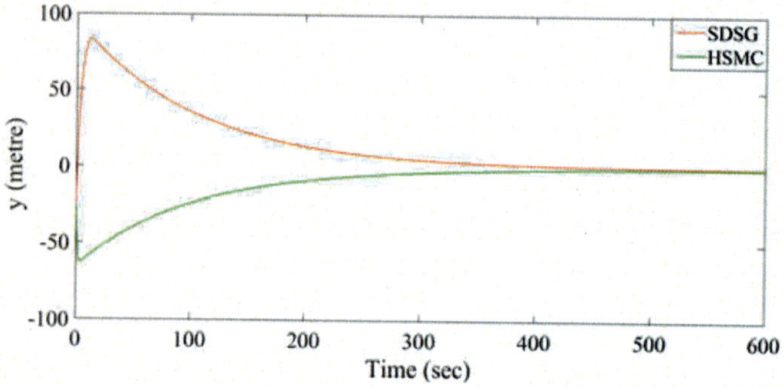

Fig. 7 Plot of y versus time

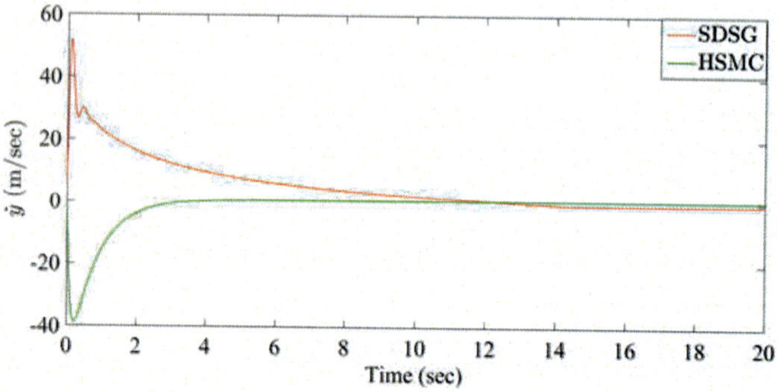

Fig. 8 Plot of $\dot{y}$ versus time

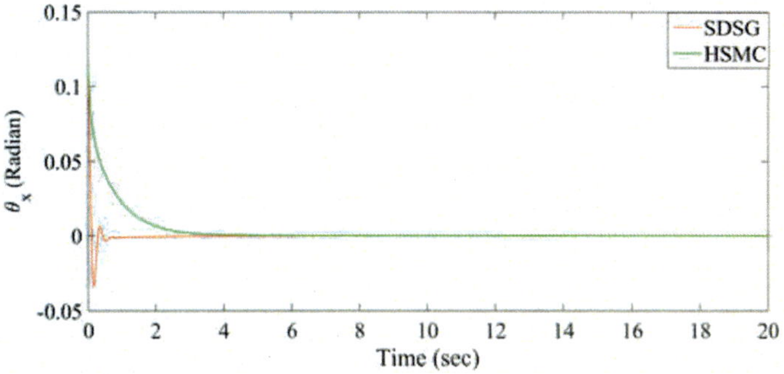

Fig. 9 Plot of θ_x versus time in the y-z plane

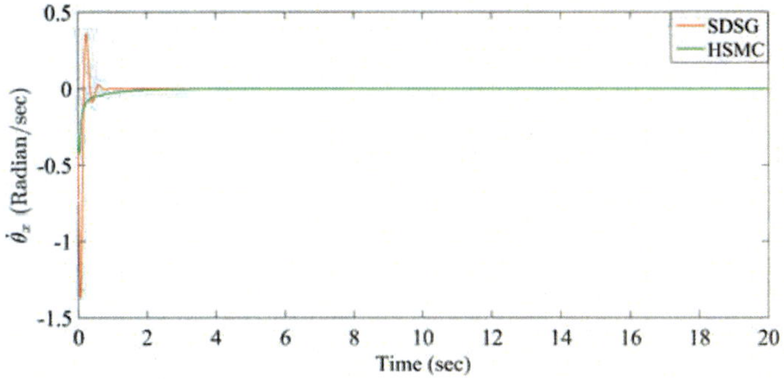

Fig. 10 Plot of $\dot{\theta}_x$ versus time in the y-z plane

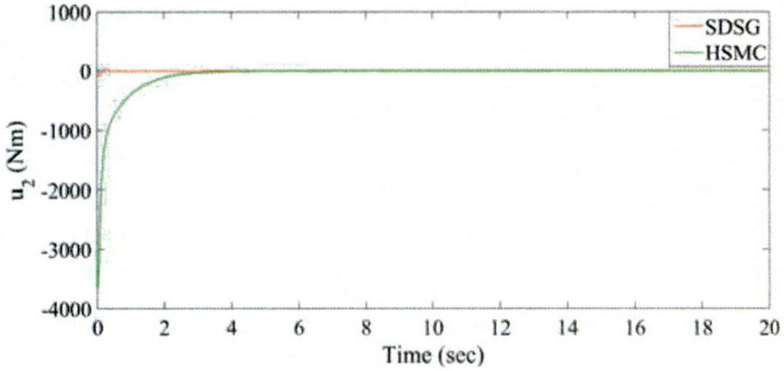

Fig. 11 Plot of u_2 versus time

5 Conclusion

State-dependent switching gain-based hierarchical sliding mode controller is proposed for the stabilization of the ball bot system. Stabilization of the ball bot is achieved using the proposed controller. The results of state-dependent switching gain-based controller and Hierarchical sliding mode controller are compared for the stabilization of the ball bot system. The proposed controller stabilizes the ball bot system in less time as compared to Hierarchical sliding mode control. The state-dependent switching gain-based controller requires less control input as compared to Hierarchical sliding mode control. Simulation results validate the efficacy of the proposed controller.

References

1. Pham DB, Lee S-G (2018) Hierarchical sliding mode control for a two-dimensional ball segway that is a class of a second-order underactuated system. J Vib Control 25(1):72–83
2. Lee SM (2020) Bong Seok park: robust control for trajectory tracking and balancing of a ballbot. IEEE Access 8:159324–159330
3. Hasan A (2020) eXogenous Kalman filter for state estimation in autonomous ball balancing robots. In: IEEE/ASME international conference on advanced intelligent mechatronics, Boston, USA
4. Hertig L, Schindler D, Bloesch M, David Remy C, Siegwart R (2013) Unified State estimation for a ballbot. In: IEEE international conference on robotics and automation. Karlsruhe, Germany
5. Nagarajan U, Kantor G, Hollis R (2014) The ballbot: An omnidirectional balancing mobile robot. Int J Robot Res 33(6):917–930
6. Nagarajan U, Kantor G, Holli RL (2009) Trajectory planning and control of an underactuated dynamically stable single spherical wheeled mobile robot. In: IEEE international conference on robotics and automation, Kobe, Japan (2009)
7. Herrera L, Hernandez R, Jurado F (2018) Control and extended Kalman filter based estimation for a ballbot robotic system. In: Robotics Mexican congress, Ensenada, Mexico
8. Kumagai M, Ochiai T (2008) Development of a robot balancing on a ball. In: International conference on control, automation and systems, Coex, Seoul, Korea
9. Lal I, Codrean A, Busoniu L (2020) Sliding mode control of a ball balancing robot. In: 21st IFAC world congress. Berlin, Germany
10. Jang H-G, Hyun C-H, Park B-S (2021) Neural network control for trajectory tracking and balancing of a ball-balancing robot with uncertainty. Appl Sci 11(11):1–12

Programmable Bot for Multi Terrain Environment

K. R. Sudhindra, H. H. Surendra, H. R. Archana, and T. Sanjana

1 Introduction

IFR (International Federation of Robotics) aims at promoting the research and development in the field of robotics, industrial robots and service robots as well as setting standards to the design and manufacturing of the robots worldwide. Development of Robotics and Automation in India is monitored by the All-India Council for Robotics and Automation (AICRA) [1]. The organization aims in making India the global leader in the field of Robotics, Artificial Intelligence and Internet of Things (IoT). It provides support to educational institutions to produce the best talents in this field [2]. An intelligent autonomous system requires accurate information about location of the vehicle and present road scenario. The system must be robust to handle adverse weather conditions. Algorithm must be designed to identify road margins with tolerably minimum error. This is possible from measurements obtained from equipment such as laser sensors and camera. Even in case of having incomplete information, Autonomous vehicles should be able to take quick decisions that mostly might not be considered by the programmer.

A miniature version of Autonomous vehicle is an Autonomous bot which is also expected to move from specified source to destination without human intervention or minimal intervention. This paper discusses about developing an autonomous bot or self-navigating bot equipped with a Kinect sensor for capturing image. It also has an IR camera which can generate a depth image. It is interfaced with a microprocessor and a Dell Vostro laptop using ROS framework on Ubuntu. Obstacles may be dynamic or static. Thus there are atleast two valid approaches to solve such problems. Ultrasonic sensors are attached to the bot to detect immediate moving objects in its path. It is controlled by an Arduino Uno. A YOLOv4 model is developed for object detection on images captured by Kinect RGB camera and bot coordinates are

K. R. Sudhindra (✉) · H. H. Surendra · H. R. Archana · T. Sanjana
B.M.S. College of Engineering, Bangalore, India
e-mail: krsudhindra.ece@bmsce.ac.in

© The Author(s), under exclusive license to Springer Nature Singapore Pte Ltd. 2023

H. Muthusamy et al. (eds.), *Robotics, Control and Computer Vision*, Lecture Notes in Electrical Engineering 1009, https://doi.org/10.1007/978-981-99-0236-1_29

collected by a GPS module. The following sections describe the development stages of the project. In Sect. 2, the block diagram of proposed solution with Hardware and Software Architecture are illustrated and described. In Sect. 3, all the implementations of the Self-Navigating bot are discussed. Section 4 discusses the results of each implementation, and finally, conclusions are given in Sect. 5.

2 Hardware and Software Architecture

The Self-Navigating bot development involves both software and hardware interfacing of different components. Raspberry Pi acts as the main processor for handling Kinect sensor, running on Ubuntu-20.04 LTS using ROS framework. Arduino Nano collects data from speed sensor for odometry of bot and sends same data to Pi and controls motors based on Pi signal. Arduino Uno collects data from GPS(Neo-6M Module) and IMU (MPU6050) and conveys it to Pi for location identification and orientation of the bot respectively. Ultrasonic sensors are connected for immediate obstacle avoidance and YOLOv4 is implemented using OpenCV and machine learning for object detection on Pi. The flow chart depicting the operation of the bot with the necessary hardware required is depicted in Fig. 1. The software packages and algorithms required for interfacing with the hardware and successful implementation of the prototype is as shown in Fig. 2.

Collision avoidance is based on reconfiguration method where the joints are made active/passive to enable collision-free tip trajectory. Previous works on collision avoidance are based on optimization approaches but with inherent limitations like not having any information about the manipulator configuration after collision avoidance [12].

3 Implementation

In this section, the integral parts of implementation such as Universal Robot Description Format (URDF) model creation, design of the hardware model, object detection module, Simultaneous localization and mapping (SLAM), path planning and interfacing of different components with Arduino are discussed.

3.1 URDF Model Creation

A 3D model of the robot will be initially designed using SolidWorks software and built using the chassis, motors, controller, and circuit connections. The design of the 3D model is shown below in Fig. 3, the chassis is made up of acrylic of 4mm in thickness. The robot has a differential drive mechanism, which is a two-stage body

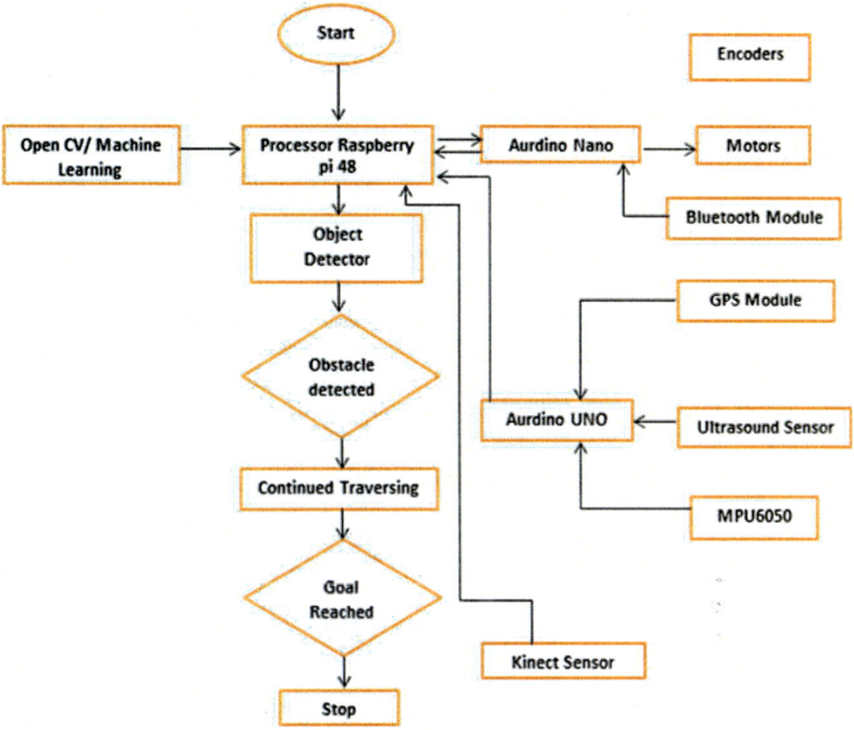

Fig. 1 Flow chart including the hardware components used

with two wheels and a castor wheel. The robot has a Kinect on top of the flat acrylic slate supported by an acrylic plate on top of spacers. Then the model is extracted into URDF (to provide the transforms between the joints for ROS integration and simulation purposes. Later the URDF is used to perform the simulation in RViz along with some ROS plugins. The robot is made to move in all possible directions and speed. Its movement is observed for any deviations due to weight distribution while both motors are given the same velocity. Figure 3a shows the robot model created in SolidWorks and Fig. 3b shows the model simulated in RViz.

3.2 *Hardware model*

As SolidWorks model depicted the hardware bot is designed with acrylic of 4mm thickness chasis and connections are made similar to block diagram and final model is shown in Fig. 4.

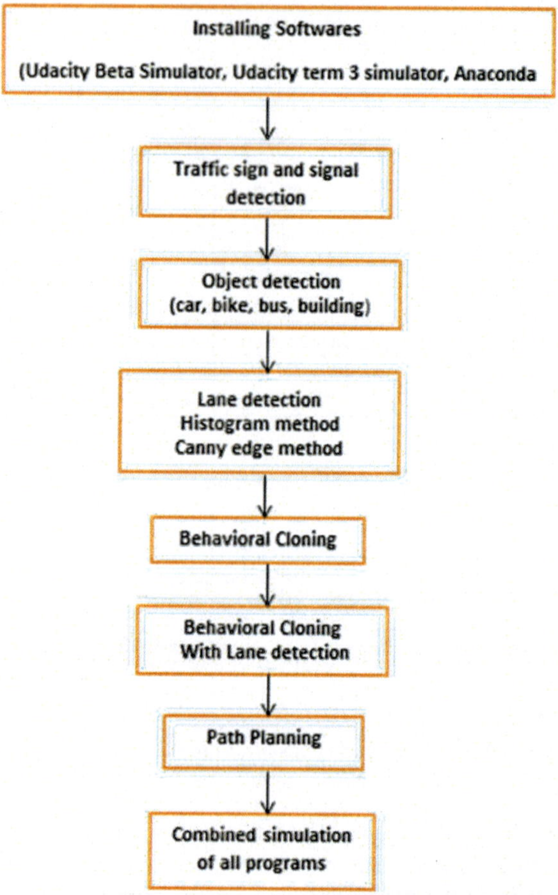

Fig. 2 Implementation flow which includes software and algorithms used

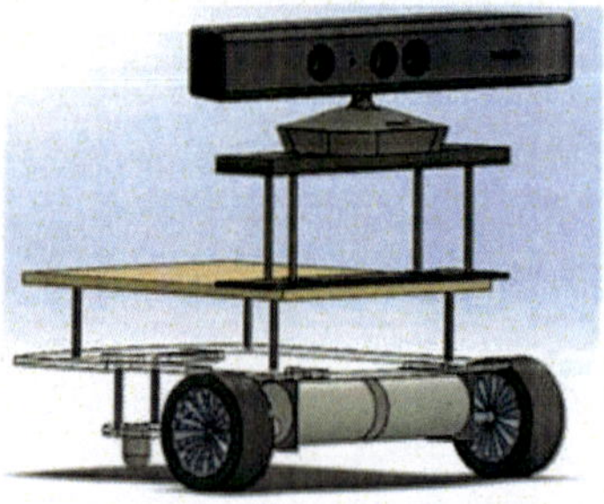

(a) Bot model created in SolidWorks

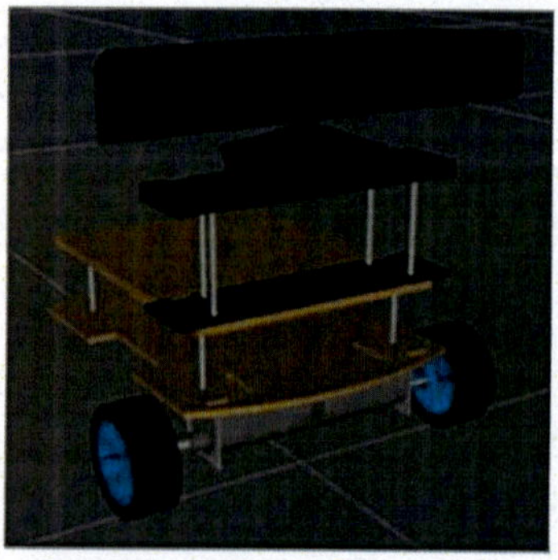

(b) Bot model simulated in rviz

Fig. 3 URDF model

Fig. 4 Hardware bot

3.3 *Object Detection Model*

Object detection is done using YOLOv4 on the Tensorflow framework. Effective implementation of the same can be done with the support of GPU and CPU. YOLO (You Only Look Once) uses one stage of detectors. It uses one neural network for the entire image. Images are divided into segments, bounding boxes are created and probabilities of match are calculated for each. Each of the bounding boxes are multiplied by the probabilities.

Input image to the YOLO is divided into grids and each grid into bounding boxes. Probability is calculated for each of the bounding boxes. The class which has probability greater than threshold is chosen. In this work, a dataset which contains images of traffic is used. Each object in the dataset (Eg. a car, a person, traffic lights) is called as a class. This data is divided for the training and testing phases.

3.4 *SLAM*

A ROS Navigation algorithm is developed using ROS framework. The map of the environment is built which acts as a reference for navigation, localization of the robot in 3D space, and path planning from the current position to the user's given destination position avoiding both dynamic and static obstacles. Localization can be achieved by a SLAM technique called RTAB-map available with ROS framework. Localization is identifying the robot's position and orientation with respect to the environment. It is an RGB-D graph SLAM method based on the global Bayesian loop closure detector. It uses an approach that, how often a new frame is captured using the Kinect sensor, from a new location or old location, which is known as loop closure detector.

IMU and encoder ticks are used to create odometrily to localize the robot in the map. Initially, sub-maps are created using the consecutive scan data from the Kinect sensor which is a probability grid (2D matrix) for a specific region of space, the values indicate the probability of grid being obstructed. After the completion of environment mapping, the map data is stored in the form of rtabmap.db database. The launch folder contains four ROS node launch configurations and the config directory contains the RViz configuration file, and a script for tele-operating the bot can be found in the script.

Path planning is performed using several functions. Move base is used for path planning, responsible for the functions like robot controlling, traversing, and trajectory planning. Given a goal in the world, move base will publish the required velocities to move the robot base towards the goal by using global plan and local plan. Cost map is a map data type that uses laser sensor data and saved maps to update the information about both dynamic and static obstacles. For instance, if it is 2 m, it means that the cost of the cells from the obstacle starts exponentially decreasing and when the distance from the obstacle is more than 2 m the cost of the cell due to this obstacle is zero. There are two types of cost maps. They are global and local cost maps where global cost map is a static map, which considers only static obstacles and local cost map accounts mainly for dynamic obstacles.

The move base path planner subscribes to the map topic along with wheel odometry and laser scan and publishes global and local plans. The planners further subscribe to their respective cost map and calculate the velocity at which the robot should move and publishes that data over the topic cmd_vel of message type geometry/Twist. The differential drive node subscribes to this twist message and calculates the velocity for two motors independently based on linear velocity in x direction and angular velocity in z direction.

It publishes two messages of float type (ex: +/−40.0). The sign indicates clockwise or anti-clockwise rotation, magnitude indicates the velocity value in m/s. With help of ROS serial, the Arduino subscribes to both the values and actuates the two motors based on the velocity commands.

3.5 Interfacing with Arduino

GPS module, Speed sensor, ultrasonic sensor, IMU, and keypad are interfaced with the Arduino. A Ublox Neo-6M GPS module is connected to the Arduino Uno. It uses serial Communication connected over Rx and Tx pins using UART protocol with a default baud rate of 9600. GPS module needs to lock on to 2–3 satellites for receiving coordinates of bot which may take upto 3–5 min. This delay is present because, the on-chip EEPROM needs to charge up to a certain level to get a lock on the satellites. A speed sensor is connected to an Arduino Nano and an encoder disk is attached to motor shaft, where the disk rotation implies motor rotation by counting ticks. The same data is used to calculate the odometry of the bot.

An HC-SR04 ultrasonic sensor is connected to the Arduino Uno and it is used for finding the distance from the bot to either a static or a dynamic object. The sensor transmits ultrasound waves, these waves hit the object and bounces back to the receiver. The distance is calcluated by measuring the amount of time the wave took to return to the receiver. The Eq. (1) is used to calculate the distance.

$$D = \frac{1}{2} \times c \times t \tag{1}$$

where D is the distance, c is the speed of sound and t is the time taken for the wave to return. A total of three ultrasonic sensors are used for three different directions.

A MPU6050 and Magnetometer (QMC833L) are connected to the Arduino Uno for orientation of bot. It is based on I2C communication and the data can be collected using the same. Rosserial communication is used to later publish the data to ROS framework.

The geo-location detected from GPS and Arduino interface is published to ROS framework, for navigation in autonomous mode towards its goal set by user. In ROS, geographiclib python library and WGS ellipsoid are used to convert the geo-coordinates into cartesian coordinates corresponding to the occupancy grid map. The location info can be sent in using either sensor msgs/NavSatFix or geometry msgs/PoseStamped message format. In pose stamped method bots desired orientation data i.e., quaternion (x, y, z, w) are sent. A launch file named initialize origin will initialize and sets origin to (0, 0, 0) and publish geometry msgs/PoseStamped message to local xy origin frame parameter of ROS coordinate frame.

A 3*4 keypad is connected to raspberry pi for insertion of security code. A security feature for delivery type bots is developed based on the Pi and keypad interface. Whenever a key on the keypad is pressed, that column gets high and pi sends high signals to each row so based on row and column combination the pressed key can be determined. A python library random is used to generate a random key code of specific length as password and using pywhatkit python library, the code can be sent to selected users.

4 Result and Discussion

In this work, indoor environment is considered for testing the bot. Results corresponding to RTAB-Map, object detection, GPS and Ultrasonic sensors are presented and discussed. SLAM is achieved on ROS framework using RTAB-map node. Results were obtained as SLAM map for the cases of absence of bot, presence of bot and bot navigating in the region of interest. These cases are depicted in Figs. 5, 6 and 7 respectively.

YOLOv4 successfully detected the objects both on webcam and a video. Figure 8 shows an example of object detection performed using YOLOv4.The encoder ticks

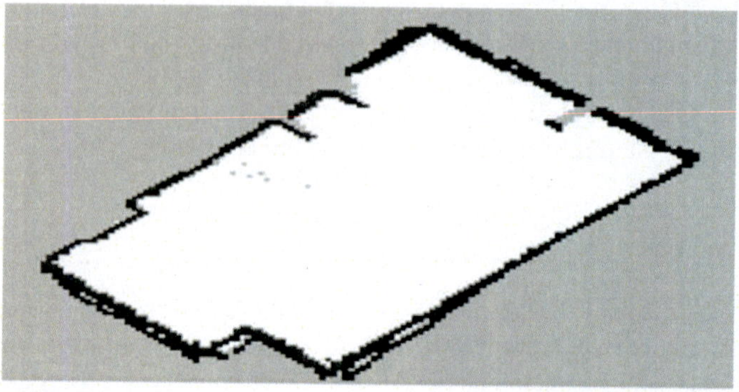

Fig. 5 SLAM map

Fig. 6 Bot shown in SLAM map

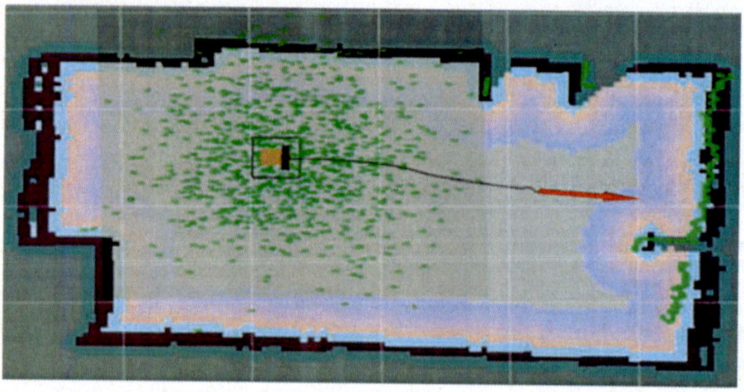

Fig. 7 Bot in auto-navigation mode

Fig. 8 Object detection using YOLOv4

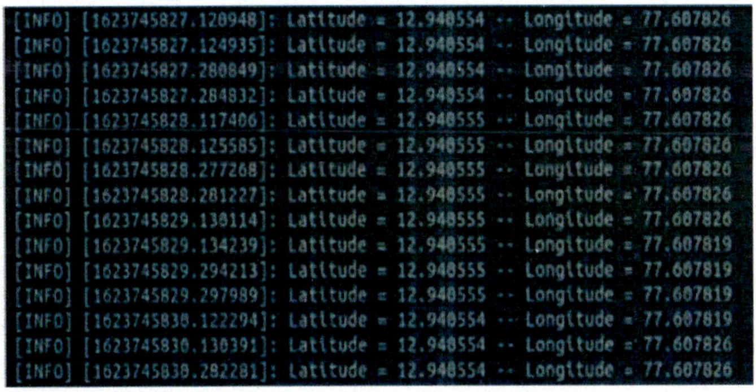

```
[INFO] [1623745827.120948]: Latitude = 12.940554 -- Longitude = 77.607826
[INFO] [1623745827.124935]: Latitude = 12.940554 -- Longitude = 77.607826
[INFO] [1623745827.280849]: Latitude = 12.940554 -- Longitude = 77.607826
[INFO] [1623745827.284832]: Latitude = 12.940554 -- Longitude = 77.607826
[INFO] [1623745828.117406]: Latitude = 12.940555 -- Longitude = 77.607826
[INFO] [1623745828.125585]: Latitude = 12.940555 -- Longitude = 77.607826
[INFO] [1623745828.277268]: Latitude = 12.940555 -- Longitude = 77.607826
[INFO] [1623745828.281227]: Latitude = 12.940555 -- Longitude = 77.607826
[INFO] [1623745829.130114]: Latitude = 12.940555 -- Longitude = 77.607826
[INFO] [1623745829.134239]: Latitude = 12.940555 -- Longitude = 77.607819
[INFO] [1623745829.294213]: Latitude = 12.940555 -- Longitude = 77.607819
[INFO] [1623745829.297989]: Latitude = 12.940555 -- Longitude = 77.607819
[INFO] [1623745830.122294]: Latitude = 12.940554 -- Longitude = 77.607819
[INFO] [1623745830.130391]: Latitude = 12.940554 -- Longitude = 77.607826
[INFO] [1623745830.282281]: Latitude = 12.940554 -- Longitude = 77.607826
```

Fig. 9 Bot coordinates

from speed sensor and orientation from mpu6050 were used to create odometry data. Figure 9 shows the latitude and longitude values obtained from the GPS module.

The current location of the bot were used as origin and the destination coordinates are to be given manually. The goal of GPS for a certain area of interest is depicted in Fig. 10.

Figure 11 shows the data gathered from all 3 ultrasonic sensors at a time.

5 Conclusion and Future Work

All the objectives including environment Map building, Autonomous Navigation, and GPS interface were successfully achieved. The design of the robot model and analysis for payload capacity was done in SolidWorks. The URDF of the robot model was extracted using the same. The Self-Navigating Bot was interfaced with RViz,

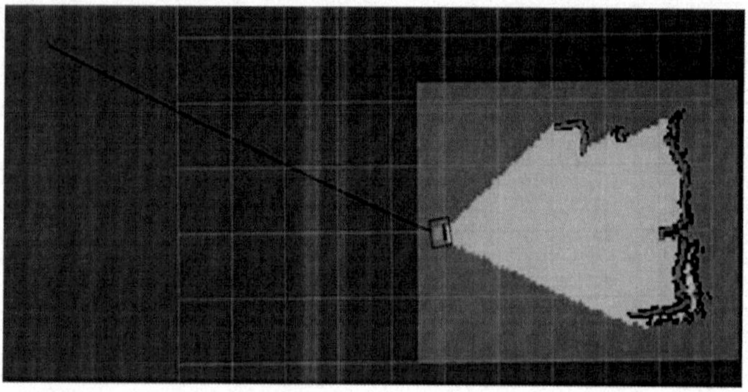

Fig. 10 GPS goal

```
left = 226 cm   center = 228 cm right = 225 cm
left = 226 cm   center = 228 cm right = 225 cm
left = 226 cm   center = 228 cm right = 225 cm
left = 226 cm   center = 227 cm right = 225 cm
left = 226 cm   center = 228 cm right = 225 cm
left = 226 cm   center = 228 cm right = 225 cm
left = 226 cm   center = 228 cm right = 225 cm
left = 226 cm   center = 228 cm right = 225 cm
left = 226 cm   center = 228 cm right = 225 cm
left = 226 cm   center = 227 cm right = 225 cm
left = 226 cm   center = 227 cm right = 225 cm
left = 227 cm   center = 227 cm right = 225 cm
left = 226 cm   center = 228 cm right = 225 cm
left = 226 cm   center = 227 cm right = 225 cm
left = 226 cm   center = 227 cm right = 225 cm
left = 226 cm   center = 227 cm right = 225 cm
left = 226 cm   center = 228 cm right = 225 cm
left = 226 cm   center = 228 cm right = 185 cm
left = 226 cm   center = 228 cm right = 220 cm
left = 226 cm   center = 227 cm right = 225 cm
```

Fig. 11 Ultrasonic measurements

which enables users to visualize the robot and the occupancy grid map in real time. The SLAM algorithm was initially developed using gmapping. To increase the speed of mapping and to navigate in unexplored areas, RTAB-map was used at the cost of system computation. The testing environment was limited to a small area of 3 m^2 due to the range constraint of the Kinect sensor. The robot was tested for different real time scenarios. In future work, we plan to increase the size of the testing area, a LIDAR or a laser scanner. A sound localization can be developed with voice-controlled based navigation, which makes the robot to identify the user's location and direction and navigate to that location when user sends the command to the robot with a unique ID, this makes easier interaction with the user.

Acknowledgements The Authors would like to thank B.M.S. College of Engineering for supporting to carry out this work.

References

1. Aziz MVG, Prihatmanto AS (2017) Implementation of lane detection algorithm for self-driving car on toll road using python language. In: 4th international conference on electric vehicular technology (ICEVT 2017). ITB Bandung, Indonesia
2. Prabhu S, Kannan G, Indra Gandhi K , Irfanuddin, Munawir (2018) GPS controlled autonomous bot for unmanned delivery. In: International conference on recent trends in electrical, control and communication (RTECC 2018), Chennai
3. Brahmanage G, Leung H (2017) A Kinect-based SLAM in an unknown environment using geometric features. In: International conference on multisensor fusion and integration for intelligent systems (MFI 2017), Daegu, Korea, 16–18 Nov 2017
4. Jape PR, Jape SR (2018) Virtual GPS guided autonomous wheel chair or vehicle. In: 3rd international conference for convergence in technology (I2CT 2018). The Gateway Hotel, XION Complex, Wakad Road, Pune, India, 06–08 Apr 2018
5. Thorat ZV, Mahadik S, Mane S, Mohite S, Udugade A (2019) Self-Driving car using Raspberry-Pi and machine learning. Int Res J Eng Technol (IRJET), Navi Mumbai 6(3)
6. Das S, Simultaneous localization and mapping (SLAM) using RTAB-Map. https://arxiv.org/pdf/1809.02989.pdf
7. ROS Noetic. http://wiki.ros.org/noetic/Installation/Ubuntu
8. rtabmap. http://wiki.ros.org/rtabmapros/Tutorials/SetupOnYourRobot
9. movebase. http://wiki.ros.org/movebase
10. IFR World Robotics. https://ifr.org/worldrobotics/
11. https://www.youtube.com/watch?v=u9l-8LZC2Dc
12. Dalla VK, Pathak PM (2015) Obstacle avoiding strategy of a reconfigurable redundant space robot. In: Proceedings of the international conferences. on integrated modeling and analysis in applied control and automation

A Computer Vision Assisted Yoga Trainer for a Naive Performer by Using Human Joint Detection

Ritika Sachdeva, Iresha Maheshwari, Vinod Maan, K. S. Sangwan, Chandra Prakash, and Dhiraj

1 Introduction

Yoga has recently gained worldwide popularity due to its physical and mental benefits. Everyone needs to practice yoga to establish a balance between themselves and their surrounding environment. The United Nations General Assembly declared June 21st as the 'International Day of Yoga' in 2014 [1]. COVID-19's ambiguity, as well as the subsequent lockdown, created a great deal of worry, tension, and anxiety and we were all compelled to remain at home, making life extremely difficult [2]. Over the last few years, yoga has received a lot of attention in the field of healthcare. Yoga helps in the reduction of stress and anxiety, as well as the improvement of physical health and the minimization of negative mental effects [3]. People who do

R. Sachdeva (✉) · I. Maheshwari
Department of Electronics and Communication Engineering, Indian Institute of Information Technology Kota, Kota, India
e-mail: 2018kuec2044@iiitkota.ac.in

I. Maheshwari
e-mail: 2019kuec2041@iiitkota.ac.in

V. Maan
Mody University of Science and Technology, Laxmangarh, Raj, India
e-mail: vinodmaan.cet@modyuniversity.ac.in

K. S. Sangwan
Birla Institute of Technology and Science, Pilani, India
e-mail: kss@pilani.bits-pilani.ac.in

C. Prakash
National Institute of Technology, Delhi, India
e-mail: cprakash@nitdelhi.ac.in

Dhiraj
CSIR-Central Electronics Engineering Research Institute, Pilani, India
e-mail: dhiraj@ceeri.res.in

© The Author(s), under exclusive license to Springer Nature Singapore Pte Ltd. 2023

H. Muthusamy et al. (eds.), *Robotics, Control and Computer Vision*, Lecture Notes in Electrical Engineering 1009, https://doi.org/10.1007/978-981-99-0236-1_30

not have a clear grasp of yoga begin practicing it without proper direction, and as a consequence, they harm themselves while performing due to their incorrect posture. It should be performed under the guidance of a professional.

Human Pose Estimation is a well-studied topic with applications in a variety of fields, including human–computer interaction, virtual reality, robots, and many more [4]. A perfect blend of these techniques can create wonders. Many frameworks and keypoint detection libraries for pose estimation have been introduced which makes it easier for everyone to build AI-based applications. One of them is the Mediapipe framework by Google for solving problems such as face detection, hands, pose, object detection, and many more using machine learning [5].

The aim of our method is to correct the user's yoga asana in real-time. We have developed a user-friendly Python-Flask based web application that assists its registered users to perform every pose accurately. The user is given feedback on how to modify their incorrect posture. The name of our web application is "Yuj" which is a Sanskrit root word for yoga: meaning to join or to unite [6].

"Yuj" is currently functioning for four asanas: Adho Mukha Svanasana (downward-facing dog posture), Phalakasana (plank pose), Trikonasana (triangular pose) and Virabhadrasana II (warrior-2 pose). The rationale for selecting these four asanas comes from the ease of availability of professional videos on the web, as well as the fact that these asanas are highly popular among people and simple for those who are new to yoga or novices.

Related work, methodology, and results are discussed in the following sections, followed by concluding remarks. Section 2 provides an overview of the work that has been proposed by others in the area. In Sect. 3, data collection and methodology are described. Experimental results are discussed in Sect. 4. Sections 5 and 6 examine concluding remarks and future prospects, respectively.

2 Related Work

A plethora of work has been proposed for the identification of human posture. Chen et al. [7] proposed a yoga self-training system that uses a Kinect depth camera to assist users in correcting postures while performing 12 different asanas. It uses manual feature extraction and creates separate models for each asana.

Trejo et al. [8] suggested a yoga recognition system for six asanas using Kinect and Adaboost classification and achieved an accuracy of 94.78%. For identification, they employed a depth sensor-based camera, which may not be generally available to the general public.

Borkar et al. [9] have developed a method called Match Pose, to compare a user's real-time pose with a pre-determined posture. They employed the PoseNet algorithm to estimate users' poses in real-time. They compared and checked whether users' real-time poses were properly replicated using pose comparison algorithms. The proposed approach enabled the user to choose only the image they wanted to replicate. After then, the user's real-time postures were collected using a camera and analyzed using a human pose estimation algorithm. The same technique was used to process the image from the database that was chosen. Yoga postures with finger placements are not comparable in the system.

Rishan et al. [10] proposed a yoga posture detection and correction system, that uses open pose to detect body keypoints and a Deep Learning model that analyzes and predicts user posture or asana using a sequence of frames utilizing time-distributed Convolutional Neural Networks, Long Short-Term Memory, and SoftMax regression. OpenPose is a real-time multi-person keypoint detection library introduced by the Perceptual Computing Lab of Carnegie Mellon University (CMU) [11]. It can jointly detect a human body, hand, facial, and foot keypoints on a single image.

Islam et al. [12] used the Microsoft Kinect to capture a person's joint coordinates in real-time. This system can only detect yoga poses; it cannot, however, assist the user in correcting an incorrect yoga posture.

Hand tracking is a key component that enables natural interaction and conversation, and it has been a subject of great interest in the industry. A significant portion of previous work necessitated the use of specialized hardware, such as depth sensors. In one investigation [13], the author used Mediapipe to demonstrate a real-time on-device hand tracking system that uses a single RGB camera to identify a human hand skeleton. It also presents a unique approach that works in real-time on mobile devices and does not require any additional hardware.

3 Proposed Methodology

The overall workflow of the system is as follows: The user has to first register on "Yuj". Then he/she can select their desired asana from the following asana after logging in the system: Adho mukha svanasana (downward-facing dog), Phalakasana (plank), Trikonasana (triangular pose), and Virabhadrasana II (warrior-2 pose) as shown in Fig. 1.

As soon as the pose is selected, the webcam is activated and the user starts performing the selected pose. Once in position, he/she shows a closed fist gesture to the webcam, this starts the video recording and his/her posture is captured. After recording the pose for 5 sec duration, visual and textual feedback is generated and provided to the user. Figure 2 shows the flowchart of our implementation.

3.1 Data Collection

It is hard to find an accurate and effective yoga-pose video dataset on the web. We gathered videos of people of various age groups and genders performing four yoga asanas: Virabhadrasana II (warrior-II), Trikonasana (triangular pose), Phalakasana (plank), and Adho Mukha Svanasana (downward-facing dog) from various online sources, including video channels and websites, for training purpose. According to

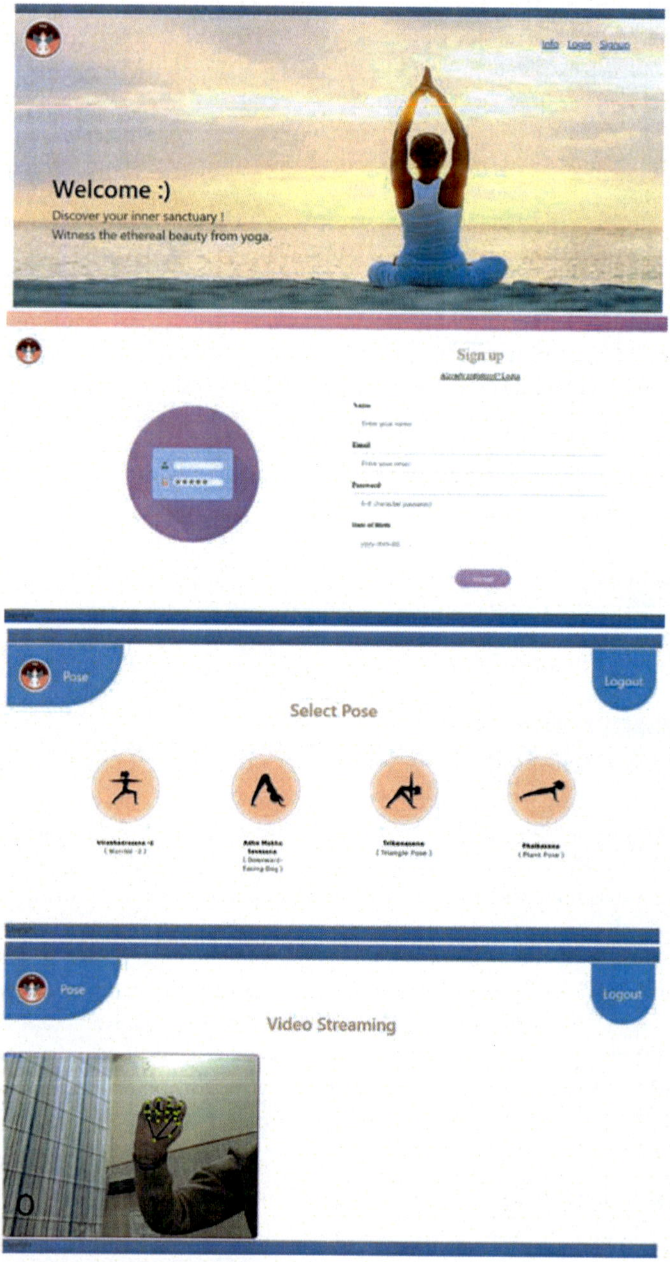

Fig. 1 Front-end designs of our web app

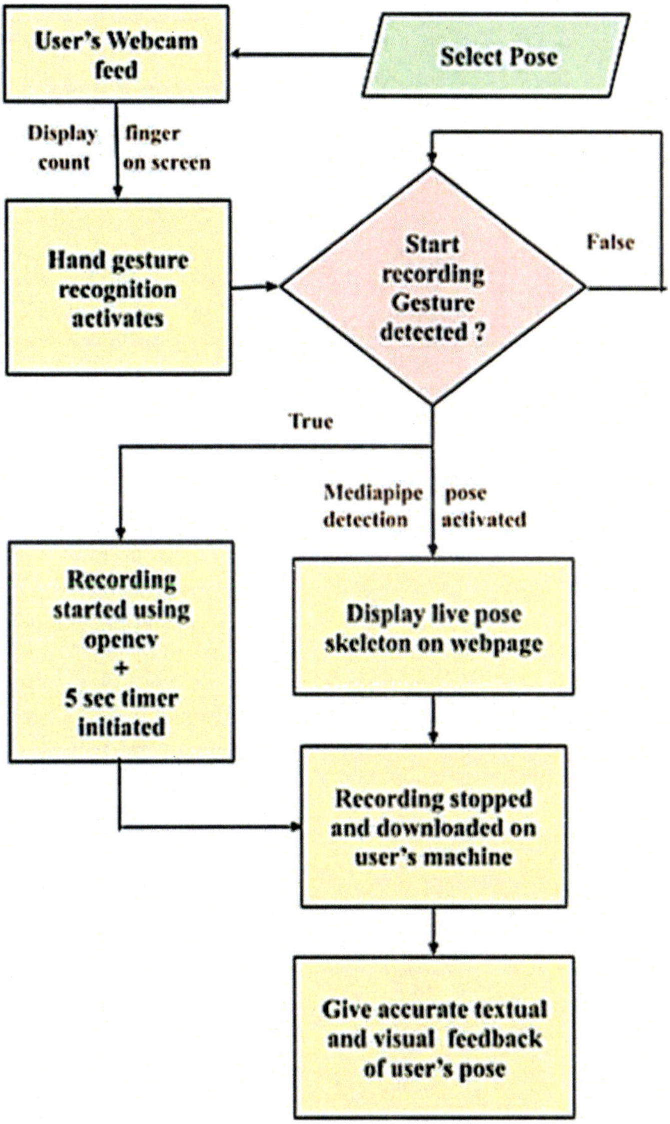

Fig. 2 Structural outline of our approach

the survey conducted by Patanjali Research Foundation [14] on 3135 yoga experienced persons. It is found that most of the people in the age group of 21–44 years, 45–59, and more than 60 years have a higher belief in the benefits of yoga and its practice. So, in our data collection of yoga videos, we have considered data ratios (shown in Fig. 3) similar to those provided in Table 2 of the survey [14].

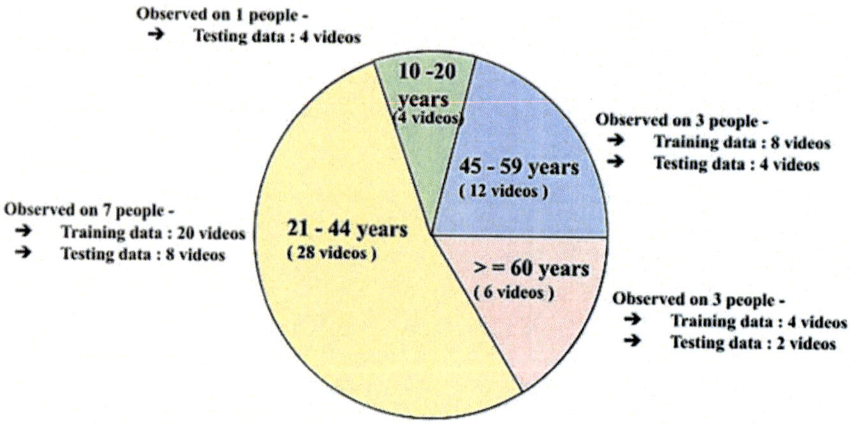

Fig. 3 Data Collection based on age group

A total of 50 videos were collected for testing and training purposes. In Fig. 3 from the term training data, we refer to those video datasets with which we determined the angle ranges for feedback generation. Whereas, the term testing data here refers to the ones we have used for observing the accuracy of our feedbacks.

For testing, all of the videos were recorded for 5 sec in an indoor as well as outdoor location at a frame rate of 20 frames per second (shown in Fig. 4). Table 1 describes the 4 poses which registered users can perform on Yuj.

Fig. 4 Row 1,2 and 3 represents testing data for age group 10–20 years, 21–44 years and > 60 years age group respectively

Table 1 Asana and Joint coordinates

S. no	Asana name	Posture
1	Virabhadrasana II (Warrior-II)	
2	Trikonasana (Triangular Pose)	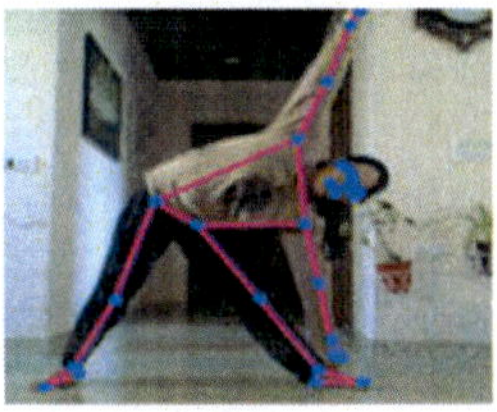
3	Phalakasana (Plank)	
4	Adho Mukha Svanasana (Downward Facing Dog)	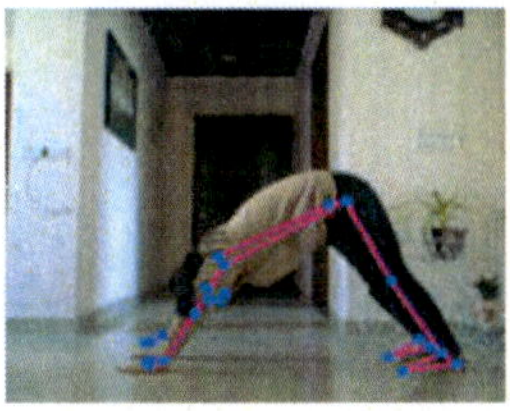

Table 2 Number of professionals' videos observed for each pose

Yoga posture	No. of videos of professionals
Virabhadrasana II	10
Trikonasana	9
Phalakasana	8
Adho Mukha Svanasana	8

3.2 *Hand Gesture Recognition*

To identify the timestamp when the user is ready in pose, we have introduced the concept of hand gesture recognition in our code. A specific gesture is defined by us, which when identified the very first time will command the machine to work on pose recognition and stop hand gesture recognition. In order to minimize latency and complexity, we have aimed to work only on those frames in which the user is ready in pose and there is a minimum deflection. To calculate minimum delflection, minimum deviation in keypoints of the user between adjacent captured frames is observed.

The Mediapipe Hands solution (initializing command: "mediapipe.solutions.hands") is used here for hand keypoints detection of the hand with detection confidence of 0.7. We have deduced 21—three dimensional landmarks of a hand from a single frame (Fig. 5b depicts all the 21 keypoints). In our approach, the mediapipe's palm detector (which has an average precision of 95.7% in palm identification) works on a full webcam captured image of 640 × 480 and locates palms via an aligned hand bounding box.

The detection of hand gestures is done with the help of finger count and frame count. A "**closed fist**" hand gesture is used as an initializing gesture to activate human pose recognition as shown in Fig. 5a. It will be identified when the finger count = 0 for continuous 50 frames (frame count = 50 f). 50 frames count means holding the closed fist gesture for 2.5 s (50 f/20fps = 2.5 s), this wait time makes sure that the triggering gesture is shown by the user when he/ she is actually ready in pose.

We have defined an array which consists of the hand landmarks of the tips of all fingers (Fig. 5b shows hand landmarks defined by Mediapipe): tips = [4, 8, 12, 16, 20].

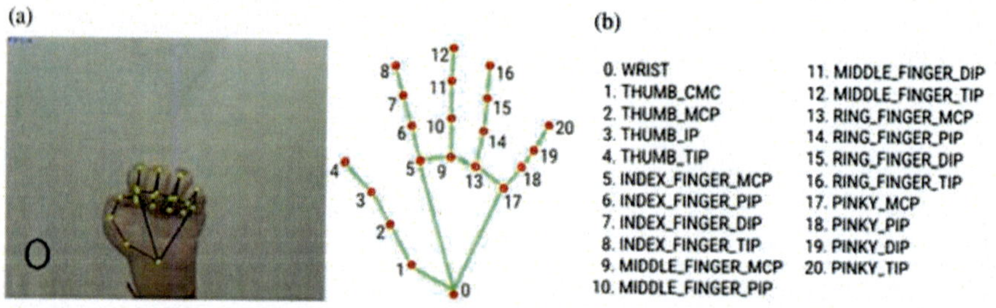

Fig. 5 **a** "Closed fist" gesture which acts as a trigger to start recording of video **b** Detailed information of Hand Landmarks in Mediapipe [15]

3.3 Human Yoga Pose Estimation

Once the closed fist gesture is detected, the incoming video stream is fed to mediapipe pose pipeline (its pose detector working is covered in brief in Fig. 6) for pose landmarks detection [16].

Upon correct detection of pose in a frame, those frames are processed in real-time to obtain the pose 33 landmarks (joint coordinates) and a live stick diagram is displayed on the web page. Only x and y coordinates of human joints, normalized to [0, 1] by the image width and height respectively, are fed in a csv file. Figure 7 depicts all the joint landmarks defined in mediapipe: it will play a vital role in our feedback mechanism.

The landmark distance from camera is represented by the z coordinate in the mediapipe, with the origin being the depth at the midway of hips, and the higher the value, the farther the joint is to the camera. The value of z is determined using a scale that is identical to that of x in the range [0, 1]. With x, y, z, and visibility the

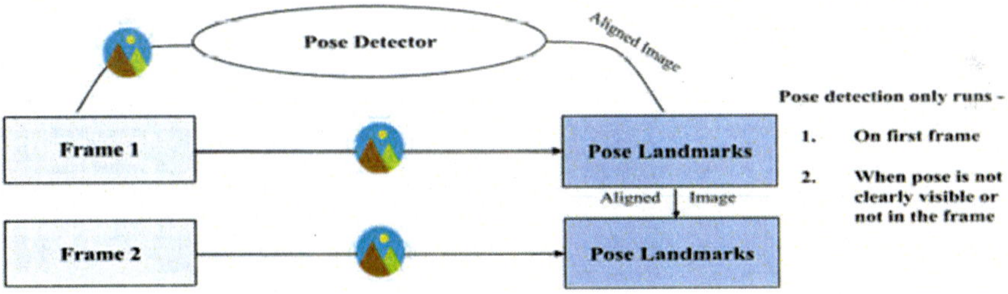

Fig. 6 Pose detection methodology of Mediapipe

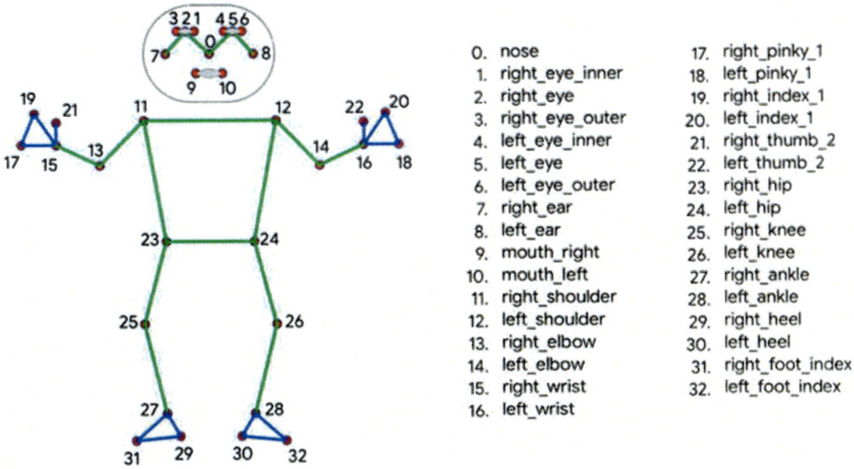

Fig. 7 Detailed information of Body Landmarks in Mediapipe [17]

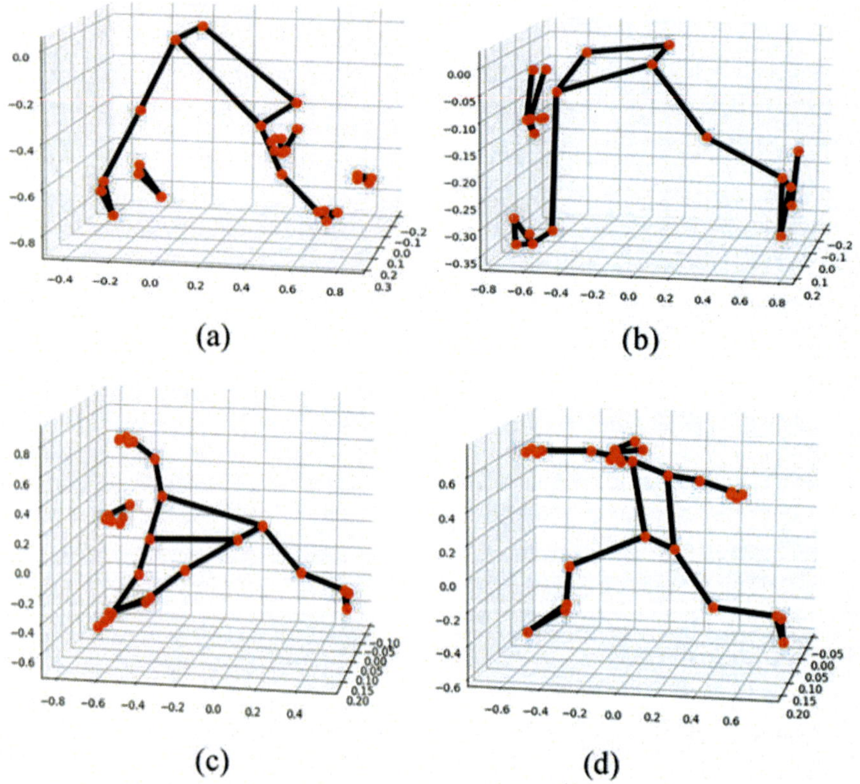

Fig. 8 3D Plot of: **a** Adho mukha svanasana (downward face) **b** Phalakasana (plank pose) **c** Trikonasana (triangular pose) **d** Virabhadrasana II (warrior-2 pose)

3D plots obtained for 4 different poses are shown in Fig. 8 but these 3D plots are not displayed on our webpage they have been shown for a better understanding of the concept.

3.4 Angle Calculation

To accurately define the angle ranges for various poses we have taken reference from 35 videos of professionals. Each frame of their video is used to determine the feasible range of angles for particular joints of specific pose. Table 2 describes the number of professional videos considered for each asana.

We have used the mathematical formulae in Eqs. (1–3) to calculate the angle between 3 joints.

Let's consider 3 joints J_1, J_2, and J_3.

To calculate angle between lines J_1 - J_2 and J_2 - J_3:

Step 1:

Using distance formula to find distances J_{12} and J_{23}

$$J_{12} = \text{sqrt}\ ((J_1(x) - J_2(x)).\ (J_1(x) - J_2(x)) + (J_1(y) - J_2(y)).(J_1(y) - J_2(y))) \quad (1)$$

$$J_{23} = \text{sqrt}\ ((J_2(x) - J_3(x)).\ (J_2(x) - J_3(x)) + (J_2(y) - J_3(y))\ (J_2(y) - J_3(y))) \quad (2)$$

Step 2:

Using "Law of Cosine" for angle calculation by taking J_2 as vertex

$$\text{angle}\ (J_{123}) = \arccos\left(\left((J_{12})^2 + (J_{13})^2 - (J_{23})^2\right)/(2 * J_{12} * J_{13})\right) \quad (3)$$

After observation, we have found that only 8 angles are sufficient to uniquely identify a particular pose as correct or incorrect. Given below is the list of angles considered for pose corrections which is also available on our website:
LH = Angle between Left_shoulder, Left_elbow and Left_wrist.
RH = Angle between Right_wrist, Right_elbow and Right_shoulder.
LU = Angle between Left_hip, Left_shoulder and Left_elbow.
RU = Angle between Right_elbow, Right_shoulder and Right_hip.
LW = Angle between Left_shoulder, Left_hip and Left_knee.
RW = Angle between Right_shoulder, Right _hip and Right_knee
LL = Angle between Left_ankle, Left _knee and Left_hip.
RL = Angle between Right_ankle, Right_knee and Right_hip.

We have further calculated the average of all the feasible angles from all the videos dataset of professionals depicted in Table 3. These values are taken as "Threshold angle" values.

For feedback purposes angle range is categorized into two categories:

Table 3 The reference angle values obtained from several professional videos

Reference values	LH	RH	LU	RU	LW	RW	LL	RL
Virabhadrasana II	178°	178°	90°	90°	135°	90°	178°	90°
Trikonasana	175°	170°	135°	85°	165°	60°	165°	170°
Phalakasana	90°	90°	90°	90°	167°	167°	178°	178°
Adho Mukha Svanasana	175°	175°	178°	178°	60°	60°	179°	179°

- Threshold angle $\pm 4°$ deviation → acceptable range (no feedback needed for that particular angle)
- Correction will be given on angles exceeding this value.

4 Experimental Results

When the trigger of the "closed fist" (depicted by 0 in Fig. 5a) gesture is provided, the code executes by parallel programming. The two side by side running processes are

- Video Recording (discussed in detail in Sect. 4.1)
- Real-Time Pose Estimation (discussed in detail in Sect. 3.3).

4.1 *Video Recording*

Recording of a 5 s video is performed using OpenCV from the very moment the trigger was captured. We have considered a 5 s timer because recording a video of more than 5 s when the user is already in pose from the very start is an increasing computational bottleneck and of no use. Once 5 s are over, the system saves that recorded video onto the user's downloads folder from the browser. The purpose of adding this feature is to provide the recorded videos to the user for his/her reference. For example, the users can compare their previously recorded videos with their latest videos (so that they can observe their improvement over time).

4.2 *Feedback Generation*

Once the 5 s timer ends, on the front-end the pose webpage directs to the feedback page, whereas backend processing is shown in Fig. 9.

The obtained csv file of joint coordinates is read and the mean of ten stillest frames is obtained. Mean of ten stillest frames for only 12 joint coordinates (mentioned in Table 4) are calculated to increase the accuracy and decrease the complexity thereby reducing latency of code. The mean coordinates obtained from the csv file are the most precise coordinates (for 5 s recorded video duration). The purpose of selecting only 12 joints is that these joints are sufficient to calculate the 8 angles for feedback generation.

Before moving to the angle approach, let's compare the deviation of 4 poses performed by the user with the reference pose i.e., professional's pose. This deviation approach is not a sufficient base to determine the feedback because when considering the 8 angles of user and professional are almost the same but both plots don't coincide

Fig. 9 Feedback mechanism

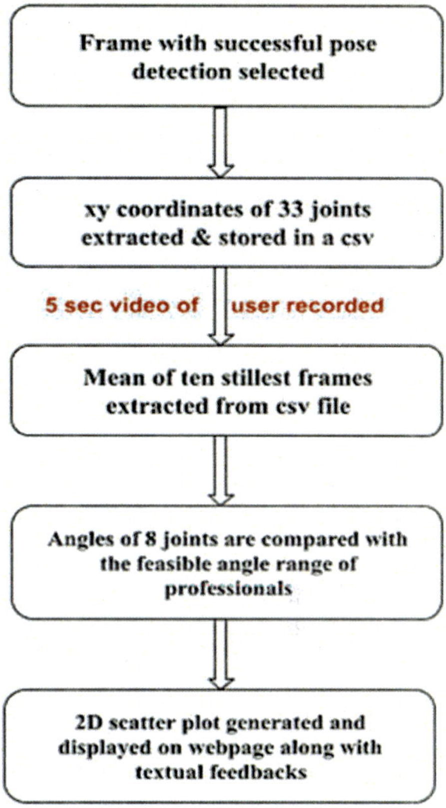

Table 4 Joint coordinates used for calculations

Mediapipe defined key points	Name given to each keypoint by us	Mediapipe defined key points	Name given to each keypoint by us
22	'L_SHOULDER_X'	46	'L_HIP_X'
23	'L_SHOULDER_Y'	47	'L_HIP_Y'
24	'R_SHOULDER_X'	48	'R_HIP_X'
25	'R_SHOULDER_Y'	49	'R_HIP_Y'
26	'L_ELBOW_X'	50	'L_KNEE_X'
27	'L_ELBOW_Y'	51	'L_KNEE_Y'
28	'R_ELBOW_X'	52	'R_KNEE_X'
29	'R_ELBOW_Y'	53	'R_KNEE_Y'
30	'L_WRIST_X'	54	'L_ANKEL_X'
31	'L_WRIST_Y'	55	'L_ANKEL_Y'
32	'R_WRIST_X'	56	'R_ANKEL_X'
33	'R_WRIST_Y'	57	'R_ANKEL_Y'

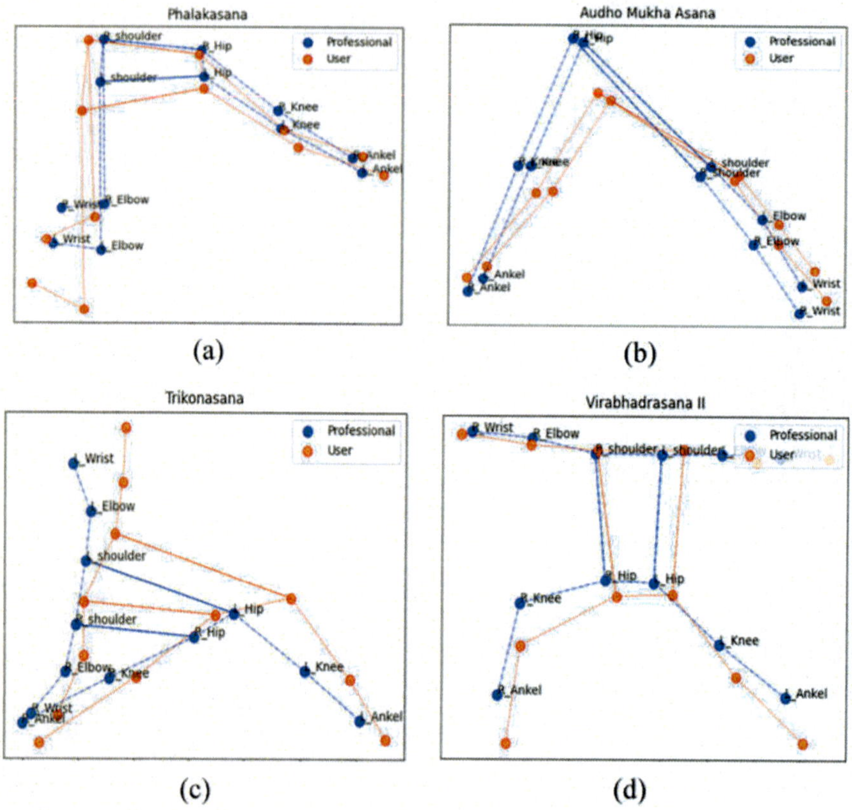

Fig. 10 Comparison scatter plot of user's pose with reference pose

as shown in Fig. 10. In the compared scatter plots (refer Fig. 10) the main cause of deviations observed is due to an individual's distance from the camera.

Figure 11 depicts a frame of the user's performed pose. All the images shown in Fig. 11 are only used for displaying the 8 angles used in the feedback mechanism but none of them are displayed on the website.

This data is used to give precise visual feedback to the user by plotting a scatter plot and displaying it on the web page. A list of strings is made to give textual feedback for all the 8 joint coordinates (if wrongly positioned). Figure 12 depicts the visual and textual feedback of one yogi performing different poses (an option to select the pose is available on our website). The feedback for the performed yoga pose is generated by the web application.

The initial word in textual feedback is categorized as

- Excellent!—When no angle exceeds beyond the acceptable range
- Good!—When 1 or more angles are beyond the acceptable range
- Oops!—When no angles are acceptable.

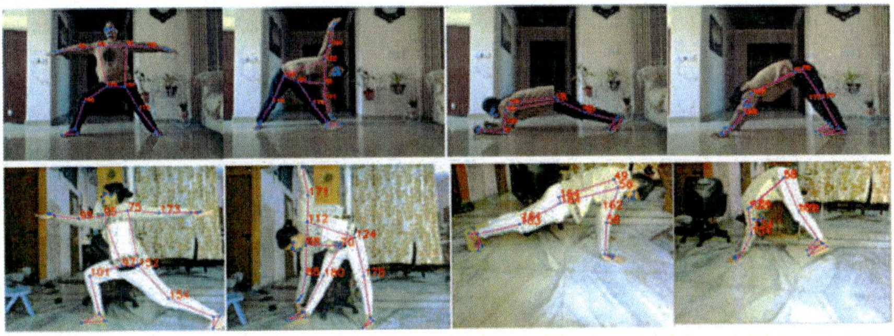

Fig. 11 Angles obtained from different users

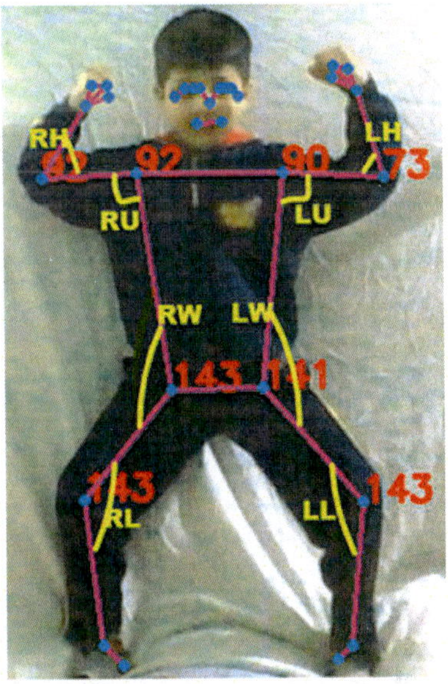

Fig. 12 Image depicting the 8 angles with their assigned names

To improve the overall visual feedback, those joint angles having a value greater than the acceptable range are highlighted with green sticks. Figure 13 shows the visual and textual feedback displayed on our webpage.

5 Conclusion

Regular yoga practice improves self-esteem and confidence. It promotes mental clarity and relaxation, which benefits people's mental health. However, yoga should be performed under professional supervision and in a regulated manner, since it can be harmful to one's health if done incorrectly. The idea behind this research is to propose

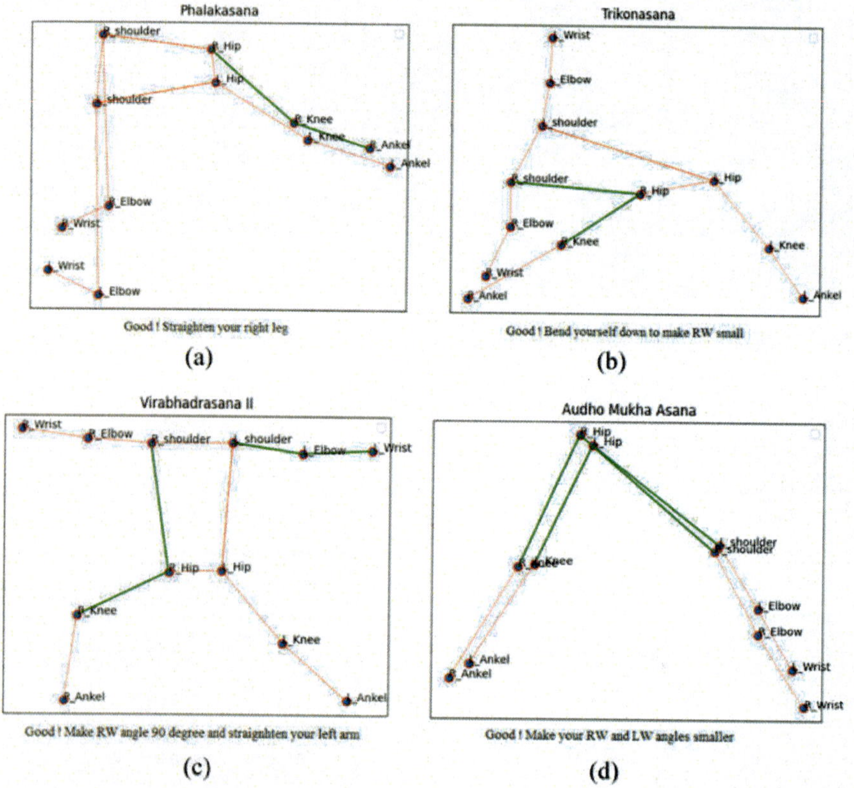

Fig. 13 Stick diagram with textual feedback **a** Phalakasana (plank pose) **b** Trikonasana (triangular pose) **c** Virabhadrasana II (warrior-2 pose) **d** Adho mukha svanasana (downward face)

a walk through application that is highly efficient in guiding a yoga-performing individual with visual and textual feedback. It utilizes cutting-edge AI-based technology to assist users in practicing correct yoga poses. In this context, a web app is developed to assist in the proper execution of postures. The user records a video of doing the pose within the app by using a webcam. Hand gesture feature is embedded in the application to start the recording of the pose. Furthermore, that recorded video is processed frame by frame through the mediapipe framework to detect body joint coordinates which further helps to calculate different body angles and then compares the different joint angles of the user with the professional's angles. At last, the user is provided guidance on how to improve their posture. With this proposal, people will be able to practice yoga anywhere, including at home. Moreover, the proposed system also takes care of the user privacy as only the calculated body joint angles are processed for final decision and feedback generation and the user videos are not stored in our database in any form.

6 Future Prospects

Further enhancements to the web app may be developed by including the concept of posture classification so that users can perform any pose they desire rather than being prompted to select a yoga pose. Data set collection is relatively small to perform this operation which can be further extended to get more accurate results. Our app is restricted to four asanas at the moment: Adho mukha svanasana (downward-facing dog posture), Phalakasana (plank pose), Trikonasana (triangular pose) and Virabhadrasana II (warrior-2 pose) which can be extended to include a variety of other yoga poses such as Suryanamaskar, Bhujangasana, Padmasana, etc. Furthermore, this can also be extended to sports-related activities. It can be applied for evaluating skating element's quality, tracking and estimating 3D human poses of the player and for estimating jumps of various types which can be beneficial for sportsmen in many ways including coordination checks and preventing injuries. The system can be improved further by incorporating voice feedback.

References

1. Guddeti RR, Dang G, Williams MA, Alla VM (2019) Role of Yoga in cardiac disease and rehabilitation. J Cardiopulm Rehabil Prev 3:146–152
2. Rodríguez-Hidalgo AJ, Pantaleón Y, Dios I, Falla D (2020) Fear of COVID-19, Stress, and Anxiety in University undergraduate students: a predictive model for depression. Front Psychol 11
3. Sharma YK, Sharma S, Sharma E (2018) Scientific benefits of Yoga: a review. Int J Multidiscip Res 03:11–148
4. Chen Y, Tian Y, He M (2020) Monocular human pose estimation: a survey of deep learning-based methods. Comput Vis Image Understand
5. Lugaresi C, Tang J, Nash H, McClanahan C, Uboweja E, Hays M, Zhang F, Chang CL, Yong MG, Lee J, Chang WT, Hua W, Georg M, Grundmann M (2019) MediaPipe: a framework for building perception pipelines
6. Yoga: Its Origin, History and Development: https://www.mea.gov.in/search-result.htm?25096/Yoga:_su_origen,_historia_y_desarrollo#:~:text=The%20word%20'Yoga'%20is%20derived,and%20body%2C%20Man%20%26%20Nature. Accessed 2021
7. Chen HT, He YZ, Hsu CC (2018) Computer-assisted yoga training system. Multimed Tools Appl 77:23969–23991
8. Trejo EW, Yuan P (2018) Recognition of Yoga poses through an interactive system with kinect device. In: 2018 2nd international conference robotics and automation science: ICRAS, pp 12–17
9. Borkar PK, Pulinthitha MM, Pansare A (2019) Match pose—a system for comparing poses. Int J Eng Res Technol (IJERT) 08(10)
10. Rishan F, Silva BB, Alawathugoda S, Nijabdeen S, Rupasinghe L, Liyanapathirana C (2020) Infinity Yoga Tutor: Yoga posture detection and correction system. In: 2020 5th international conference on information technology research
11. Cao Z, Simon T, Wei SE, Sheikh Y (2017) OpenPose: realtime multi-person 2D pose estimation using part affinity fields. In: Proceedings of the IEEE conference on computer vision and pattern recognition (CVPR), pp 7291–7299

12. Islam MU, Mahmud H, Ashraf FB, Hossain I, Hasan MK (2017) Yoga posture recognition by detecting human joint points in real time using microsoft Kinect. In: 2017 IEEE region 10 humanitarian technology conference (R10-HTC), Dhaka, pp 668–673
13. Zhang F, Bazarevsky V, Vakunov A, Tkachenka A, Sung G, Chang CL, Grundmann G (2021) MediaPipe hands: on-device real-time hand tracking
14. Telles S, Sharma SK, Chetry D, Balkrishna A (2021) Benefits and adverse effects associated with yoga practice: a cross-sectional survey from India. Complementary therapies in medicine. Elsevier
15. MediaPipe Github. https://google.github.io/mediapipe/solutions/hands. Accessed 2021
16. On-device, Real-time Body Pose Tracking with MediaPipe BlazePose. https://ai.googleblog.com/2020/08/on-device-real-time-body-pose-tracking.html. Accessed 2021
17. MediaPipe Github. https://google.github.io/mediapipe/solutions/pose. Accessed 2021

Study of Deformation in Cold Rolled Al Sheets

János György Bátorfi and Jurij J. Sidor

1 Introduction

Rolling is a commonly used method to reduce the thickness of the sheet. The generally applied parameters for rolling simulation are the radius of the rolls, roll velocity, friction coefficient, initial and the final thicknesses of a rolled sheet [1]. In general, the reference directions are indicated according to the following scheme: x, y and z correspond to rolling (RD), transverse (TD), and normal (ND) directions, respectively.

Previous studies on materials flow during cold rolling [1, 2] suggest that the displacement field across the thickness is not homogeneous and can be assessed by the function:

$$dx = \alpha \cdot z^n \tag{1}$$

where *dx* is the relative displacement in RD between, α and n are model parameters, and z represents the coordinates of points across the thickness of the rolled sheet.

It turned out that both α and n are functions of friction coefficient μ and can be accessed by using empirical expressions described in [2]. The minimum coefficient of friction (COF) required for rolling can be calculated by using Eq. 2 [3]:

$$\mu_{min} = \frac{1}{2}\sqrt{\frac{h}{R}}\frac{\ln\left(\frac{h_0}{h}\right) + \frac{1}{4}\sqrt{\frac{h_0-h}{R}}}{\tan^{-1}\left(\frac{h_0}{h} - 1\right)} \tag{2}$$

J. Gy. Bátorfi (✉) · J. J. Sidor
Faculty of Informatics, Savaria Institute of Technology, Eötvös Loránd University, Károlyi Gáspár tér 4, Szombathely 9700, Hungary
e-mail: bj@inf.elte.hu

J. Gy. Bátorfi
Doctoral School of Physics, Faculty of Natural Sciences, Eötvös Loránd University, Pázmány Péter sétány 1/A, Budapest 1117, Hungary

© The Author(s), under exclusive license to Springer Nature Singapore Pte Ltd. 2023
H. Muthusamy et al. (eds.), *Robotics, Control and Computer Vision*, Lecture Notes in Electrical Engineering 1009, https://doi.org/10.1007/978-981-99-0236-1_31

where μ_{min} is the minimum friction coefficient necessary for cold rolling, h is the sheet thickness of the deformed sheet, h_0 is the thickness prior to rolling, and R is the radius of the rolls.

In numerous studies [2–6], the value of friction coefficient is estimated either by analytical approximations or results of finite element modeling, however, the exact quantity remains unknown. In this view, this contribution presents a way that allows asses the friction coefficient based on experimental evidence and finite element calculations.

2 Modeling Methods

The rolling process was modeled by DEFORM 2D software, which is specifically designed for simplified 2D modeling of plastic deformation [7]. There are additional modeling approaches that can be employed for the simulation of rolling, such as the Flow Line Models (FLM) [2, 4, 8] or analytical approximations [5, 6]. These methods require minimum computational capacity, but they can be applied to specified processing routes and particular material models. On the other hand, Finite Element Method (FEM) can be used for a wide range of processes, while technological and material parameters should be precisely defined. The major disadvantage of FEM is that this method is time-consuming.

Three assumptions were used in the FEM simulation of the rolling process: (i) the 3D rolling geometry was approximated by a 2D geometry with a plane strain mechanical model (this simplification is generally employed for the rolling process [4]); (ii) the effect of temperature rise in deformation is neglected (similarly to [1]); (iii) an isotropic material model is assumed [5, 6].

As it was already shown [2], the yield strength and material models have a slight effect on the displacement patterns, so the elastoplastic properties of the Al-6063 material model incorporated in the software were used for modeling. The geometric setup of the FEM model used in our investigation is shown in Fig. 1. The rolls were considered as rigid bodies and therefore the meshing procedure of rolls was not necessary. To investigate the flow of a workpiece (sheet) in the rolling gap, the mesh consisting of square-shaped elements was used, as is shown in Fig. 2. The displacement of the initially horizontal line (see Fig. 3) was examined to determine the quantitative indicators of the deformation process. The points along the line are marked with P1 to P11 and the displacement of these points was determined by tracing their coordinates along the x and z axes.

Equation 2 can be employed to assess the lower bound of friction coefficient, however, the real value of μ might be higher, and therefore, the finite element calculations were performed for a spectrum of friction coefficients ranging between the μ_{min} and $5\mu_{min}$.

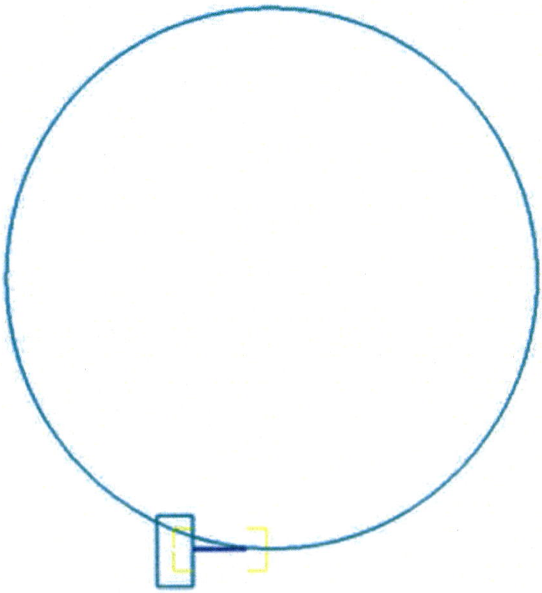

Fig. 1 Geometric setup employed in FEM simulations of rolling

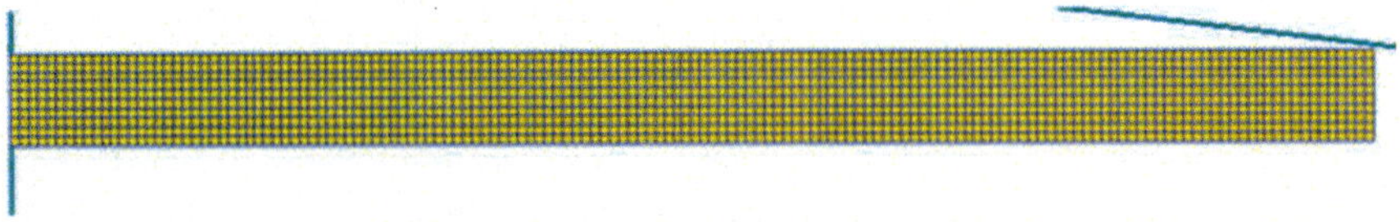

Fig. 2 Mesh applied to the rolled sheet

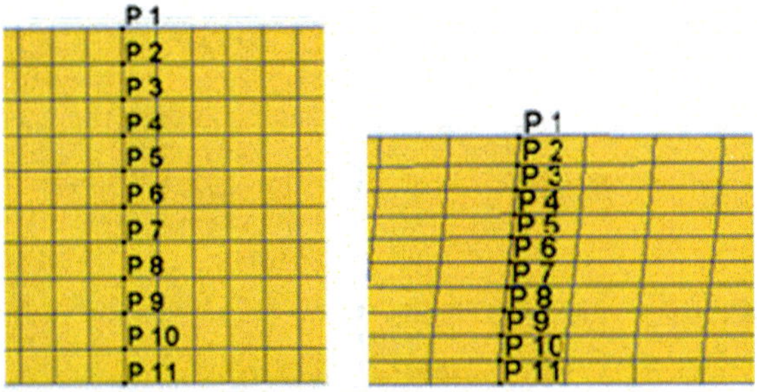

Fig. 3 Virgin material and distorted mesh after rolling

3 Model Parameters

The model parameters used to simulate the rolling process are presented in Table 1. The minimum value of μ, calculated with Eq. 2, is 0.048 and therefore the following COF values were used for the simulation: 0.05, 0.06, 0.07, 0.08, 0.10, 0.15, 0.20 and 0.25.

In order to examine the deformation flow in the rolled Al sheet, the TD plane of a virgin (deformation-free) material was marked by the microindentation technique, and as a result, rectangular patterns were created (see Fig. 4). The distortion of initially straight lines (perpendicular to RD) after 30% reduction (with a roll diameter of 150 mm) is shown in Fig. 5. The displacement values can be determined using the function expressed by Eq. 3. This equation is a polynomial approximation of Eq. 1, and the advantage of expression 3 is that it can be used for the nonmonotonic displacement patterns, which appear at high friction coefficients.

$$dx = A \cdot z^8 + B \cdot z^6 + C \cdot z^4 + D \cdot z^2 \tag{3}$$

where coefficients A, B, C, and D are fitting parameters and their values are listed in Table 2 for various friction coefficients.

Table 1 Parameters used in FEM simulation

Parameter	Notation	Unit of measurement	Value
Radius of roll	R	mm	75
Initial thickness	h_0	mm	2
Final thickness	h	mm	1.4
Coefficient of friction	μ	1	0.05–0.25
Velocity of rotation	ω	rad/s	1.1

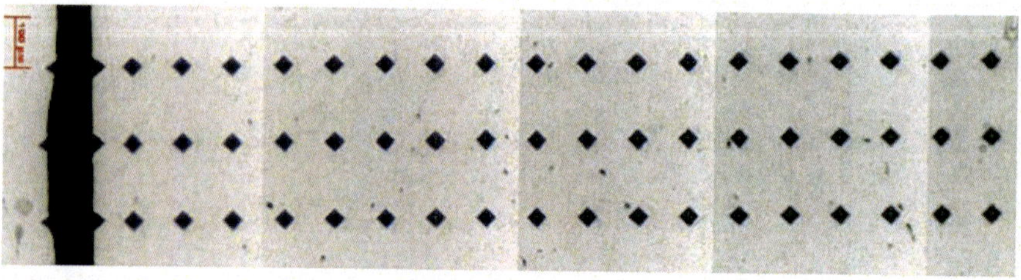

Fig. 4 Reference patterns, made by microhardness indentation on the plane perpendicular to the TD prior to rolling (rolling direction is parallel to the scalebar)

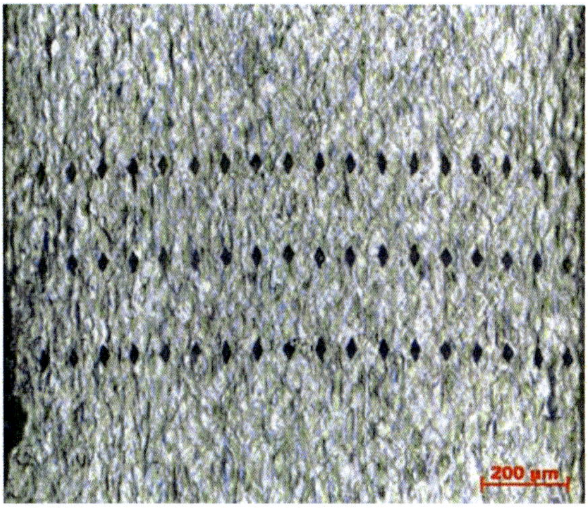

Fig. 5 Displacement of microhardness patterns after 30% thickness reduction (rolling direction is perpendicular to the scalebar)

Table 2 Parameter values for different coefficients of friction (COF)

COF	A	B	C	D
$\mu = 0.05$	−1.3145	1.7795	−0.9034	0.1961
$\mu = 0.06$	−1.3196	1.7775	−0.9038	0.2113
$\mu = 0.07$	−1.3248	1.7755	−0.9041	0.2264
$\mu = 0.08$	−1.3300	1.7735	−0.9045	0.2416
$\mu = 0.10$	−1.3403	1.7696	−0.9051	0.2719
$\mu = 0.15$	−1.3661	1.7597	−0.9068	0.3476
$\mu = 0.20$	−1.3918	1.7498	−0.9085	0.4233
$\mu = 0.25$	−1.4176	1.7399	−0.9102	0.4991

Analyzing the data of Table 2, one can conclude that the fitting parameters A-D are functions of friction coefficient μ and can be calculated by employing Eqs. 4–7. The corresponding displacement patterns for various COFs are shown in Fig. 6.

$$A = -1.2887 - 0.51565 \cdot \mu \tag{4}$$

$$B = 1.7894 - 0.19771 \cdot \mu \tag{5}$$

$$C = -0.9018 - 0.03375 \cdot \mu \tag{6}$$

$$D = -0.1204 - 1.51475 \cdot \mu \tag{7}$$

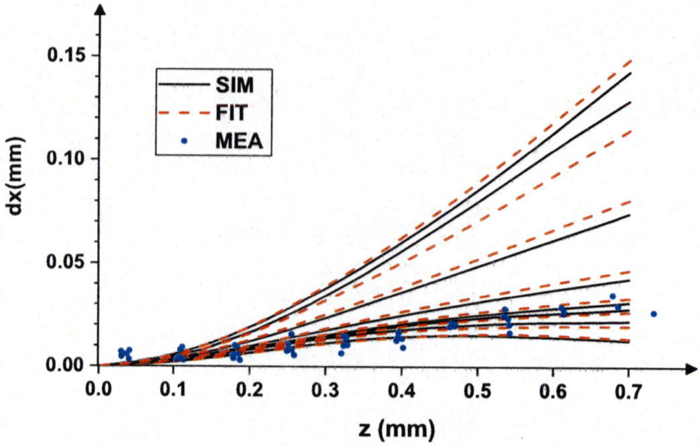

Fig. 6 Experimentally observed (MEA) and calculated displacement patterns by FEM (SIM) and analytical expression 3 (FIT)

As Fig. 6 reveals, the displacement lines obtained by means of FEM (continuous line, SIM) can be reproduced by the analytical expression (3) (dashed line, FIT) and this, in turn, allows for the assessment of friction coefficient. In the present investigation, the assessment of μ was performed by comparing the simulated and measured displacement patterns (see Fig. 5), observed on the plane perpendicular to the transverse direction of a rolled sheet. As it is shown in Fig. 6, the experimentally measured displacement (MEA) can be successfully fitted by Eq. 3 (FIT). The best fit suggests that the rolling was carried out with a friction coefficient of 0.068. Comparable value ($\mu = 0.07$) was reported elsewhere [9]. The correlation coefficient between the measured and simulated values is estimated to be 0.871.

The flowchart of the simulation used in the present study is shown in Fig. 7.

4 Calculating the Strain Values

The strain values can be subdivided into two groups: normal and shear components [9]. The normal strain can be computed by using Eq. 8 [10, 11], while the shear component can be estimated by Eqs. 9 and 10 [12, 13]. Once both components are known, the value of equivalent strain can be determined by Eq. 11 [13].

$$\varepsilon = \varepsilon_x = -\varepsilon_z = \ln\left(\frac{h_0}{h}\right) \tag{8}$$

$$\varepsilon_s = \frac{2(1-\varepsilon)^2}{\varepsilon(2-\varepsilon)}\gamma \ln\left(\frac{1}{1-\varepsilon}\right) \tag{9}$$

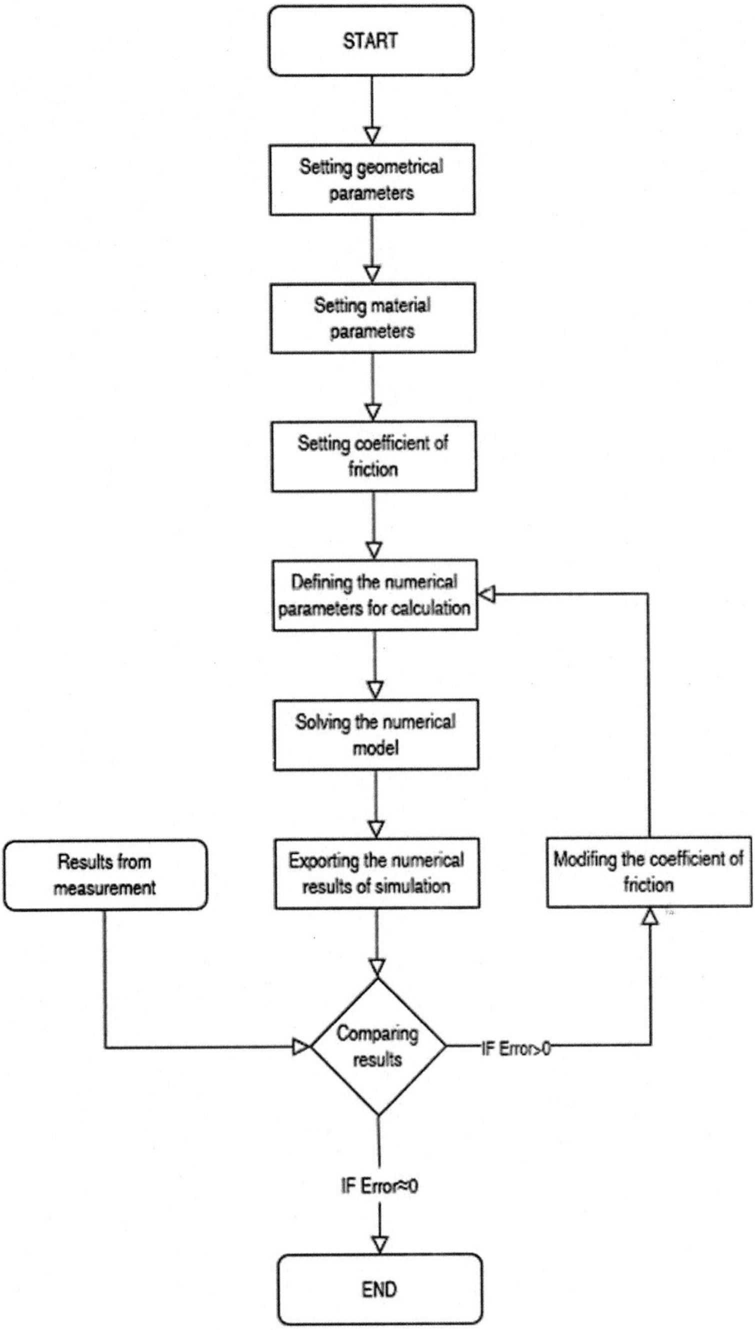

Fig. 7 Flow chart of calculations employed in the current study

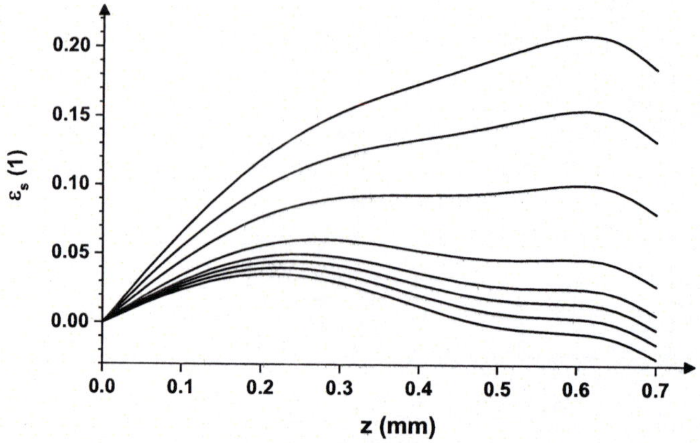

Fig. 8 Shear strain values computed for different friction coefficients (μ from bottom to top: 0.05, 0.06, 0.07, 0.08, 0.10, 0.15, 0.20 and 0.25)

$$\gamma = \frac{dx}{dz} \tag{10}$$

$$\varepsilon_{VM} = \sqrt{\frac{4}{3}\left(\ln\left(\frac{1}{1-\varepsilon}\right)\right)^2 + \frac{\varepsilon_s^2}{3}} \tag{11}$$

where ε and γ are normal and shear strain components.

Knowing the correlation between the friction coefficient and γ, one can estimate the evolution of strain during cold rolling. The simulated shear strain components and corresponding equivalent strain values are shown in Figs. 8 and 9, respectively. It is obvious that the strain flow is very heterogeneous across the thickness due to inhomogeneous shear strain distribution, while the character of strain distribution depends on the friction conditions. The smallest difference between the maximum and minimum strain values across the thickness is observed for so-called wet rolling, i.e. low μ. In the mid-thickness plane (z = 0), the shear strain is negligibly small implying that the material experiences majorly plane strain deformation, whereas both surface and sub-surface regions are subjected to complex straining, characterized by normal strain component and extensive shear.

5 Summary

In this study, the friction coefficient was determined for a given roll gap geometry based on both experimental evidence and numerical simulations. It was shown that rolling of Al sheet with 30% thickness reduction with a roll diameter of 150 mm

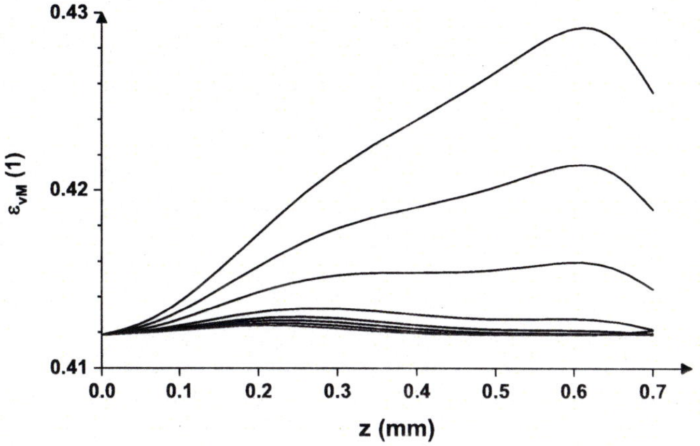

Fig. 9 Equivalent strain values calculated for different friction coefficients (μ from bottom to top: 0.05, 0.06, 0.07, 0.08, 0.10, 0.15, 0.20 and 0.25)

accounts for the friction of 0.068 and this value correlates well with the one reported in literature sources.

A new polynomial function was developed for the estimation of displacement fields during cold rolling. The model parameters for the polynomial equation were determined by analyzing the data obtained from finite element calculations. It was shown that the analytical expression developed is capable of reproducing the FEM outputs with high accuracy.

The measured displacement profile values were used for validation of the simulated data. The newly developed model accurately reproduces the experimentally observed deformation flow profile. The correlation coefficient between the measured and simulated values is estimated to be 0.871.

The model parameters of the polynomial function developed can be determined for various rolling conditions by the algorithm described in the current study. The analytical model can also be extended to other materials.

Acknowledgements Project no. TKP2021-NVA-29 has been implemented with the support provided by the Ministry of Innovation and Technology of Hungary from the National Research, Development, and Innovation Fund, financed under the TKP2021-NVA funding scheme.

References

1. Bátorfi JGy, Chakravarty P, Sidor J (2021) Investigation of the wear of rolls in asymmetric rolling. eis 14–20. https://doi.org/10.37775/EIS.2021.2.2
2. Sidor JJ (2019) Assessment of flow-line model in rolling texture simulations. Metals 9:1098. https://doi.org/10.3390/met9101098

3. Avitzur B (1980) Friction-aided strip rolling with unlimited reduction. Int J Mach Tool Des Res 20:197–210. https://doi.org/10.1016/0020-7357(80)90004-9
4. Decroos K, Sidor J, Seefeldt M (2014) A new analytical approach for the velocity field in rolling processes and its application in through-thickness texture prediction. Metall Mat Trans A 45:948–961. https://doi.org/10.1007/s11661-013-2021-3
5. Cawthorn CJ, Loukaides EG, Allwood JM (2014) Comparison of analytical models for sheet rolling. Procedia Eng 81:2451–2456. https://doi.org/10.1016/j.proeng.2014.10.349
6. Minton JJ, Cawthorn CJ, Brambley EJ (2016) Asymptotic analysis of asymmetric thin sheet rolling. Int J Mech Sci 113:36–48. https://doi.org/10.1016/j.ijmecsci.2016.03.024
7. Fluhrer J DEFORM(TM) 2D Version 8.1 User's Manual
8. Beausir B, Tóth LS (2009) A new flow function to model texture evolution in symmetric and asymmetric rolling. In: Haldar A, Suwas S, Bhattacharjee D (eds) Microstructure and texture in steels. Springer, London, pp 415–420
9. Bátorfi JGY, Sidor J (2020) Alumínium lemez aszimmetrikus hengerlése közben fellépő deformációjának vizsgálata. eis 5–14. https://doi.org/10.37775/eis.2020.1.1
10. Pesin A, Pustovoytov DO (2014) Influence of process parameters on distribution of shear strain through sheet thickness in asymmetric rolling. KEM 622–623:929–935. https://doi.org/10.4028/www.scientific.net/KEM.622-623.929
11. Inoue T (2010) Strain variations on rolling condition in accumulative roll-bonding by finite element analysis. In: Moratal D (ed) Finite element analysis. Sciyo
12. Ma CQ, Hou LG, Zhang JS, Zhuang LZ (2014) Experimental and numerical investigations of the plastic deformation during multi-pass asymmetric and symmetric rolling of high-strength aluminum alloys. MSF 794–796:1157–1162. https://doi.org/10.4028/www.scientific.net/MSF.794-796.1157
13. Inoue T, Qiu H, Ueji R (2020) Through-Thickness microstructure and strain distribution in steel sheets rolled in a large-diameter rolling process. Metals 10:91. https://doi.org/10.3390/met10010091

Modelling and Control of Semi-automated Microfluidic Dispensing System

M. Prabhu, P. Karthikeyan, D. V. Sabarianand, and N. Dhanawaran

1 Introduction

Nowadays, in the field of the syringe dispensing system, the development of the high precision device is a challenging task that is achieved using the proposed design. The author developed the syringe injection rate detection system based on two Hall-effect sensors in the differential mode of operation. From tests conducted on the prototype developed, the worst-case error in p was found to be less than 1:2% and the error in the determination of the rate of injection to be less than 2:4%. This is within clinically acceptable limits since the rate of injection in practical scenarios rarely exceeds 15 ml/s [1]. The electronic technique uses a needling instrument for the purpose of detaching the needle automatically, i.e. an action that can detach the used needles from the syringe and then collect them respectively The caliber of the developed design aims at the common 10 and 20 ml syringe practice in the hospital [2]. A novel machine-driven injection device is bestowed specifically designed for correct delivery of multiple doses of product through a variety of adjustable injection parameters, as well as injection depth, dose volume and needle insertion speed. The device was originally planned for the delivery of a cell-based medical aid to patients with skin wounds caused by epidermolysis bullosa [3]. Consequently, there's a robust demand for machine-controlled liquid handling strategies like sensor-integrated robotic systems. The sample volume is at the micro- or nanoliter level, and therefore the variety of transferred samples volume is immense once work in large-scope combinatorial conditions. Below these conditions, liquid handling by hand is tedious, long, and impractical [4].

Some of the patents related to the microfluidic dispensing systems are the technical field of cell culturing for the production of in-vitro tissues and provides a device for

M. Prabhu · P. Karthikeyan (✉) · D. V. Sabarianand · N. Dhanawaran
Department of Production Technology, Madras Institute of Technology (Campus), Anna University, Chennai, India
e-mail: pkarthikeyan@mitindia.edu

© The Author(s), under exclusive license to Springer Nature Singapore Pte Ltd. 2023

H. Muthusamy et al. (eds.), *Robotics, Control and Computer Vision*, Lecture Notes in Electrical Engineering 1009, https://doi.org/10.1007/978-981-99-0236-1_32

dispensing a suspension of biological cells into culture vessels for culture, comprising mean for re-suspending cells among the suspension [5]. The extremely machine-driven, high volume multichannel pipetting system transfers liquid from mother plates to daughter plates, or from a fill station to daughter plates [6].

2 Structure Design and Analysis of Dispensing Mechanism

The proposed model of the semi-automated microfluidic dispensing system with syringe actuation with plunger's movement is automated through actuator [7]. The actuator is controlled through microcontroller. The objective of this design is to automatically control the amount of fluid that is sucked in or dispensed through the syringe. The methodology of the syringe dispensing system is shown in Fig. 1. The process of needle dispensing system is to get the volume 'V' in ml, then calculate stroke length "L" in mm i.e. $V/0.035$ mm, after that calculate the angle of stepper motor in 'θ' is given by $L°/360$. Then the number of steps 'n' is derived by $\theta/1.8$, finally send 'n' pulse to the stepper motor and stop the sequence.

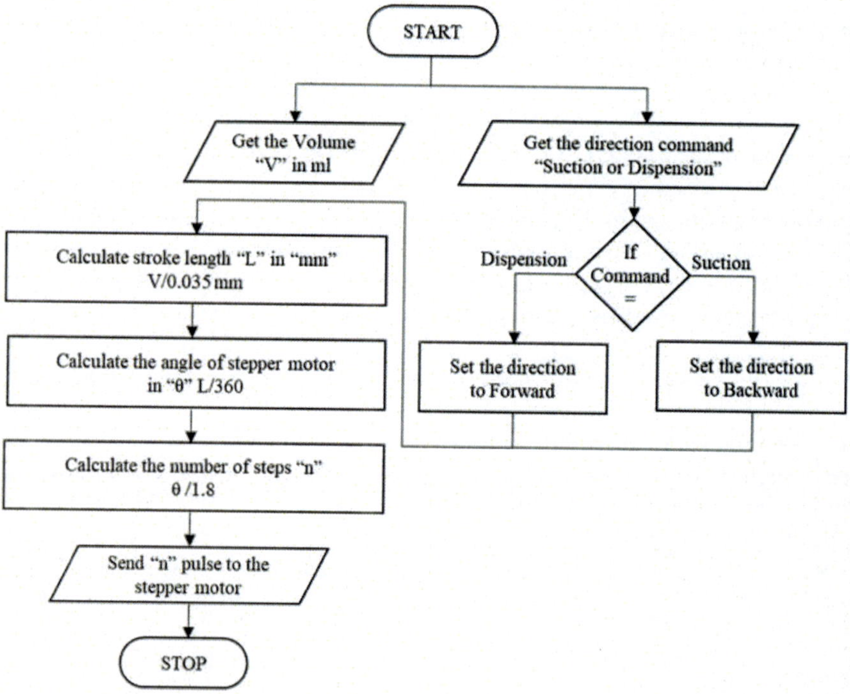

Fig. 1 Syringe dispensing system

2.1 Design Calculations

Stepper Motor: A motor has to be coupled with lead screw to give the rotary motion. As mentioned above the pitch of the screw does not meet the required minimum movement. Hence the actuator is selected such that it can achieve the minimum required movement. In this case, the high-resolution stepper motor can be used as the stepping angle of the motor can be controlled to the required position. The specifications of the stepper motor were tabulated in Table 1.

Torque Calculation of the Microfluidic Dispensing System.

Required minimum volume to be manipulated 25 μl.

Pitch of the screw 1 mm.

Coefficient of friction between nut and screw, μ_f—0.73.

Volume displaced by the syringe per mm stroke 0.035 ml or 35 μl.

Required minimum movement of plunger—25/35 = 0.714 mm.

Peak Load P—100 g or 0.1N.

Angle made by stepper motor per mm-360°

Angle required to make 0.714 mm – 0.714/360 = 257°

For a stepper 1.8° Resolution Stepper motor number of steps required to make 257° is 142.8 step. So, the stepper motor with 1.8° resolution can be chosen. For a screw having a pitch (p) of 1 mm and Diameter (D) of 4 mm,

$$\text{Thread Angle, } \alpha = \frac{p}{\pi D} \tag{1}$$

$$\text{Thread Angle, } \alpha = 0.079 \text{ rad}$$

$$\tan \phi = 0.73$$
$$\phi = 0.63\text{rad}$$

Torque Equation is given by

$$\text{Torque, } \tau_{GZ} = \frac{PD}{2} \tan(\phi + \alpha)$$

Table 1 Specification of components used in the proposed model design

Components	Motor type	Step angle	Holding torque kg-cm	Rated torque kg-cm	Diameter (D) (mm)	Screw pitch (p) (mm)	Capacity (ml)	Capacity per (mm) stroke
Stepper motor	Bipolar	1.8°	4.2	2.2	5	–	–	–
Lead screw	–	–	–	–	4	1	–	–
Syringe	–	–	–	–	10	–	2.5	0.035

$$= 0.1715\,\text{Nmm or } 0.0001715\,\text{Nm} \quad (2)$$

Therefore, the torque required is 0.1715 Nmm or 0.0001715 Nm.

Lead Screw: The minimum volume the syringe has to manipulate is 2.5 μl. The lead screw is chosen to give minimum volume of actuation i.e. the minimum stroke by the syringe must be equal to the Highest common factor of 2.5 μl. The lead screw is attached to the body of the syringe whereas the nut follower is attached to the plunger of the syringe. The diameter of the screw is 4 mm and screw pitch are of about 1 mm. Here the pitch of the screw is 1 mm. Although the pitch of the screw is not a factor of 0.2, it can be adjusted through the actuator. While rotating the lead screw with the rotation of the nut is fixed, the nut follower can be able to move upwards and downwards to give the stroke for the plunger as shown in Fig. 2a.

Selection of Syringe: In the Antibiogram process at any point a minimum of 0.2 ml and a maximum of 2.5 ml in volume is sucked and dispensed during operation. So, taking the maximum volume as the syringe volume, a commonly available 2.5 ml DISPOVAN® syringe is used in the system. Barrels are made of non-toxic, medical-grade polypropylene compatible with any medication. Gaskets are made of natural rubber which is chemically inert and compatible with any medication as shown in Fig. 2a.

3 Simulation of Semi-automated Syringe Dispensing System

The proposed model of syringe dispensing system is simulated using Matlab-Simulink as shown in Fig. 3. In order to calculate the position value, total force and total torque generated by the stepper motor with lead screw setup for the microfluidic dispensing system and also the drug delivery system to deliver the fluids in micro to nanometers.

4 Results and Discussions

The desired position signals were given to stepper motor to achieve the various stroke length. Figure 4 shows the input position value of the stepper motor with lead screw.

The signal input represents the continuous suction and dispensing operations for five cycles without break in operation. Each cycle consists of 10 s plunger movement of the syringe from bottom to top and vice versa for five times with some random direction change within each cycle (Fig. 5).

The total torque required by the motor were simulated and shown in Fig. 6. The graph shows recorded maximum torque value in positive half is 0.00008 Nm

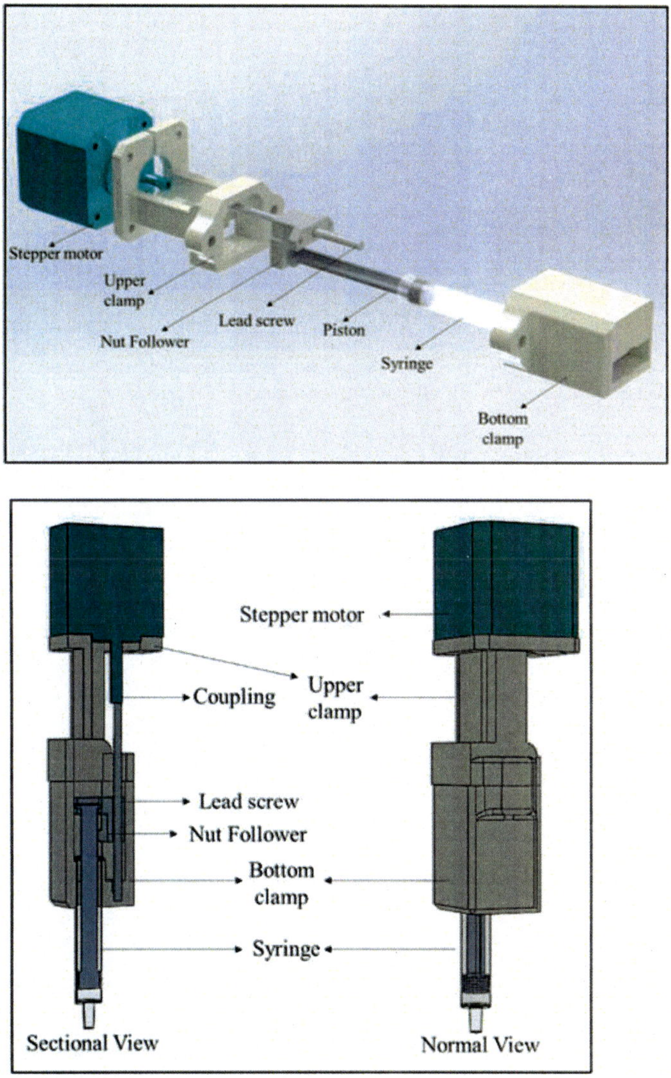

Fig. 2 Semi-automated syringe dispensing system. **a** Automated syringe. **b** Exploded view. **c** Cut section of syringe dispensing system

representing the motor rotating in clockwise direction and maximum torque value in negative half is 0.00012 Nm representing the motor rotating in anti-clockwise direction. The stepper motor which is used in the assembly has 0.4609 Nm or 4.7 kgcm torque. The calculated theoretical torque value is 0.0001715 Nm.

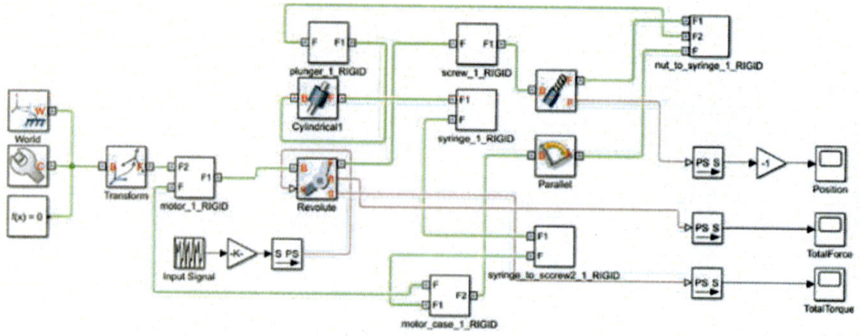

Fig. 3 Matlab simulink diagram for syringe dispensing system

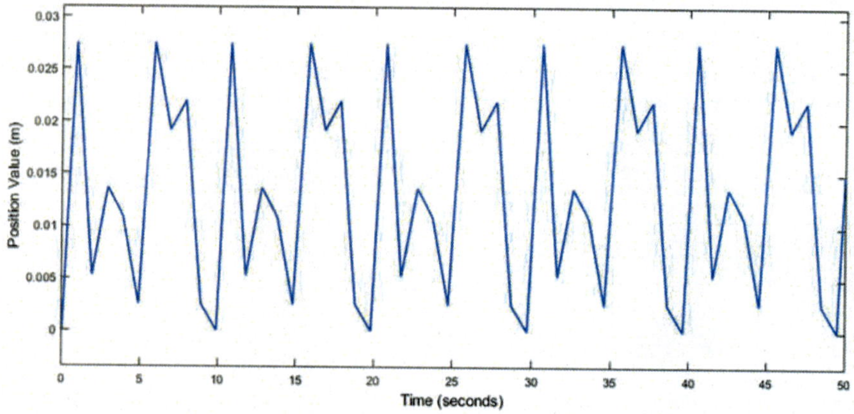

Fig. 4 Time versus position value (m)

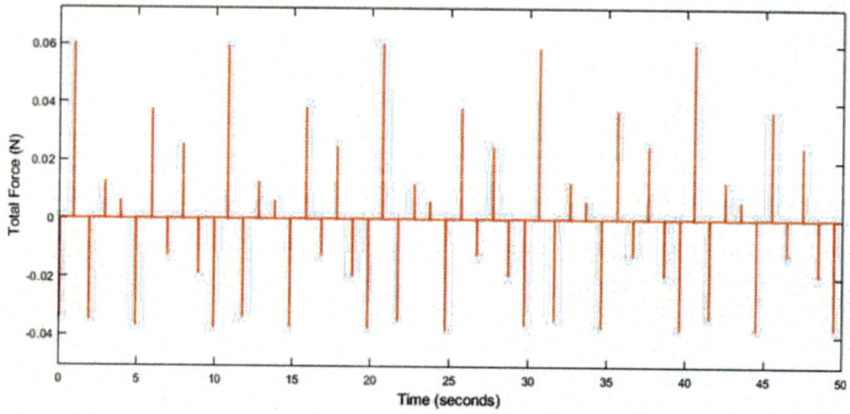

Fig. 5 Time versus total force (N)

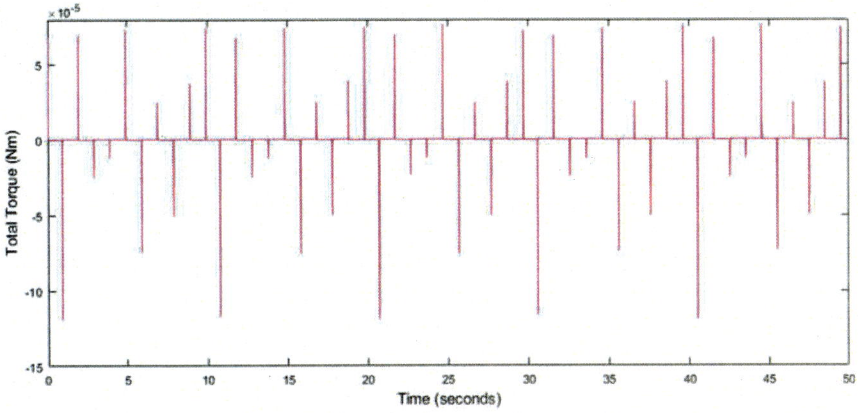

Fig. 6 Time versus total torque (Nm)

5 Experimental Validation

Once the pipette tip is used, it cannot be used again for processing another sample. It must be detached to trash. The pipette tip attaches and clamps itself to the syringe by means of frictional force between the outer face of the syringe tip and the inner face of the pipette tip. A simple push operation between these contact faces is enough to detach the pipette tip from the syringe. An actuator which moves relatively fixed to the syringe is required. So, a cam and follower mechanism, as the cam is driven by a motor and the follower moves and pushes the pipette tip is deployed. Servo motor can be used for this purpose as they can make a full or half step rotation precisely [8]. Hence there needs a clamp to hold the servomotor in a fixed position. The simple control algorithm is used to run the stepper motor precisely in micrometres while using piezo-stepper motors it is possible to achieve the motion in nanometers range by utilizing the appropriate high precision control algorithms explained [9]. The semi-automated microfluidic dispensing system is shown in Fig. 7. The flow rate of the sample were shown in Fig. 8.

6 Conclusion and Future Works

Thus, the Position, Total force and Total Torque for the proposed model design of the syringe dispensing system is theoretically calculated and simulation results were carried out using the software. The syringe system is found to have to move 0.714 mm to dispense 35 μl. The torque required by the motor to dispense the sample is 0.00012 Nm which is lesser than the calculated theoretical value of 0.0001715 Nm. The future work is to ensure the fluid flow will travel in high precise movements especially for

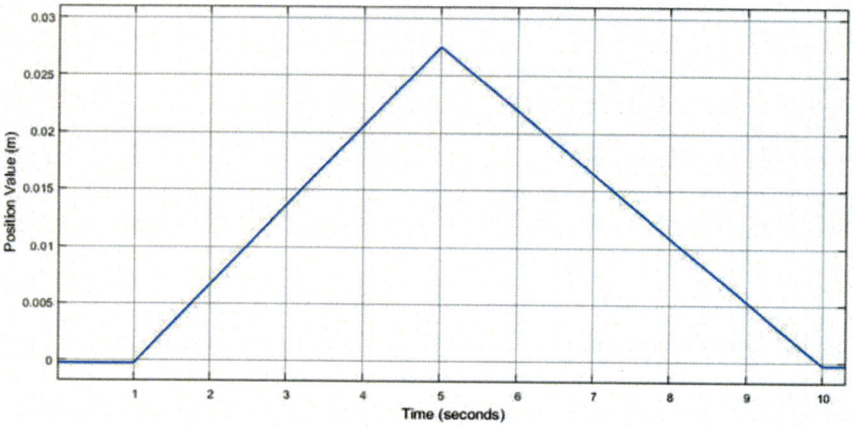

Fig. 7 Experimental Setup **a** Stepper motor with lead screw setup **b** Microfluidic dispensing system

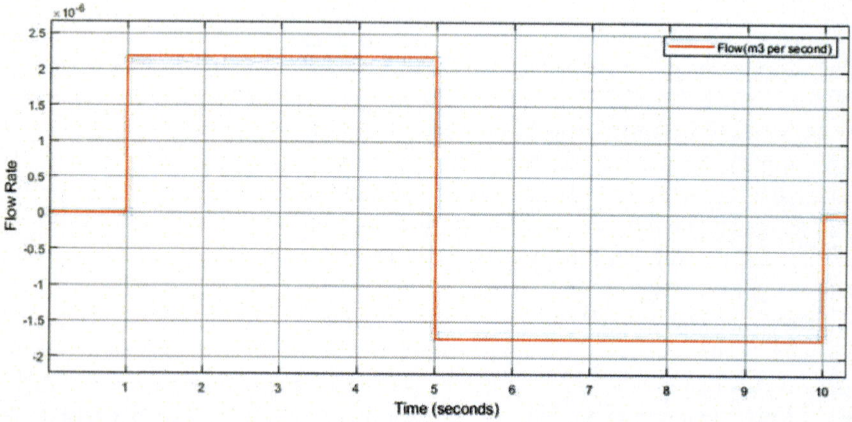

Fig. 8 Flow rate (m^3/s)

the following application such as the drug delivered system, cell injection and cell piercing using the developed dispensing system.

Acknowledgements I would like to express my deep and sincere gratitude to my former research super-visor, late Dr. R. Sivaramakrishnan, Ph.D., Anna University, Chennai, for giving me the opportunity to do research and providing invaluable guidance throughout this work.

It was a great privilege and honor to work and study under his guidance. I express my heartfelt thanks for his patience during the discussion I had with him on this work and many other research activities. In addition to that I sincerely thank him for establishing advanced facilities and equipments in the Mechatronics lab under lab modernization scheme of University.

References

1. Mukherjee GB, Sivaprakasam M (2013) A syringe injection rate detector employing a dual Hall-effect sensor configuration. Annu Int Conf IEEE Eng Med Biol Soc
2. Chen CSC, Shih YY, Chen YL (2011) Development of the syringe needle auto-detaching device. In: 5th international conference on bioinformatics and biomedical engineering, pp 1–4
3. Leoni LAG, Ginty P, Schutte R, Pillai G, Sharma G, Kemp P, Mount N, Sharpe M (2017) Preclinical development of an automated injection device for intradermal delivery of a cell-based therapy. Drug Deliv Transl Res 7:695–708
4. Kong YLF, Zheng YF, Chen W (2012) Automatic liquid handling for life science: a critical review of the current state of the art. J Lab Autom 17:169–185
5. Andreas T (2015) Cell dispensing system. In: WIP Organization (Ed), pp 1–18
6. Walter Meltzer NM (2006) Conn, Automated Pipetting System, Matrix Technologies Corp, Hudson, NH (US); Cosmotec Co, Ltd, Tokyo (JP), US, pp 1–20
7. Sabarianand DV, Karthikeyan P (2019) Nanopositioning systems using piezoelectric actuators, In: Kamalanand K, Jawahar DNJAPM (eds) (2019) Advances in nano instrumentation systems and computational techniques. Nova Sci
8. Sabarianand DV, Karthikeyan P, Muthuramalingam T (2020) A review on control strategies for compensation of hysteresis and creep on piezoelectric actuators based micro systems. Mech Syst Signal Process 140:1–17
9. Sabarianand DV, Karthikeyan P (2022) Duhem hysteresis modelling of single axis piezoelectric actuation system. In: Suhag MCS, Mishra S (ed) Control and measurement applications for smart grid. Springer, Singapore

Im-SMART: Developing *Im*mersive *S*tudent Participation in the Classroom *A*ugmented with *M*obile *T*elepresence *R*obot

Rajanikanth Nagaraj Kashi, H. R. Archana, and S. Lalitha

1 Introduction

1.1 The Need for an Effective Mobile Telepresence Robot (MTR)

The COVID-19 pandemic which completely disturbed the entire world has caused an enormous and long-lasting impact on day to day lives. Several sectors of the economy have taken massive hits and are working relentlessly for coming back on track as soon as possible. Specifically, the education sector has taken a massive hit due to the ongoing pandemic and the emergent 'not-so' promising scenario. The educational sector faces a large number of hurdles with the delivery of knowledge and skills with educational institutions grappling with alternative and efficient ways to match the efficiency and effectiveness of an offline classroom. Studies and various surveys show that even though classes have taken a virtual route through online platforms, they have failed to provide engagement and an environment similar to an offline classroom [1]. The connectedness, interaction, and engagement that exists between a faculty and a student in offline classes is something that online classes have failed to replicate. With the pandemic still not completely over, educational sector bears a large burden and therefore there is a cogent need to ensure that aspects of student attention, inclusion, and participation levels do not drop while at the same

R. N. Kashi (✉) · H. R. Archana · S. Lalitha
Department of Electronics & Communication Engineering, B M S College of Engineering, Bangalore, India
e-mail: rajanikanthnkashi.ece@bmsce.ac.in

H. R. Archana
e-mail: archanahr.ece@bmsce.ac.in

S. Lalitha
e-mail: lalithas.ece@bmsce.ac.in

© The Author(s), under exclusive license to Springer Nature Singapore Pte Ltd. 2023
H. Muthusamy et al. (eds.), *Robotics, Control and Computer Vision*, Lecture Notes in Electrical Engineering 1009, https://doi.org/10.1007/978-981-99-0236-1_33

time maintaining similar experiences that were prevalent with traditional offline classrooms. Several projects and research works have been reported in this area and we propose a framework and platform that enables development of an MTR that meets the need for ensuring an environment which is similar to conventional offline classes and also addresses the engagement aspect through Virtual Telepresence. The novelty of our approach is a scalable platform for the provision of 'build-as-you-go' platform that will incrementally add features taking into account cost aspects too. The prototype also serves as a research test bed for future work.

Mobile Robot Presence (MRP) and Mobile Telepresence Robot (MTR): The terms MRP and MTR are used interchangeably in literature. Telepresence is the ability to provide a context and an environment wherein the user of the system is empowered with both tasks and the environment that are present at a remote location. There are two components: (a) The presence of an automated system which can be a Mobile Robot which enables task accomplishment (MRP) and (b) resources on the robot which extend the remote environment to the user so that the user feels he is 'situated' and contextually aware of the environment; this would require providing necessary and sufficient information about both tasks and the environment using the resources on the Robot (MTR).

Telepresence has wide ranging applications: Healthcare, Security and Surveillance, Business Meetings, Mining in remote areas, and medical applications [2] describes an application of MTR deployed in the Healthcare Sector for remote disinfection while [3] provides an example of training nurses using Virtual Telepresence. MTR also find use in training medical students prior to them performing real surgeries or operations. MTRs also find use in hazardous or inconvenient environments and [4] describes an application in the Mining Sector, involving a bot that is used in mixed-presence teleoperation. A novel mobile robot in a Search and Rescue Operation is detailed in [6] and serves to save a lot of human efforts by employing Virtual Telepresence Robots. The area of MTR provides a fertile ground for research and advancement with the possibility of many applications and the challenges.

1.2 Related Work

Several initiatives and projects have sprung up during the pandemic times that propose to tackle the issues and challenges of home bound pupils. In this section, we provide information related to MTR in learning environments. [5] details work that emphasizes the initial challenges of the use of telepresence robots and provides some quantifications to measure the psychological efficacies of using such approaches, while detailing supporting infrastructure like student and teacher preparation. [7] discusses experiences with two MTR systems in the area of academics and stresses the need for the requirement of stable network coverage and power sources. Reference [8] provides research findings in the context of virtual transnational education scenarios. Some inputs from this research are to address the technology aspects using appropriate hardware, software and their integration. This research also suggests that one needs to look into specific solutions for a particular context since technology is evolving. A review of system design aspects is collated in [9]. Application area challenges are discussed and user evaluation studies are performed. Reference [10] provides the concept of visibility checks and guiding remote users for enabling visual access to materials in classrooms teaching foreign languages. Reference [11] describes 'Professor Avatar', a telepresence robot using a human scale holographic projection. A good review of available telepresence robots in education is provided in [12] and analyzes responses from students, teachers, and also parents. [14] outlines a web-based framework for providing a robotic telepresence environment. In [13] a framework that utilizes seven identified dimensions for designing telepresence robots is provided.

Outline of the paper: In Sect. 2, we formulate the system requirements and provide the overall system design. Hardware architecture is dealt with in Sect. 3 with allocation of functions to system components. Section 4 provides an overview of the Software aspects. Section 5 discusses implementation and system use cases. Section 6 provides data collected from experiments with the system and also provides insights into future work that has been planned with the platform.

2 Problem Context

2.1 *Requirements for an MTR and Proposed System Overview*

The most important requirement of an MTR system is the need to ensure the "connectedness" of the remote student with the classroom. The introduction of new technologies into the learning environment will address the necessity of social presence for the remote users or individuals. An attendant requirement is to provide opportunities that enhance the learning skills in a participative environment. These opportunities can become more immersive with the use of appropriate sensory inputs connected to the user or remote location. The immersive aspects are enabled by interfaces that are amenable for adaptation based on the remote individual needs. An associate aspect is the requirement for providing flexible movement of the Mobile Robotic Presence system and its subsystems. A goal often cited in this context is the reduction of transactional distance, 'psychological and communication space to be crossed, a space for potential misunderstanding between the inputs of instructor and those of the learner'. It follows that the learning experience of the student is increased with smaller measures for the transactional distance. Considering an MRP system, this transactional distance is dependent on providing the right technological base for the robot so that the user experience meets the need for social presence, and providing the right controls for the user so that the learning and immersive experiences are enhanced. In order to meet these design aspects, we have conceived Im-SMART (***Im****mersive* ***S****tudent participation in the classroom* ***A****ugmented with* ***M****obile* ***T****elepresence* ***R****obot*). Considering the aspects of connectedness, immersive experience and prior work in this area (outlined in Sect. 1), we envisaged to develop the MTR system in two phases and captured requirements in a structured manner:

Phase-1 system requirements:
[R1]: The MTR system shall be capable of being controlled by a simple low-cost open-source platform [15].
[R2]: Since Visual Interactions assume importance, the MTR system shall be capable of capturing the classroom scene effectively.
[R3]: The video capture system on the MTR shall provide controls to make appropriate adjustments to enhance the visual scene. This entails providing a mechanism that slaves the camera movement on the MRP system to the user head movement on the remote site where a virtual reality headset is being used.
[R4]: The MTR system shall provide a mobile platform that can be moved within the classroom environment.
[R5]: The MTR system shall provide a microphone system on the Mobile platform that will provide an effective communication from the classroom to the student.

Phase-2 system requirements:
[R6]: The MTR system shall provide the ability to connect any remote user to Mobile platform over the internet.
[R7]: Considering the environment in which the remote user operates, the MTR system shall provide a Voice Commanded Control system.
[R8]: The MTR system shall be commanded from a smartphone at the remote user end.
[R9]: The MTR system shall provide the ability to capture screenshots and record the sessions.
[R10]: The MTR system shall provide strong authentication via email mechanism.

The MTR system that we propose, consists of two subsystems: The Remote User subsystem and the Virtual Telepresence Bot subsystem. The remote user subsystem is hosted on a smartphone integrated with a VR headset and comprises of the ‘Mail App’, ‘Access Mechanisms for Screenshots and recordings’, ‘Main Script’ that invokes ‘Sensor Value Stream’ service, ‘Video Stream’ service, ‘Audio Stream’ service, ‘VLC Media Player’, and a ‘Voice Command’ stream service. In order to provide an orderly access to the bot infrastructure, all services on the remote user subsystem will need the Virtual Telepresence bot credentials to unlock the various services on the remote end.

The Virtual Telepresence bot hosts the computing and control platform. The platform processes remote user requests for access via an email login. Upon successful authentication, the platform sends back credentials of the bot via an email mechanism for subsequent access. These credentials are used with the Main script on the remote user subsystem. The computing and control platform is also responsible for providing the necessary control signals to the camera on board the bot. The signals are derived after processing the raw commands coming over the internet and processing them through filtering, estimation and conversion algorithms. The computing and control platform also serves to convert the commands for movement of the bot to signals that drive the bot's locomotory motors. The computing and control platform incorporates the Server which is responsible for the audio and video streaming functions, along with the microphone integration. Figure 1 indicates the proposed MTR System block diagram with various components of the Remote user subsystem. Figure 2 shows the hardware architecture as a block representation.

A modular design approach is employed in the incremental development of the prototype, driven by requirements in the two phases. The top-level modules are the 'Android Application', 'Camera Module', 'Video and Audio Feed Module', 'Bot Locomotion Controller module', and 'Microphone Control Module'. The 'Android Application module' generates the manual (phase-1) and voice commands (phase-2) for locomotion, camera control, and screenshot capture. The Camera

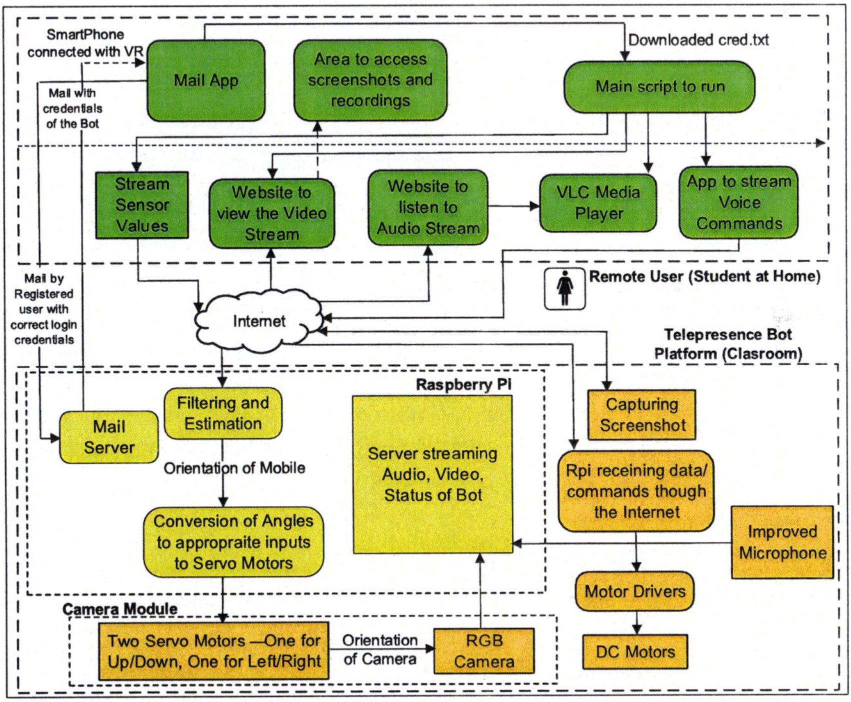

Fig. 1 The proposed Im-SMART system block diagram, outlining the various components of the Remote User subsystem and the Virtual Telepresence bot

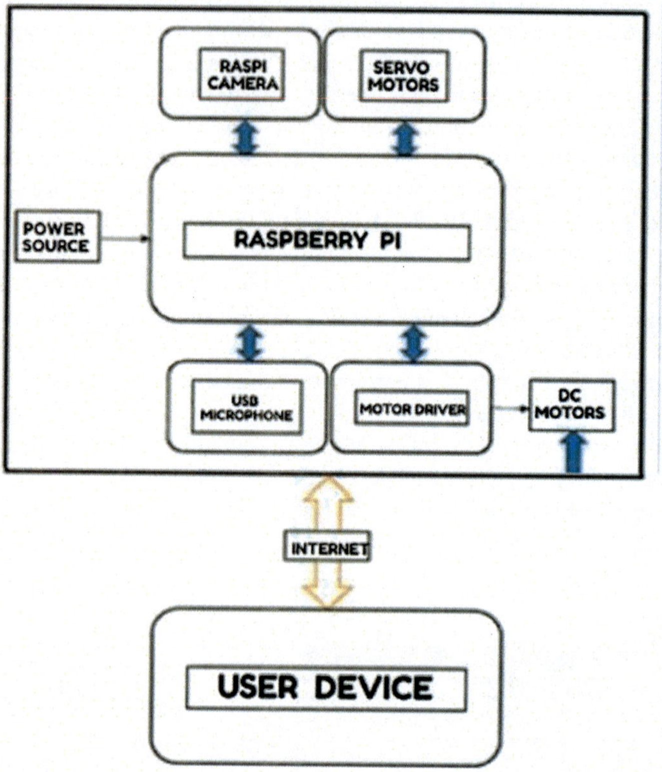

Fig. 2 Hardware architecture of the Im-SMART system (Remote End and MTR)

module processes all movements and orientations of the camera in synchronism with the user's orientation. The Bot Locomotion Controller module drives the motors using the traditional PWM technique and is integrated with the ability to receive voice commands from an android application hosted on the remote user end via a Speech Recognizer. The 'Microphone Control Module' integrates with the Liquid Soap encoder client on the Bot to interface with an IceCast Streaming server on remote user end.

3 Hardware Architecture

3.1 System Functional Allocation

Table 1 provides the allocation of requirements to Hardware elements and serves as the base framework from which the overall hardware architecture emerges. This table

Table 1 Allocation of hardware elements

Req. Id	MTR design aspect	Hardware element
[R1]	Affordability	All elements
[R2]	Immersive experience	Camera
[R3]	Telepresence	Camera motor
[R4]	Connectedness	MTR platform motor and driver support
[R5]	Interaction, mobility	Microphone
[R6]	Immersive experience, communication	Mobile device, raspberry Pi platform
[R7]	Connectedness, connectivity	Mobile Device
[R8]	Ease of use, user interface	Mobile Device, Raspberry Pi Platform
[R9]	Flexibility	Mobile device, raspberry Pi platform
[R10]	Extensibility	Raspberry Pi platform

also serves as a checklist to ascertain whether all appropriate MTR design aspects are met.

3.2 Hardware System Block Diagram, Components and Their Functions

The Hardware System block diagram is provided in Fig. 2. The key components are the Raspberry-Pi platform, servo motors for robot camera movement, power source, USB microphone, and motor driver circuitry for driving the robot's locomotion motors.

Assembly of the robot frame: The chassis of the bot is designed and assembled in a two-tier fashion. The bottom storey houses the servo motors and brackets used for it, batteries and the motor driver, whereas the top storey houses the microphone, Raspberry Pi and the Power Bank. The bot has two wheels controlled by DC motors and one caster wheel in the front. Locomotion of the Studo Bot.

Movement of the Bot's Camera based on sensor readings from smartphone of the user: We use the concept of Socket Programming to communicate between the bot and the user. Here, the Raspberry Pi acts as a Server, and the User Device acts as a Client. The sensor values needed are of accelerometer, gyroscope and magnetometer to determine the orientation of the phone because using only one of them will cause integration error or noise due to the movement of the bot. Smartphone sends sensor readings to the bot using a reliable TCP protocol. The PWM values are obtained on the Raspberry Pi on the bot and are mapped to appropriate pulse signals and fed to the servo motors which moves the camera to the orientation of the head of the user.

Based on the commands received at the motor driver enable pins, the studo bot moves in the required directions. The control logic is as shown in the tabulation in Fig. 3 and the image in Fig. 4 is the test set up of the bot.

Obtaining video feed and streaming it to the user: The camera is tested and configured in order to be able to send the video stream. A web interface is designed using PHP to create the User Interface for the Camera. The video is streamed through this web interface and can be viewed on the phone in VR. The latency of the stream is extremely low and the quality is great over the previous version. Using the inbuilt features on the web interface, we can control camera settings like brightness, contrast, camera scheduling, motion detection etc. The features of taking a screenshot and starting the recording of an ongoing session are added on the Web Interface. Two buttons have been added, which store the recorded files on the server which can be

Command	Logic			
Pins	20	21	23	24
Forward	T	F	T	F
Stop	T	T	T	T
Backward	F	T	F	T
Left	T	T	T	F
Right	T	F	T	T

Fig. 3 Enable pin configurations in the motor driver

Fig. 4 Bot camera movement test setup

downloaded on to the User's Device. Options are also included to delete any file, if needed. The Web Interface is hosted on a Server at Port 80. Voice Commands have been integrated to enable the Screenshot button when needed, significantly reducing the user intervention. We can combat the negative effects of prolonged screen time by using reading mode, night mode or blue light filter on the phone by software.

Integration of microphone on the bot: There is availability of a USB port on the Raspberry Pi which is used to connect the USB microphone with ease. A soundcard can be employed to reduce the low frequency noises by the microphone due to low proximity of distance with the Pi's circuitry. We make use of the Icecast Streaming Server and the Liquidsoap Client to send the low latency live audio stream from the microphone on the bot. Liquidsoap is an encoder client and a scripting language which reads the microphone input and encodes it to the format required by the user. In our case, we have deployed the.opus encoding due to its extremely low latency and amazing usability in live audio streaming. To reduce the effect of microphone noise, Liquid soap provides inbuilt filter functions which have been employed to reduce noise and obtain a better streaming function. Icecast is a streaming server which is hosted on Port 8000. It can automatically stream the audio data incoming from the client, Liquidsoap and play it on the server.

Locomotion of the Studo Bot: Locomotion information is obtained from the user as Voice Commands, through an Android Application specifically designed to suit the application. The Android Application is designed to run in the background continuously, and on receipt of a trigger command, in our case: "Start", starts running the Speech Recognizer and picks up commands like, "Forward", "Backward" etc., which are converted to text and sent to the Studo Bot through the reliable TCP protocol with the help of sockets. The Motor Driver is mapped accordingly and the wheels are actuated as per the commands by the User. A mechanism is provided to notify the teacher if a student has any questions or doubts through an LED on the bot.

4 Software Architecture

Figure 5 lays out the software architecture of the system and is partitioned into two parts, the user side subsystem and the MTR subsystem.

4.1 System Threads

On the Bot platform, the 'main.sh' spawns five threads namely ngrok tunnels, camera web interface, Liquidsoap streamer, locomotion and servo control python scripts. A mail server thread is distinct and created separately to handle the initial registration and setup process. This thread is responsible for generating the credentials for a remote user. On the user side the 'main.py' spawns the four threads Camera feed

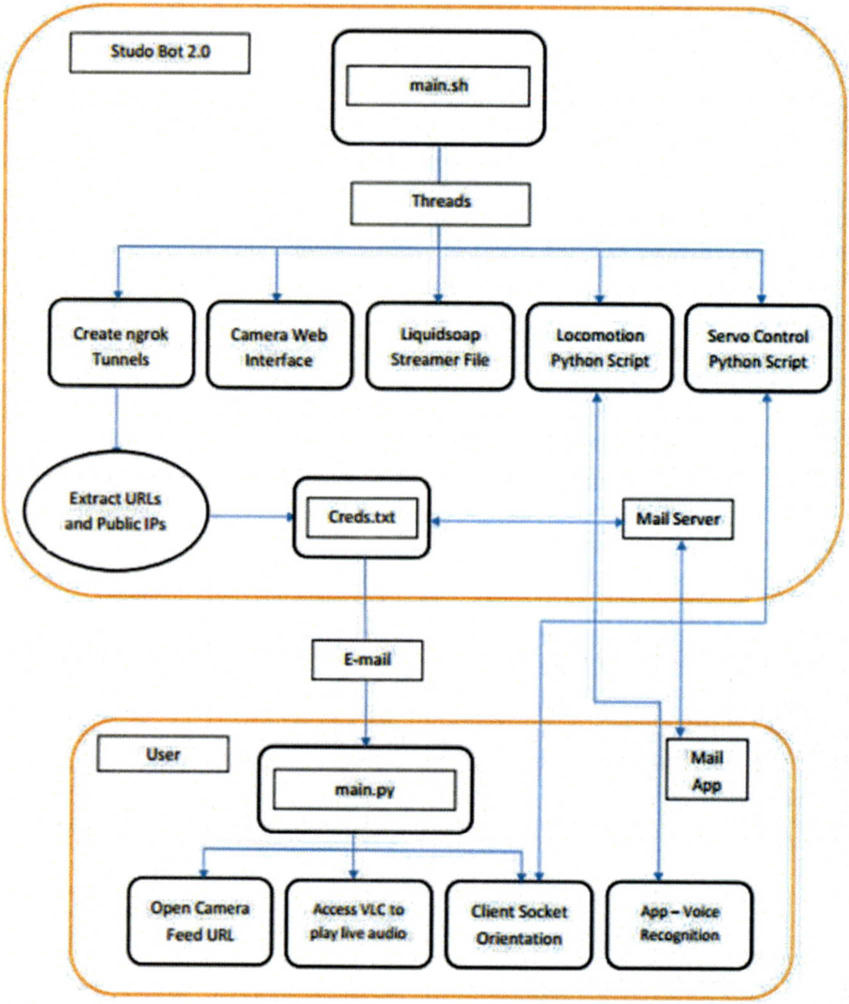

Fig. 5 Im-SMART system software architecture

URL, access VLC, Client socket and app voice recognition which form the complementary components of their counterparts on the BOT platform. Figure 5 indicates the Software Architecture of the system, depicting the key software components.

4.2 *Moving the Entire Setup to the Internet*

In order to be able to minimize User Interventions, and to make the Bot accessible truly remotely and from anywhere in the world, it is necessary for the entire setup

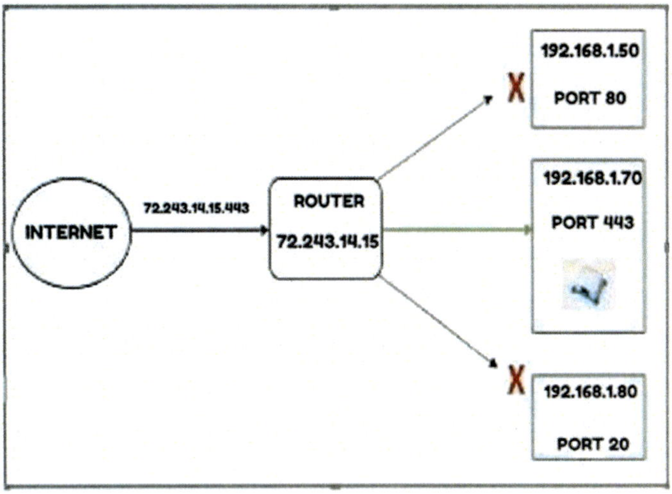

Fig. 6 Port forwarding in Im-SMART

moves to the internet. One of the simplest mechanisms to ensure a Server and a Client stay connected remotely is through the concept of Port Forwarding. To make use of a safe, reliable and cost friendly option, we opted for the services of ngrok, a platform which creates secure tunnels, enables accessing of local websites from anywhere, and also enables port forwarding to easily send data packets through TCP from anywhere in the world which is shown in Fig. 6. ngrok creates tunnels, and provides secure URLs to view the camera feed and stream the live audio from the Bot as represented in Fig. 7. It also provides access to a Public IP and enables port forwarding on local ports to send in data through sockets from User Device seamlessly.

Port forwarding is very useful in preserving public IP addresses. It can help in protecting servers and clients from unwanted access, hide the services and servers available essentially on a network and can also limit access to a network. Port forwarding adds an extra layer of security to networks.

5 Implementation

5.1 System in Work Scenarios

The main working scenarios consist of registration and subsequent authentication, connecting from the remote user end to the Bot for a camera/audio feed, and controlling the Bot's camera and motion using voice commands. These scenarios can be better understood using a system sequence diagram shown Fig. 8.

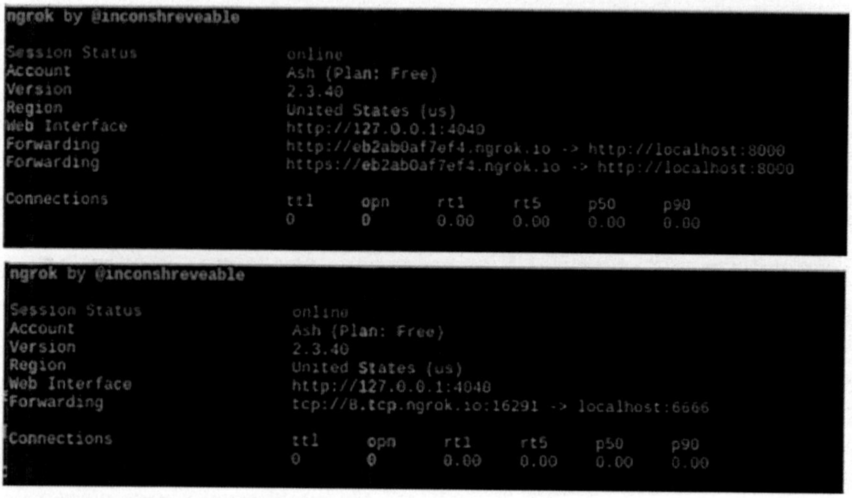

Fig. 7 Ngrok Port forwarding and secure public URLs in Im-SMART

The Bot is switched on and connects to the network at the remote location. As soon as the bot turns on, it starts scanning for emails using the email server and if a new one is received from a registered user with the right login credentials (email and password), it sends the credentials of the bot using which the user can connect to the bot. The bot is marked busy and no other requests are entertained until the bot is free again. The User downloads the file received on the E-mail. A python script reads the recently downloaded file and extracts necessary information needed to establish a connection with the Studo Bot, the Camera stream opens on the browser and the Audio Stream begins on the VLC Player and the user can now use the VR headset to control the orientation of the virtual environment. If it's needed to move the Bot around the physical location, the user speaks the trigger command, "Start" to start the Speech Recognizer in Android Application which runs in the background. On trigger, the control commands, as spoken by the user are converted to text and are sent to the Bot over the Internet. While in the session, if the user feels a need to record and store the session for future use, or needs to take a snap of something useful, the "Click" voice command is spoken and the Camera web interface clicks a snap. This way, the user with minimum interventions, controls and experiences a physical environment, virtually at the comfort of their own place and surroundings.

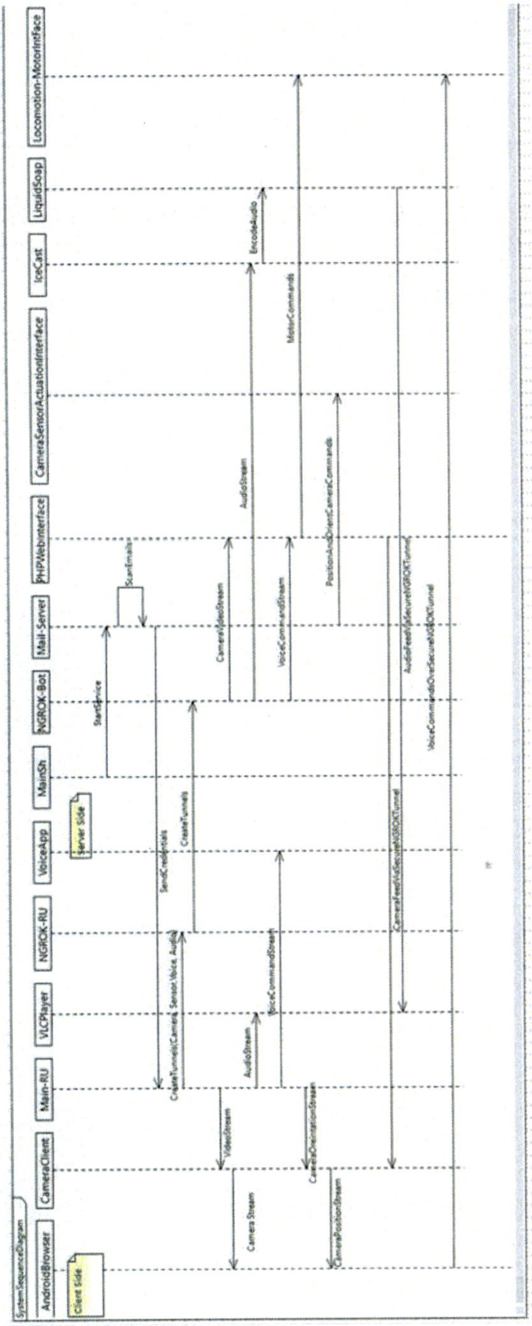

Fig. 8 Sequence diagram outlining the Im-SMART system operational scenarios

6 Results and Discussion

6.1 Experimental Work

A user-friendly platform with added features like Voice Recognition, Screen Recording and Screenshots with improved streams for Virtual Telepresence Robot was designed and the implementation of a prototype was successfully completed. The entire setup of the work from phase-1 is moved to the internet in phase-2 and the Bot can be accessed and controlled from anywhere.

Figure 9 shows images from the Bot Testing and its environment. The top portion of the diagram shows the scenario when credentials are sent to the remote user on authentication, while the bottom portion shows the video feed obtained on the remote user system. The various options available (Eg: Buttons to download a screenshot of the classroom and video recording for the classroom session can be seen). The bottom right portion shows the Mobile Robotic platform.

Table 2 shows the data recorded from experiments conducted with an Android smartphone (remote user subsystem) and the Bot. Mean values are indicated under

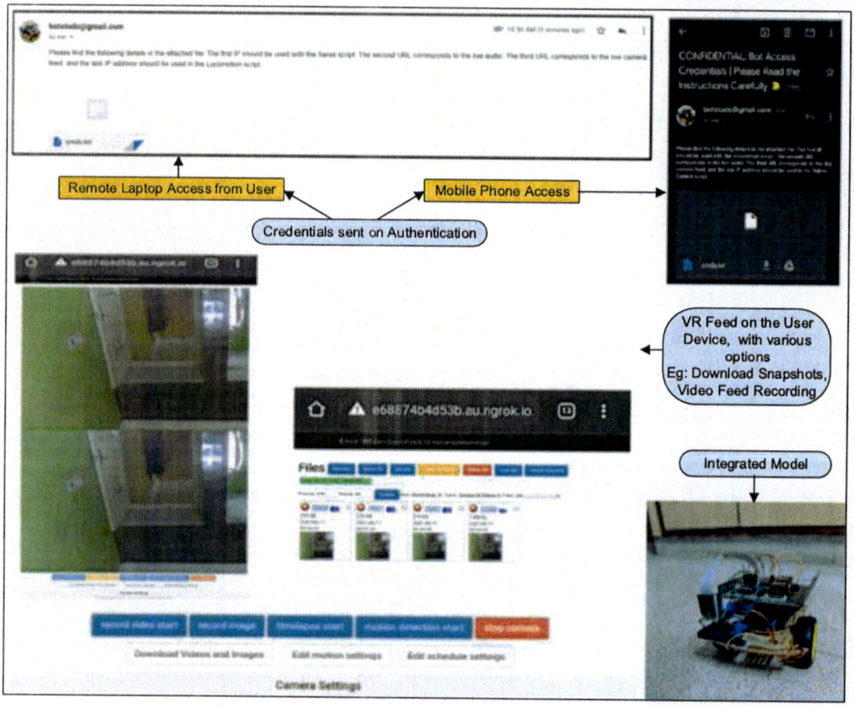

Fig. 9 Im-SMART Bot Testing and its environment

column 'μ' and Standard Deviation under 'σ'. The Camera Web Interface provides extremely low latency and higher quality video feed and is upgraded with new features including brightness control, contrast control etc. We found that the average latency for the video stream was about 0.89 secs and this was acceptable at the remote user end. The remote user did not perceive any difficulties with the video feed and was able to participate in the classrooms effectively. Features like taking a screenshot and starting the recording of an ongoing session are added on the Web Interface. Two buttons have been added, which store the recorded files on the server which can be downloaded on to the User's Device. Options are also included to delete any file, if needed. These features have proven to be useful as utilities. The audio streaming interface which was enhanced with Liquid soap encoder client and Ice Cast streaming Server has provided a delay of about 0.59 s on the average and the remote user did not find any appreciable delays that usually accompany audio–video synchronization, providing a seamless integration. The bot mobility was also measured in the context of time delay for controls to take effect at the bot platform from the moment a voice command appeared at the remote user end. The average delay was slightly more and was measured to be within about 2 s. This aspect is being investigated, since the parameters communicated are very few. The camera deployed on the bot moves according to the orientation values of the remote end user device as commanded, in real time accurately (within about 2 degrees of actual position) with very less delay of approximately 1 s on the average. In Phase-2, major improvement was obtained in the servo's movement with the orientation values with a minimal delay after a smoothing filter was added to the PWM output. Currently, there is ongoing work to measure the control aspects of the camera movement like time for camera to settle down in the locked position. The entire Im-SMART bot platform was built with a budgeted cost of Rs 8000 and actual expenditure was limited to Rs 6500.

6.2 Future Scope

We are currently working to improve the system platform and exploring the usage of Machine Learning algorithms to provide adaptable aspects related to the video and audio functions. We are also exploring aspects related to the efficient usage of the bot in educational context by examining usage scenarios more closely. User Interfaces are another area where we are currently examining the integration of Virtual Reality aspects into the platform. One area of active research is repurposing the platform for other applications like medical education, industry, survey operations, and surveillance.

Table 2 Im-SMART prototype experimental values

Test no	Video streaming latency (Secs)	μ	σ	Audio streaming latency (Secs)	μ	σ	Locomotion latency (Secs)	μ	σ
1	0.80	0.89	0.15	0.5	0.557	0.093	1.4	1.715	0.15
2	0.75			0.6			1.5		
3	1.00			0.55			1.7		
4	0.85			0.45			1.65		
5	0.65			0.65			1.75		
6	0.80			0.40			1.8		
7	0.90			0.70			1.85		
8	0.95			0.55			1.5		
9	1.05			0.50			1.6		
10	1.00			0.75			1.75		
11	1.10			0.66			1.8		
12	0.85			0.69			1.9		
13	0.75			0.57			1.6		
14	0.70			0.54			1.55		
15	0.69			0.44			1.9		
16	1.00			0.49			1.85		
17	1.20			0.49			1.75		
18	1.10			0.56			1.7		
19	0.87			0.5			1.85		
20	0.85			0.55			1.9		

References

1. Ying L, Jiong Z, Wei S, Jingchun W, Xiaopeng G (2017) VREX: Virtual reality education expansion could help to improve the class experience (VREX platform and community for VR based education). In: 2017 IEEE frontiers in education conference (FIE), Indianapolis, IN, USA, pp 1–5. https://doi.org/10.1109/FIE.2017.8190660
2. Potenza A, Kiselev A, Saffiotti A, Loutfi A, An open-source modular robotic system for telepresence and remote disinfection. arXiv:2102.01551. [cs.RO]
3. Mudd SS, McIltrot KS, Brown KM (2020) Utilizing telepresence robots for multiple patient scenarios in an online nurse practitioner program. Nursing Edu Perspect 7/8 41(4):260–262
4. James CA, Bednarz TP, Haustein K, Alem L, Caris C, Castleden A (2011) Tele-operation of a mobile mining robot using a panoramic display: an exploration of operators sense of presence. In: 2011 IEEE international conference on automation science and engineering
5. Gallon L et al (2019) Using a Telepresence robot in an educational context. In: Proceedings of the international conference on frontiers in education: computer science and computer engineering FECS
6. Ruangpayoongsak N, Roth H, Chudoba J (2005) Mobile robots for search and rescue. In: Proceedings of the 2005 IEEE international workshop on safety, security and rescue robotics Kobe, Japan, June 2005
7. Herring SC (2013) Telepresence robots for academics. In: Proceedings of the American society for information science and technology 50(1). https://doi.org/10.1002/meet.14505001156
8. Khadri HO, University academics' perceptions regarding the future use of telepresence robots to enhance virtual transnational education: an exploratory investigation in a developing country. https://doi.org/10.1186/s40561-021-00173-8
9. Kristoffersson A, Coradeschi S, Loutfi A (2013) A review of mobile robotic telepresence. Hindawi Publishing Corporation Advances in Human-Computer Interaction, vol 2013, Article ID 902316, 17 pages. http://dx.doi.org/https://doi.org/10.1155/2013/902316
10. Jakonen T, Jauni H, Mediated learning materials: visibility checks in telepresence robot mediated classroom interaction. https://doi.org/10.1080/19463014.2020.1808496
11. Belmonte LEL (2018) "Professor avatar: telepresence model" IACEE world conference on continuing engineering education (16TH MONTERREY 2018)
12. Velinov A, Koceski S, Koceska N (2021) Review of the usage of telepresence robots in education. Balkan J Appl Math Inf 4(1) (2021)
13. Rae I, Venolia G, Tang JC, Molnar D (2015) A framework for understanding and designing telepresence, CSCW '15, 14–18 Mar 2015
14. Melendez-Fernandez F, Galindo C, Gonzalez-Jimenez J (2017) A web-based solution for robotic telepresence. Int J Adv Robot Syst November-December 2017: 1–19[a]
15. Kachach R, Perez P, Villegas A, Gonzalez-Sosa E (2020) Virtual tour: an immersive low cost telepresence system. In: 2020 IEEE conference on virtual reality and 3D user interfaces abstracts and workshops (VRW), Atlanta, GA, USA, pp 504–506

Architecture and Algorithms for a Pixhawk-Based Autonomous Vehicle

Ankur Pratap Singh, Anurag Gupta, Amit Gupta, Archit Chaudhary, Bhuvan Jhamb, Mohd Sahil, and Samir Saraswati

1 Introduction

The research in autonomous driving is complicated because many interlinked modules are involved. Thus, to properly analyze the performance of a local planning algorithm in real life, we first need to develop a minimal working model of perception and control also.

This work presents our approach to an autonomous electric vehicle research platform, enabling the researcher to focus on one area while exploiting the rest of the modules. We also share how we plan to deploy this system on a full-size golf cart to carry out more realistic experiments. The platform can also serve as an educational tool in robotics, control, and computer vision courses.

A. P. Singh · A. Gupta · A. Gupta (✉) · A. Chaudhary · B. Jhamb · M. Sahil · S. Saraswati
Motilal Nehru National Institute of Technology, Allahabad, India
e-mail: amit.20194090@mnnit.ac.in

A. P. Singh
e-mail: ankur.20191020@mnnit.ac.in

A. Gupta
e-mail: anurag.20195168@mnnit.ac.in

A. Chaudhary
e-mail: archit.20173053@mnnit.ac.in

B. Jhamb
e-mail: bhuvan@mnnit.ac.in

M. Sahil
e-mail: mohd.20195148@mnnit.ac.in

S. Saraswati
e-mail: samir@mnnit.ac.in

© The Author(s), under exclusive license to Springer Nature Singapore Pte Ltd. 2023

H. Muthusamy et al. (eds.), *Robotics, Control and Computer Vision*, Lecture Notes in Electrical Engineering 1009, https://doi.org/10.1007/978-981-99-0236-1_34

Prior work in this direction includes platforms like MuSHR, AutoRally, and MIT Racecar. While the MuSHR Platform focuses on multi-robot systems, it does not have any functionality related to GPS Navigation, which makes prior mapping a necessity for outdoor navigation. It thus restricts its use in a featureless or external environment which is our primary use case. The AutoRally platform comes with an RTK (Real-time kinematic positioning) corrected GPS (Global Positioning System) module but added complexity because of the need to set up a ground station and the necessity to deploy sensor fusion as GPS alone is not accurate enough for reliable navigation. Further, the high cost of the AutoRally platform might be a bottleneck to student researchers. MIT Racecar also does not provide GPS-based navigation facilities. In our approach, we exploit the accurate localization and control capabilities of Pixhawk, an open-source flight controller popular in the UAV (Unmanned Aerial Vehicle) Industry. Pixhawk already provides the capability of precise localization using EKF (Extended Kalman filter) and GPS-based waypoint navigation using cheap sensors like Neo7m GPS-Compass Module and inbuilt IMU (Inertial Measurement Unit). We combine this with our perception and planning modules. This enables one to get started with the platform quickly without extensive calibration and tuning. Our GUI Part fetches global path from google maps and passes these to Pixhawk as a mission using Mavlink protocol. Hence, there is no need to create an SD (Standard Definition) or HD (High definition) Map beforehand to start autonomous navigation using our approach. Further, we also share the possible ways to surpass dependency on google maps and pixhawk using Open Street Maps.

Our overall high-level architecture can be represented by this diagram (Fig. 1).

1. The user first selects the start point and end point
2. We fetch the global path from the google maps API or our own SD Map

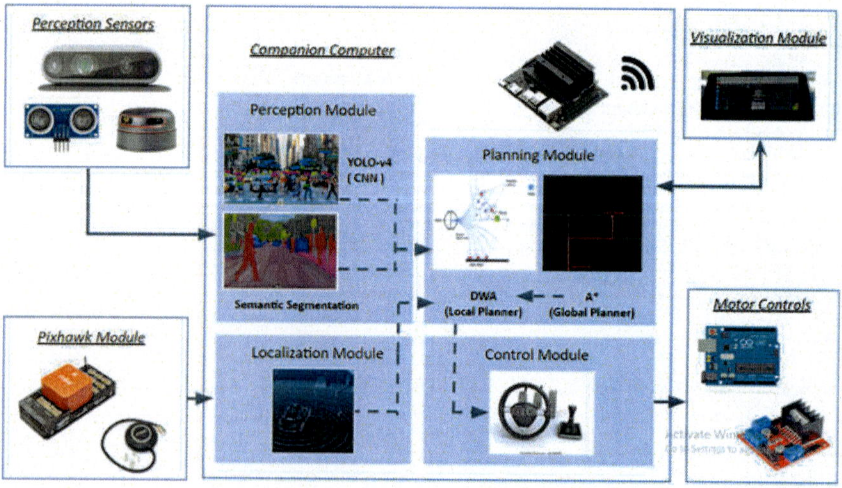

Fig. 1 Overall high level architecture

3. The Global Map, Localization from Pixhawk and Output of Perception Sensors (Camera/LIDAR/Ultrasonic Sensor Etc.) reaches companion computer (Jetson Nano)
4. The perception module in Companion computer processes data from the perception stack and extracts data useful to the local planner, like type and position of obstacles, drive-able region, etc.
5. Based on the output of perception stack, Global Map, and current state of the vehicle, the local planner decides the trajectory for the next step (using DWA)
6. The trajectory is executed using the control module directly or passed to pixhawk to implement it.
7. User sees all this in real time on screen and also this info is live streamed to ground station through radio telemetry/4G Communications and data is logged.

We first elaborate on our perception and planning stack. Our perception stack uses YOLOv4 for object detection and drivable region estimation using semantic segmentation, RANSAC, edge detection, and filtering. We demonstrate all these algorithms using the SOA CARLA Simulators. In the planner part, we first develop an OSM Format map for CARLA towns and use A* to find the global path given a start and endpoint. Then our local planner uses the global path and input from our perception module to calculate the local path using the Dynamic window avoidance algorithm. We demonstrate the accuracy of our planner through our simulation, where the car is able to reach the goal point through 3 cars in between. Finally, we present the overall architecture approach to scale it to real size golf cart and future upgrades.

All the Videos[1] and Codes[2] of simulation are released as open-source.

2 Object Detection

Object detection is the task of detecting instances of objects of a certain class within an image. The state-of-the-art methods can be categorized into two main types:

1. One-Stage Methods
2. Two-Stage Methods.

One-Stage methods prioritize inference speed, for example, YOLO, SSD, etc. Two-Stage methods prioritize detection accuracy, for example, Mask R-CNN, Faster R-CNN, etc. With this kind of identification and localization, object detection can be used to count objects in a scene and determine and track their precise locations, all while accurately labeling them (Fig. 2).

For our project, we have used "You Only Look Once" or YOLO, the family of Convolutional Neural Networks that achieve near state-of-the-art results with a single end-to-end model that can perform object detection in real time and can

[1] https://youtube.com/playlist?list=PL3HszLlqYTxCdmZk7xqaDreLpCilpBEyz.

[2] https://github.com/AmitGupta7580/Static_vehicle_avoidance_carla.

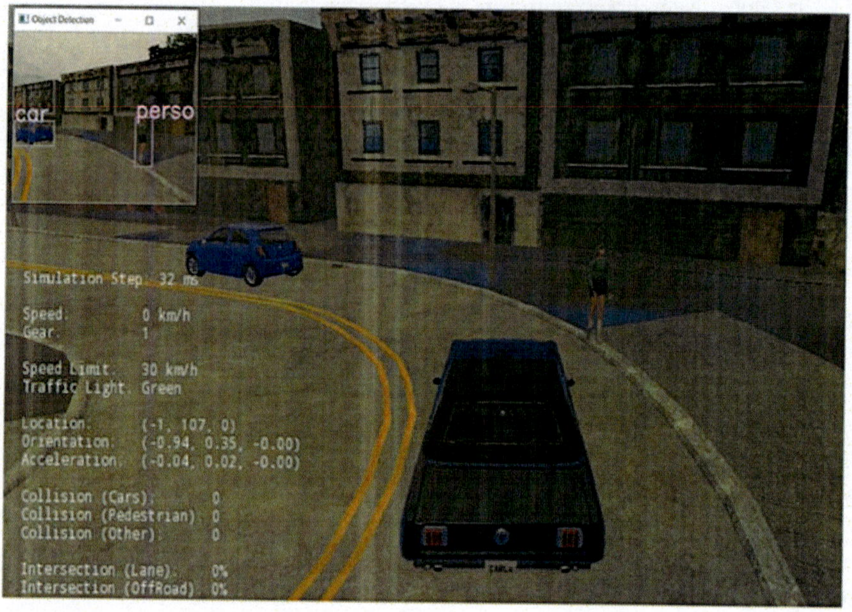

Fig. 2 Output of YOLOv4 model in carla simulator

identify multiple objects in a single frame with high precision and is faster than other models. Its implementation is based on Darknet, an Open-Source Neural Network in C. Compared to other Region Proposal Classification Networks (e.g., Faster R-CNN) which perform detection on various region proposals and thus end up performing prediction multiple times for various regions in an image thus takes more time for predictions.

3-4fps is achieved on the live feed of camera fetching from Carla Simulator on a computer with 8GB of RAM, 2GB of Nvidia GPU, and intel i7 processor. The confidence threshold is set to 0.5 for detecting the object. Classes and weights of the model that we use can be found here.[3]

3 Lane Detection (Estimating Drivable Space)

Detection of Drivable space is a very important task to calculate the possible trajectories in which our agent can move. For this task, the RGB and DEPTH camera feed is used as input and return the equation of lanes in the real-world 3-D coordinate system. This Algorithm consists of 5 steps which are explained below in a sequential order (Fig. 3).

[3] https://github.com/AlexeyAB/darknet/releases.

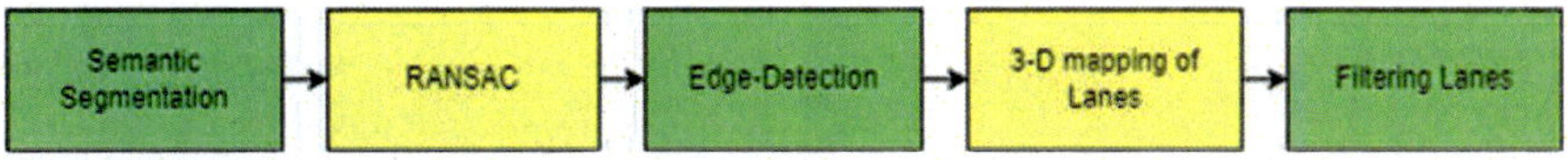

Fig. 3 Block diagram of lane detection algorithm

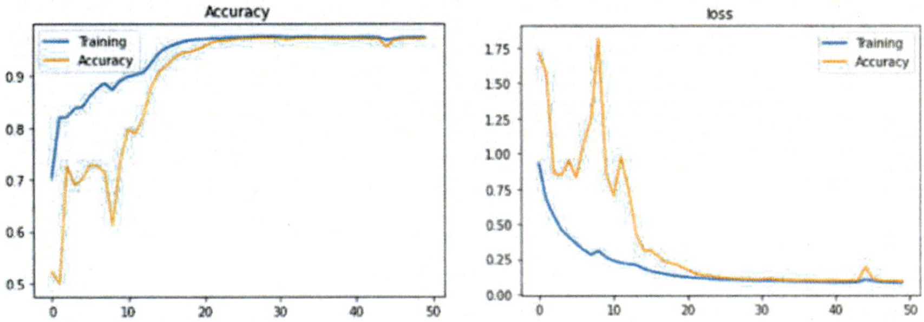

Fig. 4 Accuracy and loss of semantic segmentation model

3.1 Semantic Segmentation

It is a computer vision task in which it labels specific regions of an image according to what's being shown. More specifically, the goal of semantic segmentation is to label each pixel of an image with a corresponding class of what is being represented. As it is predicting for every pixel in the image, this task is commonly referred to as dense prediction (Fig. 4).

Using the inbuilt feature of CARLA for generating semantic segmentation of scenes, we have trained our semantic segmentation model with the help of the Lyft-PerceptionChallange dataset which includes over 1000 scenes of both the semantic segmentation and RGB images.

The architecture of the model: Our model consists of 11 Conv2D layers along with LeakyReLU layer for encoding the image into (17, 25, 192) shape matrix, then it consists of UpSampling2D and concatenate layer for decoding layers. Complete model architecture can be found here[4] Fig. 5 shows the accuracy and loss during the training of our Semantic segmentation model (Fig. 6).

3.2 Random Sample Consensus (RANSAC)

The RANSAC algorithm is a learning technique to estimate the parameters of a model by random sampling of observed data. Given a dataset whose data elements contain both inliers and outliers, RANSAC uses the voting scheme to find the optimal fitting

[4] https://www.kaggle.com/ammmy7580/lane-road-detection.

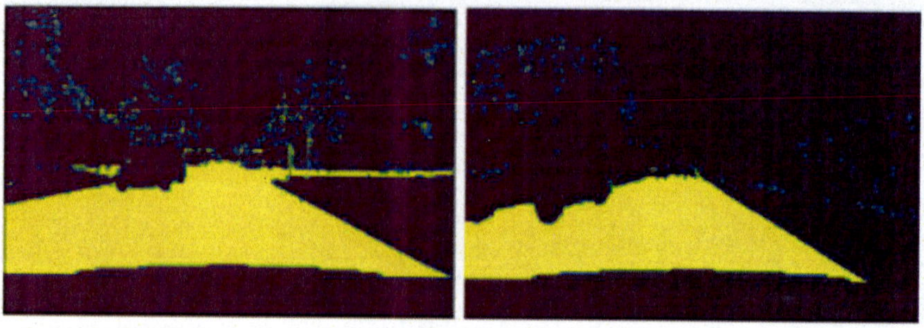

Fig. 5 Prediction of road mask using semantic segmentation model

Fig. 6 Overlapping the calculated road mask over original RGB image

result. Our model uses the RANSAC algorithm to reduce the noise coming from the Machine Learning model. Parameters for RANSAC (Fig. 7)

1. **No. of Iterations: 100**
2. **Initial max No. of inliers: 50% of total points**
3. **Threshold Distance for inliers: 0.01**.

3.3 Canny Edge Detection

Canny edge detection is a technique to extract useful structural information from different vision objects and dramatically reduce the amount of data to be processed. It has been widely applied in various computer vision systems (Fig. 8).

Hysteresis is a filter used in Canny Edge Detection to remove noise in the edge detection process. Each edge is given a score based on the Sobel Filter and a lower and an upper threshold is decided. All edges above the upper threshold are accepted and the ones with scores below the lower threshold are rejected. The edges with scores in between the two are accepted only if they are connected to an edge with a score greater than the upper threshold. This removes the noise while avoiding streaking (discontinuous edges).

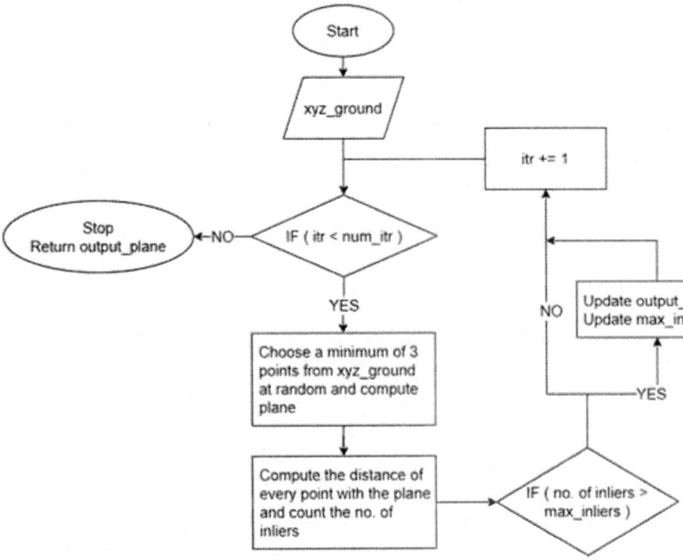

Fig. 7 FlowChart of RANSAC algorithm

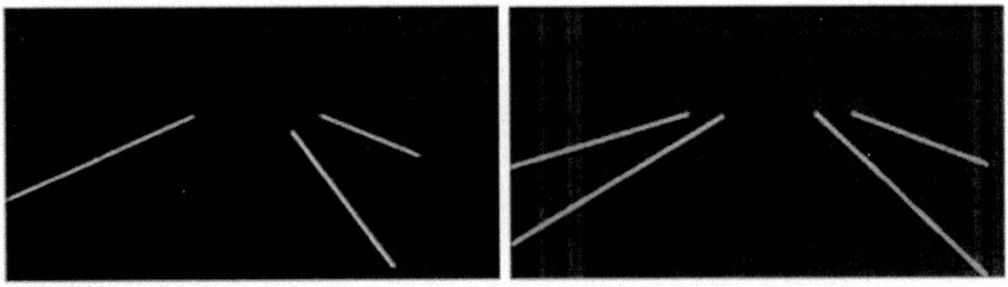

Fig. 8 Canny Edge detection over the RANSAC output

In this step, canny edge detection is applied over the road-masked image to get the edges of lanes as their starting and ending points.

Parameters used for Canny Edge Detection:

1. **First threshold for the hysteresis procedure: 0**
2. **Second threshold for the hysteresis procedure: 150**.

Filters are used to rule out noise in edge detection like merging almost similar lanes.

1. **Threshold for slope similarity: 0.1**
2. **Threshold for intercept similarity: 40**
3. **Threshold for minimum slope: 0.3**.

3.4 3D Mapping and Store Lanes as Lines

This step is about the casting of RGB image pixels into a 3D coordinate system. Estimating the x, y, and z coordinates of every pixel in the image.

With the help of the starting and ending pixels of lanes in the image, it can be easily cast into a 3D coordinate system with the help of a depth camera feed. Now for storing and visualizing the lane data, our code calculates the equation of the line using its starting and ending point (Fig. 9).

3.5 Filtering Lanes

By using the equation of lanes, the similarity of the currently encountered lane with the previously encountered lane equations is checked, and if the similarity is less than some threshold, then this lane gets merged into the previous one by calculating the collective lane equation. The similarity is checked based on their slopes and intercept in the equation of lane (Fig. 10).

1. **Threshold for slope similarity: 4**
2. **Threshold for intercept similarity: 2.**

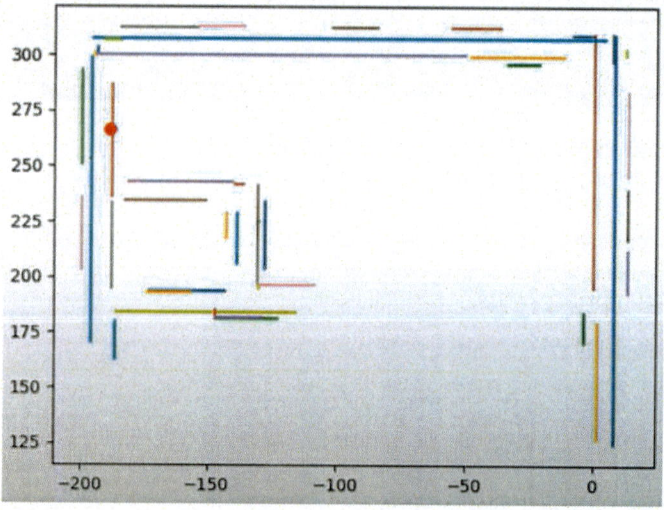

Fig. 9 Projecting detected lanes in 2D plane

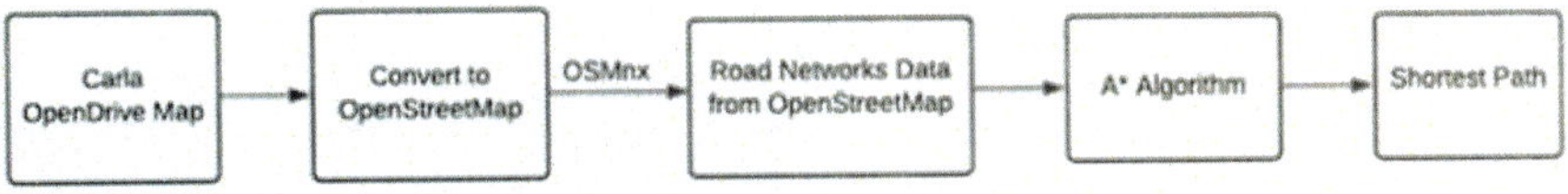

Fig. 10 Flow chart of global path planning

4 Path Planning

4.1 Global Path Planning

For our project, the Town02 map of the Carla Simulator is used. Carla maps are in OpenDrive format, so we converted the Town02 map into OpenStreetMap format. With the help of the Python library OSMnx, we extracted the road network data from the converted Town02 map and selected two nodes of the road network as the start node and goal node. Then, we applied the A* algorithm to find the shortest path (Fig. 11).

A* search algorithm approximates the shortest path in real-life situations, like in maps where there can be many obstacles. It is a popular technique used in pathfinding. A* uses a heuristics function that estimates the cost of the shortest path from the start node to the goal node (Fig. 12).

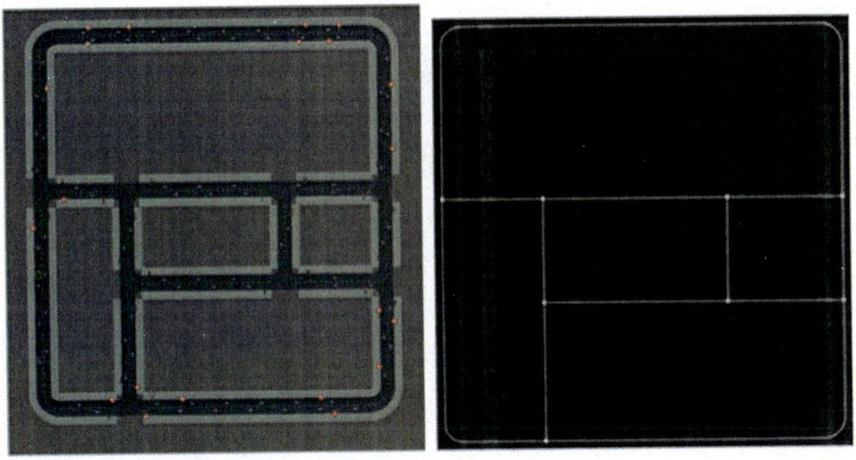

Fig. 11 Carla Town02 map|Streets network data

Fig. 12 Map with shortest path data plotted

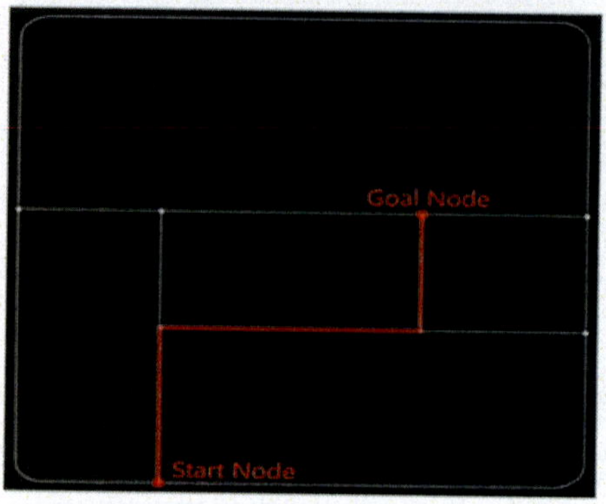

4.2 *Local Path Planning*

Once the global path of the mission is fetched, it can be divided further into small straight paths. For example, if the global path is ABCD, then the small missions would be A to B, B to C, and finally, C to D. Now, to follow these sub-paths, DWA algorithm is used as our local path planner which provides the path that meets certain criteria (Fig. 13):

1. Minimum distance to our goal.
2. Avoiding obstacles comes into the path.
3. Follow the lanes and do not cross them.
4. Smoothness of the motion avoids absurd turns.

DWA (Dynamic Window Approach) is an algorithm used to find the best collision-free trajectory among all the possible trajectories. Figure 11 shows the complete flowchart of the DWA algorithm (Fig. 14).

The Current state (position, orientation, linear velocity, angular velocity) of the car, the position of obstacles, lane equations, and goal position is provided as input to our DWA model. Using these values it returns the next optimal state having the minimum cost. It computes 4 different costs for optimal path:

1. Goal Cost (Calculates the distance of the next possible state with the goal)
2. Speed Cost (For smoothness of the motion)
3. Obstacle Cost (Calculates the distance of the next possible state with all the obstacles)

Fig. 13 Local path planning using DWA

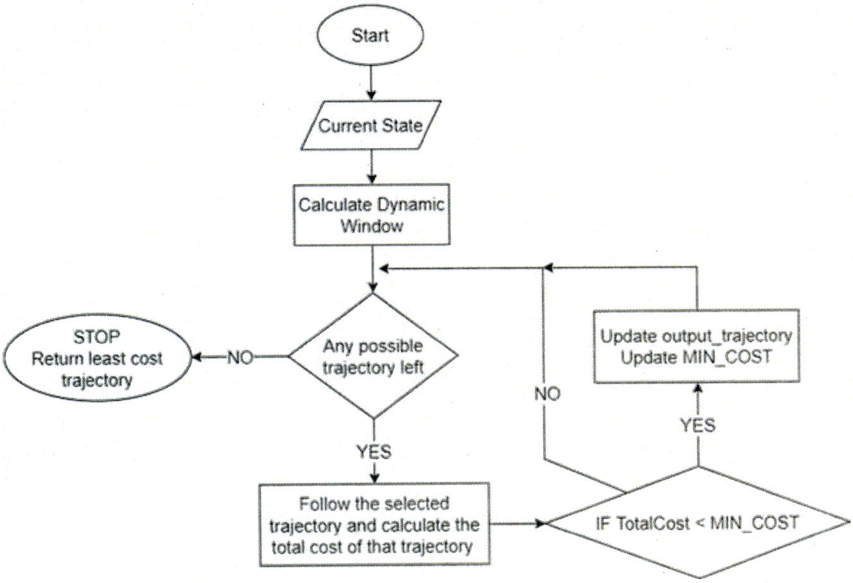

Fig. 14 DWA algorithm flowchart

4. Lane Cost (Calculates the perpendicular distance of the next possible state with the lanes)

$$\text{Total Cost} = \text{Goal} + \text{Lane} + \text{Speed} + \text{Obstacle} \tag{1}$$

The above figure shows the predicted trajectory using DWA, where red dots are displaying static objects and orange and blue lines are representing lanes of the road.

Complete implementation of Dynamic Window Approach in Carla Simulator can be found here.[5]

5 Controllers

The controller is an essential task for deciding the inputs for steer and throttle to efficiently move the vehicle from the starting coordinate to the final destination given by the path planning module. There are various Controllers (e.g., Stanley, Pure Pursuit, etc.), but for our project, a controller that is more responsive to the change in the path and also has fewer errors compared to the actual path is used.

[5] https://github.com/AmitGupta7580/Static_vehicle_avoidance_carla/blob/master/DWA.py.

The controlling part is divided into two sub parts which are:

1. Lateral Control—For Controlling the steer of the vehicle
2. Longitudinal Control—For Controlling the Speed of the vehicle.

5.1 Lateral Control

The lateral control part is the most important task for navigating the vehicle on the actual path by deciding the steering value, i.e., how much the vehicle should have to turn to follow the path. For selecting the best lateral controller, a comparison is made between three different Controllers, Pure Pursuit, Stanley, and Model Predictive Controller (MPL), in the CARLA simulator on the same track with the same inputs of coordinates.

The first two Controllers are geometric path tracking controllers. A geometric path tracking controller is any controller which uses the vehicle kinematics and the actual path to decide the steering value (Fig. 15).

Pure Pursuit controller uses a look-ahead point which is a fixed distance on the actual path ahead of the vehicle. The vehicle needs to proceed to that point using a steering angle which we need to compute. In this method, the center of the rear axle is used as the reference point on the vehicle. The target point is selected on the actual path. And the distance between the rear axle and the target point is calculated to accordingly determine the steering angle of the vehicle. Our target is to make the vehicle steer at a correct angle and then proceed to that point (Fig. 16).

Pure Pursuit controller ignores dynamic forces on the vehicle and also the limitation of the vehicle to steer at such high angles. One improvement is to vary the look-ahead distance based on the current speed of the vehicle to fine-tune the steering angle. For lower speed, it should be small so that the vehicle can steer at high angles, and for higher speed, it should be large to limit the steering changes.

Stanley Controller is also a geometric path-tracking controller. The Stanley method uses the front axle as its reference point. Meanwhile, it looks at both the

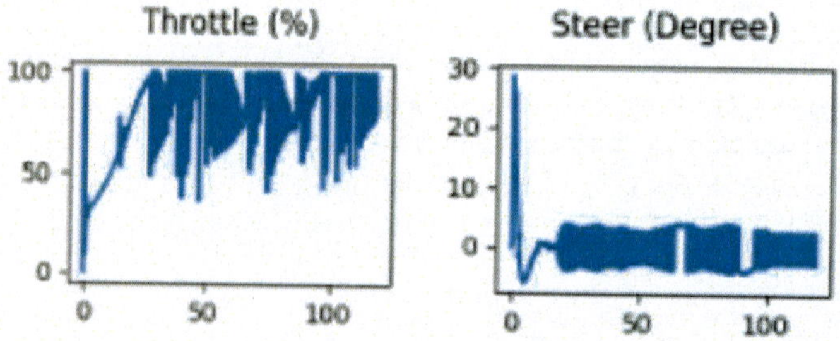

Fig. 15 Throttle value ans Steer value by the pure pursuit controller

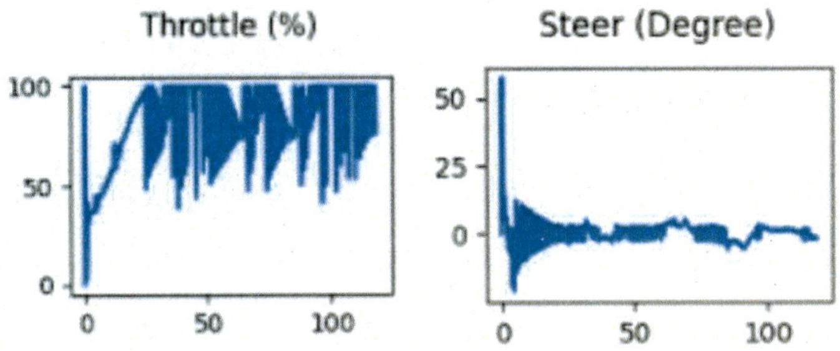

Fig. 16 Throttle value and Steer value by the Stanley controller

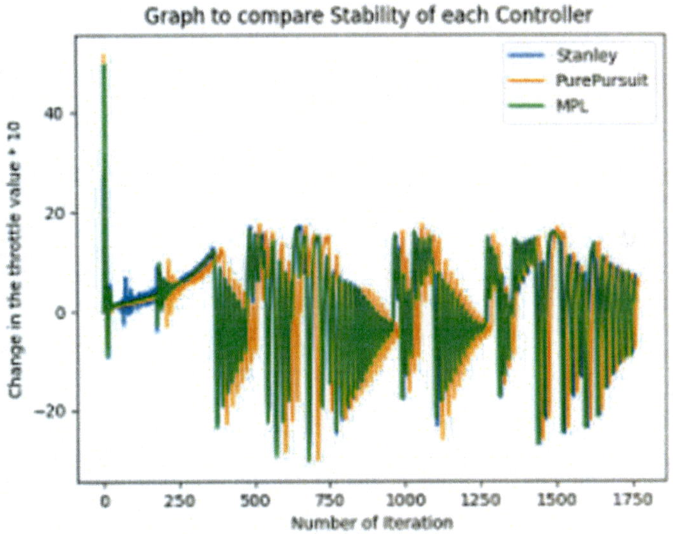

Fig. 17 Comparison of three different controllers on the basis of change in throttle in the CARLA simulator

heading error and cross-track error. In this method, the cross-track error is defined as the distance between the closest point on the path with the front axle of the vehicle (Fig. 17).

Model Predictive Controller is not a geometric path-tracking controller. Model Predictive Controller uses cost function and predictive model to output the steering values. Cost function contains the deviation from the reference path, smaller deviation better results. Meanwhile, minimization of control command magnitude to make passengers in the car feel comfortable while traveling, smaller steering better results (Fig. 18).

By plotting the change in the steering value compared to the previous value of the steer multiplied by 10 (so that we can observe small changes also), we simply observe the sudden change that happens. Because sudden changes in the steering

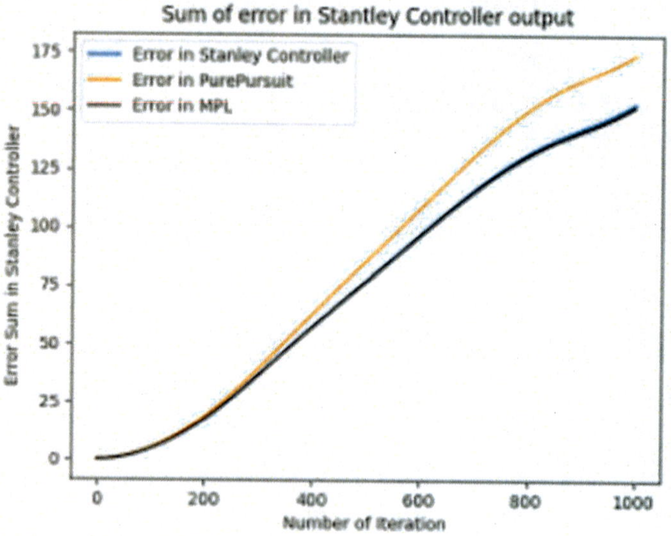

Fig. 18 Comparison of three different controllers on the basis of sum of error in the CARLA simulator

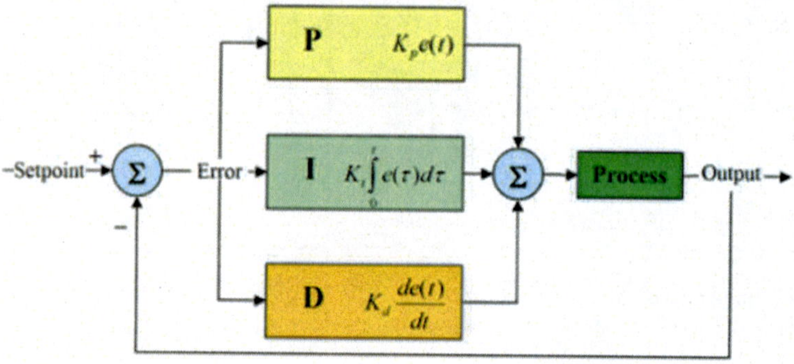

Fig. 19 Longitudinal control

will make the vehicle unstable at a higher speed. In the plot Stanley, MPL has better resistance to the sudden change that happens in the actual path so that it slowly turns the vehicle to the actual trajectory but the pure pursuit method makes sudden changes in the steer. Also, Stanley, MPL has less error compared to pure pursuit.

By comparing Stanley and MPL controller, it was found that the MPL controller is more stable than Stanley. Mainly when the road ahead is straight, Stanley controllers still have variation, but MPL (from 0–250 range in the plot) is stable in that range (Fig. 19).

5.2 *Longitudinal Control*

For longitudinal control, PID (Proportional Integral Derivative) controller is used. PID controllers use a control loop feedback mechanism to control process variables and are the most accurate and stable controllers. The feedback mechanism has a sensor that continuously provides information about the vehicle's speed, which is then compared to the actuator signal, and accordingly, the signal is varied to attain desired speed.

6 Overall Architecture

Firstly, the user inputs start point and end point through the screen using our GUI. Then, we fetch the path as a series of GPS waypoints through google maps or our global planner implemented on OSM. We feed this global plan as well as the output of the perception module to our local planner which calculates the local plan to reach the closest waypoint on the global path. We can pass this local plan either to our Control module or to the pixhawk rover firmware's controller. We use ROS for all the communications between different nodes and processes. We present the communication between different nodes of our system as follows (Figs. 20 and 21).

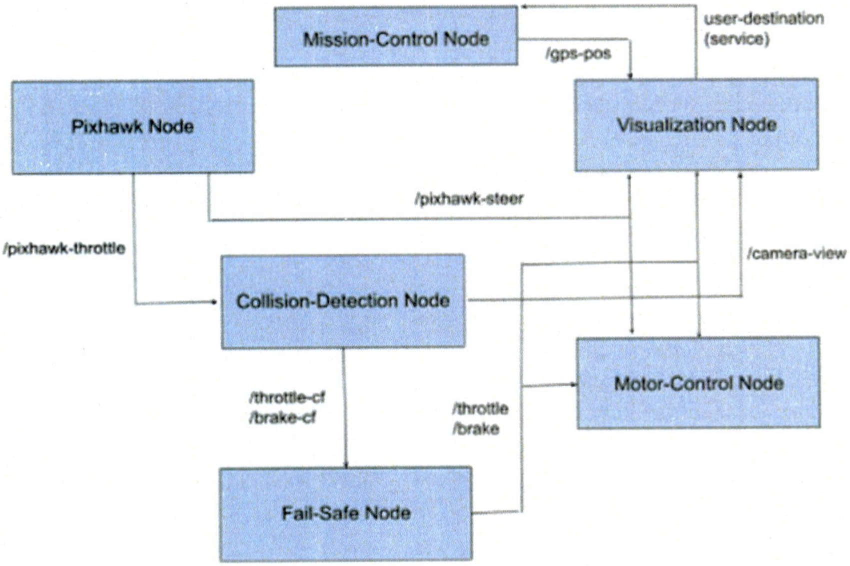

Fig. 20 Flow Chart of ROS nodes and their topic

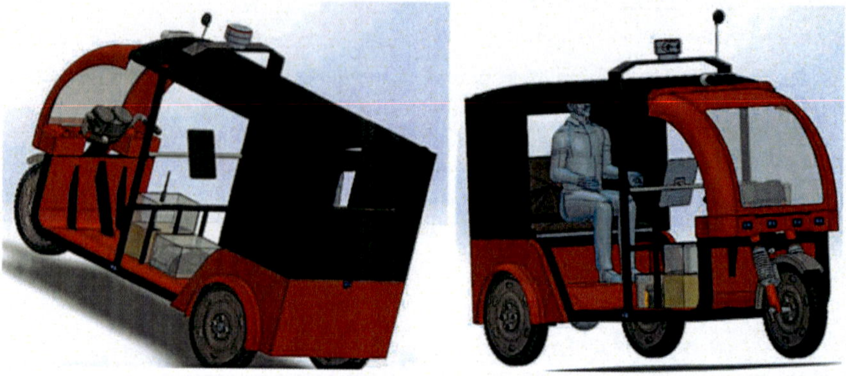

Fig. 21 CAD model of self-driving cart

7 Future Work

1. Prepare a hardware prototype as a proof of concept
2. Scale up the model by implementing it on a full-size golf cart. Further, we present how our work can be scaled to a real-size golf cart through the following CAD File.
3. Add LIDAR and other functionalities like Optical flow, navigation against HD Map in addition to GPS to further initiate algorithmic research
4. Integrate Autoware functionalities in the software stack
5. Convert our solution into a box that can be easily deployed on any e-vehicle to prepare the platform in a DIY manner with a supported Simulation Stack.

8 Conclusion

We present our study and implementations on perception, planning, and control aspects of an electric autonomous vehicle research platform for educational and research purposes. We also present an architecture to merge our modules with the accurate localization and control capabilities of pixhawk that allows the researcher to quickly get started with his work without rigorous calibrations. In the end, we share how this work can be scaled on a real-size golf cart and present our future work plans.

References

1. Srinivasa S, Lancaster P, Michalove J, Schmittle M, Rockett C, Smith J, Choudhury S, Mavrogiannis C, Sadeghi F (2019) MuSHR: a low-cost, open-source robotic racecar for education and research
2. Goldfain B, Drews P, You C, Barulic M, Velev O, Tsiotras P, Rehg J (2018) AutoRally: an open platform for aggressive autonomous driving
3. Dosovitskiy A, Ros G, Codevilla F, Lopez A, Koltun V (2017) CARLA: an open urban driving simulator. In: Proceedings of the 1st annual conference on robot learning. http://proceedings.mlr.press/v78/dosovitskiy17a/dosovitskiy17a.pdf
4. Bochkovskiy A, Wang C-Y, Mark Liao H-Y (2020) YOLOv4: optimal speed and accuracy of object detection. https://arxiv.org/abs/2004.10934
5. Kumaresan (2017) Semantic segmentation for self driving cars. In: Dataset with semantic segmentation labels generated via CARLA simulator, version 1 from https://www.kaggle.com/kumaresanmanickavelu/lyft-udacity-challenge
6. Blaga B, Nedevschi S (2019) Semantic segmentation learning for autonomous UAVs using simulators and real data. In: 2019 IEEE 15th international conference on intelligent computer communication and processing (ICCP) 2019 ICCP48234.2019.8959563
7. Fischler M, Bolles R (1981) Random sample consensus: a paradigm for model fitting with applications to image analysis and automated cartography
8. Canny J (1986) A computational approach to edge detection. IEEE Trans Pattern Anal Mach Intell PAMI-8(6):679–698. https://doi.org/10.1109/TPAMI.1986.4767851
9. Hart P, Nilsson N, Raphael B (1968) A formal basis for the heuristic determination of minimum cost paths. IEEE Trans Syst Sci Cybern 4(2):100–107. https://doi.org/10.1109/tssc.1968.300136
10. Boeing G (2017) OSMnx: new methods for acquiring, constructing, analyzing, and visualizing complex street networks. Comput Environ Urban Syst 65:126–139. https://doi.org/10.1016/j.compenvurbsys.2017.05.004
11. Fox D, Burgard W, Thrun S (1997) The dynamic window approach to collision avoidance. IEEE Robot Autom Mag 4(1):23–33. https://doi.org/10.1109/100.580977
12. Wang W, Hsu T, Wu T (2017) The improved pure pursuit algorithm for autonomous driving advanced system. In: 2017 IEEE 10th international workshop on computational intelligence and applications (IWCIA), 2017, pp 33–38. https://doi.org/10.1109/IWCIA.2017.8203557.
13. AbdElmoniem A, Osama A, Abdelaziz M, Maged SA (2020) A path-tracking algorithm using predictive Stanley lateral controller. Int J Adv Rob Syst. https://doi.org/10.1177/1729881420974852

3D Obstacle Detection and Path Planning for Aerial Platform Using Modified DWA Approach

Ankur Pratap Singh, Amit Gupta, Bhuvan Jhamb, and Karimulla Mohammad

1 Introduction

Obstacle detection and avoidance are crucial for modern-day drone applications like drone delivery, surveillance, mapping, etc. We present our novel approach for this task. We first detect the type and location of obstacles through CNN. Once the obstacles are detected, we divide the field of view of the RGB-D camera into a 12*16 grid. We find a cost value for each grid based on factors like proximity to the goal, distance from obstacles, smooth motion of drone, etc. Our cost function is built on the concept of the DWA algorithm for 2d path planning. Based on the cost distribution and type of obstacle, drone maneuvering takes place.

We first present details on dataset and model to train obstacle detection in AirSim, followed by our overall approach for obstacle avoidance and then elaboration on each component of our cost function and calculation of optimal velocity. We also present the implementation and results of our approach using AirSim Simulator.

Ankur Pratap Singh, Amit Gupta, Bhuvan Jhamb, Karimulla Mohammad—These authors contributed equally.

A. P. Singh · A. Gupta (✉) · B. Jhamb · K. Mohammad
Motilal Nehru National Institute of Technology, Allahabad, India
e-mail: amit.20194090@mnnit.ac.in

A. P. Singh
e-mail: ankur.20191020@mnnit.ac.in

B. Jhamb
e-mail: bhuvan@mnnit.ac.in

K. Mohammad
e-mail: karimulla.20198060@mnnit.ac.in

© The Author(s), under exclusive license to Springer Nature Singapore Pte Ltd. 2023
H. Muthusamy et al. (eds.), *Robotics, Control and Computer Vision*, Lecture Notes in Electrical Engineering 1009, https://doi.org/10.1007/978-981-99-0236-1_35

2 Dataset for Object Detection

YOLOv4 is trained over different types of classes but the YOLO is never used for Aerial object detection. To fill this void in YOLO we have build our own dataset which helps YOLO for detecting the specific Aerial Obstacles like the current version of our dataset provides data of these 4 classes (Figs. 1 and 2)

1. Bird
2. Drone
3. Building
4. Blocks (Big structures in AirSim Blocks Environment).

Complete Dataset can be found here.[1]

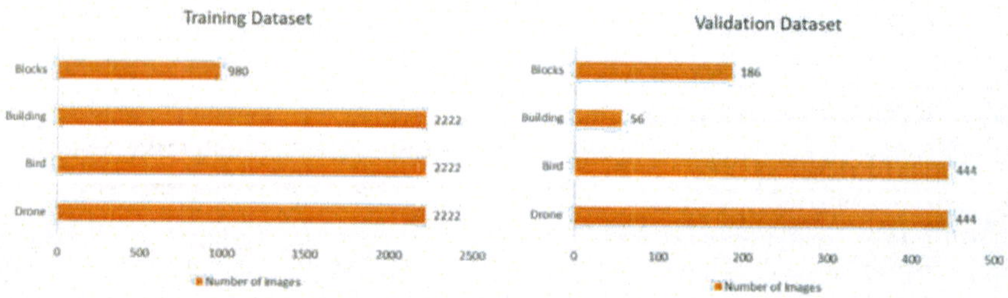

Fig. 1 Training and validation dataset images

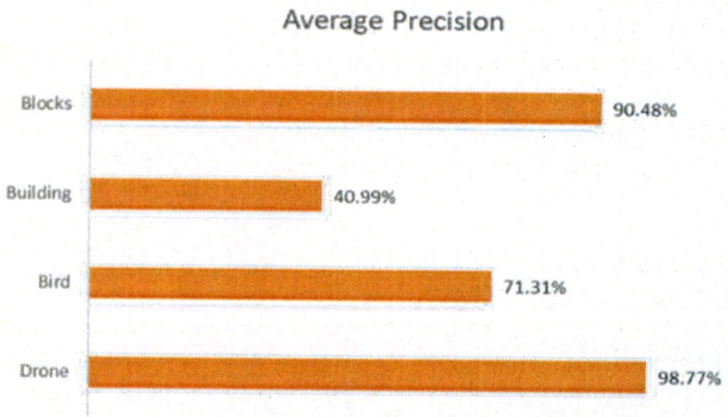

Fig. 2 Average precision for each class

[1] https://drive.google.com/drive/folders/1mfD6Pdkb4Y8l9C4ksE0ZVxfuQJXgTOeG.

3 Obstacle Detection Model

In this chapter, we aim to design an aerial object detection system for autonomous drones in the AirSim simulator using YOLOv4 object detector.

In object detection, the task is to detect instances of all the objects of a certain class within an image. The state-of-the-art methods can be categorized into two main types:

1. One-Stage Methods
2. Two-Stage Methods.

One-Stage methods prioritize inference speed, example YOLO, SSD, etc. Two-Stage methods prioritize detection accuracy, example Mask R-CNN, Faster R-CNN, etc. With this kind of identification and localization, object detection can be used to count objects in a scene and determine and track their precise locations, all while accurately labeling them.

For our project we have used "You Only Look Once" or YOLO, family of Convolutional Neural Networks that achieve near state-of-the-art results with a single end-to-end model that can perform object detection in real-time and can identify multiple objects in a single frame with high precision and is faster than other models. Its implementation is based on Darknet, an Open-Source Neural Networks in C (Fig. 3).

To perform the task of obstacle detection we employed the concept of transfer learning. Transfer Learning is when existing models are reused to solve a new challenge or problem. Transfer learning is a technique or method used while training

Fig. 3 Blocks detection in airsim environment

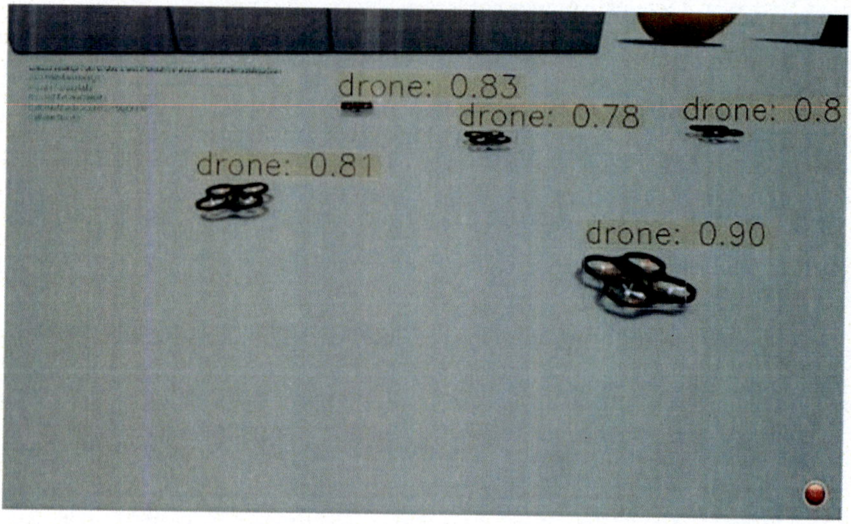

Fig. 4 Drones detection in airsim environment

models. The knowledge developed from previous training is recycled to help perform a new task. The new task will be related in some way to the previously trained task, which could be to categorize objects in a specific file type. The original trained model usually requires a high level of generalization to adapt to the new unseen data.

We used pretrained YOLOv4 model and used its convolutional layers weights which help our custom object detector to be way more accurate and not have to train as long and it will converge and be accurate way faster. We are using the Blocks environment of AirSim because it consumes less resources of our computers, so we can get better FPS (Frames Per Second) for testing algorithms and implementing future work. We have trained YOLOv4 to detect custom objects (Fig. 4).

Developing and testing algorithms for autonomous vehicles in the real world is an expensive and time-consuming process. Also, in order to utilize recent advances in machine learning and deep learning we need to collect a large amount of annotated training data in a variety of conditions and environments. AirSim is a new simulator built on Unreal Engine that offers physically and visually realistic simulations for both of these goals. The simulator is designed from the ground up to be extensible to accommodate new types of vehicles, hardware platforms, and software protocols. In addition, the modular design enables various components to be easily usable independently in other projects.

YOLOv4's architecture is composed of CSPDarknet53 as a backbone, SPP (spatial pyramid pooling) additional module, PANet path-aggregation neck, and YOLOv3 head. **Our custom trained model achieves 75.39% mAP@0.5–59.585 BFlops.**

Simulation Video[2] of Aerial Obstacle Detection using YOLOv4 in Airsim. Weights of our customized YOLOv4 model[3] are open-source.

3.1 Our Approach

In this approach we take evenly distributed points in the field of view of the camera. And then calculate the Obstacle cost, Smoothness Cost, and Goal Cost of each path present in the FOV, and then we add up all the costs to get the total costs of the path. The path with the minimum total cost is selected (Fig. 5).

As our depth camera gives a feed of 144*256 array so we decided to divide out fov into 12*16 paths/points that a drone can take so as to maintain the symmetry. Then for each of these paths we find out the obstacle cost, smoothness cost, and the depth cost of that path. The obstacle cost is proportional to the proximity of the obstacles, the smoothness cost is proportional to the sudden changes in velocity and the goal cost is inversely proportional to how close a path takes the drone toward goal. This is how we calculate the costs (Figs. 6 and 7):

Obstacle Cost

1. We first fetch the camera feed of the depth image, which contains the distance of the object present at a certain pixel. Airsim returns the feed in form of a 144*256 array.
2. We then pass the array through the average 12*16 pooling with the stride of 12 in the horizontal direction and a stride of 16 in the vertical direction such that no two layers overlap. The pooling will result in a 12*16 array. And each of these points denotes the 12*16 paths of fav which a drone can take.

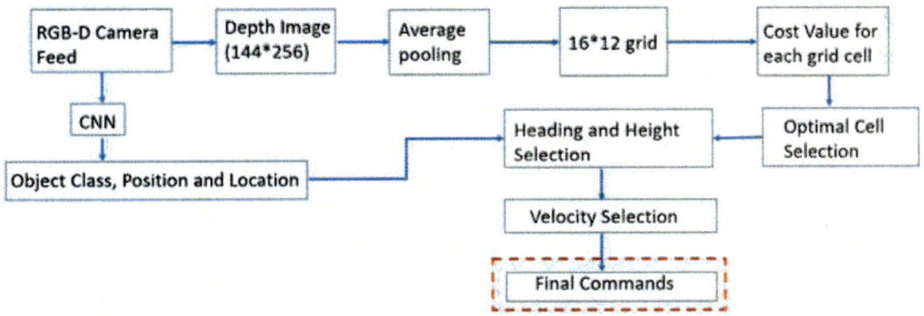

Fig. 5 Our approach

[2] https://drive.google.com/file/d/1v5KT0cw5LgAQFfhb\discretionary-4VtaZJBodEPZaEf/view.

[3] https://drive.google.com/file/d/1OkrreuxpYbSFslZ3irBKa48P7X9gpYxq/view.

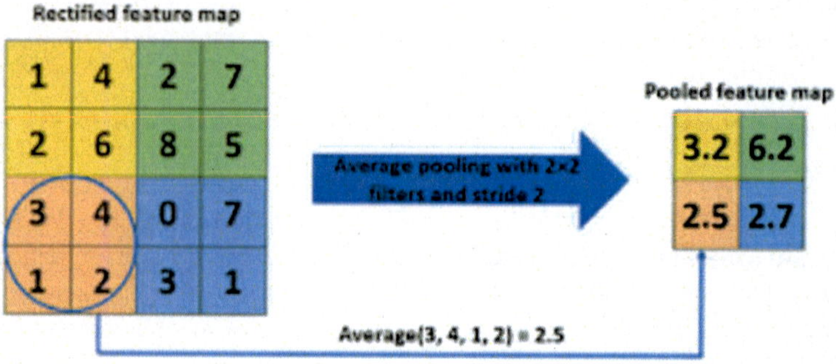

Fig. 6 Example of average pooling

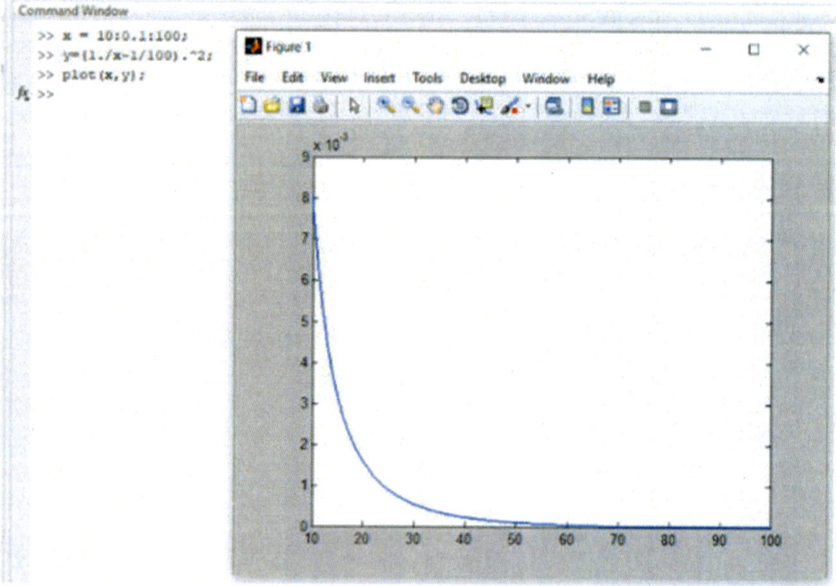

Fig. 7 Obstacle cost versus obstacle distance with effective distance 100

3. We pooled it down to a 12*16 array as it is costly to do operations on the 144*256 array. It is the very reason why we selected 12*16 evenly distributed points rather than 144*256 points.
4. We also define effective distance for the drone. It is the distance from the drone beyond which the obstacles are ignored.
5. We want the function to change highly at a smaller value of distance as compared to the higher values. This is because we would highly prefer a path with 15 units distance over 10 units distance but it would not make much difference selecting a path with 70 units distance and 75 units difference. So at a large distance, the

optimal path to select depends upon smoothness and goal cost as the obstacle cost doesn't change highly.

6. So we pass each value in the distance matrix to get the cost matrix. The function being:

$$Cost_{i_j} = W_g * ((1/effective\_dist) - (1/dist_{i_j}))^2$$

For all $dist_{ij}$ less that effective distance;
$Cost_{ij} = 0$ for all $dist_{ij}$ greater that equal to effective distance
Where W_g is the goal weight
$Cost_{ij}$ is the cost of ith row and jth column
$Dist_{ij}$ is the distance of ith row and jth column

7. It will not be the best choice to select the block with the minimum cost (highest obstacle distance) from the cost array as we may end up selecting a path which has an obstacle in the path right next to it.
8. Here it would be much better to select the block with 30 costs as compared to 20 costs as the column next to 20 has a very high obstacle cost. The high cost implies that there is an obstacle very close to the drone in that path so it would be better to go with cost 30.
9. To overcome this problem, we take the exponential average of the rows and the column before moving ahead so that the cost of the block is increased /decreased according to the closeby blocks. This is how we take the exponential average:
Let there be an array of n numbers a_1, a_2, a_3 ..
Then the exponential average numbers e_1, e_2, e_3 ..
Will be $e_i = (a_i * (1 - \beta) + (e_i - 1) * \beta)/1 - \beta^i$
where $e_0 = 0$;
$\beta = 0.3$; (In;our;case;which;can;be;modified)
10. At the start of the array, a1 = e1, i.e., the value of e1 is unaffected by other values while en is affected by all values a1, a2, a3, ..., that comes before an. So to maintain consistency we will do exponential averaging in all 4 directions: top to bottom, bottom to top, right to left, and left to right.
11. Let the sum of the values over all 4 directions be Eij for the block at the i th row and jth column. Eij is the Exponentially averaged cost of the ith row and jth column which is also the final obstacle cost of the ith row and the jth column.
Obstacle cost (i, j) = E_{ij}
Where E_{ij} is averaged value

Smoothness Cost

1. The best smoothness can be achieved by the drone if it moves with the same velocity acceleration, i.e., with the same change in displacement velocity. In reference to the 12*16 blocks we mentioned above for constant velocity acceleration, we need constant need a constant rate of change in the rows and the columns selected.
2. The smoothness cost is broken down into two parts: vertical and horizontal smoothness so that we can have more control over the flight.
3. The smoothness cost of the block will be the distance of the block from the most preferred block, in the vertical direction for vertical smoothness and vice versa.

Fig. 8 Example of 3*5 grid containing cost

20	100	45	35	55
20	100	48	30	37
20	100	42	32	39

4. Let in x-direction, change in velocity be dvx, change in a position be dx1, and last selected position be x. Now the most preferred velocity is dx + dvx and the position is x + dx1 + dvx. Vice versa in y-direction.
5. For best smoothness the change shall remain the same, i.e., the vel should remain the same. So the most preferred block will be (x + dx + dvx, y + dy + dvy).
6. For each block (i, j) the smoothness cost is referred to as

$$Smoothness\_cost(i, j) = abs(x + dx - i) * W_{sh} + abs(y + dy - j) * W_{sv}$$

where

W_{sh} is smoothness weight in horizontal direction
W_{sv} is smoothness weight in vertical direction (Figs. 8, 9 and 10).

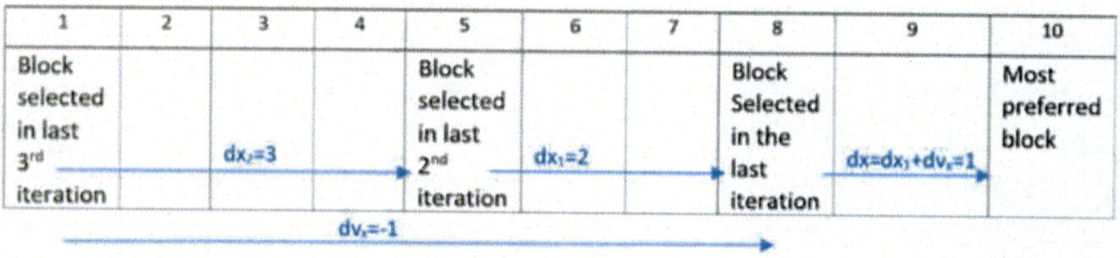

Fig. 9 Depicting next best preferred column

Fig. 10 Angle which each of the column directs

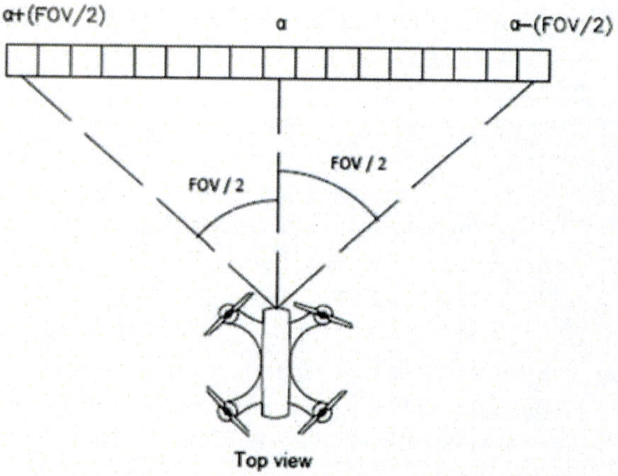

Goal Cost

1. Goal cost helps to determine whether the path moves toward the goal or away from the goal. The more the path directs toward the goal the lesser will be its Goal cost.
2. Goal cost is broken down into two parts:
   – Due to the difference in the angle the drone will face if it selects the block and the angle in which the drone should go to reach the goal.
   – Due to the difference in the height of the drone currently and the preferred height at which we want our drone.
3. We get to know about the current facing of the drone using the compass. Let the current facing angle be alpha.
4. Now we will find the angle in which the drone will move if it selects any of the boxes. Let the field of view of the camera be fov so the rightmost block will be (alpha+fov/2) degrees and the leftmost will be (alpha-fov/2) degrees. And these values change linearly.
5. We can find the goal angle, gamma, using the current location and final location by

   $$\gamma = ((G_y - y)/(G_x - x))$$

   where G_x is x-coordinate of goal
   G_y is y-coordinate of goal
   x and y are the current coordinates of the drone
6. To find the goal cost due to the second part we shall also predict the height, hi, of the drone after selecting the certain row.
7. So the total goal cost will be :

   $$Goal\ cost = (a_j - \gamma) * W_{o_a} + (h_i - G_h) * W_{o_h}$$

   where:
   A_j = angle of the drone if jth column is selected
   γ = angle:toward:the:goal
   W_{oa} = obstacle weight for angle
   h_i = height of drone if ith row is selected
   G_h = height of the goal
   W_{oh} = obstacle weight due to the height of the drone

The total cost of each block at ith row and jth column will be:

$$Total\_cost(i, j) = Obstacle\_cost(i, j) + Smoothness\_cost(i, j) + Goal\_cost(i, j)$$

The block containing the minimum cost is selected. Let the selected optimal block be in ith row and the jth column then we calculate the effective angle aj of the column j and the height hi for the ith row similar as we have calculated in the Goal cost. In airsim, to move the drone we give velocity in x-direction, velocity in y-direction, and destination height. So with ht help if aj and hi we can easily determine them,

$$X\_velocity = v * cos(a_j)$$

$$Y\_velocity = v * sin(a_j)$$

$$Destination\ height = h_i$$

where v is the velocity of the drone

We decide the velocity of the drone according to the obstacles cost and the smoothness cost of the block. We will loop over again and again to find the best block with the timestep of 0.2 s until we reach the destination.

3.2 *Selecting Velocity of the Drone*

Let the optimal path we get by adding up the costs be of ith row and jth column. The distance that the drone can move in that path safely will be the value of the ith row and jth column of the depth image(distance of obstacle) which we got after applying average pooling (refer obstacle cost section). We then take the minimum of this distance with the effective distance, distance from the drone beyond which obstacles are ignored.

$$d = pooled\_depth\_image(i, j);$$

$$d = min(d, effective\ distance);$$

We must stop the drone if it reaches near the goal so we shall limit the distance the drone can move in that path with the distance of the drone from the goal.

$$d = min(d, goal\_distance);$$

So we want our drone to be fastest when d is equal to effective distance and zero when d is equal to zero.

We can change velocity linearly but it would be better to have the rate of change in the velocity of the drone to be low at low values of d as we have obstacles nearby. So we change velocity as a quadratic function of d. This also ensures that the velocity of the drone is less at lower values as compared to that of when we change velocity linearly (Figs. 11 and 12).

So our velocity will be:

$$v = max\_velocity * (d/effective\_dist)^2;$$

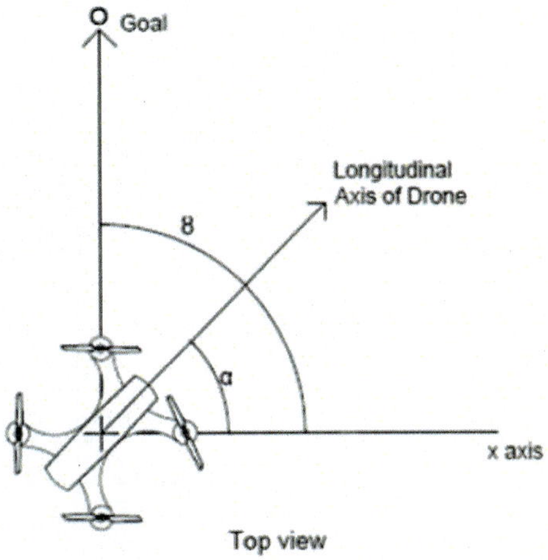

Fig. 11 Top view of goal angle (gamma) and drone angle (alpha)

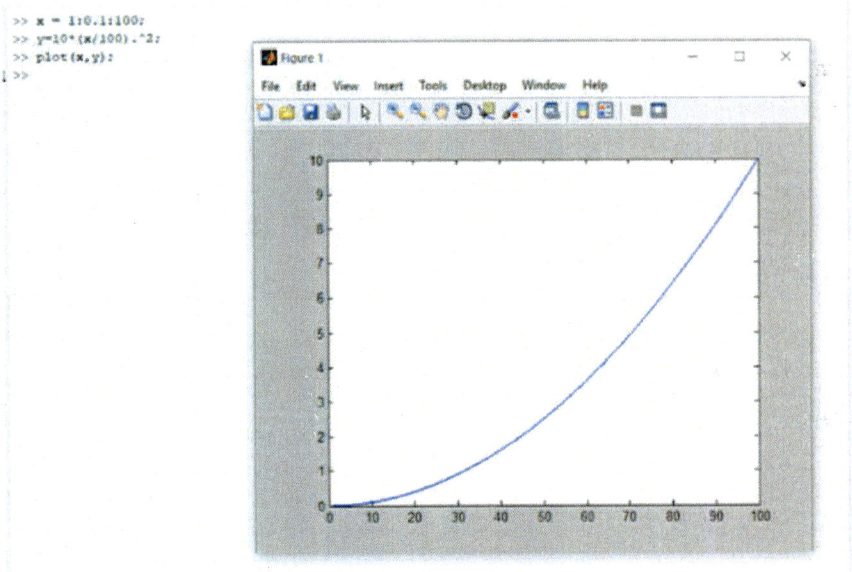

Fig. 12 Velocity versus safe distance with max velocity set to 10 and effective distance 100

We also don't want the velocity to highly increase as the safe distance may increase suddenly in just a turn so we will always store the previous velocity of the drone and make sure the current velocity should not be more that the previous velocity summed up with v, where v is the maximum change in velocity which we want to allow so

$$v = min(v, prev\_vel + \mathrm{v});$$

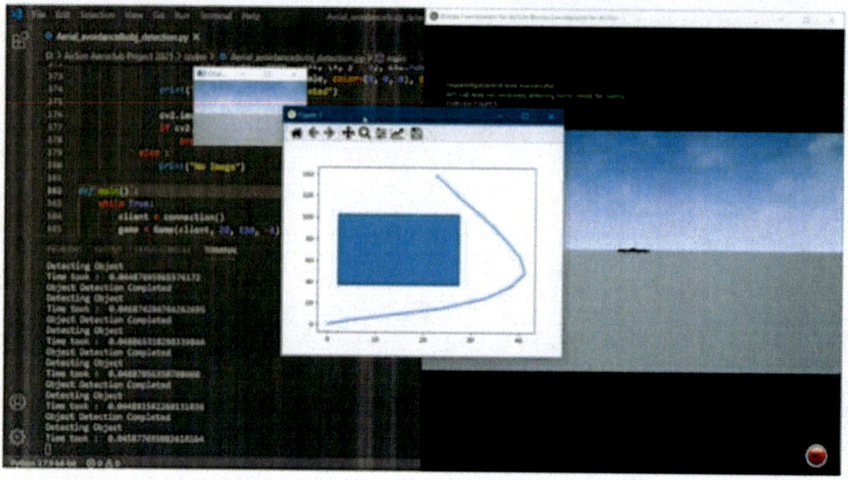

Fig. 13 Output of path planning

We will not put restrictions while decreasing velocity as if we will not decrease velocity as much as it demands then we may end up colliding with the obstacle.

3.3 Combining Object Detection and Modified DWA Algorithm

The weights we defined above in the algorithm that determines which of the factors, obstacles, smoothness, or goal, will be contributing more in the selection of the path changes with different types of obstacle.

This has been done as we can see that in the case of birds, it's more preferable to move from the top so we can reduce vertical smoothness weight and increase horizontal smoothness weight. Whereas in the case of poles, it is better to go sideways so we can increase vertical smoothness weight and reduce horizontal smoothness weight

In case, if there are multiple types of obstacles in the view then the general weights are selected (Fig. 13).

Simulation Videos[4] are released publicly.[5]

Acknowledgements This work was carried out as an intern project for TSAW We would like to thank Mr. Kishan Tiwari and Mr. Rimashu pandey for their guidance, mentorship, and valuable inputs.

[4] https://drive.google.com/drive/folders/1eSa_CJ5WKoi4o3tcwirDeDtM1xMRdsQv.

[5] https://www.tsaw.tech/.

References

1. Shah S, Dey D, Lovett C, Kapoor A (2017) AirSim: high-fidelity visual and physical simulation for autonomous vehicles
2. Bochkovskiy A, Wang C-Y, Mark Liao H-Y (2020) YOLOv4: optimal speed and accuracy of object detection. arxiv:abs/2004.10934
3. Fox D, Burgard W, Thrun S (1997) The dynamic window approach to collision avoidance. IEEE Robot Autom Mag 4(1):23–33. https://doi.org/10.1109/100.580977
4. Borenstein J, Koren Y (1991) The vector field histogram-fast obstacle avoidance for mobile robots. IEEE Trans Robot Autom 7(3):278–288. https://doi.org/10.1109/70.88137
5. Zhuang F et al (2021) A comprehensive survey on transfer learning. Proc IEEE 109(1):43–76. https://doi.org/10.1109/JPROC.2020.3004555

Vibration Suppression of Hand Tremor Using Active Vibration Strategy: A Numerical Study

Anshul Sharma and Rajnish Mallick

1 Introduction

Tremor is a neurological disorder that develops physical disabilities due to involuntary oscillatory movements of various human body parts, particularly the upper limbs. It is reported that approximately 6.3 million people around the globe have Parkinson's disease, which causes numerous difficulties to perform daily tasks; therefore, the study of both treatment and preventive methods is highly significant in order to suppress the tremors and to provide comfort to the patients [1, 2]. In recent times, the suppression of tremors using less disturbing non-pharmacological or non-surgical practices, which include limb cooling [3], tremor-suppressing orthoses [4], vibration therapy [5], and transcranial magnetic stimulation [6] has been experimented to reduce the effect of Parkinson's disease in people.

Mechanical vibration control techniques based on passive or active strategy may be adopted for hand tremor suppression. Passive vibration equipment, which includes mass–spring–damper vibration absorbers, being bulky, may cause muscle fatigue [7]. Moreover, passive technique can simultaneously suppress involuntary and voluntary motions, instead of involuntary motions alone. Therefore, active vibration strategy which includes various types of sensors and actuators is capable of suppressing involuntary motions alone and has been deployed successfully for tremor suppression [8, 9].

Active vibration control (AVC) mechanism involves the detection of change in equilibrium using sensors, amplification of sensor signal, implementation of control algorithm, and generation of actuator force. For AVC, smart transducers, such as piezoelectric materials can be utilized as both sensors and actuators for AVC owing to excellent electro-mechanical properties. It is essential to design

A. Sharma (✉) · R. Mallick
Mechanical Engineering Department, Thapar Institute of Engineering and Technology, Patiala, Punjab 147004, India
e-mail: anshul.sharma@thapar.edu

© The Author(s), under exclusive license to Springer Nature Singapore Pte Ltd. 2023

H. Muthusamy et al. (eds.), *Robotics, Control and Computer Vision*, Lecture Notes in Electrical Engineering 1009, https://doi.org/10.1007/978-981-99-0236-1_36

a precise electro-mechanical interface between the host structure and piezoelectric materials. Numerous studies have been successfully carried out to investigate the active vibration control of smart piezoelectric structures and different control methods [10–16].

In the present study, the human forearm is covered with a cylindrical shell panel with specific boundary conditions to actively suppress the hand tremor using a closed loop. The cylindrical shell panel is modelled using degenerated shell element, and further finite element-based Hamilton principle is used to obtain the dynamic response corresponding to hand tremors experienced by patients suffering from Parkinson's disease. Harmonic force is applied to simulate the hand tremors. This article emphasizes on hand tremor suppression using the concepts of smart structures which includes the host structure integrated with a pair of piezoelectric sensor and actuator layers. The influence of control gains on vibration suppression using active control is investigated.

2 Mathematical Modelling

In the present study, a cyylinderical shell device is integrated with piezoelectric sensor-actuator layers. The piezo-laminated cylindrical shell is supposed to be mounted on the wrist of the human hand as illustrated in Fig. 1.

The cylindrical shell used to control the hand tremors in patients suffering from Parkinson's disease is modelled using finite element formulation incorporating the first-order shear deformation theory. Figure 2 illustrates a degenerated shell element with four nodes; used to capture the desired curvature of the cylindrical shell such that it may be easily mounted on the human forearm. η–ξ–ζ represents the local coordinate system while x–y–z represents global coordinate system. This section presents the finite element formulation for vibration response of piezo-laminated shells mounted on human forearm. The geometry of a layered shell of a general shape is presented and equations of motion are obtained. Thereafter, the equations of

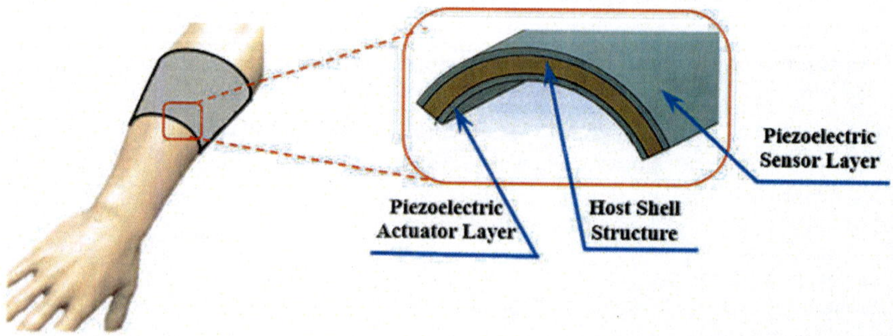

Fig. 1 Human forearm covered with piezo-laminated cylindrical shell model

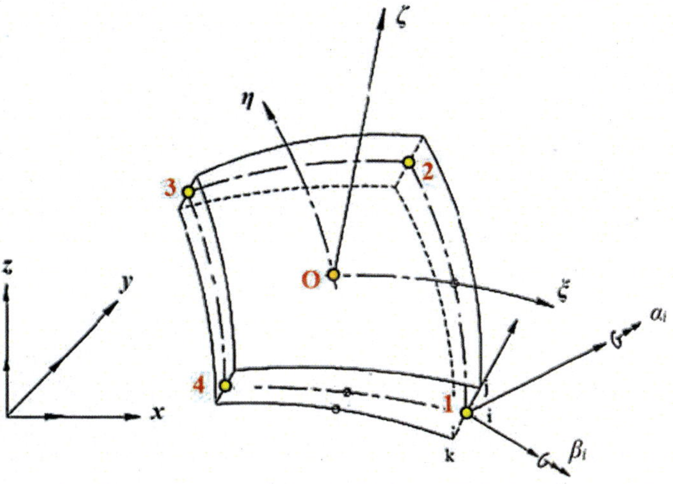

Fig. 2 Degenerated shell element used to model cylindrical shell mounted on human forearm

motion are then decoupled into sensor and actuator parts for active vibration control of hand tremors.

2.1 Geometric and Displacement Field

The shell element under study is capable of capturing five degrees of freedom per node. In addition, at element level, the potential difference through the piezoelectric thickness is incorporated. The coordinates of any location within the structure under study is represented as

$$\begin{Bmatrix} x \\ y \\ z \end{Bmatrix} = \sum_{i=1}^{nnel} N_i \left\{ \begin{Bmatrix} x_l \\ y_l \\ z_l \end{Bmatrix} + \frac{1}{2} t h_l \begin{Bmatrix} l_{3l} \\ m_{3l} \\ n_{3l} \end{Bmatrix} \right\} \tag{1}$$

where h_{i} is the thickness of node l, *nnel* is no. of nodes per element, t is thickness of shell element, and N_{i} is the shape function.

The displacement within the element may be calculated as

$$\begin{Bmatrix} u \\ v \\ w \end{Bmatrix} = \sum_{i=1}^{nnel} N_i \left\{ \begin{Bmatrix} u_l^0 \\ v_l^0 \\ w_l^0 \end{Bmatrix} + \frac{1}{2} t h_l \begin{Bmatrix} l_{1l} & -l_{2l} \\ m_{1l} & -m_{2l} \\ n_{1l} & -n_{2l} \end{Bmatrix} \begin{Bmatrix} \alpha_l \\ \beta_l \end{Bmatrix} \right\} \tag{2}$$

α_l and β_l represents the rotational degree of freedom.

2.2 Piezoelectric Constitutive Equations

In order to include the multi physics for electro-mechanical analysis, piezoelectric constitutive equations are represented as

$$\{D\} = [e]\{\varepsilon\} + [b]\{E\} \tag{3}$$

$$\{\sigma\} = [Q]\{\varepsilon\} - [e]^T\{E\} \tag{4}$$

where $\{D\}$ is electrical displacement, $\{\sigma\}$ is stress vector, $\{E\}$ is electric field, $\{\varepsilon\}$is strain vector, $[e]$ is piezoelectric coefficient, $[Q]$ is elastic stiffness coefficients, and $[b]$ is dielectric constant.

2.3 Electric Field

With the assumption of electric field acting in transverse direction of piezoelectric layer and constant electric effect in the piezo layer, the electric field can be given as

$$\{E\}_k = -\left[\frac{1}{t_{piezo\ k}}\begin{Bmatrix} l_3 \\ m_3 \\ n_3 \end{Bmatrix}\right]\{\phi_{piezo\ k}\} \tag{5}$$

$\phi_{piezo\ k}$ represents electric potential *in* piezoelectric layer of thickness $t_{piezo\ k}$.

2.4 Equation of Motion

By Hamilton's principle, the equation of motion of cylindrical shell structure is written as

$$[M_{uu}]\{\ddot{q}\} + [C_{uu}]\{\dot{q}\} + [K_{uu}]\{q\} + [K_{u\phi}]\{\phi\} = \{F_m\} \tag{6}$$

$$[K_{\phi u}]\{q\} - [K_{\phi\phi}]\{\phi\} = \{F_q\} \tag{7}$$

where $[M_{uu}]$ is the mass matrix which includes mass of cylindrical shell structure and piezoelectric layers, $[K_{uu}]$ is the elastic stiffness matrix which includes elastics stiffness of host shell structure, piezoelectric layers, and torsional stiffness of elbow joint motion as torsional spring, $[K_{\phi\phi}]$ is the electric stiffness matrix and $[K_{u\phi}]$ is coupled elastic-electric stiffness matrix. $\{F_m\}$ is mechanical force and $\{F_q\}$ is applied electrical charge, respectively.

As cylindrical shell is piezolaminated, the upper layer of the host structure is modelled as sensor, while lower layer is modelled as actuator. During deformation, the sensor senses the change in equilibrium resulting in generation of input voltage which is fed to the controller. As per the pre-defined control algorithm, control voltage is supplied to the actuator for control action. Therefore, the total voltage in Eqs. (6) and (7) can be split into sensor and actuator voltage as

$$[M_{uu}]\{\ddot{q}\} + [C_{uu}]\{\dot{q}\} + [K_{uu}]\{q\} + [K_{u\phi_s}]\{\phi_s\} = \{F_m\} - [K_{u\phi_a}]\{\phi_a\} \quad (8)$$

$$[K_{\phi_s u}]\{q\} - [K_{\phi_s\phi}]\{\phi_s\} = \{F_{q_s}\} \quad (9)$$

$$[K_{\phi_a u}]\{q\} - [K_{\phi_a\phi}]\{\phi_a\} = \{F_{q_a}\} \quad (10)$$

From Eq. (9), the open circuit sensor voltage may be predicted as

$$\{\phi_s\} = [K_{\phi_s\phi}]^{-1}[K_{\phi_s u}]\{q\} \quad (11)$$

Using $\{\phi_s\}$ in Eq. (8)

$$[M_{uu}]\{\ddot{q}\} + [C_{uu}]\{\dot{q}\} + ([K_{uu}] + [K_{u\phi_s}][K_{\phi_s\phi}]^{-1}[K_{\phi_s u}])\{q\} = \{F_m\} - [K_{u\phi_a}]\{\phi_a\} \quad (12)$$

In Eq. (12), the actuator voltage (ϕ_a) is determined by the controller.

2.5 Active Vibration Controller

The crucial objective of the controller design is to regulate the hand tremor to a desired level by driving an actuator using control force. The cylindrical shell is sandwiched among piezoelectric layers. The output voltage form piezoelectric sensor subjected to external force is predicted using Eq. (11). After filtration and magnification, the output sensor voltage is send to the controller for analysing the input voltage. The controller, thereafter, delivers control voltage (ϕ_a) as output to piezoelectric actuator which in resultgenerates control force as represented in Eq. (12). In the present study, negative velocity feedback controller is used for active vibration control of hand tremor. The closed loop active vibration control strategy is illustrated in Fig. 3 and is mathematically represented as

$$\{\phi_a\} = -Gain_V\{\dot{\phi}_s\} \quad (13)$$

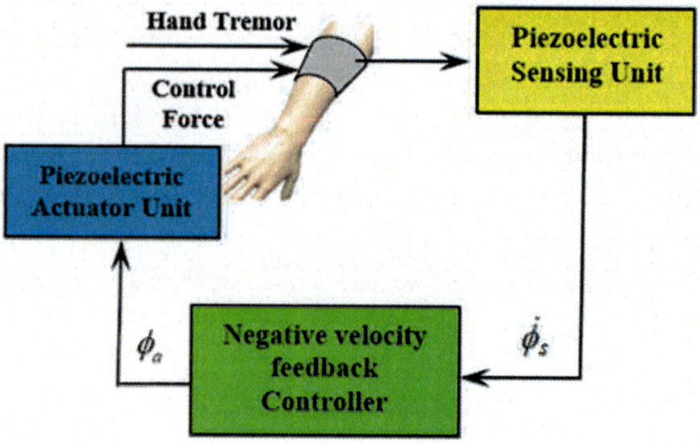

Fig. 3 A cylindrical shell panel mounted on forearm in close loop with negative velocity feedback controller

3 Validation

3.1 Static Analysis of Piezo-Laminated Cylindrical Shell

A simply supported composite cylindrical shell integrated with collocated piezoelectric material (PZT-4) is considered. Degenerated shell element (presented in Sect. 2) is utilized to discretize the structure. The present formulation is validated with the results reported by Balamurugan and Narayanan [12]. All the geometric and material properties; loading and boundary conditions are kept same as reported in [12]. Two laminate stacking configurations of [p/0/90/90/0] and [0/90/90/0/p] are considered. The actively induced radial deflections along the axial midspan of the panel are shown in Fig. 4. The results are in very good agreement with the reference results.

3.2 Natural Frequencies of Cylindrical Shell

The natural frequencies of composite laminated cylindrical shells (without piezoelectric layers) are compared with results presented by Saravanan et al. [13]. The geometric properties and material properties are considered same as mentioned by Saravanan et al. For the boundary conditions, both the curved edges of cylindrical shell are clamped and other edges are kept free. The lowest natural frequencies for different orientations are listed in Table 1. The results of present modelling is in very good agreement with the reference results.

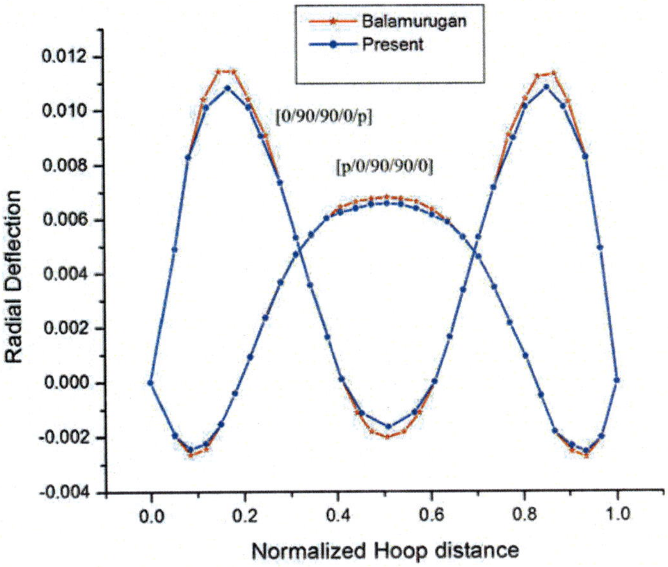

Fig. 4 Radial deflection versus normalized hoop distance of simply supported cylindrical shell

Table 1 Validation of natural frequencies of cylindrical composite shell (in Hz)

Orientation	Saravanan et al. [13]	Present
[0/0/0]	262	263.44
[30/0/30]	380.5	382.7
[45/0/45]	422.6	423.9
[60/0/60]	430.7	433.3
[90/0/90]	370.5	377.1

4 Numerical Analysis and Results

In this section, numerical study on active vibration control of hand tremor in the patients suffering from Parkinson's disease is confronted. A cylindrical shell sandwiched between piezoelectric sensor and actuator layers is modelled using finite element formulation presented in Sect. 2. The upper piezoceramic layer acts as sensor and lower piezoceramic layer acts as actuator as shown in Fig. 1.

As the human hand tremor is a kind of sinusoidal movement, the external force that is applied to the cylindrical shell model as a tremor is a harmonic force. Mathematically, the applied load is applied as

$$f(x,\ t)\ =\ F(x)\ \sin(\omega t) \tag{14}$$

Table 2 Physical properties of cylindrical shell and piezoelectric ceramics [15]

Physical property	Cylindrical shell	Piezoelectric layer
Elastic modulus		
E_{11} (GPa)	181	61
E_{22} (Gpa)	10.3	61
G_{12} (Gpa)	7.17	23.64
Density, (kg/m^3)	1600	7700
Poisson's ratio	0.31	0.29
Piezoelectric properties		
$e_{33} = e_{31}$ (C/m^2)	–	14.69
$\zeta_{11} = \zeta_{22} = \zeta_{33}$	–	16.5×10^{-9}

To model the human forearm motion, the pined boundary conditions are incorporated at one curved face of the cylindrical shell to capture motion about elbow joint. The material properties used for cylindrical shell and piezoelectric layers are listed in Table 2.

The effect of active vibration suppression of hand tremor using different control forces is presented in Fig. 5. With the increase in control gain (Gain $_v$), the vibration due to hand tremor can be damped out quickly. The observed damping ratio subjected to the control gain of 0.05, 0.1 and 0.5 is 0.0047, 0.0092, and 0.0295, respectively. With the increase in the value of control gain, the damping ration increases. However, due to the hardware limitations, the maximum value of control gain must be restricted. It should be noted that there might be a restriction on the maximum control gain which necessitates the use of optimum combination of other parameters as well. The numerical results show that the active vibration control strategy with collocated piezoelectric sensor and actuator pair can efficiently suppress the hand tremor.

5 Conclusion

This paper numerically investigates the active vibration suppression of hand tremor in patients suffering from Parkinson's syndrome. For the same, forearm is covered with cylindrical shell panel sandwiched between piezoelectric sensor and actuator layers. The cylindrical shell is modelled using degenerated shell element with four nodes. Hamilton principle is used to capture the dynamic response of forearm subjected to harmonic tremor. Harmonic force is applied to simulate the hand tremors. This article emphasizes on hand tremor suppression using the concepts of smart structures. The effect of control gains on active vibration suppression are investigated. Numerical simulations reveals that the active vibration control strategy with collocated piezoelectric sensor and actuator pair can efficiently suppress the hand tremor. The observed damping ratio subjected to the control gain of 0.05, 0.1, and 0.5 is 0.0047, 0.0092, and 0.0295, respectively.

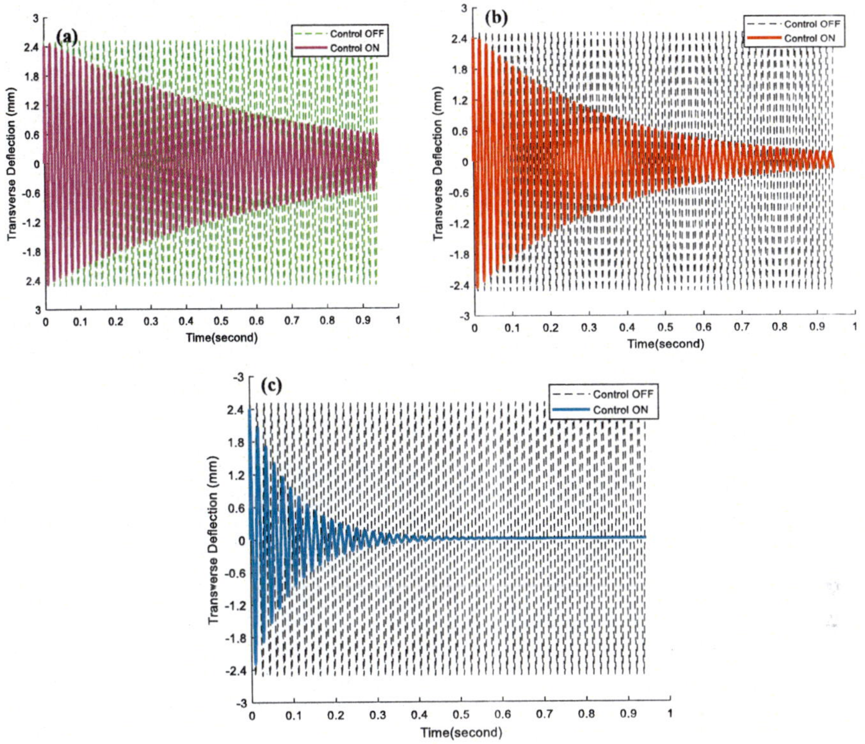

Fig. 5 Active vibration suppression of hand tremor subjected to harmonic motion corresponding to **a** Gain $_v = 0.05$, **b** Gain $_v = 0.1$ and **c** Gain $_v = 0.5$

References

1. Abbasi M, Afsharfard A, Safaie RA (2018) Design of a noninvasive and smart hand tremor attenuation system with active control: a simulation study. Int Fed Med Biol Eng 56(7):1315–1324
2. Su Y, Allen CR, Geng D, Burn D, Brechany U, Bell GD, Rowland R (2003) 3-D motion system data-gloves application for parkinsons disease. IEEE Trans Instrum Meas 52(3):662–674
3. Cooper C, Evidente VGH, Hentz JG (2000) The effect of temperature on hand function in patients with tremor. J Hand Ther 13(4):276–288
4. Zhou Y, Naish MD, Jenkins ME (2017) Design and validation of a novel mechatronic transmission system for a wearable tremor suppression device. Robot Auton Syst 91:38–48
5. Haas CT, Turbanski S, Kessler K (2006) The effects of random whole-body-vibration on motor symptoms in Parkinson's disease. Neuro Rehabil 21(1):29–36
6. Filipovic SR, Rothwell JC, Bhatia K (2010) Low-frequency repetitive transcranial magnetic stimulation and off-phase motor symptoms in Parkinson's disease. J Neurol Sci 291(1):1–4
7. Kotovsky J, Rosen MJ (1998) A wearable tremor-suppression orthosis. J Rehabil Res Dev 35:373–387
8. As'arry A, Zain MM, Mailah M (2011) Active tremor control in 4-DOFs biodynamic hand model. Int J Math Models Methods Appl Sci 5:1068–1076 (2011)
9. Kazi S, Mailah M, Zain ZM (2014) Suppression of hand postural tremor via active force control method. Manuf Eng Autom Control Robot 12(6):76–82

10. Sharma A, Kumar R, Vaish R, Chauhan VS (2016) Experimental and numerical investigation of active vibration control over wide range of operating temperature. J Intell Mater Syst Struct 27(13):1846–1860
11. Mallick R, Ganguli R, Bhat MS (2015) Robust design of multiple trailing-edge flaps for helicopter vibration reduction: a multi-objective bat algorithm approach. Eng Optim 47(9):1243–1263
12. Sharma A, Kumar A, Susheel CK, Kumar R (2016) Smart damping of functionally graded nanotube reinforced composite rectangular plates. Compos Struct 155:29–44
13. Balamurugan V, Narayanan S (2001) Shell finite element for smart piezoelectric composite plate/shell structures and its application to the study of active vibration control. Finite Elem Anal Des 37:713–738
14. Saravanan C, Ganesan N, Ramamurti V (2000) Analysis of active damping in composite laminate cylindrical shells of revolution with skewed PVDF sensors/actuators. Compos Struct 48:305–318
15. Sharma A, Kumar R, Vaish R, Chauhan VS (2014) Lead-free piezoelectric materials' performance in structural active vibration control. J Intell Mater Syst Struct 25(13):1596–1604
16. Mallick R, Ganguli R, Kumar R (2017) Optimal design of a smart post-buckled beam actuator using bat algorithm: simulations and experiments. Smart Mater Struct 26(5):055014

Design of a Self-reconfigurable Robot with Roll, Crawl, and Climb Features for False Ceiling Inspection Task

S. Selvakumaran, A. A. Hayat, K. Elangovan, K. Manivannan, and M. R. Elara

1 Introduction

False or suspended ceilings are favorable for rodents to seek refuge and build their habitat. These pests can wreak havoc in the buildings, whether residential, commercial, or industrial. Pests infestation is a significant health hazard as well as [1]. For example, pests such as rats, cockroaches, and mosquitoes spread asthma, allergy, and food contamination illnesses. Rats damage building structures, chew electrical wires, and transmit diseases. The false-ceiling environment and the manual inspection process are shown in Fig. 1.

The requirement of smoothly and implementing the autonomous task in uncertain environments with robust adaptive autonomous features is vital for developing next-generation robots. Legged robots have higher adaptability to the different conditions of ground [2, 3]. However, they are more complex and require high torque and power. On the other hand, a wheeled robot is comparatively simpler in structure, easier to control [4], and is efficient on moving a plane surface. Nevertheless, it is inferior to adapt to obstacles or rough terrain. Track wheels can overcome irregularities in the terrain with limited height, and it was used in the design of a false-ceiling robot named Falcon reported in [5]. However, the track wheels have limitations in overcoming and accessing the vertical surfaces such as sidewalls and ducts, in the false ceiling. Therefore, we propose a novel robot design with roll, crawl, and climb capabilities referred to here as *FalconRCC*, i.e., Falcon with Roll Crawl and Climb (RCC) features.

The mobility of a wheel-legged type device can be used to negotiate obstacles. This system combines the benefits of both a leg and a wheel mechanism. With the disadvantage of high power consumption, track wheel robots can overcome obstacles and operate on unstructured ground. The evolvability, multi-functionality, and

S. Selvakumaran · A. A. Hayat (✉) · K. Elangovan · K. Manivannan · M. R. Elara
Engineering Product Development Pillar, Singapore University of Technology and Design (SUTD), Singapore, Singapore
e-mail: abdullahaamir@sutd.edu.sg

© The Author(s), under exclusive license to Springer Nature Singapore Pte Ltd. 2023
H. Muthusamy et al. (eds.), *Robotics, Control and Computer Vision*, Lecture Notes in Electrical Engineering 1009, https://doi.org/10.1007/978-981-99-0236-1_37

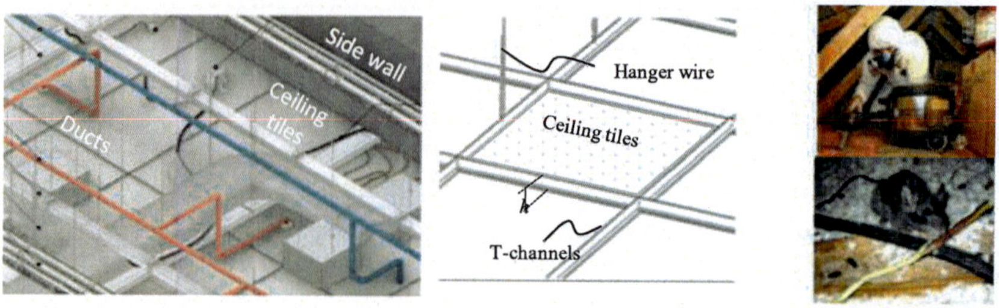

(a) False ceiling environment (b) Section of false-ceiling (c) Manual inspection and hazards

Fig. 1 Manual inspection and hazard of pests on false ceiling

survivability in reconfigurable robots [6] are useful for challenging terrains. Several robotic architectures based on the reconfigurable design principles proposed in [7, 8] were implemented in width changing pavement sweeping robot [9], Tetris inspired floor cleaning robots [10, 11] staircase accessing robot [12], rope climbing robot [13], drain inspection robots [14, 15], among others. Quattroped [16] was designed with a unique transformation technique from wheeled to legged morphology. It includes a transformation mechanism that allows them to convert the morphology of the driving mechanism between wheels (i.e., a full circle) and two degrees of freedom legs (i.e., combining two half circles as a leg). In [17], a robot with a unique claw-wheel transformation design is described. Moreover, mobile robots with differential wheel action were used in the area coverage strategy for false ceiling in [18]. However, the robot cannot self-recover, and the dimension restricts it from accessing cluttered regions in false ceiling. In this work, we present the novel design of the reconfigurable robot with the ability to switch between crawl and roll mode and the modular attachment that can aid in climbing walls.

The rest of this paper is organized as follows. Section 2 explains the requirements and considerations for the design, mechanical layout, and system architecture of the false-ceiling robot *FalconRCC*. The mechanical design of the quadruped robot with the ability to crawl and roll, along with the modular attachment for wall climbing is detailed in Sect. 3. Section 4 explains the components for the system architecture, and experimental results for the transition and climbing of the wall by the robot are shown in Sect. 5. Finally, Sect. 6 concludes the paper.

2 Operational and Design Requirements

The existing inspection and surveillance task of the false ceiling is done manually (Fig. 1b) and is tedious. The environmental scenario and design requirements are discussed here.

Environmental requirements: Figure 1 shows the typical situation which exists with the false ceiling. For designing a robot to inspect inside the false ceiling the typical interior of the false ceiling is shown using the CAD model in Fig. 1b. There are positive obstacles in the form of the T-section of the supporting frame and negative obstacles in the form of open ends (AS/NZS 2785:2000 [19] suspended ceilings—Design and installation standard). The environment inside is mostly dark, ceiling tiles conceal the heating ventilation, air-conditioning (HVAC) ducts, fire safety pipes, cables, etc.

Mobility requirements: The set of requirements for the mobility features desired in the false-ceiling inspection robots are: (a) Able to access uneven terrain and overcome obstacles such as wires and channel height, (b) Able to detect an obstacle and significant height drop, (c) Able to orientate itself to be upright, (d) Able to access vertical surfaces.

Functionality requirements: The functionality features desired in the false-ceiling inspection robots are: (a) Access and map the area, and view false-ceiling environments to detect potential rat pathways (e.g., holes, gnaw marks), (b) Autonomous complete area coverage for cleaning and path planning for inspection, (c) Conduct visual inspection, preferably with its own light, (d) Detect a different kind of pest droppings. Capture images with location tagging, date/time stamp, and collect samples, (e) Estimate the density of the pest.

Operability requirements: The operability features desired in the false-ceiling inspection robots are: (a) Small in the dimension that can fit into a cube of $25 \times 25 \times 25$ cm, (b) Lightweight, typically less than a kilogram, (c) Single charge lifetime of 2 h and more in running state, (d) Ability to operate independently with minimal human intervention (e) Support first person view (FPV) and be remotely controlled by the operator.

2.1 Design Considerations

By these observations, the following features in the robot designed for the inspection and surveillance task will be of help:

- From the false-ceiling standards and observations, it was concluded that the channel height h (Fig. 1b) varies as $30 < h < 90$ mm. Hence the robot design should overcome this obstacle height.
- The platform must be lightweight and should generate less noise while moving over the false ceiling.
- The platform should be able to recover itself from the fall.
- It can climb over vertical surfaces.
- Night vision camera mounted for the inspection task in the dark environment.

The transformation design principles [20] were utilized to cater to the need for crawl, roll, and climb the wall by designing the subsystems accordingly. The detailed aspects of utilizing the design principles and facilitators with the mechanisms presented in [8] were utilized in this work to come up with a system facilitated by roll/wrap/coil, modularity, shared transmission, furcate, and fold.

3 Mechanical Layout

In the robotics area, reconfiguration refers to a system's ability to change its configuration to fulfill the required task by reversibly changing its mechanism type, mobility, gaits, architecture (say, serial to parallel), and so on. In this work a self-reconfigurable robot designed using transformation principles [20] aimed at a false-ceiling inspection task. The scale of the designed robot is depicted in Fig. 2a, b which also adheres to the system requirement. The two configurations for the crawling and rolling are also shown along with the exploded view of the system showing it components and the symmetry in design. The crawl and roll capabilities over the false ceiling are discussed next.

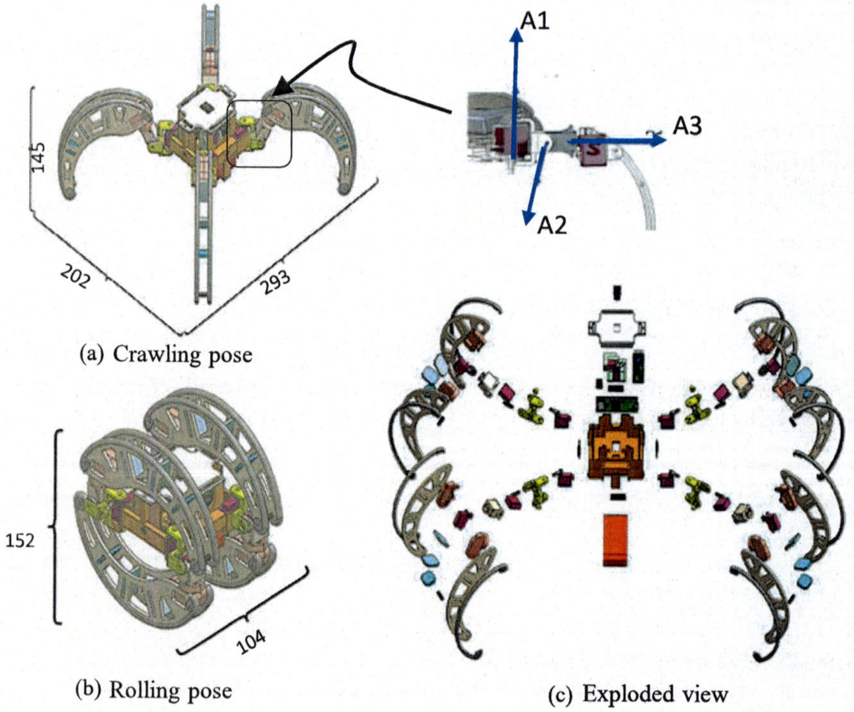

Fig. 2 Dimensions of *FalconRCC* (in mm) and its exploded depicting the components

3.1 Crawling and Rolling Mechanisms

The primary mechanism for the crawling and rolling locomotions of the robot are its four semi-circular limbs connected using the spherical joint with the body. Each limb has an active spherical joint that is providing three Degrees of Freedom (DoF) and is made possible with three perpendicular revolute joints that are powered by micro servo motors. The robot has a total of 12 such servo motors to control the movement of its four limbs. The servo motors on each limb are arranged such that the joint proximal to the robot body, i.e., Axis A1 (Fig. 2a) controls the legged locomotion, the joint in the middle with A2 helps to control the rolling and the third motor connected the arched leg helps in lifting the leg and transition of the position of the leg. HS-35HD HiTec Ultra nano servos motors are placed to give the spherical joint using three servo motors in each leg. The same joint configuration is provided for the four legs attached to the main body, and as a result, twelve units of the servo motors are used.

The home position state for *FalconRCC* is when its limbs face diagonally outward at an angle of 45° from its body. Crawling begins from this state, and periodic drags are created by each leg, one leg at a time. Figure 3a, b shows the home position of the robot and the cawing gait pattern over the false-ceiling environment. Forward translation, backward translation, clockwise rotation, and anticlockwise rotation are the four basic locomotion patterns in the crawling state. This enables the maneuverability of the robot to be used for the inspection task. With the change in the spread angle of each leg or by increasing the leg footprint the height of the robot can be varied from 135 to 155 mm. This enables lower change of the configures the height as per the obstacle. Figure 3c shows the *FalconRCC* reconfiguring its go beneath the duct pipe.

The reconfigurability of the limbs also plays a crucial role in enabling the robot to transition from its crawling state to the rolling state, as shown with the sequence of leg transformation in Fig. 4a. The advantage of rolling is observed with ten times higher speed than during the crawling, and in the false-ceiling environment, the small height (<90 mm) obstacle can be overcome during rolling. Both forward and

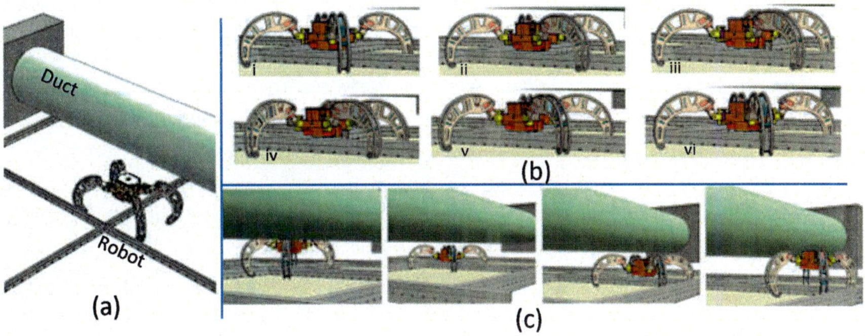

Fig. 3 Crawling gaits and the ability to go beneath the ducts

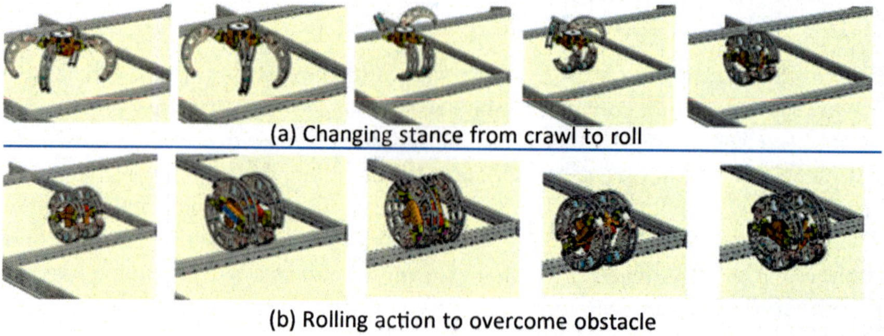

Fig. 4 Changing the state from crawl to roll and overcome obstacles

backward rolling can be achieved. The feedback from the time of flight sensor (ToF) and inertial measurement units (IMU) attached in the center helps avoid the fall due to depth and regulates the rolling action. The robot has self-recovery mode upon falling, since the design is made symmetric, the upside-down recovery can be easily achieved.

3.2 *Transition and Wall Climbing Mechanism*

The exploded view of the *FalconRCC* robot is shown in Fig. 5 with the subsystems. The main subsystems shown are the Transition wheel, chassis, and bipedal. The four limbs play a crucial part in enabling the robot to transition from its crawling to climbing state. Also, this transition mechanism consists of a small pair of wheels to enable a seamless transition from the floor to the wall after achieving the desired climbing configuration. These wheels are driven by an 8V DC motor, which is engaged to the wheel shaft using a pair of bevel gears as shown in Fig. 5a.

The robot uses micro-suction tape from AirStick. This tape establishes stable bonds between robot and wall, similar to a gecko or spider forming Van der Waals forces between its feet and the wall surface. The sticky surface of tape contains microscopic air pockets that create partial vacuums between the tape and wall surfaces. It leaves no residue behind, similar to suction cups. Thus it can be used repeatedly without losing its adhesive holding power. However, it is not pressure-sensitive, a property lacking in regular suction cups that rely on ambient pressure. These micro-suction tapes are designed such that it is hard to pull off the entire tape in the direction perpendicular to the attachment surface, but they are easy to peel off.

A pair of pedals move synchronously with the central trunk in a periodic fashion. This periodic motion is made possible with the cam mechanisms incorporated within the bipedal mechanism. The pair of pedals is similar to the long limbs on either side of an ape, and the central trunk of the robot is like the trunk of the ape. The climbing motion is achieved using a single motor. Before climbing, the transition from the

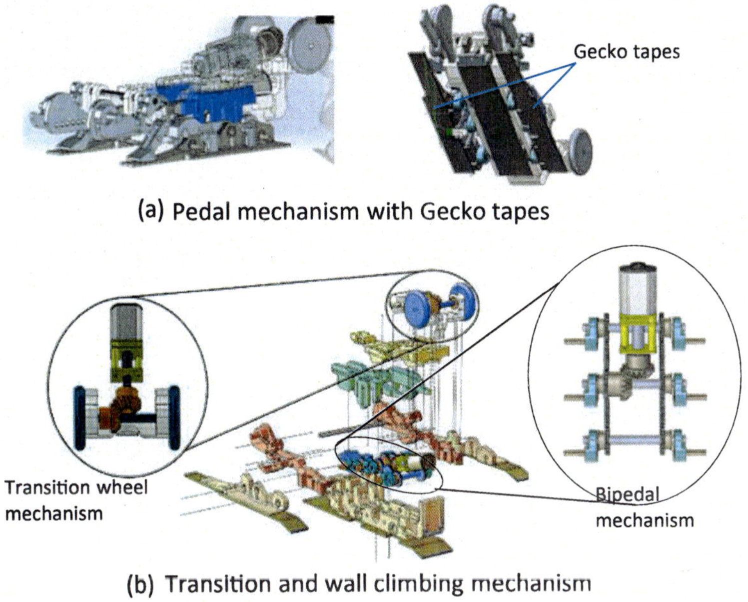

Fig. 5 Mechanisms helping in the transition from flat to a vertical surface and pedals with Gecko tape actuated using transmission from gears for vertical climbing

floor to the wall is assisted using the pair of wheels, as shown in Fig. 5. The assembly is moved toward the vertical surface with the help of these wheels to make contact of sticky pedals to the vertical surface.

4 System Architecture

The system architecture of the designed robot is depicted in Fig. 6. It gives an overview of the electronics and hardware components within the robot. We have selected Arduino Pro Mini as the microcontroller, which is the central unit that directly connects to the sensor, actuator, and communication units, as well as the other components within the controller unit. Communication between the server and the robot is made possible with a Bluetooth device, Adafruit EZ-Link Bluetooth Transceiver. This enables the robot to receive commands wirelessly from external devices. The robot's motion is actuated by DC gear motors (DCM) and servo motors (SM). Twelve HS-35HD HiTec Ultra nano servo motors in total are used to control the four limbs, with three servo motors controlling 3-DoF of a single limb. Two Pololu micro metal geared motors are used as the DC motors to control the bipedal mechanism and transition wheel mechanism. The sensor unit consists of three types, an Inertial Measurement Unit or IMU (Pololu minIMU-9 v5) that sends gyroscope

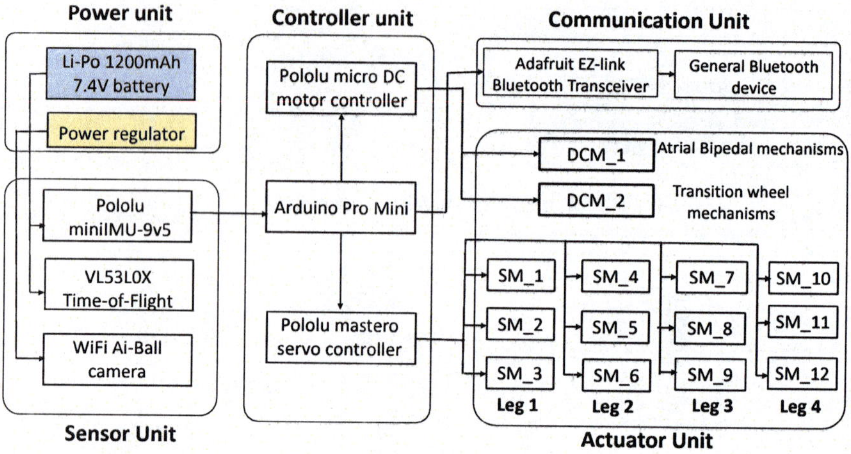

Fig. 6 System architecture

and accelerometer readings to the microcontroller, two Time-of-Flight laser sensors for obstacle detection (VL530L0X), and a camera for vision (WiFi Ai-Ball Camera). A Li-Po 1200 mAh 7.4 V battery is used as a power source.

5 Experiments

The transition of the robot from its crawling state to the climbing state is shown in a series of images in Fig. 7a. In this transition, the robots' legs use two out of their three Degrees of Freedom (DoF). When the robot has completed its crawling motion, the four legs return to their default state. The configuration is achieved by rotating the four legs, such that there are now two legs facing left and another two facing the right side of the robot. After which, all four legs rotate about A3, away from one another. As two of the legs contact the wall the other two legs continue rotating about the A2-axis. The two legs furthest from the wall then rotate about the A1-axis, until the joints of the two legs are parallel to the X-axis. Next, the legs closest to the wall rotate until they are right above the robot's body and parallel to the wall. Once this configuration is achieved, the robot rolls forward on the two wheels in contact with the floor until the micro-suction tape (MST) face pedals contact the wall.

While moving on a vertical or steep inclined surface, the modular attachment to *FalconRCC* utilizes a bipedal gait. The figure shows one cycle of the bipedal motion, from the moment where the center foot is about to begin detaching from the wall surface to the same moment one cycle later. Some vertical distance is covered as a result, which is represented by the dotted lines Fig. 7b.

The limitations in the current design observed are, namely, (a) a Higher number of actuators used for each leg movement. This can be overcome by studying the

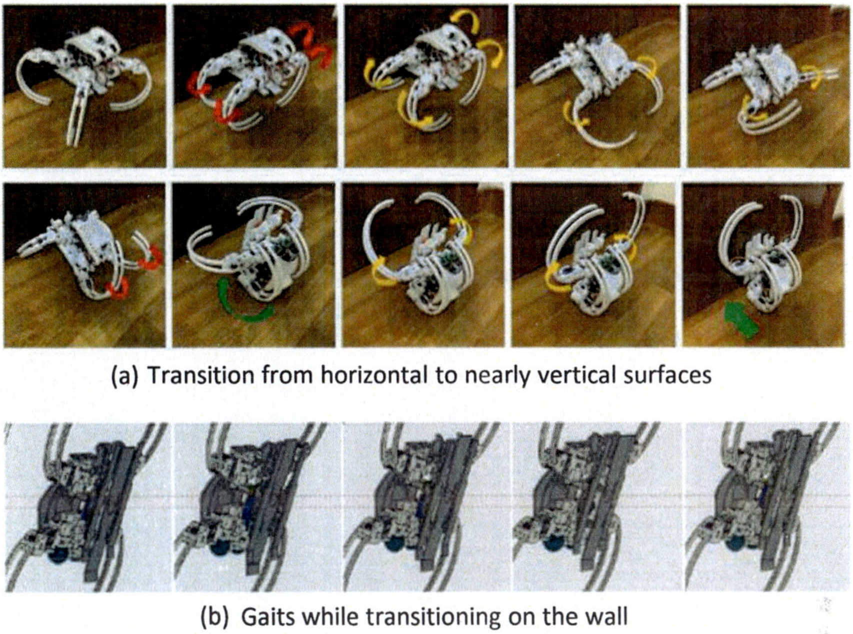

Fig. 7 Experiments showing the gaits for the transitions

kinematic behavior and replacing a joint with a flexural joint as in [21] to reduce the number of active actuation, (b) The passive suction unit to adhere to the vertical surface works only for the clean and flat surface, like glass, mica, etc., (c) The localization of the robot on the false ceiling is another limitation in the current design, (d) The kinematic model and identification [22] of the geometric parameters of the assembled robot along with the dynamic model is not incorporated in the control scheme, and, () The noise with the current servo motors due to the gearbox is more than the acceptable limit to operate in the false-ceiling environment.

6 Conclusions

In this paper, we conceptualize the design of the *FalconRCC* robot with the capability to roll, crawl, and climb inclined surfaces. It was shown that the design of the robot is suitable for the false-ceiling environment and its inspection task due to the self-reconfigurable and modular design. The mechanisms for each subsystem were selected according to the transformation design principles. The modular attachment of the climbing and transitioning mechanism using spatial transmission of motion from motors to the pedals was demonstrated using the experimental transition. The viability of the flexible joint for each leg to improve leg movement and reduce the number of actuators is being carried out. Future work includes the design optimization and control of the robot.

Acknowledgements This research is supported by the National Robotics Programme under its Robotics Enabling Capabilities and Technologies (Funding Agency Project No. 192 25 00051), National Robotics Programme under its Robot Domain Specific (Funding Agency Project No. 192 22 00058), National Robotics Programme under its Robotics Domain Specific (Funding Agency Project No. 192 22 00108), and administered by the Agency for Science, Technology and Research.

References

1. List of common household pests. https://en.wikipedia.org/wiki/list-of-common-household-pests. Accessed 20 Apr 2019
2. Clark JE, Cham JG, Bailey SA, Froehlich EM, Nahata PK, Full RJ, Cutkosky MR (2001) Biomimetic design and fabrication of a hexapedal running robot. In: Proceedings 2001 ICRA, IEEE international conference on robotics and automation (Cat No 01CH37164), vol 4. IEEE, pp 3643–3649
3. Jha RK, Singh B, Pratihar DK (2005) On-line stable gait generation of a two-legged robot using a genetic-fuzzy system. Robot Autonom Syst 53(1):15–35
4. Caracciolo L, De Luca A, Iannitti S (1999) Trajectory tracking control of a four-wheel differentially driven mobile robot. In: Proceedings 1999 IEEE international conference on robotics and automation (Cat No 99CH36288C), vol 4. IEEE, pp 2632–2638
5. Muthugala M, Apuroop KGS, Padmanabha SGA, Samarakoon S, Elara MR, Wen RYW (2021) Falcon: a false ceiling inspection robot. Sensors 21(16):5281
6. Tan N, Hayat AA, Elara MR, Wood KL (2020) A framework for taxonomy and evaluation of self-reconfigurable robotic systems. IEEE Access 8:13969–13986
7. Hayat AA, Yi L, Kalimuthu M, Elara M, Wood KL (2022) Reconfigurable robotic system design with application to cleaning and maintenance. J Mech Des 144(6):063305
8. Kalimuthu M, Hayat A, Elara M, Wood K (2021) Transformation design principles as enablers for designing reconfigurable robots. In: International design engineering technical conferences and computers and information in engineering conference, vol 85420. American Society of Mechanical Engineers, p V006T06A008
9. Hayat AA, Parween R, Elara MR, Parsuraman K, Kandasamy PS (2019) Panthera: design of a reconfigurable pavement sweeping robot. In: 2019 international conference on robotics and automation (ICRA). IEEE, pp 7346–7352
10. Hayat AA, Karthikeyan P, Vega-Heredia M, Elara MR (2019) Modeling and assessing of self-reconfigurable cleaning robot hTetro based on energy consumption. Energies 12(21):4112
11. Prabakaran V, Elara MR, Pathmakumar T, Nansai S (2018) Floor cleaning robot with reconfigurable mechanism. Autom Constr 91:155–165
12. Prabakaran V, Shi Y, Prathap KS, Elara MR, Hayat AA (2022) s-sacrr: a staircase and slope accessing reconfigurable cleaning robot and its validation. IEEE Robot Autom Lett (2022)
13. Ratanghayra PR, Hayat AA, Saha SK (2019) Design and analysis of spring-based rope climbing robot. Machines. Mechanism and robotics. Springer, Singapore, pp 453–462
14. Hayat AA, Elangovan K, Rajesh Elara M, Teja MS (2018) Tarantula: design, modeling, and kinematic identification of a quadruped wheeled robot. Appl Sci 9(1):94
15. Parween R, Hayat AA, Elangovan K, Apuroop KGS, Heredia MV, Elara MR (2020) Design of a self-reconfigurable drain mapping robot with level-shifting capability. IEEE Access 8:113429–113442
16. Chen SC, Huang KJ, Chen WH, Shen SY, Li CH, Lin PC (2014) Quattroped: a leg-wheel transformable robot. IEEE/ASME Trans Mechatron 19(2):730–742

17. Chou JJ, Yang LS (2013) Innovative design of a claw-wheel transformable robot. In: 2013 IEEE international conference on robotics and automation. IEEE, pp 1337–1342
18. Pathmakumar T, Sivanantham V, Anantha Padmanabha SG, Elara MR, Tun TT (2021) Towards an optimal footprint based area coverage strategy for a false-ceiling inspection robot. Sensors 21(15):5168
19. AS/NZS 2785:2000. Suspended ceiling design and installation. https://www.shop.standards.govt.nz/catalog/2785. Accessed Apr 2019
20. Singh V, Skiles SM, Krager JE, Wood KL, Jensen D, Sierakowski R (2009) Innovations in design through transformation: a fundamental study of transformation principles. J Mech Des 131(8)
21. Hayat AA, Akhlaq A, Alam MN (2012) Design of a flexural joint using finite element methods. Mach Mech 198–205
22. Hayat AA, Chaudhary S, Boby RA, Udai AD, Dutta Roy S, Saha SK, Chaudhury S (2022) Identification. Springer Singapore, Singapore, pp 75–113

Smart Technologies for Mobility and Healthcare

Review Paper on Joint Beamforming, Power Control and Interference Coordination for Non-orthogonal Multiple Access in Wireless Communication Networks for Efficient Data Transmission

Leela Siddiramlu Bitla and Chandrashekhar Sakode

1 Introduction

In conjunction with successive interference cancellation (SIC), researchers have studied the concept of superposition coding (SC) for calculating the capacity of many access channels [1]. The experimental evaluations of several SC approaches are put to the test in [2]. SC can increase spectral efficiency by using various signal power levels in a multiple-access channel. To integrate NOMA into current standards, [3] recommends the use of multi-user superposition transmission (MUST) techniques. [4] looks into a realistic MUST system's power allocation algorithm. Researchers have looked at multiple input multiple output (MIMO) for NOMA in [5, 6] to better understand how it can be used in MIMO systems. Downlink coordinated two-point systems, which make use of NOMA, are also investigated in [7]. There is a performance study in [8], and a power allocation issue is dealt with in [9]. Even without SIC in [10], NOMA can still be used with only minimal impact on performance as may be seen in various examples. In a multiple user model, multiple user downlink beamforming have been investigated to improve spectral efficiency. Different multi-user downlink beamforming systems have used uplink-downlink duality to overcome multi-user downlink beamforming challenges with individual QoS limitations utilising the signal-to-interference-plus-noise ratio (SINR). Semidefinite programming (SDP) has been utilised to solve other beamforming problems, such as multicast

L. S. Bitla · C. Sakode (✉)
Department of Electronics and Communication Engineering, Indian Institute of Information Technology, Nagpur, Maharashtra, India
e-mail: chandrashekharsakode@gmail.com

L. S. Bitla
e-mail: leelabitla@gmail.com

© The Author(s), under exclusive license to Springer Nature Singapore Pte Ltd. 2023

H. Muthusamy et al. (eds.), *Robotics, Control and Computer Vision*, Lecture Notes in Electrical Engineering 1009, https://doi.org/10.1007/978-981-99-0236-1_38

beamforming [11–13]. There aren't many multi-user downlink beamforming techniques that utilise NOMA, save from a few research efforts. While random beams are used in [14, 15], researchers are also looking at different approaches in [5, 16]. A memorisation- maximisation strategy (MMA) is used in [17] to maximise the NOMA sum rate. [18] looks into beamforming with a small amount of data to investigate optimal NOMA beamforming problems and utilise the power and spatial domains. Although this problem is similar to others, it varies from them, in that, it has additional SINR constraints. For specific QoS requirements, we are constrained by SINR, and this article's problem is distinct from that [17]. Even when the correlation between channels is inadequate, NOMA beamforming may perform worse than traditional multi-user beamforming at ideal levels (without NOMA). The performance benefit from using SC and SIC in conjunction with power and space domains may be less when interference is modest due to poor spatial correlation than t vectors. To get around this issue, we suggest using simple space domain for channel addition to NOMA beamforming that makes use of variable SIC sets to account for channel vector spatial correlation.

In scattering scenarios like those found in millimeter-wave cellular networks, extended NOMA beamforming outperforms traditional multi-user beamforming, according to this study's analysis and simulation results [19–21]. Therefore, generalised NOMA beamforming may be well-fitted to mmWave cellular network beamforming in order to improve spectral efficacy. Generalised NOMA beamforming has also been demonstrated in [22]. In addition to NOMA beamforming, there may be other generalisations. This and other radio resource allocation issues for multi carrier systems could be investigated more in the future. Other contributions include formulating an SDP problem for optimal NOMA beamforming and generalising NOMA beamforming with a low complexity approach based on channel vector correlations to decide on the SIC sets via an analysis of how the correlation affects generalised NOMA beamforming. This paper makes both of these contributions, among others. In scattering scenarios like those found in millimeter-wave cellular networks, extended NOMA beamforming outperforms traditional multi-user beamforming, according to this study's analysis and simulation results [19–21]. Cooperative NOMA systems were initially developed in Ref. [4] as an alternative to traditional NOMA systems for expanding coverage and improving reception reliability. Weak users sharing a subchannel can employ SC and SIC techniques to help themselves by passing the signal to the strong user, which is already present in this cooperative scenario. As a result, the weak user will experience better performance and have their QoS requirements met. Since improving energy efficiency is a fundamental goal of 5G in conjunction with cooperative NOMA. As a result, the powerful user can double as a helper and an energy collector by drawing power from its own batteries. This scenario (cooperation between users) takes into account the integration of cooperative networks with NOMA as well as the use of dedicated relays.

2 Related Works

Liu et al. [23] suggested a method to combine the Reconfigurable Intelligent Surfaces (RIS) with Unmanned Aerial Vehicle (UAV) in order to enhance the service quality of UAV. The method effectively allowed roaming of the mobile users coupled with increased spectrum efficiency even when roaming. The method developed the decaying deep Q-network (D-DQN) algorithm to reduce the energy utilisation of the system. The UAV-enabled status wireless network was monitored periodically and measures were made appropriately to make the system adapt to the changing environment. The interference arising in the NOMA was removed via the usage of linear precoding technique. Upon simulation, the methods demonstrated to decrease the energy usage substantially in the UAV.

Ashok and Sudha [24] proposed a beamforming method using Conditional Time-Split Energy Extraction (CT-EE) to optimise the error rate with great energy efficiency. The MSE and attainable sum rate (ASR) were improved and a minimal MSE method was also developed to enable the system's efficiency. Based on the battery life, the time split was done between information decoding and Energy Extraction (EE) phase and the bypassing scope of EE phase was explored simply based on the power need. The method effectively manages the power depleted situation with maximum connection and increased relay lifetime. The method focuses on allowing the wireless nodes to gain complete connection even amid natural catastrophes. The method dominates the other strategies related to beamforming designs with minimal error.

Al Obiedollah et al. [25] proposed a beamforming method. In this, the trade-off between the two metrics are discovered and optimised to enhance the performance of the beamforming method. The approach had used the SCA methodology to figure out the non-convexity problem and to get an optimum solution. The integrated priori articulation method with weighted sum technique was exploited to accomplish an effective optimization of EE and SE. The simulations were carried out using the known beamforming methods relevant to power minimization.

Zhao et al. [26] proposed a method to increase the average sum rate for IRS-aided multi-user system. This method focused on optimising the passive IRS phase shifts first and then the beamforming vectors are developed with the help of minimising the channel training overhead and the design complexity compared to previous ways. The passive phase shifts are optimised based on statistics CSI (S-CSI) rather than instantaneous CSI (I-CSI) and the beamforming vectors were developed in a manner to accommodate fading channels of I-CSI of the users. The method presented the Penalty Dual Decomposition (PDD) algorithm for single-user case and generic TTS stochastic sequential convex approximation (SSCA) for multi-user scenarios. The simulations focused on the effect assessment of S-CSI and channel correlation on the performance of the system.

Ehlers et al. [27] developed a digital beamforming method to detect the spatial disparities between the nodes present in a vast network. The method focused on minimising the interference happening between the networks with the usage of

Multi-User Detection (MUD) receiver that may decrease the interference between the nodes. MUD was primarily utilised to reduce the interference that was not detected simply via the beamforming method. The overall performance of MUD was improved using the Direct Sequence Spread Spectrum (DSSS) waveform. MUD, DSSS and digital beamforming were the three main mitigation types focused on and described as MUD-Aided Multi-Beam Uncoordinated Random Access Medium Access Control (MAMBU-RAM) (MAMBU-RAM). The method proved capable of reducing the interference observed in the closely spaced nodes spanning a limited range.

Newell and Vejarano [28] proposed a method to decrease the average power consumption in the networks during data transmission. A dynamic routing method for power management during transmission was developed in the Wireless Body Area Networks (WBAN) based on the body motions of the user. The programme recorded the body movement and developed a periodic time-domain model to minimise the energy usage during transmission with assured packet delivery rate (PDR) (PDR). The algorithm also focussed on finding the shortest route towards the sink node to reduce the energy usage while reaching the destination. The movement of the sensors placed in the user's body was monitored by the algorithm utilising the inertial measurement units (IMU) (IMU). The method upon tests showed that the power consumption relies on the movement of the user and the distance between the middle of the sensor node and the destination.

Zhu et al. [29] put forward a method to jointly optimise both the power regulation and beamforming for the uplink NOMA for 2 users in mmWave communications. The method provided maximal ASR for 2 users with minimum rate restriction for both the users. The joint optimization method was addressed as two distinct issues where the first phase focused on power management and beam gain allocation, while the second phase worked on addressing analogue beamforming problem exposed to constant- modulus constraint. The tests carried out indicated that the method performed better in joint optimization than the Orthogonal Multiple Access (OMA).

Ji et al. [30] developed a method to simultaneously optimise the UAV trajectory, cache location and power needed for transmission within a limited time. The method used an alternating iterative algorithm to optimise the three stated issues in an iterative way. The method mainly focused on increasing the network throughput and full exploitation of wireless caching and mobility of UAV for multi-user content delivery. The iterative technique was based on convex approximation methods and block alternating descent. The experimental findings showed the improved convergence of the algorithm in optimising the lowest throughput for UAV users. Also, the method showed effective access latency reduction and throughput improvement when several UAVs were deployed.

Zhan et al. [31] proposed a model to enhance the spectrum sharing capabilities of the network as well as to concurrently improve the power management during transmission. To accomplish the two stated goals, the model constructed a Distributed Proximal Policy Optimization (DPPO) for efficient spectrum sharing. Based on regulating the power among the main the issue of spectrum sharing was studied. Through various power changes, the method allowed spectrum sharing between the PU and SU along with fulfilling the QoS criteria.

Lin et al. [32] proposed a method using user admission control to simultaneously optimise the base station activation, admissible users and transmission beamformers. The ultimate aim was to optimise the use of power in the networks. The author formulated the problem into a convex sparse optimisation problem to ensure proper functioning of the network and to achieve low power efficiency. To address the issue, the Alternating Direction Method of Multipliers (ADMM) was used which operates in an iterative way to find a solution. The tests demonstrated the effectiveness of the method in reducing the power in multi-cell downlink green networks.

Shen et al. [33] addressed the joint trajectory and cross-link interference problem through the use of a successive convex approximation algorithm. The author formulated the joint trajectory and power control problem (TPC) to increase the aggregate sum rate of UAV enabled Interference Channel (UAV-IC) under certain constraints. The optimal solution for TPC problem was obtained based on the fly-hover-fly technique employed in finding the optimal hovering locations of UAVs. The time complexity of the method was decreased utilising the parallel TPC technique. It jointly updated the trajectory and power variables for every iteration and the algorithm proved to dominate other algorithms in terms of mitigating the cross- link interference along with reducing the time complexity.

Li et al. [34] addressed the inter-tier interference coordination problem in heterogeneous networks (HetNet) comprising macro and small cells working under Frequency Division Duplexing (FDD) mode sharing a similar spectrum. Large-scale and multiple antenna arrays were installed in the macro and small cell, respectively. The SINR was approximated in the macro BS and 3D beamforming strategy was applied for the users utilising the macro BS. The expected SINR was derived for the small-cell users using the Wishart matrix. Then, the interference coordination algorithms were introduced based on the observations to achieve a better trade off between the network traffic and performance of both macro and small cells. Under experiments, the algorithm dominated the other algorithms in reducing the interference between the networks and the results of the experiments correlated with the Monte Carlo results.

Li et al. [35] published a technique to solve the interference problem for cell-edge users in full-dimension MIMO (FD-MIMO) systems. The authors introduced Fractional-Frequency-Reuse (FFR) method for the FD-MIMO systems allocating same frequency band for cell-centre users and the remaining frequency for the cell-edge users. Also, two joint interference coordination strategies on 3D beamforming were introduced such as the full- cooperative strategy. Wang et al. [36] developed a two- level beamforming coordination method with the help of minimising the interference happening between the wireless networks. The method operated by dividing the network into clusters and followed an inter-cluster coordination for every clusters. A dynamic time domain interference coordination method was put-forth to collect the interference information. The method assisted in decreasing the inter-cluster interference between the network clusters for the switched-beam systems (SBS) (SBS). Upon simulations, the two-level method showed to minimise the interference between the networks and also proven to achieve greater performance for the edge users than the other techniques.

Mismar et al. [37] developed a combined design of beamforming, interference coordination and power management for a 5G wireless network. The method focused on optimising the SINR function and the average sum rate. The deep reinforcement learning was used to address the issues in networks where the greedy character of the learning method predicted the future situation. Based on the estimate, an algorithm was developed for the voice and data carriers in mmWave frequency ranges. The calculations showed that the method offered maximum sum rate capacity for data bearers and outperformed the industry norms for sub-6 GHz voice bearers.

3 Motivations and Contributions

Beamforming provides flexibility in data transmission and it is mostly utilised in wireless communication. Efficient beamforming can jointly optimise the power as well as the interference occurring in the networks and when the sensors and BS are located in a sparse space. There is no known current research that has studied resources aside from that, each NOMA group will be able to share the sub-channel with several D2D pairs at once. The assignment of sub-channels will be optimised orderly to maximise the system sum rate and upgrade the performance of cell-edge users, therefore, increasing the number of assigned D2D pairs and thereby increasing the coverage.

The following are the study's major contributions.

In order to benefit from their beneficial combination in terms of increased spectrum utilisation and better cell-edge user performance, FD Cooperative NOMA technology and underpinning D2D are integrated. This interference control is systematic. The resource allocation problem must be efficiently dealt with in this circumstance. Sub-channels are allocated between NOMA groups and D2D pairs for this purpose.

4 The Concept

Non-orthogonal multiple access refers to multiple access that is not orthogonal to the rest of the network (NOMA). OFDM is used for modulation, while NOMA is used for multiple access. Orthogonal frequency division multiple access (OFDMA) is a technique that is an extension of orthogonal frequency division multiple access (OFDM). It is widely used in typical 4G networks to provide information about each user to a limited number of subcarriers. While in NOMA, each user retains full control over the subcarriers they use. Figure 1 illustrates how two users sharing spectrum in the OFDMA and NOMA bands may do so using OFDMA and NOMA.

With superposition coding and successive interference cancellation (SIC) at the transmitter and receiver, a single spectrum may be shared by a wide range of users. A single waveform is used to broadcast all the information signals, which are then decoded one at a time by SIC once they arrive at the receiver. There are three distinct

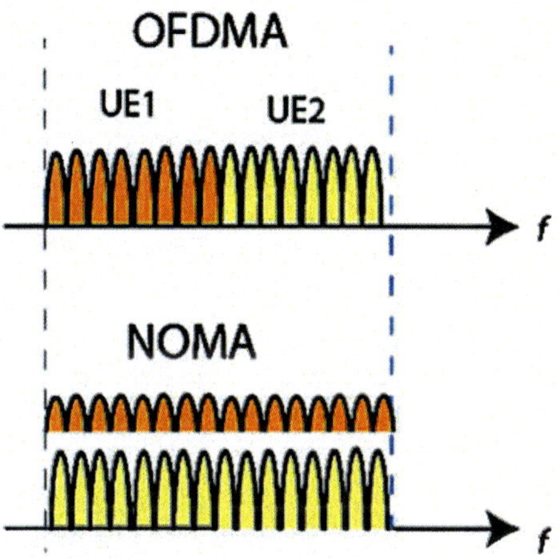

Fig. 1 Spectrum splitting for NOMA and OFDMA for two users

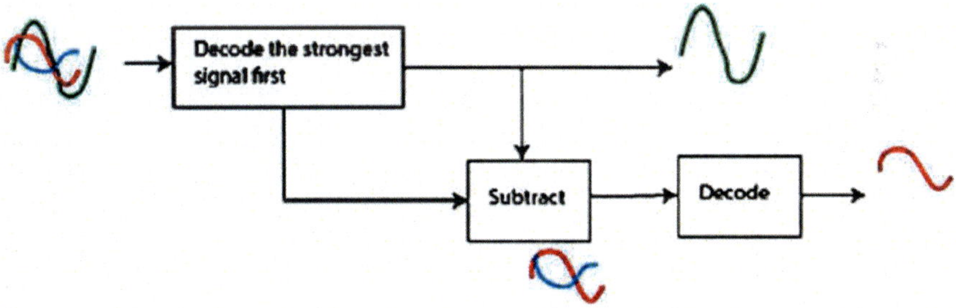

Fig. 2 Successive interference cancellation

colour-coded information signals being sent from the transmitter, as seen in the figure. All three signals are included in the signal received by the SIC receiver. The strongest signal is the first one decoded by SIC, with the others acting as noise. Decoding is performed by first subtracting the coded signal from the received signal, and if decoding is successful, a complete waveform is produced. Once SIC has found the required signal, it continues to repeat the procedure until it does

5 NOMA Downlink

The BS covers the served users information waveform on its own. SIC is used by each piece of user equipment (UE) to pick up on its own signals. Figure 3 depicts a BS and K UEs equipped with SIC receivers in a wireless network. Base stations (BS) are presumed to be nearest (UE1) and furthest (UEK), respectively.

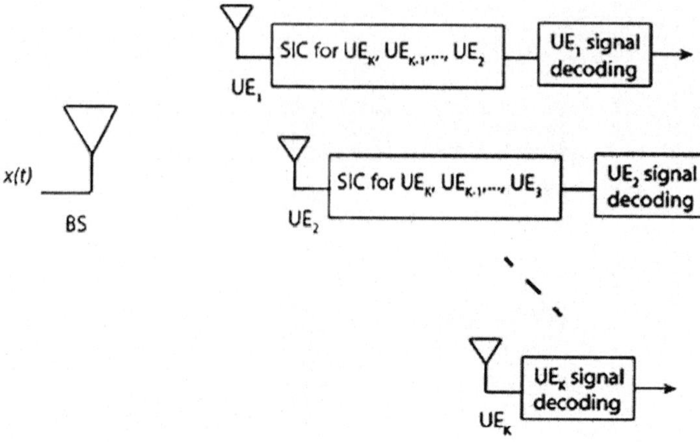

Fig. 3 Successive interference cancellation

One of the biggest problems for BS is deciding how much power to distribute across the various information waveform. When using the NOMA downlink, the UE situated further away from the base station receives more power and the UE located closest to the base station receives less power. This information is sent to all UEs in the network as a single signal. The strongest signal is decoded first by each UE, and the decoded signal is deducted from the received signal thereafter. The subtraction is repeated until the SIC receiver locates its own signal. The signals from UEs situated far away from the BS may be cancelled by UEs located near the BS. Due to the fact that the furthest UE's signal gives the most to the received signal, it will decode its own signal first.

Most of the beamforming methods rely on achieving a high data transmission rate but the interference and power consumption during transmission are not addressed. Joint optimization of all the three methods enables the network to operate effectively with maximum throughput. Interference in the network is a serious problem that confronts security leading to loss of data. Because of the development of wireless communication networks and the dense deployment of BS and antennas, interference has become a frequent issue. Also, the energy that is dissipated from the network creates the interference and capacity difficulties. Power control method assists in preventing the interference happening in the network as well as increases the capacity of the network. Efficient beamforming can jointly optimise the power as well as the interference happening in the networks and when the sensors and BS are situated in a sparse area, beamforming can operate well with maximum SINR. In a dense network, the capacity of the network and the QoS guaranteed to users are severely restricted owing to the inter-cell interference. The capacity of the network may be increased by the frequency reuse techniques, thus decreasing the interferences.

Thus, increasing the capacity of the network and effective beamforming for multiple-antennas minimises the inter-cell interference. The proactive method (prediction) of finding a solution to enhance the capacity and decrease the interference is more important than the strategies that simply respond to the issues after identification. Another most essential aspect is the power management in the network that assists in maintaining a higher battery capacity. Both the power control and interference coordination methods are tightly connected as the decrease in power consumption lowers the interference as well. In the dense deployment of BS, the dynamic flow needs to be addressed instead of static methods so that the power consumed by BS may be effectively controlled to increase the throughput. In case of static methods, the power may only be changed in the BS if the users avoid travelling and remain in a specific location for a set period of time.

There are only a few methods known related to beamforming that induce both power management and interference coordination with simultaneous optimization of SINR and network performance. Due to the urgent need for effective beamforming methods with joint optimization of power management and interference coordination, the upcoming joint beamforming strategies have been suggested. This primary goal is to accomplish efficient data transmission rate in a wireless network and to solve the faults highlighted in the current methods with decreased interference and with effective spectrum sharing viewpoint.

6 MIMO-NOMA

For the reason that it increases total capacity performance also when there are a huge number of users, MIMO systems may be able to take use of multicast beamforming. Although there are numerous applications for it, there are also numerous drawbacks. When all users share a single beam, everyone receives an identical signal [37], according to one technique. It is possible to employ numerous beams, each of which may be utilised by different groups of users to receive a different signal, as an alternate method [38]. The following studies on beamforming in MIMO-NOMA systems are examples of such investigations. [39] Proposes a downlink MIMO-NOMA system with multi-user beamforming as an alternative to the current standard. Two persons can use the same beam at the same time if they are in close proximity to one another.

Because this beam can only be shared by two users at a time, strategies like clustering and power allocation can be utilised to enhance overall capacity while simultaneously decreasing interference between clusters and between users.The effectiveness of multicast beamforming is examined in [40]. In order to distribute their information streams, broadcast system transmitters use a large number of antennas based on multi resolution broadcast ideas that only provide low priority signals for consumers who are far away from the broadcasting system or who have poor channel quality near BS in order to keep him or her connected. With the use of superposition coding and a minimal power beamforming issue, it is possible to conduct random beamforming. Due to the assumption that all users in a cluster would utilise the same beam, all transmission power is assigned to the same amount of transmission power across all beams. A spatial filter should also be used to decrease interference between clusters and between beams, according to the recommendations of the authors. The fractional frequency reuse concept is proposed as a means of optimising power distribution across a large number of beams, with consumers with varied channel conditions accepting a variety of reuse ratios. [41] describes a downlink multi-user MIMO-NOMA system that reduces interference while simultaneously increasing capacity, where mobile users receive antennas outnumber the base station's broadcast antennas. In this paper, we present a zero-force beamforming-based technique for inter-cluster interference reduction, which is particularly useful when distinguishing between users of various channel quality is predicted. To achieve the highest possible throughput while minimising disturbance, approaches such as user clustering and the minorization-maximization approach is also advised as an approximation in order to minimise the significant computational costs associated with the nonconvex optimization problem

The minorisation-maximisation approach is also advised as an approximation in order to minimise the significant computational costs associated with the nonconvex optimization problem. The primary objective of the minorisation-maximisation technique is to maximise system throughput for a given number of users that multiple beams are used, the [42] downlink MIMO- NOMA system broadcasts precoded signals to all cellular users; means every beam serves a certain number of customers; in other words, each beam serves a single consumer. It has been recommended that three distinct approaches be used in conjunction with one another in order to optimise the overall rate.

With the use of weighted sum rate maximisation, a unique beamforming matrix is created for each beam, with each beam making use of all of the CSI that is accessible at the BS. The second technique makes use of user scheduling in order to take use of super SIC for each mobile user. To realise the maximum potential of SIC, channel gains within each cluster must be considerably different, and channel correlation among mobile users must be strong in order to enjoy the benefits to the greatest extent possible. Fixed power allocation, on the other hand, attempts to optimise performance by delivering neither a higher sum rate nor also a more pleasant performance for customers who have poor channel quality. [43] investigates a layered transmission system with a two-user MIMO- NOMA maximum transmission power constraint in order to determine the most efficient power allocation strategy for the system. Because each mobile user decodes signals in SIC in a sequential manner, using layered transmission instead of non layered transmission significantly reduces the complexity of decoding signals in the SIC. As a consequence, the average sum rate and its limitations are demonstrated to have a closed-form expression in both the perfect CSI and partial CSI situations, demonstrating that the average sum rate and its limits can be expressed mathematically. The average total rate grows in tandem with the growth in the number of antennas used. It is stated in [44–46] networks in a MIMO-NOMA framework. It was also discovered that by combining two distinct power distribution systems, a reasonable balance between fairness and throughput could be achieved. Different QoS criteria may be satisfied utilising the fixed power allocation approach. Further, a power allocation approach based on cognitive radio technology ensures that the QoS needs of the end user are satisfied straight away. Also, conceivable for the open-system were the construction of exact and asymptotic equations (OP). A study published in [47] investigates the power reduction issue. According to [48], there exist precoders known as linear beamformers that provide a greater overall total throughput while simultaneously enhancing the user's throughput on channels of poor quality. These precoders also meet the requirements of the Quality of Service standard. Furthermore, it has been proven that higher distinct channel gains result in superior NOMA performance for the greatest number of users per cluster.

A superimposed pilot scheme in which the Gaussian signal prohibits the use of a pilot seems to have the greatest amount of pilot power when the power loss that may be produced by the use of a pilot appear to have zero power. It is more efficient to use the superimposed code maximisation method rather than the orthogonal method when there are more mobile users and greater mobility. Massive MIMO is distinct from massive access MIMO, which is discussed in [49], In [50], a low-complexity Gaussian message passing iterative detection technique is applied to achieve the lowest mean square error multi-user detection, and both its means and variances perfectly converge with rapid speed. MA scheme, NOMA, has also been proposed for consideration in millimeter-wave communication systems and it integrates beamspace MIMO and provides massive connectivity in situations where lot of cellular users outnumber the lot of radio frequency chains, while also achieving improved spectrum and energy efficiency performance [51]. In addition, a zero-forcing (ZF) precoding approach has been developed to reduce inter beam interference to the greatest extent possible. Another set of innovations includes a dynamic power allocation system and iterative optimization methods with higher sum rates and less complexity. The issue of energy efficiency optimization for MIMO-NOMA systems with imperfect BS CSI across Rayleigh fading channels is addressed in [52, 53]. However, there are specified limits on the total amount of money that can be spent.

7 Conclusion

Today's wireless networks make use of the Non-orthogonal multiple access (NOMA) technique in order to distribute radio resources equally among its customers devices. Because of the growth in multiple users, it is more similar to OMA-based techniques that will fall short of more stringent requirements such as high spectral efficiency, ultra-low latency and broad connectivity. Successive improve spectral efficiency while yet permitting for some multiple-access interference at the receiver end, the idea of non-orthogonal multiple access (NOMA) has evolved. Our goal is to write this instructional-style essay to provide a NOMA downlink model and to offer extensions to MIMO and cooperative communication scenarios.

References

1. Lin Z, Li Mn, Zhu W-P, Wang J-B, Cheng J (2020) Robust secure beamforming for wireless powered cognitive satellite-terrestrial networks. IEEE Trans Cognitive Communications and Networking (2020).
2. Lu Y, Koivisto M, Talvitie J, Valkama M, Lohan ES (2020) Positioning- aided 3D beamforming for enhanced communications in mmWave mobile networks. IEEE Access 8: 55513–55525
3. Papageorgiou GK, Voulgaris K, Ntougias K, Ntaikos DK, Butt MM, Galiotto C, Marchetti N et al (2020) Advanced dynamic spectrum 5G mobile networks employing licensed shared access. IEEE Commun Mag 58(7):21–27
4. Naderializadeh N, Eisen M, Ribeiro A (2020) Wireless power control via counterfactual optimization of graph neural networks. In: 2020 IEEE 21st international workshop on signal processing advances in wireless communications (SPAWC). IEEE, pp 1–5
5. Gilan MS, Maham B (2020) Virtual MISO with joint device relaying and beamforming in 5G networks. Phys Commun 39:101027
6. Choi J, Cho Y, Evans BL (2020) Quantized massive MIMO systems with multicell coordinated beamforming and power control. IEEE Trans Commun
7. Kong J, Dagefu FT, Sadler BM (2020) Simultaneous beamforming and nullforming for covert wireless communications. In: 2020 IEEE 91st vehicular technology conference (VTC2020-Spring). IEEE, pp 1–6
8. Liu Y, Li J, Wang H (2019) Robust linear beamforming in wireless sensor networks. IEEE Trans Commun 67(6):4450–4463
9. Wu Q, Zhang R (2019) Intelligent reflecting surface enhanced wireless network via joint active and passive beamforming. IEEE Trans Wireless Commun 18(11):5394–5409
10. Huang H, Peng Y, Yang J, Xia W, Gui G (2019) Fast beamforming design via deep learning. IEEE Trans Veh Technol 69(1):1065–1069
11. Ioushua SS, Eldar YC (2019) A family of hybrid analog–digital beamforming methods for massive MIMO systems. IEEE Trans Signal Process 67(12):3243–3257
12. Zhu L, Zhang J, Xiao Z, Cao X, Xia X-G, Schober R (2020) Millimeter-wave full-duplex UAV relay: Joint positioning, beamforming, and power control. IEEE J Sel Areas Commun 38(9):2057–2073
13. Peken T, Tandon R, Bose T (2020) Unsupervised mmWave beamforming via autoencoders. In: ICC 2020–2020 IEEE international conference on communications (ICC). IEEE, pp 1–6
14. AlAmmouri A, Gupta M, Baccelli F, Andrews JG (2020) Escaping the densification plateau in cellular networks through mmWave beamforming. IEEE Wirel Commun Lett 9(11):1874–1878
15. Zheng Y, Bi S, Zhang Y-JA, Lin X, Wang H (2020) Joint beamforming and power control for throughput maximization in IRS-assisted MISO WPCNs. IEEE Internet of Things J
16. Zhao C, Cai Y, Liu A, Zhao M, Hanzo L (2020) Mobile edge computing meets mmWave communications: Joint beamforming and resource allocation for system delay minimization. IEEE Trans Wireless Commun 19(4):2382–2396
17. Li X, Zhu G, Gong Y, Huang K (2019) Wirelessly powered data aggregation for IoT via over-the-air function computation: Beamforming and power control. IEEE Trans Wirel Commun 18(7):3437–3452
18. Zhu L, Zhang J, Xiao Z, Cao X, Wu DO, Xia X-G (2019) Joint Tx-Rx beamforming and power allocation for 5G millimeter-wave non-orthogonal multiple access networks. IEEE Trans Commun 67(7):5114–5125
19. Chen W-Y, Chen B-S, Chen W-T (2020) Multiobjective beamforming power control for robust SINR target tracking and power efficiency in multicell MU-MIMO wireless system. IEEE Trans Veh Technol 69(6):6200–6214
20. Mei W, Qingqing W, Zhang R (2019) Cellular-connected UAV: Uplink association, power control and interference coordination. IEEE Trans Wirel Commun 18(11):5380–5393
21. Liang F, Shen C, Wei Y, Feng W (2019) Towards optimal power control via ensembling deep neural networks. IEEE Trans Commun 68(3):1760–1776

22. Chen Y, Wen M, Wang L, Liu W, Hanzo L (2020) SINR-outage minimization of robust beamforming for the non- orthogonal wireless downlink. IEEE Trans Commun 68(11):7247–7257
23. Liu X, Liu Y, Chen Y (2020) Machine learning empowered trajectory and passive beamforming design in UAV-RIS wireless networks. IEEE J Selected Areas Commun
24. Ashok K, Sudha T (2020) Uninterrupted connectivity using conditional time split energy extraction with beamforming system for disaster affected wireless networks. IEEE Access 8:194912–194924
25. Al-Obiedollah HM, Cumanan K, Thiyagalingam J, Tang J, Burr AG, Ding Z, Dobre OA (2020) Spectral-energy efficiency trade-off-based beamforming design for MISO non-orthogonal multiple access systems. IEEE Trans Wirel Commun 19(10):6593–6606
26. Zhao M-M, Wu Q, Zhao M-J, Zhang R (2020) Intelligent reflecting surface enhanced wireless network: two-timescale beamforming optimization. IEEE Trans Wirel Commun
27. Ehlers B, Gupta AS, Learned R (2020) A MUD-enhanced multi-beam approach for increasing throughput of dense wireless networks. IEEE Sens J
28. Newell G, Vejarano G (2020) Motion-based routing and transmission power control in wireless body area networks. IEEE Open J Commun Soc 1:444–461
29. Zhu L, Zhang J, Xiao Z, Cao X, Wu DO, Xia X-G (2018) Joint power control and beamforming for uplink non-orthogonal multiple access in 5G millimeter-wave communications. IEEE Trans Wirel Commun 17(9):6177–6189
30. Ji J, Zhu K, Niyato D, Wang R (2020) Joint cache placement, flight trajectory, and transmission power optimization for multi-UAV assisted wireless networks. IEEE Trans Wirel Commun 19(8):5389–5403
31. Zhang H, Yang N, Huangfu W, Long K, Leung VCM (2020) Power control based on deep reinforcement learning for spectrum sharing. IEEE Trans Wirel Commun 19(6):4209–4219
32. Lin J, Zhao R, Li Q, Shao H, Wang W-Q (2017) Joint base station activation, user admission control and beamforming in downlink green networks. Digital Signal Process 68:182–191
33. Shen C, Chang T-H, Gong J, Zeng Y, Zhang R (2020) Multi-UAV interference coordination via joint trajectory and power control. IEEE Trans Signal Process 68:843–858
34. Li X, Li C, Jin S, Gao X (2018) Interference coordination for 3-D beamforming- based HetNet exploiting statistical channel-state information. IEEE Trans Wirel Commun 17(10):6887–6900
35. Li X, Liu Z, Qin N, Jin S (2020) FFR based joint 3D beamforming interference coordination for multi-cell FD-MIMO downlink transmission systems. IEEE Trans Veh Technol 69(3):3105–3118
36. Wang J, Weitzen J, Bayat O, Sevindik V, Li M (2019) Interference coordination for millimeter wave communications in 5G networks for performance optimization. EURASIP J Wirel Commun Netw 2019(1):1–16
37. Mismar FB, Evans BL, Alkhateeb A (2019) Deep reinforcement learning for 5G networks: Joint beamforming, power control, and interference coordination. IEEE Trans Commun 68(3):1581–1592
38. Kaliszan M, Pollakis E, Stańczak S (2012) Multigroup multicast with application-layer coding: beamforming for maximum weighted sum rate. In: Proceedings of the 2012 IEEE wireless communications and networking conference, WCNC 2012, France, pp 2270–2275. (Apr 2012)
39. Kimy B, Lim S, Kim H et al (2013) Non-orthogonal multiple access in a downlink multiuser beamforming system. In: Proceedings of the 2013 IEEE military communications conference, MILCOM 2013. San Diego, Calif, USA, pp 1278–1283. (Nov 2013)
40. Choi J (2015) Minimum power multicast beamforming with superposition coding for multiresolution broadcast and application to NOMA systems. IEEE Trans Commun 63(3):791–800
41. Ali MS, Hossain E, Kim DI (2017) Non-orthogonal multiple access (NOMA) for downlink multiuser MIMO systems: user clustering, beamforming, and power allocation. IEEE Access 5:565–577
42. Sun X, Duran-Herrmann D, Zhong Z, Yang Y (2015) Non-orthogonal multiple access with weighted sum-rate optimization for downlink broadcast channel. In: Proceedings of the 34th annual IEEE military communications conference, MILCOM 2015. Tampa, Fla, USA, pp 1176–1181. (Oct 2015)

43. Choi J (2016) On the power allocation for MIMO-NOMA systems with layered transmissions. IEEE Trans Wirel Commun 15(5):3226–3237
44. Chen C, Cai W, Cheng X, Yang L, Jin Y (2017) Low complexity beamforming and user selection schemes for 5G MIMO-NOMA systems. IEEE J Sel Areas Commun 35(12):2708–2722
45. Shin W, Vaezi M, Lee B, Love DJ, Lee J, Poor HV (2017) Coordinated beamforming for multi-cell MIMO-NOMA. IEEE Commun Lett 21(1):84–87
46. Ding Z, Schober R, Poor HV (2016) On the design of MIMO-NOMA downlink and uplink transmission. In: Proceedings of the 2016 IEEE international conference on communications, ICC 2016, Kuala Lumpur, Malaysia, May 2016
47. Cui J, Ding Z, Fan P (2017) Power minimization strategies in downlink MIMO-NOMA systems. In: Proceedings of the 2017 IEEE international conference on communications, ICC 2017, Paris, France, May 2017
48. Nguyen V-D, Tuan HD, Duong TQ, Poor HV, Shin O-S (2017) Precoder design for signal superposition in MIMO-NOMA multicell networks. IEEE J Sel Areas Commun 35(12):2681–2695
49. Liu L, Yuen C, Guan YL, Li Y, Huang C (2016) Gaussian message passing iterative detection for MIMO-NOMA systems with massive access. In: Proceedings of the 59th IEEE global communications conference, GLOBECOM 2016, Washington, DC, USA, Dec 2016
50. Liu L, Yuen C, Guan YL, Li Y (2016) Capacity-achieving iterative LMMSE detection for MIMO-NOMA systems. In: Proceedings of the 2016 IEEE international conference on communications, ICC 2016, Kuala Lumpur, Malaysia, May 2016
51. Wang B, Dai L, Wang Z, Ge N, Zhou S (2017) Spectrum and energy-efficient beamspace MIMO-NOMA for millimeter-wave communications using lens antenna array. IEEE J Sel Areas Commun 35(10):2370–2382
52. Sun Q, Han S, Chin-Lin I, Pan Z (2015) Energy efficiency optimization for fading MIMO non-orthogonal multiple access systems. In: Proceedings of the IEEE international conference on communications, ICC 2015, pp 2668–2673, London, UK, June 2015
53. Wu P, Jie Z, Su X, Gao H, Lv T (2017) On energy efficiency optimization in downlink MIMO-NOMA. In: Proceedings of the 2017 IEEE international conference on communications workshops, ICC workshops 2017. France, pp 399–404. (May 2017)

3D Reconstruction Methods from Multi-aspect TomoSAR Method: A Survey

Nazia Akhtar, Tamesh Haldar, Arindam Basak, Arundhati Misra Ray, and Debashish Chakravarty

1 Introduction

1.1 TomoSAR

The SAR system uses a radar sensor that is mounted on a satellite to synthesize an antenna for several long kilometers. When the system is placed along the path of the satellite, it accurately and continuously takes information about the particular area. After that, the captured image of a particular area is reformed by digital processing technology. The outcome of the process is a 2D high-resolution map of the image scene. The main characteristic of the process is the microwaves generated by SAR (2D). It penetrates into media like snow, cloud, rain, etc. TomoSAR (3D) system obtained from SAR (2D) uses a radar that flows along multiple trajectories or paths [1, 2]. When the TomoSAR 3D system is placed in multiple paths, it measures the distance of the target from multiple paths. For localization of SAR (2D) radar along a straight line, it only measures the distance from the target to each point of the

N. Akhtar · A. Basak (✉)
School of Electronics Engineering, KIIT-Deemed to be University, Bhubaneswar, Odisha, India
e-mail: abasakfet@kiit.ac.in

N. Akhtar
e-mail: 1808027@kiit.ac.in

T. Haldar (✉) · D. Chakravarty
Department of Mining Engineering, Indian Institute of Technology, Kharagpur, West Bengal, India
e-mail: haldertamesh@iitkgp.ac.in

D. Chakravarty
e-mail: dc@mining.iitkgp.ac.in

A. M. Ray
Space Application Center, ISRO, Ahmedabad, India
e-mail: arundhatiroy@sac.ac.in

© The Author(s), under exclusive license to Springer Nature Singapore Pte Ltd. 2023

H. Muthusamy et al. (eds.), *Robotics, Control and Computer Vision*, Lecture Notes in Electrical Engineering 1009, https://doi.org/10.1007/978-981-99-0236-1_39

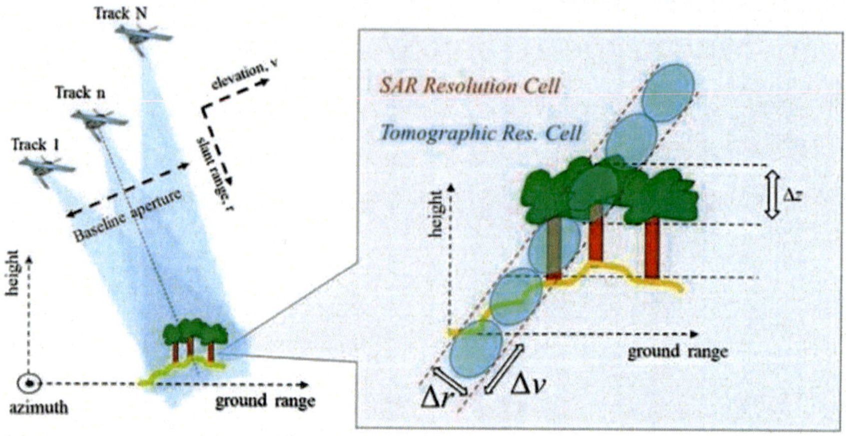

Fig. 1 Geometry of Tomosar inground plane [25]

particular line, whereas, for TomoSAR (3D) radar along multiple lines, it measures the distance from the target to multiple lines [3, 4]. The resolution of TomoSAR determines the bandwidth of the pulse along the direction of the slant range. The length of the synthetic aperture is in the azimuth direction and the baseline aperture is present in the cross range [5, 6]. The algorithm of TomoSAR is mainly classified in backward projection [7, 8], compressive detection [9] and estimation which is mainly spectral [10–12].

TomoSAR focusing is done by only one principle which is obtained by Fourier transform of the specific scene reflected in project complexity and cross range of coordinates [13, 14] (Fig. 1).

1.2 Reconstructing Building Framework in TomoSAR (3D)

Observing the side view of the target building using tomosar principle, which is marked as a red line, determines the visibility of the area, and the blue dots are the scatters [15]. These scatters will represent the structure of building and to reconstruct the same, it will return blue points to red line to improve the visibility. This process in turn provides the elimination of fake targets [16, 17] (Fig. 2).

TomoSAR has many advantages, one such advantage is the use of advanced technology by SAR (2D) as compared to TomoSAR (3D) for providing more accurate information which was not available by using SAR(2D) [1]. Another added advantage is that, for a particular given area, it covers a stack of images by flying multiple trajectories or paths [18]. Even if, it can work in all weather conditions [19].

With these added advantages the TomoSAR (3D) has some specific drawbacks also; the quality of the tomosar image deteriorates by noise and fake targets present in them [20]. Multiple methods are required to remove unwanted factors and to

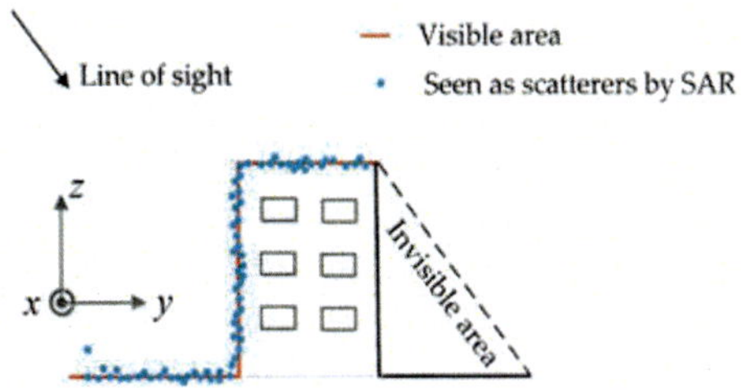

Fig. 2 Geometry of Tomosar(3D) building [17]

reconstruct the 3D cloud TomoSAR point. It takes a lot of time to provide information by flying in multiple paths and extracting the data points of a particular area [21].

In this manuscript, different approaches to the construction and reconstruction of TomoSAR images are described thoroughly. Researchers will get the advantage and flexibility of choosing different approaches very easily and in a very efficient manner.

2 Methodology

2.1 Hough Transform

As noise is present in the tomographic image, the false target is scattered in a disordered manner. To detect the outline, the motive of this transform is to find imperfect instances of an object present in a given shape class by using the procedure of voting.

It is composed of several straight lines connected in a segment. It is used in the recognition of the pattern. In this algorithm, a process of voting is held where each data point belonging to the pattern, votes for the possible pattern passing through that point. The votes are stored in an accumulator array, known as bins. The pattern which receives the maximum votes is known as the desired pattern [22].

In a N*N binary edge image equation of straight line is

$$\rho = x\cos\theta + y\sin\theta \tag{1}$$

where (x, y) is the measurement of position in coordinates of x and y

θ $(-\pi/2 \leq \theta < \pi/2)$ is the representation of the angle whose normal line makes with the x-axis.

ρ $(-N \leq \rho \leq 2-\sqrt{N})$ is the representation of the normal distance from the origin to the straight line.

In Fig. 3, (x, y) is the coordinates of point red, ρ*, and* θ is the parameter of space and a straight line is mapped at a point with reference to the parameter denoted for space. Every straight line in the (x, y) plane like a red dot in the coordinate of x and y is mapped with reference to the parameter of space. All possible straight lines through the red dot in coordinates of x and y will make a curved line, as shown in

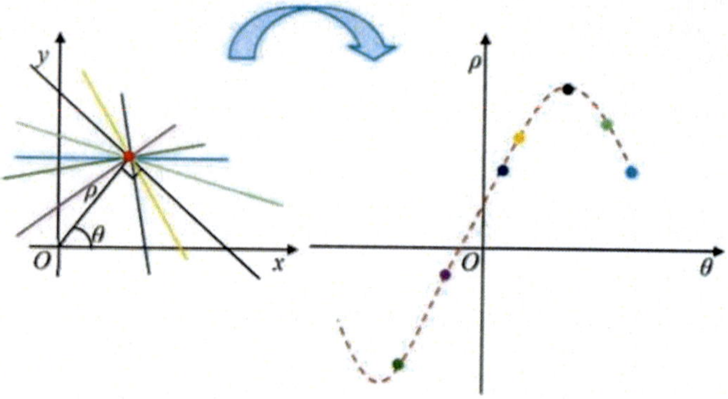

Fig. 3 Graphical representation of the relationship of (ρ, θ) and (x, y) [17]

Fig. 3; the red dashed line. So for all parameter cell, like (ρ, θ), this algorithm will calculate the value parameter and store all the pixel which lies within (ρ, θ).

(ρ, θ) is known as a straight line lying within the coordinates of x and y, but if it is not determined, then it is termed as noise. The lines which are detected may be broken due to noise present or by the density of some point clouds. Thus, some of the broken lines belong to the same segment of the outline but the parameter of the line is slightly not the same. In the K-means method of clustering, the lines detected are grouped in various clusters and parameters associated with distance are used in the clustering of the detected lines.

The computation parts of the Hough transform are as follows:

1. The calculation of value parameter and storing the pixels in the parameter of space.
2. Finding all local maximum points that represent the segment of the line.
3. The extraction of a segment of the line using maximum position.

2.2 *Facade Reconstruction*

In this method, there are three steps that are performed in TomoSAR point of extracting large datasets [23] which include the detection of facades and their extraction, segmentation, and reconstruction (Fig. 4).

In facade detection, the already existing model like DTM (DIGITAL TERRAIN MODEL) [using technical filter] is used for detection and for extraction of 2D point density in the x-y horizontal ground. Now in segmentation, the reconstruction of the individual facade is required so the point cloud which belongs to the same facade can be used. It sometimes uses unsupervised clustering technique. Lastly, for reconstruction, a facade is normally determined by a flat surface, curved surface, and edges or the boundary of facades and vertices [24]. To replace the Hough transform and Facade reconstruction, we can use some other popular techniques described below.

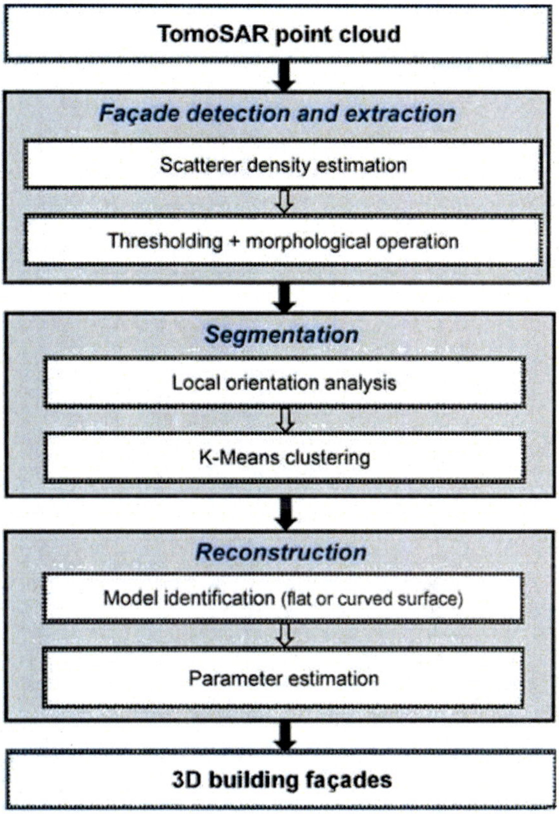

Fig. 4 Flowchart of facade reconstruction [15]

2.3 DBSCAN Method

DBSCAN METHOD is also known as DENSITY-BASED SPATIAL CLUSTERING OF APPLICATION WITH NOISE. It is used for extracting point cloud for the reconstruction of tomographic (3D) images. First, the TomoSAR point cloud is generated and then it is imputed in the DBSCAN module. This is a clustering method in which machine learning is used to separate the high-density clusters from low-density clusters (Figs. 5 and 6).

Two parameters of DBSCAN:

1. EPS: It specifies the space between two points as it has less than or sufficient "EPS" points which are normally re- referred to as adjacent points. If the selection of EPS value is too small, information is considered an outlier and if it is large, most of the data points in the group are merged into equivalent groups. To find the EPS value, a k-distance graph is used.
2. Min points: The minimum number of data points in the EPS radius. In larger datasets, a large minm point value is selected.

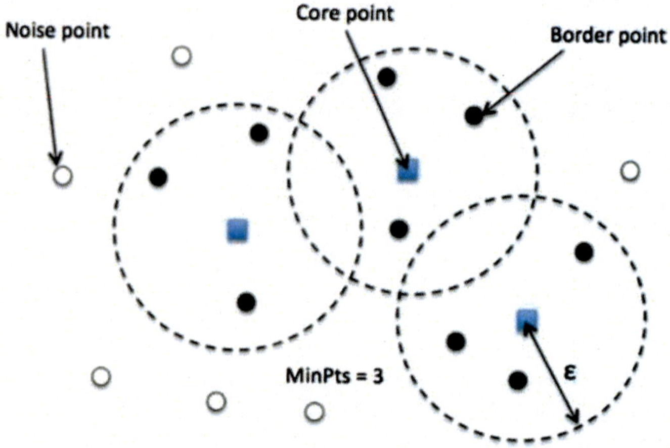

Fig. 5 DSCAN Method [*Source* https://www.kdnuggets.com/2020/04/dbscan-clustering-algorithm-machine-learning.html]

Fig. 6 DBSCAN Method Of Clustering [*Source* https://www.kdnuggets.com/2020/04/dbscan-clustering-algorithm-machine-learning.html]

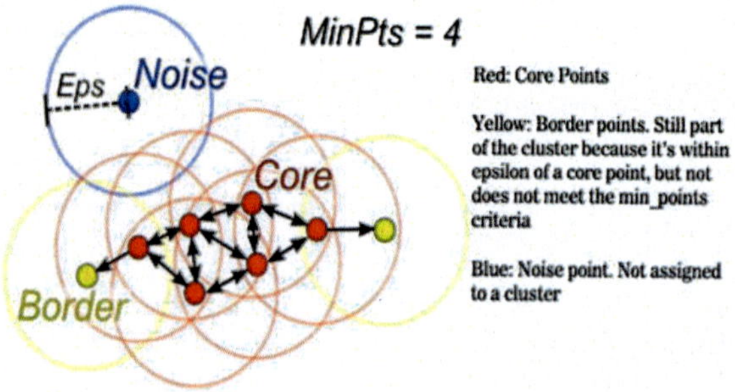

MinPts > = D + 1

Whereas MinPts > = D + 1 in dimension D of the dataset

Minm points selected = 3

In tomographic reconstruction, DBSCAN is used. Firstly the TomoSAR point cloud is generated and then it is imputed in the DBSCAN module (Fig. 7).

In the DBSCAN module, density detection is used for the separation of high-density clusters from low-density clusters and by unsupervised clustering. This process separates the data points into several groups. Having similar properties in similar groups and different properties in different groups is a common phenomenon in this type of clustering. After that, the unwanted factors such as noise and fake targets are removed. So the extraction of targeted point clouds is achieved [25].

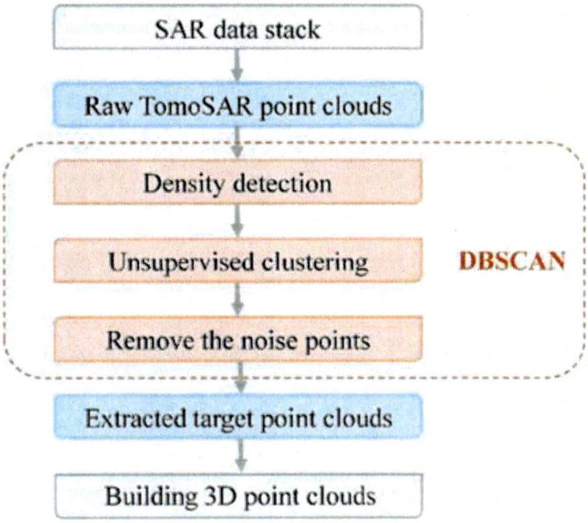

Fig. 7 DSCAN method In TomoSAR(3D) [25]

3 Result and Discussions

Using TomoSAR in Biosar (2007), the reconstructed tomographic image has analyzed the profile of large areas totally covered with forest, where HH is the dominating phase center which is locked grounded and the vegetation is almost not visible [3]. It is the same for VV, but for HV the vegetation is visible. In the campaign data of AFRISAR using the P band, the potential value which is determined by Tomography is used for the improvement in the classification of forest structures. The site was Mondah site (Gabon) and improvement was observed in classifying the forest structures [26]. In urban areas, observation of two techniques was used. One is PSI (persistent scattering interferometry) and another is TomoSAR. In comparison with the other results, TomoSAR has yielded the very best results by having four times better results in terms of the point density of PSI [27].

We can use TomoSAR (3D) in a productive manner for future scope of research as a BIOMASS [28] mission is to be launched. When the satellite will arrive, it will provide prominent layers of image by the tomographic technique [29]. There is a possibility for inspecting different types of forests. The future works of biomass mission will consider a lower spatial resolution in a more constrained way to provide a classified method. Various countries are also planning to participate in REDD (Reducing Emission from Deforestation and Degradation) program for forest biomass [30] and forest areas through which they will be benefited from monetary compensation.

4 Conclusion

Various approaches have been discussed for the reconstruction of 3D TomoSAR point clouds which is the Hough transform, Facade reconstruction and DBSCAN method. But the use of DBSCAN method is the current technique for reconstruction which gives a proper reconstructing model in urban areas, forested areas, etc. We plan to develop more reconstruction techniques in the future using TomoSAR(3D). With all these extensions, the addressed methods are presented as a feasible candidate for practical implementations in the perspective of future space missions, such as BIOMASS, aimed to estimate the global forest structure using TomoSAR data.

References

1. Ferro-Famil L, Huang Y, Pottier E (2016) Principles and applications of Polarimetric SAR tomography for the characterization of complex environments. Int Assoc Geodesy Symp 142(1–13):243–255
2. Tebaldini S, Ho Tong Minh D, Mariotti d'Alessandro M et al (2019) The status of technologies to measure forest biomass and structural properties: state of the art in SAR tomography of tropical forests. Surv Geophys 40:779–801
3. Blomberg E, Ferro-Famil L, Soja MJ, Ulander LMH, Tebaldini S (2018) Forest biomass retrieval from L- band SAR using tomographic ground backscatter removal. IEEE Geosci Remote Sens Lett 1–5
4. Frey O, Meier E (2011) 3-D time-domain SAR imaging of a forest using airborne multibaseline data at L-and P-bands. IEEE Trans Geosci Remote Sens 49:3660–3664
5. Lombardini F, Cai F (2014) Temporal decorrelation-robust SAR tomography. IEEE Trans Geosci Remote Sens 52:5412–5421
6. Aguilera E, Nannini M, Reigber A, Member S (2013) A data-adaptive compressed sensing approach to polarimetric SAR tomography of forested areas. IEEE Geosci Remote Sens Lett 10:543–547
7. Li S, Yang J, Chen W, Ma X (2016) Overview of radar imaging technique and application based on compressive sensing theory. J Electron Inf Technol 38:495–508
8. Ma P, Lin H, Lan H, Chen F (2015) On the performance of reweighted L1 minimization for tomographic SAR imaging. IEEE Geosci Remote Sens Lett 12:895–899
9. Wang Y, Zhu XX, Bamler R (2014) An efficient tomographic inversion approach for urban mapping using meter resolution SAR image stacks. IEEE Geosci Remote Sens Lett 11:1250–1254
10. Budillon A, Ferraioli G, Schirinzi G (2014) Localization performance of multiple scatterers in compressive sampling SAR tomog- raphy: results on COSMO-Skymed data. IEEE J Sel Top Appl Earth Obs Remote Sens 7:2902–2910
11. Aguilera E, Nannini M, Reigber A (2013) Wavelet-based compressed sensing for SAR tomography of forested areas. IEEE Trans Geosci Remote Sens 51:5283–5295
12. Xing SQ, Li YZ, Dai DH, Wang XS (2013) Three-dimensional reconstruction of man-made objects using polarimetric tomographic SAR. IEEE Trans Geosci Remote Sens 51:3694–3705
13. Tropp JA, Gilbert AC (2007) Signal recovery from random measurements via orthogonal matching pursuit. IEEE Trans Inf Theory 53:4655–4666
14. Xiao XZ, Adam N, Brcic R, Bamler R (2009) Space-borne high resolution SAR tomography: experiments in urban environ ment using TS-X data. J Urban Remote Sens Event 2:1–8
15. Zhu XX, Bamler R (2010) Tomographic SAR inversion by L1-norm sensing approach. IEEE Trans Geosci Remote Sens 48:3839–3846

16. Liang L, Li X, Ferro-Famil L, Guo H, Zhang L et al (2018) Urban area tomography using a sparse representation based two-dimensional spectral analysis technique. Remote Sens 10(2):109
17. Liu H, Pang L, Li F, Guo Z (2019) Hough transform and clustering for a 3-D building reconstruction with tomographic SAR point clouds. Sensors 19:5378
18. Frey O, Magnard C, Ruegg M, Meier E (2009) Focusing of airborne synthetic aperture radar data from highly nonlinear flight tracks. IEEE Trans Geosci Remote Sens 47(6):1844–1858
19. Meng M, Zhang J, Wong YD, Au PH (2016) Effect of weather conditions and weather forecast on cycling travel behavior in Singapore. Int J Sustain Transp 10(9):773–780
20. Budillon A, Crosetto M, Johnsy AC, Monserrat O, Krishnakumar V, Schirinzi G (2018) Comparison of persistent scatterer Interferometry and SAR tomography using sentinel-1 in urban environment. Remote Sens 10:1986
21. Gini F, Lombardini F, Montanari M (2002) Layover solution in multibaseline SAR interferometry. Aerospace and electronic systems. IEEE Trans Aerosp Electron Syst 38:1344–1356
22. Basca CA, Talos M, Brad R (2005) Randomized Hough transform for ellipse detection with result clustering. In: EUROCON 2005-The international conference on "computer as a tool", pp 1397–1400
23. Wang Y, Zhu X, Shi Y, Bamler R (2012) Operational TomoSAR processing using multi-track TerraSAR-X high resolution spotlight data stacks. In: Proceedings of the IEEE IGARSS, Munich,Germany
24. Zhu XX, Shahzad M (2014) Facade reconstruction using multiview spaceborne TomoSAR point clouds. IEEE Trans Geosci Remote Sens 52(6):3541–3552
25. Guo Z, Liu H, Pang L, Fang L, Dou W (2021) DBSCAN-based point cloud extraction for tomographic synthetic aperture radar (TomoSAR) three-dimensional (3D) building reconstruction. Int J Remote Sens 42(6):2327–2349
26. Bohn FJ, Huth A (2017) The importance of forest structure to biodiversity–productivity relationships. R Soc Open Sci 4:160521
27. D ˇanescu A, Albrecht AT, Bauhus J (2016) Structural diversity promotes productivity of mixed, uneven-aged forests in southwest- ern Germany. Oecologia 182:319–333
28. Toraño Caicoya A, Pardini M, Hajnsek I, Papathanassiou K (2015) Forest above-ground biomassestimation from vertical re- flectivity profiles at L-Band. IEEE Geosci Remote Sens Lett 12(12):2379–2383
29. Ho Tong Minh D, Ndikumana E, Vieilledent G, McKey D, Baghdadi N (2018) Potential value of combining ALOS PALSAR and Landsat-derived tree cover data for forest biomass retrieval in Madagascar. Remote Sens Environ 213:206–214
30. Le Toan T, Beaudoin A, Riom J, Guyoni D (1992) Relating forest biomass to SAR data. IEEE Trans Geosci Remote Sens Lett 30:403–411

Security and Privacy in IoMT-Based Digital Health care: A Survey

Ashish Singh, Riya Sinha, Komal, Adyasha Satpathy, and Kannu Priya

1 Introduction

A few decades back, there was nothing to look at or detect inside the human body because of a lack of knowledge and technology. In many cases, no one knew the cause of death of many people and the cause of the disease. People were not familiar with their bodies or which condition was inherited in their bodies. They also did not know how to overcome from the disease. But now, the scenario is different. IoMT changes the medical system. IoMT refers to the interconnection of medical devices architecture with technology. Medical sensors and wearable devices together make the IoMT. It provides better communication, remote medical assistance, management of proper medicines, tracking patients' life cycles, and many more things. The role of IoMT in human's life is people use this approach to detect different things inside the body, such as level of glucose, pulse rate, proper circulation of blood, and many more in daily life. With the help of a smart system in health care, doctors are successfully completing critical operations and saving many individuals' lives. IoMT also helps people to know and analyze their bodies. After analyzing the body, it suggests suitable yoga and exercises which keeps them fit and healthy.

In today's scenario, one-third of IoT devices are engaged in health organizations, and it is about to increase by the year 2025 [24]. Day by day, the technology of IoMT is revolutionizing. Its efficiency is also growing, and the cost is decreasing; these outcomes are far better than in the past. The data collection, transmission, and analysis of the system's raw facts and figures are speedy using IoMT tools. People can pair their devices with their smartphone applications. This makes the system keep track of the particular thing in need.

It contains different IoMT aspects from basic to advance in terms of technology and advancements. We also focus on the security system architecture, including the

A. Singh (✉) · R. Sinha · Komal · A. Satpathy · K. Priya
School of Computer Engineering, KIIT Deemed to be University, Bhubaneswar 751024, OR, India
e-mail: ashish.singhfcs@kiit.ac.in

© The Author(s), under exclusive license to Springer Nature Singapore Pte Ltd. 2023
H. Muthusamy et al. (eds.), *Robotics, Control and Computer Vision*, Lecture Notes in Electrical Engineering 1009, https://doi.org/10.1007/978-981-99-0236-1_40

device, fog, and cloud layers. Then we discussed the different communication protocols running on different IoMT protocol layers, such as link layer protocol, network layer protocols, transport layer protocols, and application layer protocols. We also discussed the requirement of security in IoMT. This survey work also covers the types of malware and mitigation techniques. The mitigation techniques include IDS, anomaly-based detection, misuse-based detection, and specification. Malware detection through blockchain is also discussed in this paper. Analysis of security attacks is done here, including eavesdropping, tag cloning, sensor tracking, etc. Different security countermeasures are explained. Applications of IoMT include fitness tracking and diagnostic, smart pills, virtual home, real-time patient monitoring, and personal emergency response system. At the end, some of the open issues and challenges are identified in this work.

The first step of the survey work is to define the research questions with different types of security attacks, security countermeasures, and applications of IoMT. The selection of accurate and concise research articles is critical in forming any research project. The research topics were addressed using "search words" or "search keyword" methodologies. Springer, Science Direct, IEEE, Elsevier, and other academic research databases return results for the search phrase. These are typical databases that cover a wide range of topics and facts that are useful. As a result, we've chosen these databases. We have provided search engines with precise search phrases. However, the results required some filtration, so the first criterion was language chosen to be English. The second criterion is to eliminate brief publications that do not adequately explain the study work. We also sought to stay away from old research publications and focus mainly on new approaches. After receiving all the required articles, we double-checked the list of all the selected works. This search procedure ensures that no crucial and relevant works are overlooked during the keyword search process. Following the discovery of relevant works, the next step is to categorize them using various criteria, including security requirements, privacy, and security aspects. The classified papers were being used to develop the sections of this paper.

The following is a list of the study's remaining sections. The existing works related to IoMT are discussed in Sect. 2. The derive security system architectural model is discussed in Sect. 3. Section 4 discusses the protocols utilized in this layered system. Section 5 discusses the security requirements for IoMT. Types of malware and mitigation techniques are discussed in Sect. 6. Security attacks and their analysis are covered in Sect. 7. In Sect. 8, security countermeasures are discussed. In Sect. 9, IoMT applications are presented, followed by challenges and open issues in Sect. 10. Finally, in Sect. 11, conclusions are discussed.

2 Literature Survey

This section discusses several previous works that are helpful to understand the IoMT phenomena from different aspects. A comparative Table 1 is included to understand all the existing works.

Table 1 A comparative analysis of previous IoMT-based works

Paper	Year	Aim of the work	Proposed approach	Advantages	Disadvantages
Joyia et al. [26]	2017	Discusses IoMT existing contributions and future scope	Developed an efficient model to reduce the threats in IoMT	Doctors and hospital workers can execute their jobs more precisely and provide better healthcare services to people	The unstructured data is handled by medical and technical expertise. Thus, chances of insider attack are very high
Aslam et al. [7]	2021	Control COVID-19 using blockchain in IOMT environment	A blockchain-based system is developed to track patients' conditions using Bluetooth-enabled cellphones	The framework provides a controlled tracking system of patient daily lives activities	The blockchain configuration and parameters have not been significantly discussed
Das et al. [12]	2019	Monitor a patient's status using wireless body area network technology (WBANT)	A system that incorporates wireless communication technology and computer science analytics to identify narcolepsy illness	Non-invasive, low cost, easy to use, and has no battery issues	Security and privacy of the patient data remain a major concern
Bibi et al. [9]	2020	An IoMT-based architecture to improve leukaemia detection	Clinical devices are linked to network resources in the proposed IoMT system using cloud computing	The technology allows patients and healthcare providers to test, diagnose, and treat leukaemia in real time	These systems should be improved in terms of correctness, learning process, and expediency
Nguyen et al. [34]	2019	Study how a mobile cloud-based IoMT system is used to track the advancement of a neurological condition	For data collection and cloud communication, an Android application is deployed	The developed application plays a critical role in offering smart and efficient medical services	The sensitive user content and data flow can be a breach in the network, leading to the loss of data

(continued)

Table 1 (continued)

Paper	Year	Aim of the work	Proposed approach	Advantages	Disadvantages
Alsubae et al. [5]	2019	Developing a web-based IoMT-SAF that helps in the selection of a solution that matches the stakeholder's security objectives and supports the decision-making process	Created a web-based IoMT Security Assessment Framework (IoMT-SAF) based on a novel ontological scenario-based approach for recommending security features in IoMT and assessing protection and deterrence in IoMT solutions	This framework can be used by solution providers to analyze and authenticate the security of their products	One of the most difficult aspects of using IoMT-SAF is the length and complexity of defining security features
Maddikunta et al. [41]	2020	Comparison of DNN with other machine learning techniques using standard intrusion detection dataset	In the IoMT context, DNN is employed to construct effective and efficient IDS	The detection accuracy of the model is good	It is not suitable for the multi-class problem
Haseeb et al. [21]	2021	Develop a machine-learning-based prediction model that predicts network resource usage and improves sensor data delivery	ML technique is used to categorize IoT nodes, and the SDN controller's configurable structure is employed for a centralized security system	It provides an unsupervised machine learning approach for IoT networks that reduce communication overheads and forecasts	Limited scalability due to the use of a single controller

(continued)

Table 1 (continued)

Paper	Year	Aim of the work	Proposed approach	Advantages	Disadvantages
Ogundokun et al. [36]	2021	Developing a CryptoStegno model to secure medical information on the IoMT environment	An amalgamated approach employing Triple Data Encryption Standard (3DES) cryptographic techniques and the steganography encoding technique Matrix XOR was deployed to safeguard medical data on the IoMT platform	User privacy, complete assurance, efficiency, and durability are all achieved using this hybrid technique	Only work on text data, not audio or video data
Doubla et al. [15]	2021	Investigate the behaviors of a two-neuron non-autonomous tabu learning model	A tabu learning two-neuron (TLTN) model with a composite hyperbolic tangent function made up of three hyperbolic tangent functions with varying offsets	Based on unpredictable sequences from the TLTN model, complicated data such as medical picture encryption is easy	It does not extract meaningful data and uses a whole chaotic sequence for encryption. Hence, taking a larger duration of time
Almogren et al. [4]	2020	Developed Fuzzy-based Trust Management System (FTM) for reducing Sybil attacks in the In FTM-IoMT	An intelligent trust management method is developed in two phases, the first phase outlines the mechanisms of processing and the second phase shows how the suggested mechanism works	It determines the trust value of a node, and then the trust traits, such as integrity, receptivity, and responsiveness, are assessed	It has high server overhead and packet delivery delay time

Vaiyapuri et al. [47] reviewed the recent developments in terms of authentication, data protection, and authorization that uses blockchain technology to share data safely in the IoMT environment. Alsubaei et al. [5] proposed an IoMT Security Assessment Framework (IoMT-SAF) that includes a unique conceptual sequence of events methodology. It is a web-based tool that proposes security mechanisms, protection, and prevention of threats in IoMT solutions. The IoMT-SAF provides security needs, granularity, flexibility, and capacity to adapt to new users. Hatzivasilis et al. [22] proposed an outline of the basic security and privacy measures that must be implemented in current IoMT environments to protect the relevant stakeholders. It provides a whole strategy that can be thought of as an ideal manual for the safe implementation of IoMT systems with the circular economy. Usman et al. [46] proposed a framework for IoMT applications that are efficient, safeguarding privacy, and data collecting and analysis (P2DCA). A fundamental wireless sensor network is partitioned into several clusters using the suggested architecture. Each cluster is responsible for protecting the privacy of individual MSNs through the accumulation of data and geographical information.

Papaioannou et al. [37] proposed a security threat in the IoMT networks based on the primary priorities and objectives that they target. Furthermore, it proposed a classification of security remedies against attacks to IoMT edge networks. Rizk et al. [40] proposed a model that focused on detecting potential security risks in the IoMT and offers security procedures for removing any potential barrier from IoMT networks. Bigini et al. [10] provide an outline of the modern blockchain-based systems for the IOMT. He also discussed possible future paths for full data control by consumers in the blockchain-based IoMT networks. Dilawar et al. [13] proposed an IoMT-based network security architecture as a method for securing the exchange of patient health records using blockchain-based technology. A blockchain-based data structure is described as a series of essentially untouchable cryptography linked blocks that can be used to hold important patient data. A decentralized blockchain-based technique would address many issues connected with the centralized cloud strategy.

Bharati et al. [8] proposed a framework for IoMT named IoT healthcare network (IoThNet), which demonstrates how hospitals at the access layer may gather user data at the information persistence layer. It presents a cloud-based IoMT framework and compares it to current frameworks in the literature. Karmakar et al. [27] proposed a security architecture for smart healthcare network infrastructures. Numerous security mechanisms or applications are designed and implemented as virtualized network functionalities in the architecture. Puat et al. [38] examine many IoMT gadgets, including implantable cardiac devices, smart pens, wireless vital monitors, and others, in terms of their working methodology and risks that could expose them to an attacker. The cardiac device, one of the IoMT devices, has been discussed further. Several security approaches and remedies are also being developed to reduce the flaws of IoMT devices. Alsubaei et al. [6] proposed a categorization of IoMT security and privacy (S&P) concerns. It also demonstrates how to analyze hazards in two IoMT devices and offers a method for quantifying IoMT risks. Its goal is to raise S&P consciousness among IoMT participants by allowing them to detect and estimate possible S&P hazards in the IoMT. Allouzi et al. [3] define a security plan for the IoMT network. Any flaws or defects in the IoMT network that could allow unauthorized users to get access and threats that could exploit these flaws are also discussed. Using the Markov transition probability matrix, the probability distribution of IoMT threats is derived. Priya et al. [41] proposed a Deep Neural Network (DNN) framework to create efficient IDS that categorizes and anticipates unexpected cyberattacks in the IoMT environment. A detailed analysis of trials in DNN with some other machine learning techniques is compared using the standard intrusion detection dataset. The Internet of Medical Sensor Data, IDS, and Intruders are the three primary components of the developed framework.

3 Security System Architecture

IoMT is an IoT-based solution that enables the construction of IoMT-enabled healthcare systems for checking vital signs such as ECG, heart rate, and blood pressure. IoMT-enabled healthcare systems aim to improve patients' quality of life by reducing the likelihood of an unpleasant hospitalization. Allowing patients to wander throughout the medical and non-medical environments while maintaining constant monitoring of their vital signs and health condition is a critical aspect of high-quality medical care [37].

Figure 1 depicts the security system architecture, which is composed of three layers—device layer, fog layer, and cloud layer.

Device layer—The lowest layer is made up of a variety of physical devices, such as implantable or wearable medical sensors that are incorporated into a small wireless module to collect contextual and medical data. Bio-medical and context signals are acquired from the body, room, sensing, and communication capability. The signals are utilized to manage medical problems' therapy and diagnosis. The signal is subsequently sent to the top layer (smart gateways in the fog layer) via wireless or wired communication protocols like IEEE 802.15.4, Bluetooth LE, Wi-Fi, etc.

Fog layer—A network of interconnected smart gates makes up the middle layer. Using the cloud computing paradigm instead of constructing and maintaining private servers and data centers is cost-effective. It increases the productivity and flexibility of online applications. Mutual authentication also takes place in this layer. With this architecture, developers and end-users can use cloud services while knowing little about the underlying hardware and infrastructure. The fog computing paradigm addresses the latency problems by extending cloud services to the network's edge.

Cloud layer—Broadcasting, data warehousing, and big data analysis servers are the parts of the cloud layer. The local hospital database is synchronized data with a remote healthcare database server in the cloud regularly. Identification, authentication, and encryption of data take place in this layer [31].

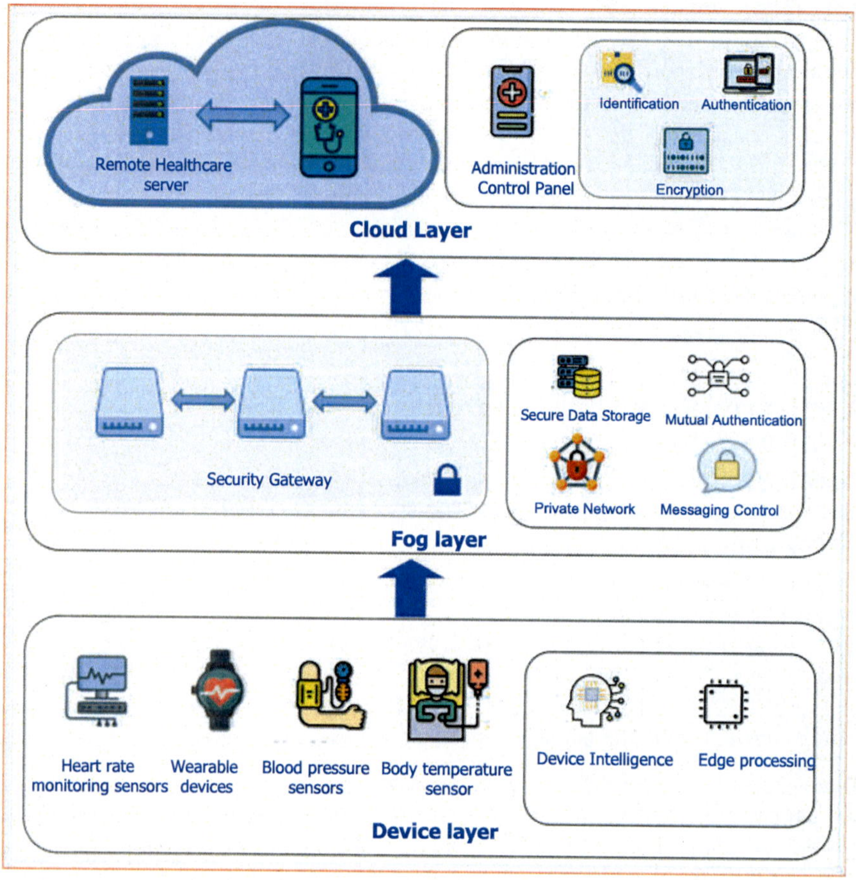

Fig. 1 Security system architecture of IoMT

4 IoMT Communication Protocol

In IoMT, many devices are linked together in a network. The communication in this network between physical objects takes place through protocols and standards. So, it is very important to use the correct protocol to make the communication secure and reliable. IoMT protocols are used in various network layers to facilitate data exchange between devices, devices to the cloud, and other interactions. This section will discuss the protocols used in various layers of IoMT. The summary of IoMT communication protocols that are running in different layers is summarized in Table 2.

- Link Layer Protocol: This layer determines how the data is physically sent over a medium. This layer uses Z-Wave, Wi-Fi, BLE, ZigBee, and NFC protocols.

Table 2 A summary of IoMT communication protocols

IoMT Communication Layers	Protocols	Remarks
Application layer	CoAP	Customized web transfer protocol for constrained devices having limited bandwidth and availability
	MQTT	Lightweight M2M chat service that uses publish-subscribe architecture. Small code footprint, low power consumption, low bandwidth, and low latency are the features of this protocol
	AMQP	Transmitting transnational messages across servers. It is quick, easy-to-use, multichannel operational protocol
	XMPP	Message-oriented middleware and application protocol based on XML intended for nearly instantaneous messaging
Transport layer	TCP	Offers end-to-end data transmission. Allows application software and computing devices to communicate over a network by establishing a reliable connection. It is a reliable communication standard that is connection-oriented, full-duplex, and stream-oriented
	UDP	Used to construct low-latency, loss-tolerant connections, and low-power applications with resource-constrained devices. Provides a simple connectionless transmission mechanism with no handshaking dialogues
Network layer	RPL	A low-power IPv6 remote vector protocol that runs on the IEEE 802.15.4 standard and includes support for the 6LoWPAN adaptation layer
	CARP	It is suitable for lightweight packets making IoT
	Ipv6	Uses 128-bit addresses and can accommodate almost 340 undecillion unique IP addresses. Auto-configuration, integrated security, complex network, and a range of new mobility capabilities are also supported by IPv6
	6loWPAN	Transporting packet data across other networks with standard IEEE 802.15.4. Compatibility with all IPv6 protocols, mesh-routing capabilities, reduced power consumption, limited footprint, and end-to-end security
Link layer	Z-Wave	Inter-operable and uses mesh topology, allowing two or more devices to connect simultaneously in the network
	Wi-Fi	Based on IEEE 802.11 standards. It is a LAN and covers a range of around 100 m. It provides high-speed data transfer with a data rate of 100–150 Mbps
	BLE	Wireless protocol based on Bluetooth v.4.0. It provides a range of up to 100 m and a data rate of up to 200 Mbps
	ZigBee	Provides low-power and low-cost wireless communications in IoMT based on IEEE 802.15.4 standards. It provides data rates up to 250 kbps and covers a range of up to 50 m indoors with a wide range of network topologies
	NFC	Very short spectrum wireless protocol. It enables simple and secure communication between devices or things. It provides a data rate of up to 400 kbps and range around 5 cm

1. *Z-Wave*: It is a wireless radio frequency protocol. Z-Wave protocol is interoperable and uses mesh topology, allowing two or more devices to connect simultaneously in the network. It has mainly been used for monitoring and controlling IoT devices.
2. *Wi-Fi*: Wireless-Fidelity (Wi-Fi) is the most common protocol used daily. It is based on IEEE 802.11 standards. It is a Local Area Network (LAN) and covers a range of around 100 m. It provides high-speed data transfer with a 100–150 Mbps data rate.
3. *BLE*: Bluetooth Low Energy (BLE) is a wireless protocol based on Bluetooth v.4.0. It reduces power consumption by ten times compared to classic Bluetooth while increasing the latency by 15 times. It provides a range of up to 100 m and a data rate of up to 200 Mbps.
4. *ZigBee*: ZigBee provides low-power and low-cost wireless communications in IoT based on IEEE 802.15.4 standards. It provides data rates up to 250 kbps and covers a range of up to 50 m indoors with a wide range of network topologies, including mesh.
5. *NFC*: Near-Field Communication (NFC) is very short spectrum wireless protocol [39]. It enables simple and secure communication between devices or things. It provides a data rate of up to 400 kbps and range around 5 cm.

– Network Layer Protocol: This layer is responsible for sending datagrams from source to destination. It performs data encapsulation, forwarding, and routing. This layer uses protocols like RPL, CARP for routing, IPv6, and 6LoWPAN for encapsulation.

1. *RPL*: Routing protocol Low-Power and Lossy Networks (RPL) is a low-power IPv6 remote vector protocol that runs on the IEEE 802.15.4 standard and includes support for the 6LoWPAN adaptation layer [17]. The protocol feature consists of the efficiency of RPL's hierarchy, timers to reduce control messages, and the flexibility of the goal function.
2. *CARP*: Channel-Aware Routing Protocol (CARP) is a network layer protocol that works in a distributed manner. It has lightweight packets, making it suitable for usage in the IoT. It carries out two distinct functions: network initialization and data transmission. It can successfully route past connectivity voids and avoid loops. Simple topology information such as hop count can be used to create shadow zones. It's also built to take advantage of power control while choosing reliable links.
3. *IPv6*: IPv6 and IPv4 can coexist without causing substantial interruption. The IPv6 system uses 128-bit addresses and can accommodate almost 340 undecillion unique IP addresses. Auto-configuration, integrated security, complex network, and a range of new mobility capabilities are also supported by IPv6. IPv6 Gateways can be fully compliant with the Internet.
4. *6LoWPAN*: IPv6-Low-Power Wireless Personal Area Network (IPv6-LoWPAN) is a protocol for transporting packet data across other networks with standard IEEE 802.15.4 [2]. Compatibility with all IPv6 protocols, mesh-routing capa-

bilities, reduced power consumption, limited footprint, and end-to-end security are the features of 6LoWPAN's wireless modules.

- Transport Layer Protocol: This layer is responsible for point-to-point communication. The transport layer protocol is critical for how one process communicates with another process. It includes two important protocols: TCP and UDP.

1. *TCP*: Transmission Control Protocol (TCP) is a fundamental Internet standard defined by the Internet Engineering Task Force's specifications (IETF). It is one of the most commonly used protocols in computerized communication networks since it offers end-to-end data transmission. It is a transport layer communication protocol that allows application software and computing devices to communicate over a network by establishing a reliable connection. It is a reliable communication standard that is connection-oriented, full-duplex, and stream-oriented.
2. *UDP*: User Datagram Protocol (UDP) is defined by RFC 768. It's mostly used to construct low-latency, loss-tolerant connections, and low-power applications with resource-constrained devices. UDP provides a simple connectionless transmission mechanism without handshaking dialogues and exposes the user's unreliability.

- Application Layer Protocol: This layer determines how the application interface communicates with lower layer protocol to send data over the network. This layer includes protocols like CoAP, MQTT, AMQP, and XMPP.

1. *CoAP*: Constrained Application Protocol (CoAP) is a customized web transfer protocol developed by the IETF group Constrained Resource Environments (CoRE) [25]. CoAP is a protocol for connecting basic, confined devices to the IoT even across constrained devices having limited bandwidth and availability. It is commonly utilized in machine-to-machine (M2M) applications. It is built on a client/server architecture in which low-overhead and low-latency applications work on a request-response basis.
2. *MQTT*: Message Queue Telemetry Transport (MQTT) is a lightweight M2M chat service that uses publish-subscribe architecture. It was created by IBM and has since become an open standard. It also supports bi-directional messaging among devices and the cloud and can scale to millions of linked devices. Small code footprint, low power consumption, low bandwidth use, and low latency are some of its features. It has three levels of Quality of Service (quality of service) to ensure reliable and consistent message delivery [44].
3. *AMQP*: Advanced Message Queuing Protocol (AMQP) is an open platform application layer protocol for transmitting transactional messages across servers. It can handle thousands of reliable scheduled transactions as a message-centric middleware [30]. Patron programs can communicate with the dealer and interact with the AMQP model via the AMQP protocol. It supports several assured messaging modes, including at-most-once, at-least-once, and exactly-once. It is a quick, easy-to-use, multichannel, and reliable application layer protocol.

4. *XMPP*: Extensible Messaging and Presence Protocol (XMPP), formerly named Jabber, is a message-oriented middleware and application protocol based on Extensible Markup Language (XML) intended for nearly instantaneous messaging and presence data. It allows for the discovery of services located locally or across a network and determining service availability. XMPP is meant to be flexible, and it has been used in embedded IoT networks for publish-subscribe systems, file sharing, and communication.

5 Security Requirements in IoMT

IoT networks enable various new services and business models for users and service providers by increasing connectivity across all markets and sectors. The better connection enables more accurate healthcare services, and faster workflows enhance operational productivity for healthcare organizations [18]. A set of security requirements is required to assure the security of IoMT sensitivity.

1. Confidentiality/Privacy: It is the capability to keep data private while collecting, transmitting, or storing it. They must also be available only to authorized users. Data encryption and access control lists are the most prevalent ways to meet this need.
2. Integrity: This refers to the ability to safeguard data during the accumulation, dissemination, and repository stages from unauthorized tampering.
3. Availability: The capability to maintain the IoMT networks operational at all times. It can be accomplished by keeping the system updated, scrutinizing any changes in performance, providing unnecessary data storage or transmission methods in the event of DoS assaults, and quickly resolving any issues.
4. Nonrepudiation: The capacity to hold each authorized user accountable for the acts they take. In other words, this criteria ensures that no system contact can be refused. Digital signature techniques can be used to accomplish this security requirement.
5. Authentication: The capacity to verify a user's identity before allowing them access to the system. Mutual authentication is the most protected form of verification since it requires both ends of the communication process to verify before any secured communication exchange occurs.

6 Types of Malware and Mitigation Techniques

Malware refers to any malicious software intended to hurt or exploit any programmable thing, application, or network. Cybercriminals often utilize it to retrieve information that they may exploit to gain financial advantage. The following malware is discussed here [48].

1. Types of Malwares:

   - *Spyware*: Spyware is a type of spyware that monitors user behavior without their permission. Such malicious actions as keylogging, activity tracking, data harvesting, account passwords, and financial data monitoring are examples of spyware. It may potentially change the software's security settings. It takes advantage of software flaws and attaches itself to a usual computer running program.
   - *Keylogger*: It is a malicious piece of code that allows a hacker to track the user's keystrokes. A keylogger [42] attack is more effective than a brute-force or dictionary-based attack. This dangerous program tries to gain access to a user's device by convincing them to download it by clicking on a link in an email. It is one of the most deadly malwares because even a strong password isn't enough to protect the system.
   - *Trojan Horse*: This malware poses as a legitimate computer program to deceive people into downloading and installing it. It enables a hacker to gain remote access to an infected system with permission. Once a hacker has access to an infected system, they can steal sensitive information. It can also install other malicious programs in the system and carry out additional destructive acts.
   - *Virus*: This harmful application can replicate itself and propagate to other computers. It infects other computers by attaching itself to different programs, and when a user runs a legitimate code, the attached infected program is also run. It can be used to steal data, cause damage to the host system, and create botnets.
   - *Worm*: It spreads across a network by exploiting flaws in the operating system. It harms their host networks by consuming too much bandwidth and overwhelming web servers. It generally contains a payload designed to harm a host system. Hackers frequently use this to steal important information, erase files, or build a botnet. In nature, worms self-replicate and spread independently, whereas viruses require human intervention to spread. Corrupted attachments transmit worms in emails.

2. Mitigation Techniques: The first step in reducing risk is to recognize the potential risk. It includes addressing main risks regularly to guarantee that your system is completely safeguarded.

   (a) Intrusion Detection System: An IDS is a part of the software that monitors and analyzes harmful activity within a network or system. It detects and protects a variety of devices (such as smart medical equipment) against potential threats and attacks [29]. The IoMT context includes the deployed IDS monitors and verifies all traffic (both usual and malicious) and looks for harmful indicators. The linked IDS component takes the appropriate action to detect any harmful behavior.

   An IDS technique can be classified into three types: anomaly-based detection, misuse-based detection, and specification-based detection. The following is a summary of these mechanisms.

i. *Anomaly-based detection*: This detection system relies on the statistical behavior of the traffic. It attempts to distinguish between usual network flow and flow under attack. It will sound an alarm if it detects any deviation from typical behavior. It has a disadvantage that we must regularly update the usual behavior database to highlight the network changes in the database.
ii. *Misuse-based detection*: It is also known as rule-based detection or signature-based detection. A signature is formed when an absurdity (such as a virus) impacts the system. The signatures of previously detected assaults are utilized to detect similar attacks in the future. The advantages of this method are that it can correctly and efficiently detect known anomalies while also having a low false-positive rate. The majority of antivirus (or anti-malware) programs in use today fall into misuse-based detection.
iii. *Specification-based detection*: This technique involves the definition of requirements and constraints to characterize the validity of the detection process. The behavior of the system or network is then monitored and analyzed following the specifications and restrictions. It also can identify unknown attacks. It uses the benefits of both anomaly and misuse-based detection systems to diagnose anomalous behavior. This method appears to act as an anomaly-based detection mechanism because it detects attacks based on deviations from typical behavior.

(b) Malware detection through blockchain: The blockchain's operations can be utilized to protect a variety of communication contexts because blockchain activities are decentralized, efficient, and transparent [16]. In the IoMT environment, blockchain processes may be used to detect malware efficiently. We can add a block containing information about harmful programs (i.e., malware) in the blockchain. We can develop such type of detection mechanism because the blockchain is accessible to all authorized parties. These parties can learn about the current malware attacks on the system. As a result, malware detection may be done efficiently.

7 Security Attacks and Analysis

The medical devices connected to IoMT-based networks over wireless networks have a huge collection of patient information, test reports, medication lists, and chronic health conditions. This leads to security breaches that adopt various malicious activities to access this information or steal them for their bad intentions [37].

1. *Eavesdropping Attack:* It's a type of attack that uses unprotected network connections to interfere with the communication of two entities without their knowledge or agreement. It usually occurs when a user connects to a network where the

traffic is not secured or encrypted. This form of attack is harmful as it is difficult to detect because it does not disrupt network communication [37].

2. *Tag Cloning*: This is a type of attack where tags clone data to gain access to sensitive information and closed areas [1]. The attacker can duplicate data acquired from a side-channel attack and can use it to access the sensitive information [14]. The cloning of a tag can be done by determining the signal that the tag transmits and building a device that mimics that signal.
3. *Sensor Tracking*: Sensors are devices widely used in electronic-based medical equipment to convert stimuli into electrical signals for the health analysis of patients. These devices include GPS sensors, fall detection, or wheelchair. The sensors attached to this medical equipment will send the patient location to the monitoring system or the doctor in an emergency. The hackers or attackers can invade this sensitive data, access the patient location, and even send inaccurate data to hinder patient health monitoring.
4. *Man-in-the-middle attack*: This attack occurs when an attacker intrudes in the communication between two authenticated entities during signal transmission [37]. The hacker intercepts and manipulates the information between the sender and receiver. The attacker can alter the communication data, leading to mistreatments such as medicinal overdosage or false reports.
5. *Denial of service (DoS)*: This is an attack where the attackers jam the system with noise interference and block radio signals. These attacks flood the device with the huge number of legitimate service requests but are sent by the attackers. This attack consumes network requests to disrupt the services [35]. The attackers can hack the IoMT devices in the botnet to infect devices with the owner's knowledge.
6. *Message Replay*: The message in the Radio-frequency identification (RFID) system is recorded and then replayed. The original message is resent later to the recipient device, which confuses the devices involved in the IoMT system. The attacker does this intending to steal information or gain access to the IoMT device [33].
7. *Malware Attack*: This attack can be stated as dangerous as it may destroy the health records or important information related to the patient in IoMT devices. This type of attack occurs when the attackers install malicious software or firmware without the user's information to harm or destroy the data and run destructive or intrusive programs. The different types of malware are viruses, spyware, ransomware, Trojan horses, and worms [37].
8. *Side-Channel Attacks*: By monitoring the electromagnetic activity near specific medical devices, attackers might utilize side-channel methods to steal and access patient records in the healthcare systems [1]. In this attack, an attacker intercepts communication between tags and a reader to extract data from various patterns using a ready-made tool. This occurs when the devices are not using wireless protocols to transfer the data.
9. *Cross-Site Scripting (XSS) attack*: These attacks are performed in IoMT apps by injecting specially crafted malicious scripts into web pages to evade access controls. The malicious scripts have accessibility to the browser's cookies, security tokens, and other essential and confidential.

10. *Impersonation Attack*: In this attack, a malicious person poses a genuine party in an authentication protocol to obtain access to resources or confidential material that they are not allowed to access [37].

8 Security Counter Measures

1. *Ensuring Confidentiality*: Confidentiality could be an important security concern when it involves IoMT. The gathering and maintenance of a clinical record must follow legal and ethical privacy guidelines stated by different organizations where only authorized persons have access to those data. It's important to secure the information associated with patient health, in addition to ensuring confidentiality [45]. Several lightweight cryptographic techniques such as symmetric-key cyphers and hash functions are available that may be used to create a secure communication between IoMT devices. Shared keys should be created to maintain the confidentiality [37].
2. *Providing Integrity*: To ensure the integrity of data transmission within IoMT devices, symmetric cryptography and attribute-based encryption (ABE) are used. The delivered messages are usually encrypted and utilize an ABE-encrypted random symmetric key (RSK). The user's privileges are represented by the secret key associated with the device set. In this scenario, properly modifying the system settings allows you to encrypt the received RSK rather than the entire message, increasing communication, and reducing encrypting costs [37].
3. *Ensuring availability*: The availability of networked medical devices should be ensured in an IoMT network due to the criticality of the data. IoMT devices have resource and processing power limits. Several research studies on jamming attacks have concentrated on centralized systems and solutions. Defending reactive jammers using a trigger identification service has also been added [32]. This approach identifies and distinguishes nodes whose transmitting patterns are identical to the jamming nodes [37]. The distributed strength of crowd (SOC) protocol may be suitable for IoMT devices with limited resources. However, a considerable percentage of the available bandwidth may be blocked. This protocol ensures message delivery to receiving nodes [43].
4. *Ensuring Authentication*: User authentication techniques are essential to access the IoMT data or device. The device authentication mechanism should communicate in a secure/encrypted manner for data confidentiality and integrity. The mutual authentication technique is a secure solution for authentication between two communicating parties.

Table 3 A summary of IoMT applications

Application	Description
Fitness tracking	Patients use fitness trackers to keep track of their progress which is especially useful during rehabilitation and recovery. Activity trackers [23], bracelets, smart wristbands [20], sports watches, and smart clothing [11] are some of the most prominent technologies for personal wellness or fitness
Smart pills	Swallowable wireless sensors and cameras are included in Smart Pills [19]. A new generation of tech-enabled pills aims to track adherence among regularly prescribed medications. Experimental sensors are also being developed to detect a disease from the inside
Virtual home wards	A virtual home system is the most important part of getting the correct therapy for homebound patients and elders in chronic conditions. It uses telemedicine apps to allow patients and doctors to speak with one another and provide long-term care from afar
Real-time patient monitoring (RPM)	RPM incorporates all home monitoring sensors and devices used for chronic disease treatment, constant monitoring of physiological signals to support long-term care in a patient's own house, and prevent re-hospitalization. This is especially useful when patients require regular monitoring using a cost-effective method
Personal emergency response systems (PERS)	PERS is concerned with emergency response times and reaching patients in a reasonable timeframe. It combines wearable device/relay units with a live medical call center service to help homebound or limited-mobility elderly become more self-reliant

9 Applications of IoMT

The IoT medical field is rapidly evolving with new developments and applications. Radical solutions are being deployed to address holistic healthcare concerns, ranging from smart monitors to patient diagnostic devices. Increased accuracy, enhanced efficiency, and lower costs are benefits of adopting IoMT into regular healthcare procedures. Table 3 gives brief information about some of the key applications of IoMT.

10 Challenges and Open Issues

This section discusses some of the open issues and challenges in the IoMT environment that is still unsolved [28]:

- *Security Concerns*: IoMT devices rely on open wireless connections. Thus, they are vulnerable to a variety of wireless/network attacks. In fact, due to a lack of security protections and security verification mechanisms, numerous IoMT devices are readily circumvented by a trained intruder. An intruder can gain access to incoming and outgoing data and information. As a result, security risks like unauthorized access can arise.

- *Privacy Issues*: Passive attacks such as traffic analysis raise privacy concerns. The majority of these attacks resulted in the intrusion of patients' privacy through data leakage, which leads to exposure of sensitive data. In this issue, the attacker can obtain and publish information about patients' identities and sensitive and secret patient data. This might create a person's medical problems, damage the patient image in the social environment, or pose a significant threat to patients.
- *Trust Concerns*: The trust of IoMT devices is another issue because device breaches may leak the patient's personal sensitive information. It might also endanger their lives and social image because hackers will access their confidential medical information.
- *Accuracy concerns*: The accuracy of IoMT devices is another concern caused by the device's malfunction. A report says more than 8061 malfunctions were reported from 2001 to 2013. These attacks lack precision and accuracy in medical robot-assisted surgeries, patient misdiagnosis, and incorrect medical prescriptions.
- *Standardization of IoT devices*: The absence of standardization of IoT devices is a vital issue. The medical devices were incorporated into IoT systems. There is a need of a standard communication protocol that will communicate in different networks or platforms. Standardization is necessary for numerous medical equipments and devices to work together. It also required manufacturers to implement the appropriate security measures to safeguard them from being attacked by hackers.

11 Conclusion and Future Scope

This paper discussed an architectural model of IoMT in terms of security and privacy. From the literature, we have identified that security and privacy are significant problems that limit IoMT usage at the consumer level, so a discussion about security system architecture is essential. The work includes different communication protocols based on the IoMT protocol stack. Security requirements, types of malware and mitigation techniques, security attacks and analysis, countermeasures, and application are important points covered in this survey work. Based on the discussed aspect, problems and open issues in the IoMT field are presented, which will assist researchers and practitioners in developing new applications securely.

Apart from this, this article has a limited number of security solutions and applications. We need to discuss application-specific security attacks and its prevention in IoMT in health care in the future. This also needed to be elaborated in the future.

References

1. Abdul-Ghani HA, Konstantas D (2019) A comprehensive study of security and privacy guidelines, threats, and countermeasures: an IoT perspective. J Sens Actuator Netw 8(2):22
2. Al-Kashoash HA, Kemp AH (2016) Comparison of 6lowpan and lpwan for the internet of things. Australian J Electr Electron Eng 13(4):268–274
3. Allouzi MA, Khan JI (2021) Identifying and modeling security threats for IoMT edge network using markov chain and common vulnerability scoring system (CVSS). arXiv:2104.11580
4. Almogren A, Mohiuddin I, Din IU, Almajed H, Guizani N (2020) FTM-IoMT: Fuzzy-based trust management for preventing sybil attacks in internet of medical things. IEEE Int Things J 8(6):4485–4497
5. Alsubaei F, Abuhussein A, Shandilya V, Shiva S (2019) IoMT-SAF: internet of medical things security assessment framework. Int Things 8:100123
6. Alsubaei F, Abuhussein A, Shiva S (2017) Security and privacy in the internet of medical things: taxonomy and risk assessment. In: 2017 IEEE 42nd conference on local computer networks workshops (LCN Workshops), pp 112–120. https://doi.org/10.1109/LCN.Workshops.2017.72
7. Aslam B, Javed AR, Chakraborty C, Nebhen J, Raqib S, Rizwan M (2021) Blockchain and ANFIS empowered IoMT application for privacy preserved contact tracing in covid-19 pandemic. Pers Ubiquitous Comput 1–17
8. Bharati S, Podder P, Mondal MRH, Paul PK (2021) Applications and challenges of cloud integrated IoMT. In: Cognitive internet of medical things for smart healthcare. Springer, pp 67–85
9. Bibi N, Sikandar M, Ud Din I, Almogren A, Ali S (2020) IoMT-based automated detection and classification of leukemia using deep learning. J Healthc Eng 2020
10. Bigini G, Freschi V, Lattanzi E (2020) A review on blockchain for the internet of medical things: definitions, challenges, applications, and vision. Futur Int 12(12):208
11. Chen M, Ma Y, Song J, Lai CF, Hu B (2016) Smart clothing: connecting human with clouds and big data for sustainable health monitoring. Mob Netw Appl 21(5):825–845
12. Das PK, Zhu F, Chen S, Luo C, Ranjan P, Xiong G (2019) Smart medical healthcare of internet of medical things (IoMT): application of non-contact sensing. In: 2019 14th IEEE conference on industrial electronics and applications (ICIEA). IEEE, pp 375–380
13. Dilawar N, Rizwan M, Ahmad F, Akram S (2019) Blockchain: securing internet of medical things (IoMT). Int J Adv Comput Sci Appl 10(1):82–89
14. Ding ZH, Li JT, Feng B (2008) A taxonomy model of RFID security threats. In: 2008 11th IEEE international conference on communication technology. IEEE, pp 765–768
15. Doubla IS, Njitacke ZT, Ekonde S, Tsafack N, Nkapkop J, Kengne J (2021) Multistability and circuit implementation of tabu learning two-neuron model: application to secure biomedical images in IoMT. Neural Comput Appl 1–29
16. Fuji R, Usuzaki S, Aburada K, Yamaba H, Katayama T, Park M, Shiratori N, Okazaki N (2019) Blockchain-based malware detection method using shared signatures of suspected malware files. In: International conference on network-based information systems. Springer, pp 305–316
17. Gaddour O, Koubâa A (2012) RPL in a nutshell: a survey. Comput Netw 56(14):3163–3178
18. Ghubaish A, Salman T, Zolanvari M, Unal D, Al-Ali AK, Jain R (2020) Recent advances in the internet of medical things (IoMT) systems security. IEEE Int Things J
19. Goffredo R, Accoto D, Guglielmelli E (2015) Swallowable smart pills for local drug delivery: present status and future perspectives. Expert Rev Med Devices 12(5):585–599
20. Grym K, Niela-Vilén H, Ekholm E, Hamari L, Azimi I, Rahmani A, Liljeberg P, Löyttyniemi E, Axelin A (2019) Feasibility of smart wristbands for continuous monitoring during pregnancy and one month after birth. BMC Pregnancy Childbirth 19(1):1–9
21. Haseeb K, Ahmad I, Awan II, Lloret J, Bosch I (2021) A machine learning SDN-enabled big data model for IoMT systems. Electronics 10(18):2228

22. Hatzivasilis G, Soultatos O, Ioannidis S, Verikoukis C, Demetriou G, Tsatsoulis C (2019) Review of security and privacy for the internet of medical things (IoMT). In: 2019 15th international conference on distributed computing in sensor systems (DCOSS). IEEE, pp 457–464
23. Henriksen A, Mikalsen MH, Woldaregay AZ, Muzny M, Hartvigsen G, Hopstock LA, Grimsgaard S (2018) Using fitness trackers and smartwatches to measure physical activity in research: analysis of consumer wrist-worn wearables. J Med Internet Res 20(3):e9157
24. Intel A (2017) Guide to the internet of things infographic. http://wwwintel.com/content/dam/www/public/us/en/images/iot/guide-to-iot-infographic.png. Accessed 11 Jan 2016
25. Jan SR, Khan F, Ullah F, Azim N, Tahir M (2016) Using CoAP protocol for resource observation in IoT. Int J Emerg Technol Comput Sci Electron ISSN 0976:1353
26. Joyia GJ, Liaqat RM, Farooq A, Rehman S (2017) Internet of medical things (IoMT): applications, benefits and future challenges in healthcare domain. J Commun 12(4):240–247
27. Karmakar KK, Varadharajan V, Tupakula U, Nepal S, Thapa C (2020) Towards a security enhanced virtualised network infrastructure for internet of medical things (IoMT). In: 2020 6th IEEE conference on network Softwarization (NetSoft). IEEE, pp 257–261
28. Kumar S, Arora AK, Gupta P, Saini BS (2021) A review of applications, security and challenges of internet of medical things. Cogn Internet Med Things Smart Healthc 1–23
29. Liao HJ, Lin CHR, Lin YC, Tung KY (2013) Intrusion detection system: a comprehensive review. J Netw Comput Appl 36(1):16–24
30. McAteer IN, Malik MI, Baig Z, Hannay P (2017) Security vulnerabilities and cyber threat analysis of the AMQP protocol for the internet of things
31. Moosavi SR, Gia TN, Nigussie E, Rahmani AM, Virtanen S, Tenhunen H, Isoaho J (2016) End-to-end security scheme for mobility enabled healthcare internet of things. Futur Gener Comput Syst 64:108–124
32. Mutlag AA, Abd Ghani MK, Arunkumar NA, Mohammed MA, Mohd O (2019) Enabling technologies for fog computing in healthcare IoT systems. Futur Gen Comput Syst 90:62–78
33. Nawir M, Amir A, Yaakob N, Lynn OB (2016) Internet of things (IoT): Taxonomy of security attacks. In: 2016 3rd international conference on electronic design (ICED). IEEE, pp 321–326
34. Nguyen DC, Nguyen KD, Pathirana PN (2019) A mobile cloud based IoMT framework for automated health assessment and management. In: 2019 41st annual international conference of the IEEE engineering in medicine and biology society (EMBC). IEEE, pp 6517–6520
35. Ni J, Zhang K, Lin X, Shen X (2017) Securing fog computing for internet of things applications: challenges and solutions. IEEE Commun Surv Tutor 20(1):601–628
36. Ogundokun RO, Awotunde JB, Adeniyi EA, Ayo FE (2021) Crypto-stegno based model for securing medical information on IoMT platform. Multimed Tools Appl 80(21):31705–31727
37. Papaioannou M, Karageorgou M, Mantas G, Sucasas V, Essop I, Rodriguez J, Lymberopoulos D (2020) A survey on security threats and countermeasures in internet of medical things (IoMT). Trans Emerg Telecommun Technol e4049
38. Puat HAM, Abd Rahman NA (2020) IoMT: a review of pacemaker vulnerabilities and security strategy. J Phys: Conf Ser 1712:012009. IOP Publishing
39. Pulipati M, Phani S (2013) Comparison of various short range wireless communication technologies with NFC. Int J Sci Res 2:87–91
40. Rizk D, Rizk R, Hsu S (2019) Applied layered-security model to IoMT. In: 2019 IEEE international conference on intelligence and security informatics (ISI). IEEE, pp 227
41. RM SP, Maddikunta PKR, Parimala M, Koppu S, Gadekallu TR, Chowdhary CL, Alazab M (2020) An effective feature engineering for DNN using hybrid PCA-GWO for intrusion detection in IoMT architecture. Comput Commun 160:139–149
42. Royo ÁA, Rubio MS, Fuertes W, Cuervo MC, Estrada CA, Toulkeridis T (2021) Malware security evasion techniques: an original keylogger implementation. In: WorldCIST (1), pp 375–384
43. Sciancalepore S, Oligeri G, Di Pietro R (2018) Strength of crowd (SOC)-defeating a reactive jammer in IoT with decoy messages. Sensors 18(10):3492
44. Soni D, Makwana A (2017) A survey on MQTT: a protocol of internet of things (IoT). In: International conference on telecommunication, power analysis and computing techniques (ICTPACT-2017), vol 20

45. Sun Y, Lo FPW, Lo B (2019) Security and privacy for the internet of medical things enabled healthcare systems: a survey. IEEE Access 7:183339–183355
46. Usman M, Jan MA, He X, Chen J (2019) P2dca: a privacy-preserving-based data collection and analysis framework for IoMT applications. IEEE J Sel Areas Commun 37(6):1222–1230
47. Vaiyapuri T, Binbusayyis A, Varadarajan V (2021) Security, privacy and trust in IoMT enabled smart healthcare system: a systematic review of current and future trends. Int J Adv Comput Sci Appl 12:731–737
48. Wazid M, Das AK, Rodrigues JJ, Shetty S, Park Y (2019) IoMT malware detection approaches: analysis and research challenges. IEEE Access 7:182459–182476

5G Technology-Enabled IoT System for Early Detection and Prevention of Contagious Diseases

Amit Saxena, Kshitij Shinghal, Rajul Misra, and Amit Sharma

1 Introduction

The outbreak of the COVID-19 virus has conveyed a message that communities, countries and civilizations are evolving and transforming due to the disease. The faster means of transport quickly convert a disease into an epidemic and then into a pandemic. Table 1 shows the global health pandemic timeline.

Table 1 clearly indicates that from time to time there has been an outbreak of a virus, and this is the right time that society should get ready for the next outbreak. An IoT-based system for early detection and prevention of the spread of contagious disease is the need of the hour. The proposed system will employ 5G wireless technologies for communication with cloud computation and storage. Figure 1 shows the death toll due to various pandemics over a century. Figure 2 shows the evaluation of wireless technologies.

The Indian Government has already started implementing 5G networks. The bands identified for 5G technology are 700 MHz, 3.5 GHz and 26/28 GHz. Table 2 gives the year-wise details of various wireless technology.

The inherent advantages of 5G technology-based IoT network over a 4G LTE-based IoT network are shown in Fig. 3. The proposed IoT-based system will be based on the latest 5G technology to take inherent advantages of 5G technology and harness higher data rates for better processing as shown in Table 2 and Fig. 3.

A. Saxena · K. Shinghal (✉)
Department of Electronics and Communication Engineering, Moradabad Institute of Technology, Moradabad, U.P, India
e-mail: kshinghal@gmail.com

R. Misra
Department of Electrical Engineering, Moradabad Institute of Technology, Moradabad, UP, India

A. Sharma
Department of Electronics and Communication Engineering, Teerthanker Mahaveer University, Moradabad, UP, India

© The Author(s), under exclusive license to Springer Nature Singapore Pte Ltd. 2023

H. Muthusamy et al. (eds.), *Robotics, Control and Computer Vision*, Lecture Notes in Electrical Engineering 1009, https://doi.org/10.1007/978-981-99-0236-1_41

Table 1 Timeline of various pandemics

Year	Virus	Death toll
165–180	Antonine Plague	5,000,000
541–542	Plague of Justinian	50,000,000
735–737	Japanese Smallpox Epidemic	1,000,000
1347–1351	Black Death (Bubonic Plague)	200,000,000
1520	Small Pox	56,000,000
1600	17th Century Great Plagues	3,000,000
1700	18th Century Great Plagues	600,000
1817–1923	Cholera 6 outbreak	1,000,000
1855	The Third Plague	15,000,000
Late 1800s	Yellow Fever	26,200
1889–1890	Russian Flu	1,000,000
1918–1919	Spanish Flu	100,000,000
1957–1958	Asian Flu	4,000,000
1968–1970	Hong Kong Flu	4,000,000
1981-Present	HIV/AIDS	35,000,000
2002–2003	SARS	774
2009–2010	Swine Flu	284,000
2014–2016	Ebola	11,323
2015-Present	MERS	886
2019-Present	Novel Coronavirus (COVID-19)	3,840,000

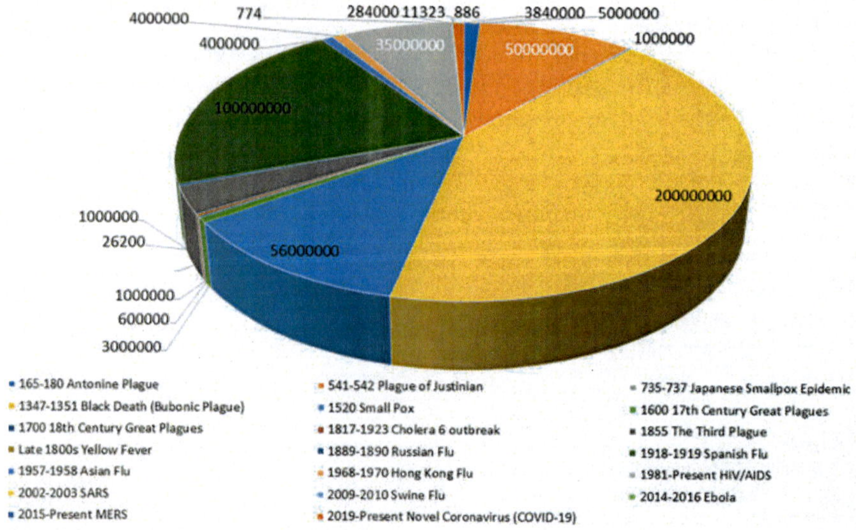

Fig. 1 Death toll due to various pandemics

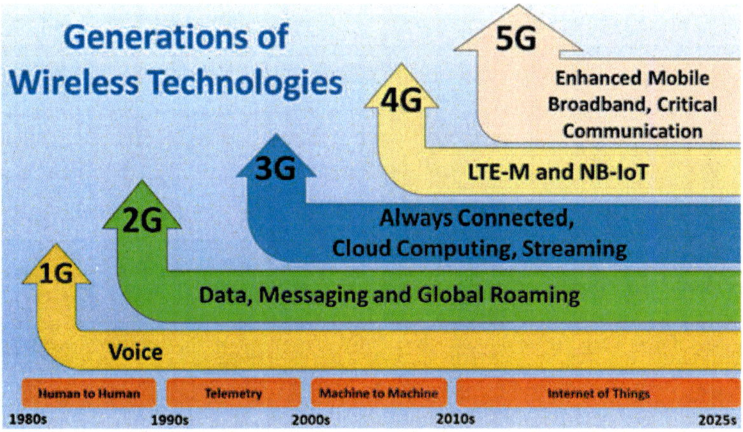

Fig. 2 Evolution of wireless technology

Table 2 Timeline depicting various generations and features of wireless technology

Year	Generation	Maximum data speed
1991	2G	14.4 kbps
2001	3G	384 kbps
2010	4G	100 Mbps
2020	5G	1 Gbps

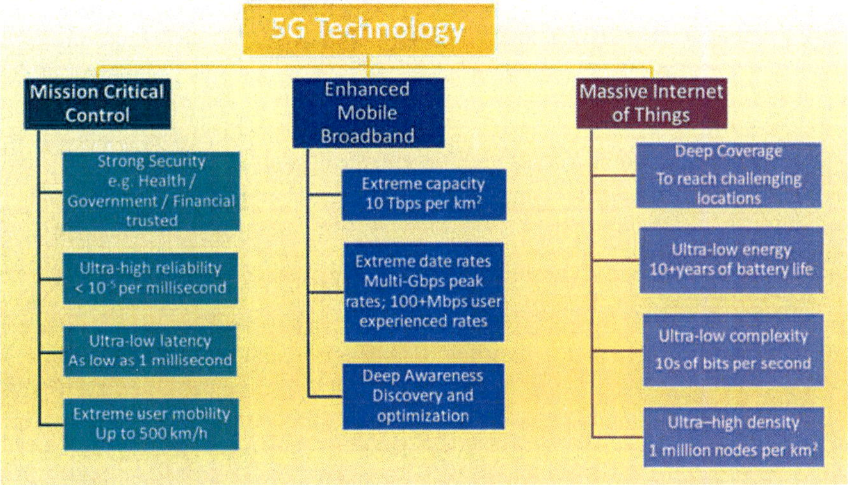

Fig. 3 5G technology features

The rest of the paper is organized as follows: literature review, problem identification and gap in existing technology are carried out in Sect. 2, the proposed system architecture is presented in Sect. 3 followed by implementation details in Sect. 4 and hardware description of the proposed work in Sect. 5. The results are discussed in Sect. 6 and finally the conclusion and future work of the proposed work in Sect. 7.

2 Related Work

U. Varshney in his paper on health monitoring of disabled patients using wireless technology proposed a health monitoring system that uses wireless and mobile networks. The proposed system operated autonomously without patient intervention, which is generally not possible with patients suffering from one or more disabilities. However, the system didn't address the issue of the detection of disease [1]. V. Sharma et al. in their paper on low-energy health monitoring for patients based on the LEACH protocol proposed a health monitoring wireless device that contains good range and capability and improved the performance of the health monitoring network by the Low Energy Adaptive Clustering Hierarchy (LEACH) protocol. The proposed system had the limitations of portability and easy implementation [2]. M. Baswa et al. in their paper on e-health monitoring architecture proposed health monitoring architecture using GSM based upon the communication devices like mobile phones and wireless sensor networks for the real-time analysis of the patient health condition. The main focus of the paper was on developing a model that can facilitate doctors through tele-monitoring. The device failed to address the health monitoring of a large number of people; it was suitable for individuals who were at home or at the hospital [3]. M. S. Uddin et al. in their paper on IoT-based patient monitoring system proposed a remote monitoring system which includes vehicle or assets monitoring, kids/pets monitoring, fleet management, parking management, water and oil leakage, energy grid monitoring, etc. They have proposed an intelligent patient monitoring system for monitoring the patients' health condition automatically through sensor-based connected networks. However, the system had severe limitations in monitoring patients who are suspected of contagious diseases [4]. A. Bhatti et al. in their paper on economical patient tele-monitoring system for remote areas proposed a novel, rapid and cost-effective tele-monitoring architecture based on an Arduino device hardware system. Their prime goal was to design a prototype that could serve as a reliable patient monitoring system, so that healthcare professionals can monitor their patients in real time, who are either hospitalized in critical conditions or unable to perform their normal daily life activities. The system was not designed for the early detection of disease nor it had any feature to check the spread of disease once it is detected [5]. T. Erlina et al. in their paper on patient's smart health system proposed a system that monitors the number of heartbeats and respiratory rate, and detects eyelid opening using a pulse sensor, thermistor and Infrared Light Emitting Diode (IR LED), respectively. Still, the system severely suffered from the limitation of continuous unattended monitoring and alarm generation on detection of symptoms of infection in monitored subjects [6]. Shahbaz Khan et al. in their paper on COVID-19 patients monitoring using a health band proposed a health band that is developed for monitoring the patients sent to quarantine, or under medical treatment. The novel COVID-19 created a time of pandemic as large crowds of people were sent to either isolation or quarantine centers; their health monitoring is a challenge for today's medical team as well as patients under observation. This health band is developed to provide quality monitoring without spreading the virus among the patients and medical staff. However, the implemented system requires some necessary changes in terms of parameters monitored, response time and reliability [7]. Otoom M. et al. in their paper on identification and monitoring of COVID-19 using IoT proposed a system that collects real-time symptom data from users using an IoT framework for early identification of suspected coronavirus cases. The system also monitors the treatment and response of those who have already recovered from the virus; thus, it tries to understand the nature of the virus by collecting and analyzing relevant data. The proposed system severely suffered from the limitation of continuous unattended monitoring and alarm generation on detection of infection in monitored subjects [8].

3 Proposed System Architecture

The proposed system consists of four parts: the sensors, the data aggregator, application and the cloud server. The health of persons needs to be monitored, and the sensors deployed should be able to detect any deviation from normal values and send an alert message to responsible persons such as government authorities, doctors, hospitals and family members. Several sensors can be deployed to measure and monitor various physiological changes. The sensors can be deployed in jackets, wristbands, watches, clothes, shoes, jewelry, handbag, etc. in order to monitor various parameters like heart rate, blood pressure, body temperature, oxygen level in body, pulse rate, etc. Figure 4 depicts various possibilities for deploying the proposed system.

3.1 The Sensors

The number of sensors used can be changed depending on the parameters sensed. The system is fully customizable. In the present paper, pulse sensor, heart rate, SPO2, temperature sensor, heart ECG monitoring sensor and PIR motion sensors are used.

Fig. 4 Various apparel for deploying proposed system

3.2 Data Aggregator

The sensor continues to sense various physical parameters and sends the data to Node MCU for aggregation, analysis and monitoring purposes.

3.3 Application

The application part of the proposed system continuously checks the aggregated data for any unusual and abnormal activity which means the acquired data crosses the required pre-set values.

3.4 Cloud Server

The analyzed data along with any alert signal in case of abnormal reading is sent to the cloud server for sending an alert signal as decided by the user to the hospital, to relatives or to his own smartphone.

4 Implementation Details

The proposed system was implemented and a hardware prototype was prepared for testing. The hardware details of the proposed system are shown in Fig. 5. Figure 6 gives the complete hardware implementation of the proposed system and its experimental setup.

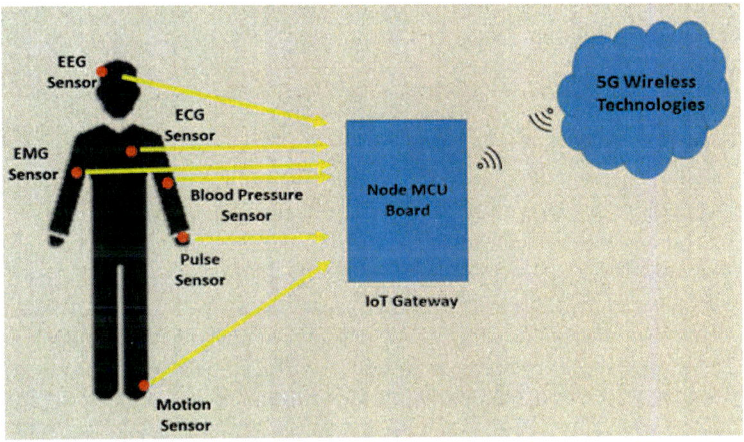

Fig. 5 Block schematic of proposed system

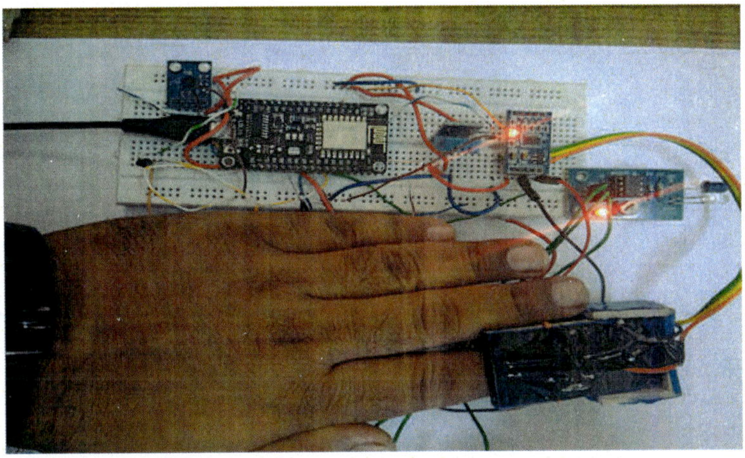

Fig. 6 Hardware implementation of proposed system

5 Hardware Description

The different sensors used in designing the system prototype (shown in Fig. 6) are listed below.

1. Node MCU ESP8266
2. Pulse Sensor (SKU-835048)
3. Heart rate, SPO2, Temperature sensor (SKU-845800)
4. Heart ECG Monitoring Sensor (AD8232)
5. PIR Motion Sensor.

Node MCU ESP8266 is the main controller that is used in this IoT application as shown in Fig. 7a. Its high processing power and low operating voltage of 3.3 V with in-built Wi-Fi/Bluetooth and Deep Sleep Operating features make it ideal for the present application [9]. Pulse Sensor used in the proposed circuit is SKU-835048 shown in Fig. 7b. The used sensor is compatible with most of the microcontrollers such as Arduino and Node MCU. The output of the pulse sensor is digital, therefore it can be directly interfaced with MCU. The sensor works on 5VDC [10]. Heart rate, SPO2, Temperature sensor used in the proposed circuit is SKU-845800 shown in Fig. 7c. The used sensor is compatible with most of the microcontrollers such as Arduino and NODE MCU. The output of Heart rate, SPO2, Temperature sensor is digital, therefore it can be directly interfaced with MCU. The sensor is compatible with 3.3 and 5 V logic levels. This sensor has three LEDs green, red and infrared. The amount of light reflected back to the sensor can be detected by these LEDs in combination with the photodetectors. Photoplethysmography (PPG) is a technique that is used to detect the patient's heart beat. When the patient's fingertip is pressed against the sensor, the change in color of the patient's skin with each beat of his/her heart is detected. This sensor measures the amount of light bounced back to the sensor by the particles and thus can also be used to detect particles in the air, like smoke [11]. Heart ECG Monitoring Sensor used in the proposed circuit is ECG Module AD8232 shown in Fig. 7d. The used sensor is compatible with most of the microcontrollers such as Arduino and NODE MCU. The output of the pulse sensor is analog, therefore it cannot be directly interfaced with MCU; it needs connection through ADC. The sensor works on 5VDC. The sensor is designed to extract, amplify and filter bioelectric signals in the 0.1–10 mV range. The sensor can measure signals in the presence of noisy conditions, such as those created by motion or remote electrode placement. It is a cost-effective board for measuring the ECG of the patient. Body movement sensor used in the proposed circuit is the SeeedStudio Grove Mini PIR Motion sensor shown as shown in Fig. 7e. The used sensor is compatible with most of the microcontrollers such as Arduino and NODE MCU. The output of the pulse sensor is digital, therefore it can be directly interfaced with MCU. The sensor works on 5VDC [12]. Human body movement sensor Grove Mini PIR Motion Sensor

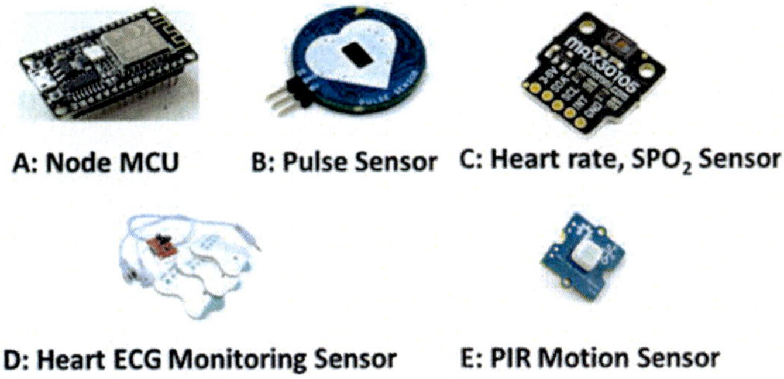

Fig. 7 Sensors used for designing prototype of proposed system

v1.0. is ideal for the present application. PIR stands for Passive Infra-Red. PIR sensor measures infrared (IR) light radiating from objects in its field of view. This sensor can be easily used in various things with the proposed design. This sensor is compact, cost-effective and has low power consumption; moreover, this sensor has adjustable sensitivity, and there is a reserved pin out on the back of the board so that it can be soldered to a slide rheostat to adjust the sensitivity [13].

The features and specifications of the above components are given in appendix.

6 Results and Discussions

The proposed system prototype was implemented and evaluated for performance. The proposed system is working as per theoretical predictions. With the help of the sensors, it is able to predict and send timely alerts in case any of the parameters sensed by the sensors indicate chances of infectious disease. Table 3 gives the various sensor output, condition suspected, alert message sent or not and the response time of the system. Figure 8 shows the alert message on a smartphone.

7 Conclusion and Future Work

This research found that the spreading of any contagious disease may quickly turn as an epidemic and then as a pandemic if not checked timely. So, the timely detection and control of the spread of infectious disease is a much-waiting kind of research investigation. This paper has proposed an IoT-based and 5G technology-based automatic system to mitigate the impact of contagious diseases like COVID-19.

Table 3 Sensor status, response time and alert generation of proposed system

S.no	Sensor data				Condition status	Response time (ms)	Alert message
	Pulse sensor	Heart rate, SPO2, Temp. sensor	Heart ECG sensor	PIR motion sensor			
1.	BT	BT	BT	BT	OK	52	Not generated
2.	AT	BT	AT	BT	Alert	59	Generated
3.	AT	AT	AT	BT	Alert	64	Generated
4.	BT	BT	AT	AT	Alert	62	Generated
5.	BT	AT	AT	AT	Alert	68	Generated
6.	AT	BT	AT	AT	Alert	67	Generated
7.	AT	AT	BT	AT	Alert	67	Generated
8.	AT	AT	BT	BT	Alert	60	Generated
9.	BT	BT	BT	AT	Alert	57	Generated
10.	AT	AT	AT	AT	Alert	72	Generated

BT = Below Threshold, AT = Above Threshold

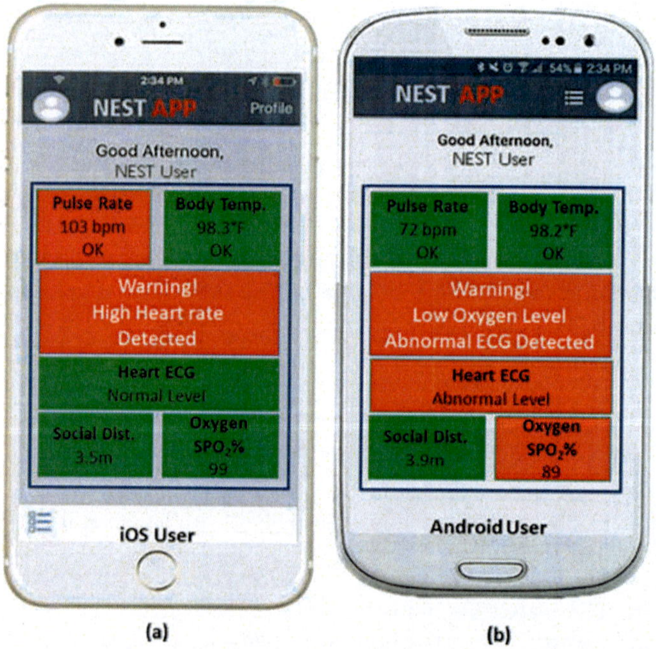

Fig. 8 App showing normal and abnormal parameters of iOS user and Android user

An experimental prototype was developed and tested, the results showed that the prototype developed achieved desired accuracies of more than 90%, and its response time confirmed the theoretical results. Using the proposed design, end user will be equipped with an effective and accurate system to fight against the spread of COVID-19 and other such contagious diseases. Employing the proposed system in day-to-day life usage could potentially reduce the impact of pandemics, as well as mortality rates through early detection of cases. The proposed system will also provide the ability to follow up on recovered cases, and a better understanding of the disease. The system will take to its leverage the inherent properties of 5G and IoT for its benefit and to overcome the limitations posed by the 4G/LTE technologies. IoT-based and 5G technology-based automatic system was able to utilize 5G-enabled IoT technologies ensuring reduced date delay and increased reliability in terms of quality of service. It has been suggested to deploy the system in various wearable apparel. There has been extensive study of this work to provide the best performance of the device by comparing the existing domains. The new features of this design accomplish different objectives to measure the health symptoms, track and monitor the patient during quarantine and maintain the data to predict the situation. As future work, and due to the unavailability of the required data and testing on real subjects, the system will be tested in hospitals and nursing homes for field testing and its performance established in real-time operations.

Acknowledgements The authors are thankful to Prof. Rohit Garg, Director MIT and the Management of MITGI for their constant motivation and support.

Appendix

See (Tables 4, 5, 6 and 7).

Table 4 Pulse Sensor SKU-835048 specifications

Sl.no	Parameters	Output
1.	Operating voltage	3–5 VDC
2.	Operating current	4 mA
3.	Sensor output	Digital
4.	Sensor weight	0.03 kg
5.	Sensor size	5 × 3 × 1 cm

Table 5 Specifications of PIMORONI MAX30105 Heart rate, Oximeter, temperature sensor SKU-845800

Sl.no	Parameters	Output
1.	Operating voltage (VDC)	5
2.	Interface	I2C
3.	I2C Address	0 × 57
4.	Compatible with	All Models of Raspberry Pi and Arduino
5.	Sensor length (mm)	19
6.	Sensor width (mm)	19
7.	Sensor height (mm)	3.2
8.	Sensor weight (gm)	10
9.	Sensor weight	0.015 kg
10.	Sensor dimensions	5 × 5 × 1 cm

Table 6 Specifications of ECG Module AD8232 Heart ECG Monitoring sensor

Sl.no	Parameters	Output
1.	Operating voltage (VDC)	3.3
2.	Operating temperature (°C)	−40 to 90
3.	Sensor length (mm)	36
4.	Sensor width (mm)	30
5.	Sensor height (mm)	18
6.	Sensor weight (gm)	5
7.	Sensor weight	0.01 kg
8.	Sensor dimensions	7 × 5 × 2 cm

Table 7 Specifications of body movement Sensor, i.e. SeeedStudio Grove Mini PIR Motion sensor

Sl.no	Parameters	Output
1.	Input supply voltage (VDC)	3.3 ~ 5
2.	Working current	12 ~ 20 μA
3.	Sensitivity	120 μ–530 μV
4.	Sensor Max. detecting range	2 m
5.	Sensor length (mm)	24
6.	Sensor width (mm)	20
7.	Sensor height (mm)	12
8.	Sensor weight (gm)	8
9.	Sensor weight	0.012 kg
10.	Sensor dimensions	6.8 × 4.3 × 1.2 cm

References

1. Varshney U (2006) Managing wireless health monitoring for patients with disabilities. In: IT professional, vol 8, no 6, pp 12–16, Nov–Dec 2006. https://doi.org/10.1109/MITP.2006.139
2. Sharma V, Sharma S (2017) Low energy consumption based patient health monitoring by LEACH protocol. In: International conference on inventive systems and control (ICISC) 2017, pp 1–4. https://doi.org/10.1109/ICISC.2017.8068632
3. Baswa M, Karthik R, Natarajan PB, Jyothi K, Annapurna B (2017) Patient health management system using e-health monitoring architecture. In: International conference on intelligent sustainable systems (ICISS) 2017, pp 1120–1124. https://doi.org/10.1109/ISS1.2017.8389356
4. Uddin MS, Alam JB, Banu S (2017) Real time patient monitoring system based on Internet of Things. In: 2017 4th international conference on advances in electrical engineering (ICAEE), 2017, pp 516–521. https://doi.org/10.1109/ICAEE.2017.8255410
5. Bhatti A, Siyal AA, Mehdi A, Shah H, Kumar H, Bohyo MA (2018) Development of cost-effective tele-monitoring system for remote area patients. In: International conference on engineering and emerging technologies (ICEET) 2018, pp 1–7. https://doi.org/10.1109/ICEET1.2018.8338646
6. Erlina T, Saputra MR, Putri RE (2018) A smart health system: monitoring comatose patient's physiological conditions remotely. In: International conference on information technology systems and innovation (ICITSI) 2018, pp 465–469. https://doi.org/10.1109/ICITSI.2018.8696094
7. Khan S, Shinghal K, Saxena A, Pandey A (2020) Design and development of health band for monitoring of novel covid-19 under medical observation. Int. J. Adv. Eng. Manag. (IJAEM) 2(1):332–336. (June 2020)
8. Otoom M, Otoum N, Alzubaidi MA, Etoom Y, Banihani R (2020) An IoT-based framework for early identification and monitoring of COVID-19 cases. Biomed Signal Process Control 62:102149. https://doi.org/10.1016/j.bspc.2020.102149
9. Datasheet of Node MCU ESP8266. https://www.espressif.com/sites/default/files/documentation/0a-esp8266ex_datasheet_en.pdf
10. Datasheet of Pulse Sensor SKU-835048. https://robu.in/wp-content/uploads/2020/10/Pulse-Sensor.pdf
11. Datasheet of Heart rate, SPO2, Temperature sensor (SKU-845800). https://datasheets.maximintegrated.com/en/ds/MAX30102.pdf
12. Datasheet of Heart ECG Monitoring Sensor (AD8232). https://www.analog.com/media/en/technical-documentation/data-sheets/ad8232.pdf
13. Datasheet of Grove–Mini PIR Motion Sensor v1.0. https://www.mouser.com/datasheet/2/744/Seeed_101020020-1217525.pdf

A Brief Review of Current Smart Electric Mobility Facilities and Their Future Scope

Darbhamalla Satya Sai Surya Varun, Tamesh Halder, Arindam Basak, and Debashish Chakravarty

1 Introduction

The usual petrol-diesel type vehicles are driven by Internal Combustion Engine/Vehicle (ICE/ICEV). With an ever-increasing automobile industry, EV's or electric vehicles are becoming the most promising dream to come true in the future. Electric Vehicles are the future of mobility systems and it's very much evident as being one of the most sprighting and controversial topics of discussion [1, 2].

Climatic change is another topic of concern for scientists. A study claims that Automobile pollution contributes to over 40% of total global pollution [3, 4]. And hence, this topic is "long in the tooth" for auto-mechanical-researchers and has been under the constant focus of development and research and has been an ambitious topic, to achieve real-life implementation [5]. Though there exist papers discussing brief and classified types of EV's, not all of them have been mentioned all together yet [6, 7], and hence we shall discuss them as follows.

1.1 Wheeler-Based EV Types

(a) Two-Wheeler Electric Vehicle (E2V's)
(b) Three-Wheeler Electric Vehicle (E3V's)
(c) Four-Wheeler Electric Vehicle (EV's).

D. Satya Sai Surya Varun · A. Basak (✉)
School of Electronics Engineering, KIIT-Deemed to be University, Bhubaneswar, Odisha, India
e-mail: abasakfet@kiit.ac.in

T. Halder · D. Chakravarty
Department of Mining Engineering, Indian Institute of Technology, Kharagpur, West Bengal, India
e-mail: haldertamesh@iitkgp.ac.in

© The Author(s), under exclusive license to Springer Nature Singapore Pte Ltd. 2023

H. Muthusamy et al. (eds.), *Robotics, Control and Computer Vision*, Lecture Notes in Electrical Engineering 1009, https://doi.org/10.1007/978-981-99-0236-1_42

1.2 Charging-Station-Based-EV Types

(a) Normal-EV (NEV)
(b) Super-EV (SEV).

1.3 Component-Wise EV's Classifications

- Battery (size, capacity, type, packs)
- Internal motor
- Reducer
- PCU or Power Control Unit
- Power conditioner type
- Humidifier
- Fuel processor and its reliability
- Fuel stack.

As per the previously reported studies, there are research papers available on electric vehicles, but elaborated review on smart electric vehicles is rare. This paper discusses elaborated classifications and the current status of the sales of electric vehicles in the market, hence it is necessary to classify EV's in terms of various performance features, current trends and additional developments necessary to overcome the disadvantages and challenges faced by modern society for adapting to complete electric mobility systems from, senior type vehicles relying on an ever-depleting source of energy. It is also necessary to develop structured details and collective types of EV's in the current market with either minor or major Structural/Manufactural/Performance-Based as well as pre-installed attributes and feature-equipped types of EV's into the light of the complete picture, for researches to understand/better organize and develop the possibilities of above-mentioned features for the future of mobility systems.

2 Internal Structure and Design/Architecture Classified Types of EV's

2.1 Hybrid EV (HEV): (Based on Degree of Hybridization)

These types of EV's are usually powered by both electricity as well as gasoline/petrol/diesel and are driven mainly by the engine (ICE/IC) and electric motor. These types of EV's are further classified as follows:

(a) ***Series Hybrid EV/EREV's/REEV's (Range extended)***

These kinds of hybrid EV's are usually equipped with similar batteries as in battery electric vehicles (BEV). The ICE is utilized for power generators as well as the battery. For high power requirements, the combined power from both batteries as well as the generator is used. Since petrol/diesel is only used to drive the electric motor, these are range-extended EV's.

(b) ***Parallel Hybrid EV***

These kinds of hybrid EV's are powered by both ICE as well as electric motor/generator. The varying power distribution system allows both components to work simultaneously. Unlike Series Hybrid EV's, having a separate generator, for this case, isn't required.

(c) ***Parallel Mild Hybrid EV***

The components for this type of EV are the same as that of a Parallel Hybrid EV, but the only disadvantage of this is being unable to be driven purely using electric power. The motor is turned on only when the extra boost is required for the EV, under extreme situations of need. As these types are unable to provide individual functioned engine system's deployment, either of ICE or Electric Motor Engine, they are termed as "Mild Hybrids".

(d) ***Parallel Split Hybrid EV/Through-the-Road (TTR) HEV***

This type of HEV is usually equipped with both ICE as well as an electric motor (in-wheel motor (IWM)) just as the above-mentioned EV's. The electric motor, however for this case, is capable of providing propulsion power to different axles [8]. This type of HEV doesn't consist of any kind of mechanical system to drive the wheels of the EV; rather, the coupled power of the two systems is used to move the wheels. These types of EV's are equipped with power-split devices for the driver/customer to opt for either of both mechanical as well as electrical operation of driving. These types of HEV's are capable of zero-emission driving, generally 20–30 miles.

(e) ***Series–Parallel Hybrid EV***

This type of HEV could be driven with the help of petrol/diesel or by completely reliance on electric motors or with the help of both the components to get optimum performance. While both of them could be utilized, the engine is given higher priority of performance and power input than that of the motor as it is the main component that drives the whole system and also gets a maximum operating range for the same.

(f) ***Micro HEV's***

This kind of HEV is equipped with an integrated alternator/starter-type electric motor to start or stop the engine. The ICE system for this is utilized when the EV starts moving.

(g) ***Mild HEV's***

This type of HEV is mostly similar to that of Micro HEV in terms of components. The integrated alternator/stator for this is designed larger and is more efficient as compared to that of the Micro HEV's component. A battery is equipped for the same, which is only utilized for propulsion while EV is under cruising mode.

(h) ***Full HEV's***

This type of HEV consists of a large battery that could be powered from either a grid system or from home. FHEV's aren't completely emission-free vehicles yet are the best-known option for environmental control.

(i) ***Dual HEV's***

These types of HEV's usually are sophisticated type vehicles currently used only for racing and testing purposes. These are the EV's equipped with Hybrid powered 4-stroke piston otto-cycle engine along with highly efficient REE-based ICE components capable of producing high power emissions and an engine burning system. However, these types of EV's are unreliable keeping environmental & affordability concerns of such EV by a common citizen, of any country alongside.

(j) ***Plug-in Hybrid Electric Vehicle (PHEV)***

These types of EV's usually are provided with a Large Battery (larger than usual HEV battery) and larger discharge rate, but it is able to charge from CS's from time to time. These are again usually powered by ICE and electric motors. These have the ability to operate in all-electric modes, i.e. either in charging or depleting energy mode. Studies related to the prospective CO_2 emission reduction model specifically for PHEV-type vehicles suggest that this type of EV, though having less automotive CO_2 and NOx emissivity, could be highly reliable to better countries suffering high pollutive crises [9, 10].

2.2 Plug-In Electric Vehicle (PEV)

As the name suggests, these types of EV's are equipped with a plug-in facility for the electric motor/engine/battery to be charged with the help of grid-connected wall sockets.

2.3 All-Electric Vehicle (AEV)

These types of EV's are equipped with one or more than one electric motors to power the engine. These are also equipped with exceptional batteries to get powered from grid systems directly. They don't use any form of gasoline. These types of EV's include the following types:

(a) ***Battery Electric Vehicle (BEV)***

Propulsion is provided by an electric motor and the rest is powered by the power storage unit. These types of EV's are solely driven by batteries. There is zero emission claim for these types of EV's.

(b) ***Fuel Cell Electric Vehicle (FCEV)***

As the name suggests, these kinds of EV's are powered directly or indirectly by a powerful fuel cell consisting of various powering ingredients or gases (according to availability).

2.4 *Extended Range Electric Vehicle (EREV)*

This type of EV is equipped with a medium-sized battery in comparison to other, above-mentioned EV's. This type of EV has the same system, i.e. battery storage (ESS) and ICE. These EV batteries are to be charged directly from the grid system/CS's.

3 Interactive Powering Mode's Classification for EV's

(a) ***Grid-to-Vehicle (G2V) Mode***

In a Grid-to-Vehicle mode, the EV could be considered as an electrical load as it consists of a series of electrical components. The basic concept of this system involves the power being drawn from the grid system to the EV [11].

(b) ***Vehicle-to-Grid (V2G) Mode***

In a Vehicle-to-Grid mode, the power is provided to the grid system by the EV while under steady-state (parked) condition [11–13]. The system's performance is independent of the type of EV under consideration. These are most commonly applicable for fuel, battery or hybrid-type EV's. The subclassifications for this are as follows:

(c) ***Vehicle-to-Building (V2B) Mode/V2H***

This type of mode is a system especially designed for EV's equipped with large batteries with exceptional storage capacities. This type of system can communicate directly with the building itself to sell demand response [14].

(d) ***Vehicle-to-Vehicle (V2V) Mode***

This type of mode is a system capable of sharing and transferring/deriving and gaining electric charge from another EV. Hence, it is also referred to as "bi-directional conductive charging" EV's [15]. This type of facility is essential for vehicles as EV's far away from CS's cannot get access to them everywhere and under stranded conditions, they could avail electricity enough to reach the CS under emergencies. Various studies and research have been conducted on the same as well [15, 16].

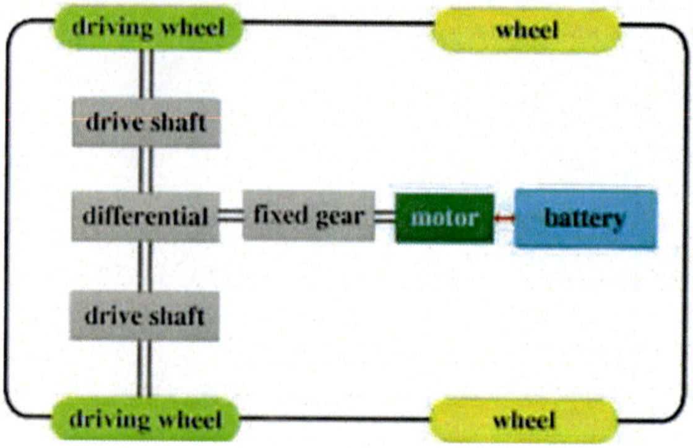

Fig. 1 Study of different topologies of EV's

(e) ***V2X Mode***

This type of model could also facilitate bi-directional charging [15]. This is a Vehicle to Infrastructure (X) type of charging facility. And it gives great future scope for smart grid facilities of CS's for EV's. **Large Capacitive Batteries (LCB)** are deployed for this type of EV's, and the common users for the same are HEV- and PHEV-type EV's. The LCB's could be used to drive the whole vehicle.

4 Topology of EV's

See Fig. 1.

5 Current EV Trends and Expectations

The EV industry was always under constant progression for a decade and yet continues to develop to this day. The ever-growing research gives greater opportunities for implementing better components of replacement. A study states that in the year 2020, the global EV stock hit the 10 million mark and with a 43% increase, more than that of 2020's mark. As technology keeps developing, new models and designs for EV's keep changing/evolving. The efficiency of batteries also keeps advancing. As technologies as such continue to emerge to meet the growing needs of the Electric Vehicle industries, one of the trends to look forward to is changing customer sentiments. As the fuel rates keep skyrocketing, the demands and expectations for alternatives for the customers keep increasing towards the automobile industry. Some more of the important trends are listed below (Figs. 2 and 3).

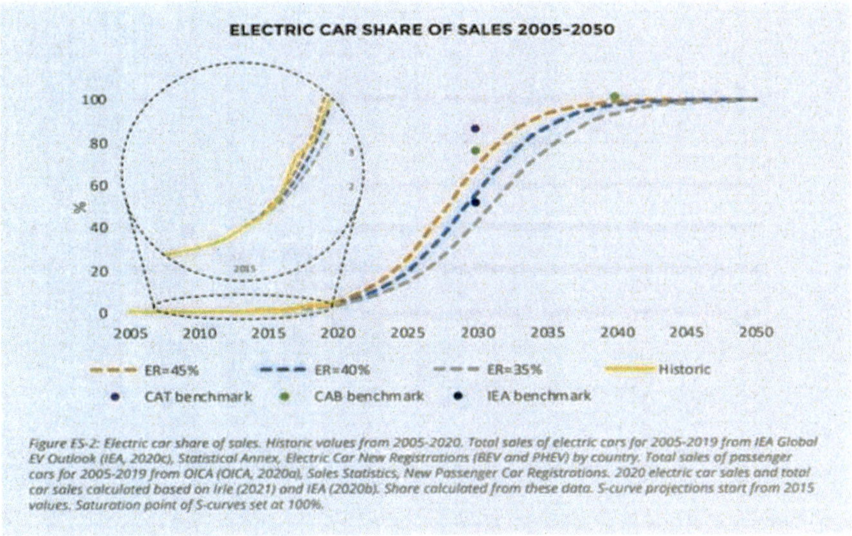

Fig. 2 S-curve, IEA report for EV sales. (*Image Source* https://thedriven.io/2021/05/27/electric-vehicle-s-curve-puts-global-uptake-in-line-with-paris-goals/)

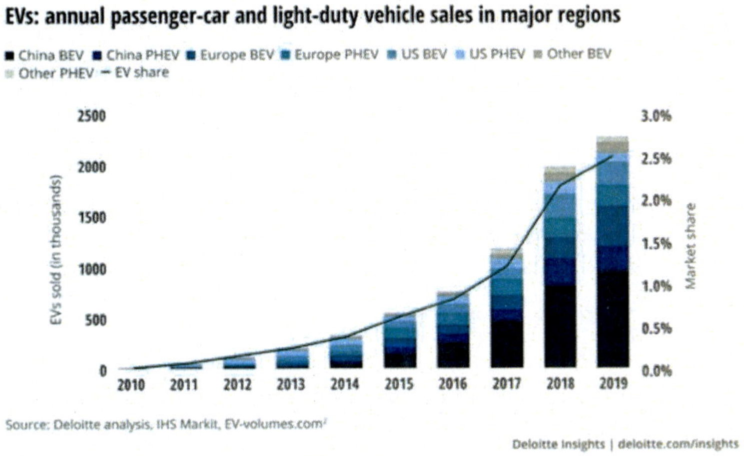

Fig. 3 Annual Passenger-car and Light duty vehicle sales analytic (2010–19). (*Image Source Electric vehicles.* (2020, July 28). Deloitte Insights. https://www2.deloitte.com/us/en/insights/focus/future-of-mobility/electric-vehicle-trends)

(a) ***Better Automotive Design/Comfort***

The components or elements used inside the vehicle such as the dashboard and touchscreen are some of the futuristic designs and a symbol of luxury and comfort. Customers expect the cruising journey in an EV to be comfortable and rather better

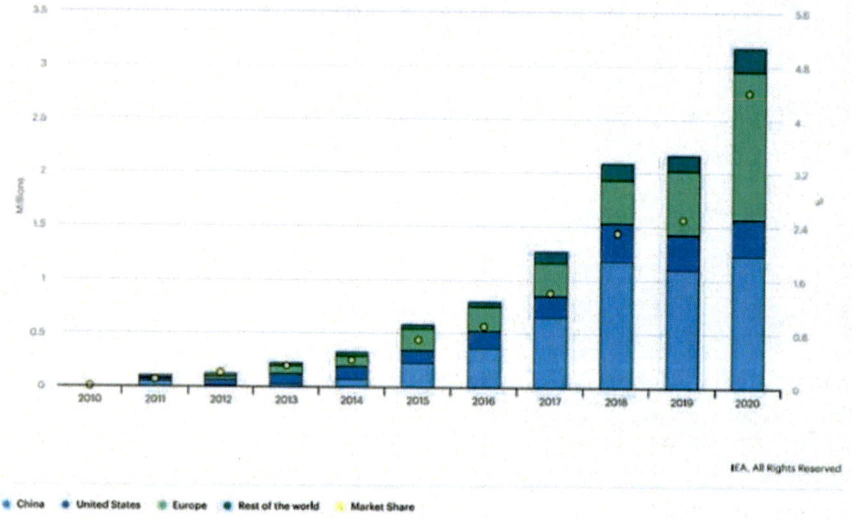

Fig. 4 Annual Passenger-car and Light duty vehicle sales Analytic (2010–20). (*Image Source* https://www.iea.org/commentaries/how-global-electric-car-sales-defied-covid-19-in-2020)

than what they've been experiencing in usual ICE-based vehicles. Comfort and design play a crucial role in getting better sales in the automotive industries. The Utility Vehicle (UV) design keeps getting popular being the most suitable design for middle-class customers. The EV's exterior design has become some sort of competitive art form. The aerodynamics of the exterior design plays a crucial role, especially for the exterior elements of manufacture. The EV in comparison with usual ICE vehicles has no front area of occupation, i.e. separate crash absorption system is uniquely designed for the same. This type of trend gives greater scope of marketing in the automobile industry and yet is under constant evolution (Figs. 4 and 5).

(b) ***Demand for Autonomous Facilities***

Harmonized charging standards are very important especially for cities with the requirement to achieve zero emissions. Development and research of achieving ultra-fast charging facilities are booming in the industry [17]. V2G research with better equipment is also a topic under development for the same. The electrification efficiency of the battery affects the grid system, and hence smart charging facilities are to be developed. Autonomous EV's have the potential of replacing traditional ICE-type vehicles. Advanced charging and connective solution facilities would create better business opportunities for the industry to excel [18].

(c) ***Demand for better Life Cycle Assessment (LCA)***

A LCA is a method of study for investigating certain components/entire machine's ecological Assessments. A variety of studies exist for different types of EV's as mentioned above. Each form of EV has been given individual importance and study

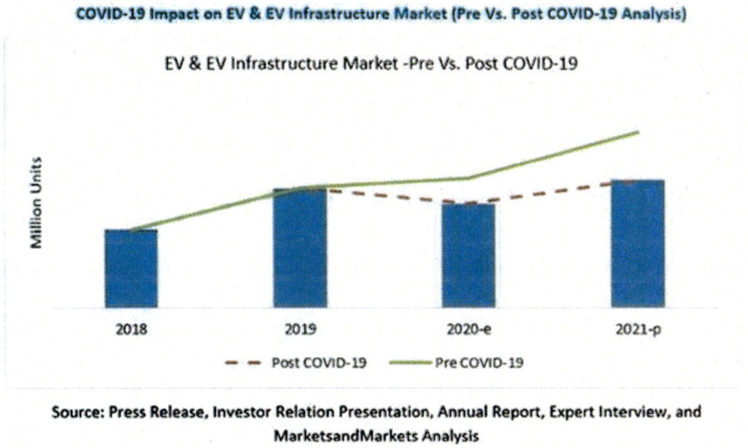

Fig. 5 EV sales review pre- and post-COVID-19 pandemic. (*Image Source* https://www.marketsandmarkets.com/Market-Reports/covid-19-impact-on-electric-vehicle-market-81970499.html)

cases [19]. However, it's difficult to find LCA statements of each type as very short review papers for each type exist. EV LCA performance analysis and literature review are increasing constantly as being an important subject of concern, especially for the customers. Most studies, yet, majorly consider only Well to Wheel performance of EV's while neglecting factors such as battery production. A brief comparison of specific types of EV's was conducted and framed with over 79 study cases [19–22]. The Wheel-to-Wheel (WTW) study highlights the amount of carbon emission intensity and amount of electrification that could be assessed for the specific type of vehicle. The study states that full EV emits over ~ ½ of the amount of CO_2 emission in comparison to that of an ICE-based common conventional vehicle [21].

The study suggests the average CO_2 emission produced by an EV over a common ICE-based vehicle is over 25% less in percentage [20].

The prognosis of EV's carbon footprint study also suggests that the life cycle performance and efficiency of EV's are going to increase in upcoming years. For better performance, the demand for better metals are increasing [23]. For example, Tesla utilizes metals like Lithium, Aluminium Oxide, Manganese, Nickel and Cobalt. Rare Earth Elements (REE) are used for manufacturing electric motors for greater performance. Electrification of vehicles with larger sizes and weights has been constantly criticized (e.g. SUVs) for them requiring larger battery sizes and storage capabilities which is yet hard to achieve in a complete (Full-EV)-type system. Yet the same study enlightens upon the fact that the batteries manufactured with REE's would create better and efficient batteries which would also be capable to drive large-sized-SUV-type vehicles as well. It is a true fact that the accountability of the LCA's influencing factors into consideration can be taken up by individual Automobile industries and depends upon their individual goals and aims pre-set and yet it is also the veracious fact that modern society relies on believes and trusts new technology and ever-expanding scientifically efficient devices and auto-mobility as well.

Conclusively, the expanding research efforts and studies for the subject of EV's keep generating better chances of decreasing Carbon emissions in comparison to that of the utilization of the conventional fuel-based form of vehicles. The life cycle analysis of the previous subject research papers shows that the carbon footprints emitted by EV's are way lower, justifying the replacement of conventional ICE-based vehicles with electrified EV's for good. With the increase in studies for developing means for generating/harvesting electricity from renewable sources of energy, the hazardous climatic-carbon effects are expected to diminish rapidly. The technological improvements not only in the field of energy harvesting systems but also in the field of improving battery chemistry, battery-efficient-materials chemistry and battery storage capacity will contribute to the same goal of achieving a carbon-free environment.

(d) ***Demand for Price reductions***

EV's are viewed as the ultimate problem solution for many types of issues. For the same, economic value and for them to be the replacement for common ICE-based vehicles, the price should be such that middle-class people from different countries could afford to buy them.

There are many strategies that would work to help improve this situation of affordability by common members of society to rely on EV's. EV's are costly/expensive mainly for their batteries. Batteries of different types have different lifetime and energy storage capabilities with which one's owner would have to worry about its "health" (Lifeline). The battery has to be such that it would not have materials of Extreme Rare Earth Elements (EREE's) and wouldn't consume more electricity while manufacturing (depending on the individual manufacturing device's capability). With improving technology, EREE's are being utilized for better quality. Cobalt-based batteries are cheap and affordable in comparison to other forms of recent-type batteries such as Lithium-Titanate and Lithium-Iron-Phosphate [24].

Falling battery prices would solve 25% of the price demand issue. But what other types of factors could be done to reduce the EV's battery value? The performance. Performance of the battery is essential and is something people/customers would ask about before even deciding to buy a certain type of EV.

EV's design optimization play's a crucial role to reduce the price as well. For this case, it isn't the vehicle's exterior design but rather mostly for its battery and other components' compatibility design that we are going to focus on. Having LCB in a luxury type V2X EV's, it is hard to keep their state as such without increasing the height of the vehicle. If so, this type of vehicle would consume a lot of energy even if it were manufactured with an ICE-based engine. Its design could be easily compared to an SUV and hence a complex-internal design has to be taken into account for the same. There shall be fewer compromises and higher flexibility for these types of EV's design. The electric cable design slots have to be pre-designed using computerized software to avoid mistakes and to save space.

The estimation for total electric vehicle battery manufacturing includes almost 50–40% of the total vehicle cost. Investing in new upraising companies is hence essential to let new technologies with better ideas of replacement to give a chance of development.

Electric Vehicles are expected to get cost-effective with time as better sources and materials for manufacturing the batteries are yet under constant development and research. A study suggests that 77% per cent of EV battery cost will be reduced in the 2016 to 2030 time frame [25]. The continued efforts given by current researchers in the automotive industry are evident for the same.

(e) ***Demand for better Wireless systems***

The "Wireless Charging Facilities" is an enormous topic of discussion, improvement and research. Optimizing such a facility is a necessity for getting better performance output from the EV. Various studies for the same are under great interest of automobile/EV researchers [26, 27].

There are many factors affecting the charging facilities and major ones include the charging time and charging location. This type of facility challenges the current grid systems available. And the types of grid-facilitated charging system types have been discussed above. When it comes to charging location, CS's should be abundant in certain areas of cruising. In fact, the current gas stations should be deployed with a CS facility for electric vehicles [3]. The wireless charging system is quite a new concept of origin and yet under major construction of government policies for different countries, technological development and manufacturing developments. With the progressive market competition, the wireless EV charging facility development is increasing.

Types of wireless systems for power transfer (WPT's):

- Near-Field-WPT
- Inductive-WPT-System
- Capacitive-WPT-Systems.

Current Trends in WPT's:

- Reducing the component sizes and increasing spaces
- Achieving high power transfer with high efficiency
- Achieving variable compensations
- Multi staging Matching Network Systems in the EV's
- Phased array Field Focusing.

(f) ***Demand for better accessibility***

Irreversible climatic conditions have been and are affecting the environment as well as different types of flora and fauna unnoticeably yet noticeable under years of inspection. The greenhouse effect is real and yet people don't seem to bother/get concerned about it. Even if people agree to replace their fuel-type vehicle with a modern EV-type vehicle, they have their own budgets which are hard to compensate for newly developing technology to be owned. People even take their own time, for "the development of tech" before buying it to avail extremely advanced tech possible. Emerging technologies are always financially inaccessible at the initial stages of development, and it's true for every one of them. Though EV's reduce carbon emissions and take the financial perspectives of individuals to afford the EV to cause immense and immediate effects of CO_2, greenhouse gas emission reduction is impossible to be achieved within even 20 years of time. Even though EV evolutionary development might take years of time to achieve the ultimate EV, the controlling of the carbon emissions is yet if not better than utilizing ICE Based EV's. Accessing EV's is not only difficult but also might cause a burden to people if any sort of equipment repeatedly needs to be replaced or to be changed, financially for the owners. Equipment failure plays a crucial role in customers' interest in owning EV's. Bad reviews Might cause serious issues for the individual automobile industry, and hence every piece of equipment needs a thorough examination of life assessment and lifetime warranty facility to be provided for the customers.

Countries like Africa, India, Bangladesh and so on yet suffering from immense poverty issues are far from achieving the goal of full EV replacement modernization even after a complete century. Yet people of modernizing places in such countries should get access to technologies as such to avoid further owning of ICE-based vehicles. A study suggests over 96% of people in India might not access such features even after 5 decades of development [28]. Accessibility of such technologies to be showcased at various places has to be ensured by the automobile industries; the question of affordability depends upon the people and customers.

(g) ***Demand for complete Electric Facilities***

With the increased number of electric vehicles, the electrical bills are highly expected to be the same as the usual fuel available in the market. This sounds remote in reality but is also true to be achieved with the reports and suggestions for the EV futuristic research reviews. Electrical facilities are associated with many components of the vehicle in a usual ICE-based vehicle/car. But in the case of EV's, the customers expect every component to be driven with only electricity which isn't true in the current situations of EV's. EV's are manufactured into different models and ideas and with different ranges and even with internal component differences. This type of analysis is hard to be taken into account because the performance and maintenance for the same do not stay the same for even individual types of EV's as discussed above. Although most of the components could be just driven by electricity, the challenge is only "Batteries" with a great capacity to store energy and to be having as high mileage as for usual fuel-type vehicles. Hence, it is the common perspective of the people who desire to invest their money in something of such "nearly emerging technology".

EV's account for a significant load capacity in most of the country's as of 2021. For EV future to flourish, new technology and innovative ideas have to be taken into consideration by any minor or major companies working for the same [4]. This is time taking, and the evolution of EV's is also expected to escalate same as in 1900's usual car evolution.

Though electrical facilities will be available, it doesn't ensure complete customer interest as electrical prices and production rates at different places vary and also significantly depend on the economic situation of the place's location. In places like India, it is highly expected to get only around 2–3% of power supply requirements by the end of 2030 as the customer's reliability of such is very low [29].

(h) ***Demand for Quality & Eminence***

EV is yet an emerging and developing type of technology with high aspirations/expectations and demands increasing among the general public. The quality and expectations from the exterior to the interior design of the vehicles are expected to be better than that of the previous ICE-based mobility systems used by the general public. It is the same case of expectation from battery performances and the usual facilities to be provided in the EV as well.

The quality of EV's would determine the effectiveness of the expectations to be achieved for the customer's "quality of living". Early accessed individuals with EV's may face significant constraints under the conditions and places or regions they live in as earlier said; every technology has immense disadvantages while they might look completely reliant. EV's on the other hand could significantly affect the reliance of ICE-based vehicles on foreign oil production which would be a humongous loss for several oil/fuel-producing countries. EV eminence would result in decreased utility prices in the market as well. The utility rates relying on distributive power consumptions would affect EV's. The EV's are expected to be charged mostly at night times or at steady-state conditions, while at night the electricity is the cheapest peak as the demand for the same at that time period being less would result in a decrease of overall rates of electricity.

6 Demand for Financial Incentives

Governments of many different countries support the ideology of switching to a completely electric facilitated mobility system and also offer enthralling financial inducements for the same [30]. The government of different countries can take initiatives such as.

- Reducing taxes on EV's
- Increasing maintenance and repairing taxes on ICE-based vehicles.

This cost of maintenance and taxes have to be carefully examined by individual governments as this might cause price parity between the ICE and EV-type mobilities.

7 Recent Machine Learning Study Based on EV Charging Behaviours

Renewable sources of energy are hard to extract or harvest, and it's challenging to get better efficiency from them as well. However, these types of energy harvesting systems are the future of powering sources for people in decades to come. With challenging environmental hazards created by us, humans keep evolving; repairing the same gets more difficult with time. With greater provisions and availability of renewable energy resources, the utilization of grid systems is constantly decreasing. But it is a matter of fact that people have to rely on the grid systems to get fuel/energy for their EV's.

There has been a significant amount of studies to improve the driving range of EV's and accuracy predictions of the electric motors equipped in the vehicles. BEV types of EV's are given greater importance than any other type as this would be the test module of real-life implementation for further improvements. However, just as there are many differences between all types of EV's discussed above, each and every type also has its own advantages and disadvantages over the other. As BEV's are equipped with sophisticated battery systems, it usually takes a lot of time to charge and hence are unreliable under emergency situations if the battery runs out. To count and analyse this range issue, conventional multiple linear regression methods could be used [31]. And hence it is one of the recent innovative applications for ML-based EV development. Batteries are the main source of development for EV's and that is to ensure a high range of mobility and deployment.

AI is under great expansion and is one of the leading subjects to contribute to the field of machine learning for the future of electric battery and automobile battery development and research [32]. Recent studies conducted by "Stanford" are innovative and may also be one of the branch of study choices that may be under development for the future of EV's. The study claims it to be helping future automobile batteries to have long-lasting charges and fast charging facilities with fast charging powering grid systems [17].

Machine Learning in the field of EV's can be considered a type of "hit and trial" method to achieve successive outcomes. The patterns of failure from previously examined and tested batteries could be observed and solutions for the same with thoughtful and scientific concepts by current researchers could lead to the future of this study. Storage facilities for not only EV's but also for a wide range of applications such as House-Inverters, wind and solar energy harvesting systems would lead to more efficient utilization of renewable power resources as well.

Predicting the EV driving range using the Machine Learning concept is a fairly recent topic of discussion [31]. Charge scheduling and manufacturing and designing the cable structure for charging the EV with minimum waiting time is a challenge facing ML in recent times, and as discussed above it is one of the main focuses of study for a decade [33].

Conclusively one of the main factors that affects the study of machine learning in the field of Electric vehicles, Powering systems and Battery chemistry includes the following obstacles:

- Battery enhancement difficulty
- Battery storage capacity
- Battery's physical dimensions for space and adjustments
- Charging equipment's modelling
- Charging port's design for PEV-type EV's
- Charging time efficiency
- Grid system enhancements for CS's.

And hence the current focus of studies and research under ML for futuristic development and to achieve goals of comfort for the customers also by maintaining the environmental issues into consideration.

8 Recent Deep Learning-Based Study for EV's Behaviours

Deep learning is a sub-subject of machine learning involving deeper analysis of systematic calculations and algorithmic analysis of any specific subject, and it is true also for the study of electric vehicle futuristic plans under study by many research groups across the world. Deep learning allows one to exactly predict the behaviour and nature of the Machine or any other device under specific conditions with a little bit of mathematics. Deep Learning (DL) methods have various branches of undergoing studies, for example, to estimate the power requirements and power consumption space for batteries to be designed for various types of EV's, as mentioned above.

Optimizing the power distribution and estimating the power requirements of the battery is one of the most recent papers [34]. This paper predicts the power requirements of a specific EV-type under consideration using a DL algorithm based on Modular-Recurrent-Neural-Network (MRNN).

The paper successfully suggests using the DL algorithm; the power requirement along with the driving range could be simultaneously predicted. Upon the advantages of finding these two data together, the jitteriness of the EV under the training phase can be avoided, and the results predicted will be a lot smoother in comparison to the ML study of predicting the driving range.

In the whole study, BEV-type vehicles are the most commonly found study subject as these are the most prominent future of EV's in the coming years with great business and tests in the market value shown in recent years. BEV-type vehicles are also of great interest for the scientists and researchers of the modern era of EV-development programs to work on for the same reason. Eco-driving systems are the greatest advantage of this type of EV and are also encouraged by the governments of different countries taking environmental concerns seriously. As discussed above, BEV types of EV's have the least carbon and NOx emission record and so have been said [35], though it is not certain that it is a complete non-emitting greenhouse gas vehicle either.

The study of deep learning utilities isn't just confined to the engine and electrically driven components of the EV, but also to the growing demand for grid system facilities for building CS's [3]. Grid systems as mentioned above have different types of subclassifications and types and methods of their own systems of charging the EV. However, in recent times the V2G strategy for deploying in public buildings has gained a lot of popularity [36]. Deep learning-based charging port detection facility along with location detection facility based on machine and deep learning technology came into existence a decade ago [37]. This topic is convenience-based technology for the customers to locate the charging port more easily, avoiding charge leakages while the EV is being charged, Low space management and increasing the charging efficiency for the customers as well. A brief theory for image sensing and filtering technology for the same is also discussed in the paper [37].

Demand-Side Management (DSM)-based electric vehicle system is another DL-type learning program that proposes a smart way of managing the influence energy consumption patterns of the EV to make the best use of the electricity stored in the battery [38].

With the demands of new technology, the deep learning methods of improvising the performance of the EV's is expected to keep getting better; it is for certain that ultra-low emission vehicles (ULEV) are the future of mobility systems. And hence we can say that deep learning is a crucial structure of study that is necessary for building logical algorithmic plans for the EV architecture and also working principles with promising future implications and development for upcoming transportation and grid facilities.

8.1 Deep Learning Study Based on Road Lane Detecting Facilities in EV's

As discussed above, deep learning facilitates and provides basic infrastructural development plans for futuristic EV's each and individual types. In ever-growing populations such as India and China, Road Mapping applications in the EV dashboard feature demand is increasing with greater dependency and efficiency expectations [39]. Driving assistance and dependence on the automated facility in these countries is a challenge of deployment and requires complex algorithmic codes and developments for the same as no risks of life-threatening circumstances should be avoided in the same. The automated forms of such programs have already been seen in ICE-based vehicles [39, 40]. The High-Definition-Road-Network (HDRN) for the road mapping facility provides the greatest, to date, solution for self-driving vehicles. The power requirement for the same requires a complex and separate battery or powering system in which, BEV-type vehicles are the best for deployment having high-capacitated Lithium-ion batteries to facilitate both, engine/wheels as well as the

Automotive driving instrument/component. It is evident from previously seen works to this date that very little importance has been given to improving road mapping, and Automotive driving systems have been given for the last 5 years [40]. Though the automotive systems to this date have advanced, they aren't reliable for the assured safety of the customers/passengers yet.

Though not many efforts are made to improvise the automated obstacle-avoiding facility using deep learning in the field of electric vehicles for self-driving vehicle systems, the warning systems are under constant development [41].

The facility of offline and online modes of mapping systems should be provided for the driver's convenience. Yet offline mode is considered the most reliant as no data waiting has to be dependent, while for the online mode, the system has to be connected with the satellite system which needs network access. Though this in-built facility of deep learning technology is still under constant development, it might take years for complete online-mode dependency without any waiting period of access. In the case of offline mode, the sensor data is accumulated in the pointers or central location of the vehicle which involves pre-installed software of either satellite or plain road map imagery. SD and HD mapping facilities are provided in the same software according to the convenience of the driver's choice while buying the vehicle, and the only difference between them would be centimetre to meter-type accuracies and better definition of imagery. With better development of this technology, such innovative projects would be worth implementing for EV's of the future with higher precision, accuracy and comfortable automated self-driving electric vehicle systems [4]. Road Mapping and obstacle-detecting facility involve deep learning technology and algorithmic logic which in fact are the future of EV's of the future generations and modules of Automobile industries.

Some of the articles involving in-depth studies of different types of mapping systems of automated technology have been referred to in the bibliography/references section [42–49].

8.2 Basic DL-ML Algorithmic Study-Type Classifications for EV's

1. Decision Tree (DT)
2. Random Forest (RF): Subclass of DT
3. Modular-Recurrent-Neural-Network (MRNN)
4. Support Vector Machine (SVM)
5. Demand-Side Management (DSM)
6. Naïve Bayes (NB)
7. K-Nearest Neighbours (KNN)
8. Deep Neural Networks (DNN)
9. Long Short-Term Memory (LSTM)
10. Short-Term Load Forecasting (STLF).

9 Diagnostic and Prognostic Analysis for Battery Managements in EV's

Let's brief here about lithium-ion battery management systems for EV's. In EV's, the Battery Management Unit (BMU) is a small part of the system that stores and converts energy from the battery-stored electricity into motion and vice versa. As we have discussed above, the different types of EV's have different battery requirements and also space for deployment. Modern-type electrical impact of the battery system vehicle impacts the overall performance of the EV and is expected to be very highly efficient than the hydraulic-based (ICE) type vehicle, and it is the same for every type of EV discussed above. Battery management is the most important aspect and concept of concern in the overall EV system as discussed above, modern type EV major deployment issue being battery systems, their performance and storage capabilities. Battery system being the most expensive component of the vehicle, the type of battery of deployment and its design make a great impact on the overall performance of the vehicle. Care and the feeding pack of the battery system is a great deal of focus to ensure the best performance and avoid any damages in the future along with the longevity of the battery's life. There are several factors that are taken into account while designing the battery pack as well as the BMU. In ideal conditions, the service of the battery pack and its performance outlast that of the overall life span of the vehicle driven by itself and is highly expected as well. The safety and efficiency of the same have to be ensured as well. Some of the variations in the BMU available in the market are as follows.

Increased capacity of the battery pack than that of the targeted range of the same. Despite the overall capacity of the battery getting diminished over time, the overall performance of the vehicle is retained over a longer period of time (in years). Some of the diagnostics and prognostics of the EV battery system along with other types have been studied as well [50].

Some of the BM-Fuel Gauging techniques are as follows:

- Monitoring Cell Voltage
- Hydrometer Analysis
- Coulomb Counting.

The overall aim of the BMU monitoring system is to carefully update and note the capacity of the battery being charged and discharged frequently and how it's affecting the battery's performance which in turn affects the vehicles' performance range, efficiency optimization and battery's life span. Usually, the stacked battery system consisting of cells having one lesser amount of charge being charged entirely may cause damage to the entire system. This discharge and charged levels of the BM being damaged depend upon various factors of the situation. Hence, we can say that the situation of damage to the Battery Management System (BMS) doesn't only get affected by external influences but also by internal structure and the level of charges in each and every stacked cell as well. Hence, Cell Charge level balancing and equalization provide a mechanism for all of the stacked cells in the BMU to be maintained at the almost identical level of charge and hence maintaining the overall performance of the Battery pack over a long period of time of being charged and discharged.

The long period of charging and discharging cycles affects the battery performance and has been monitored by many papers as well [50]. Strategic propositions to maintain these levels are necessary and have to be monitored for individual EV battery health.

Some of the strategies include [51]:

- Un-Coordinated Direct Charging (U-Di-C)
- Un-Coordinated Direct Charging and Discharging (U-Di-CD)
- Un-Coordinated Delayed Charging and Discharging (U-De-CD)
- Un-Coordinated Delayed Charging (U-De-CD)
- Un-Coordinated Random Charging (U-R–C)
- Un-Coordinated Delayed Charging and Discharging (U-R-CD)
- Continuous Coordinated Direct Charging (CC-Di-C)
- Continuous Coordinated Direct Charging and Discharging (CC-Di-CD)
- Continuous Coordinated Delayed Charging (CC-De-C)
- Continuous Coordinated Delayed Charging and Discharging (CC-De-CD).

For a detailed analogy of these strategic propositions, it is highly recommended to endure the paper [51].

10 EV's Future Scope

The growth of EV's is expected to follow the S-curve trend.

Factors powering constant growth of EV's:

- Developing Customer's Sentiments for New technology
- Improving EV Policy and Legislation
- OEM (Original Equipment Manufacturers) EV Strategy Implementation [30]
- Involvement and Interest given by Corporate Companies.

This is most common for any type of newly emerging technology. EV's have a bright future and yet would take a lot of time for complete reliance until the technology is efficient enough to be replaced with usual fuel ICE-based vehicles. In this case, the EV battery is expected to decrease while the technological development for the same increases. Though a completely reliant type of EV would successfully make it to the market, the same would have energy and transport policies. Customer preferences, demands and infrastructural expectations play a huge role in the futuristic development and scope of EV's. A study suggests that over 36% of the future and model depend upon expectations and influences of public charging infrastructure [52]. Some of the key factors and challenges to overcome the current expectations of the general public/customers that affect the same are mentioned, and are as follows:

- Affordability of EV's
- Charging and electric facilities to be robust
- Exterior design and architectural design
- Quality design and comfortable interior for opulence travel experiences
- Manufactured with quality material resources
- Less likelihood of repairs
- Affordable charging (to charge) facilities
- Greater mileage
- Fast charging facility
- Home charging facility and CS's deployment.

With all of these factors taken into consideration, the expectancy of the goal to be achieved is prominent and would definitely take years of development and research.

Though the emission expectations might not be as high as expected, the production of gases such as NOx and CO_2 is expected to be lesser in comparison to the situations as recorded in 2019 at various regions with high population and pollution rates globally. Electrification of transportation systems is one of the major steps that has to be undertaken within this century to maintain the balance of the environment and humanity on this planet.

This paper has discussed various types of possible EV's available in the market and has given a brief review of the type of batteries that are in the market and what new developments are being undertaken by automobile industries to achieve this goal.

EV's create a major role in the power sector and especially for the futuristic reliance on power grid systems according to this yet emerging technology. Energy conservation and Harvesting systems and Environmental consciousness are two different factors that are taken into account for EV development. Various papers have suggested that this could be achieved hand-in-hand sooner or later. Vast-ranging types of EV's in the market show a greater potential to achieve the goal with time.

Futuristic innovations such as metal intensity batteries apart from usual high-capacitive yet expensive batteries made of Nickel, Cobalt or Lithium could be a potential future of batteries and power systems as well apart from EV's [25].

Innovative interior and exterior designs and architectural development of modern EV designs could attract people's attention towards improvising technology of EV's. Slow-powering different components via electrification with slowly developing battery systems with slight or higher voltage outputs could be utilized to power complete vehicles on electrified systems as a whole by years of studies and burgeoning. In the case of interior design, better and wireless systems equipped could attract customers' interest to invest as well.

Tablet features and automated driving systems with automated parking systems are the future of electric vehicles.

The grid system technology is a complicated powering system that has to be developed and would provide great opportunities and employment for the people as well. Wireless or contact mode transmitted Energy systems are yet a research topic that might also be the future of charging facilities in EV's. Better and fast charging facilities are also an enormous topic of debate in recent years for PEV-type EV's.

Conclusively, it's an undeniable fact that EV's have a scintillating future with an appreciable scope of deployment globally in the coming decades.

11 Pandemic Situation EV Updates

As Covid-19 struck the world in 2020, while the whole world went under isolative conditions, the global market not for EV's but for usual vehicles dropped drastically within a very short time [53]. In 2019, the integrated annual sales of BEV- and PEV-type electric vehicles reached over the 2 Million mark and yet are expected to increase till 2030 [30].

The Sources suggest that overall 15% of the vehicle sales dropped on a year-on-year basis. Though the effect could be observed for EV's as well, the expected sales for EV's in 2019 couldn't hit the mark in 2020 since the pandemic became an obstacle, though it has been observed that the sales of EV's rather increased slightly. For a detailed analysis of the same over that affected the market in different countries, it is highly suggested to refer to the paper [53].

The impact of the sales of EV's is likely to increase rather than that of fossil-fuel-powered vehicles across the world. There have been various investigative reports on the affective impact of the coronavirus over sales of EV's as well as normal ICE-based vehicles [53, 54]. While some of the reports have suggested the decline of Charging Station Implementations to be decreased to over 70–75% in regional basis areas, the overall EV demand is yet increasing which is evident from IEA reports [53]. It's very much evident that other than EV's, all forms of transportation systems in the market have majorly got impacted/affected by the same. The EV after the Covid-19 situation has a bright future and has to be taken into account by different countries; the reasons are as follows [55]:

To Incentive the Economy of Individual economic situations of different countries.

- Cost Saving for EV's.
- At times of Emergencies and low demand in the market, to increase new revenue streams.
- To encourage people to retain local air quality using EV's.

According to the article [55], the following steps have to be undertaken by the government to encourage the supplement of EV's in the market:

- By Increasing Studies and Research for Charging infrastructure
- By Encouraging and Supporting the people for purchasing EV's
- By Implementing Emission Standards and EV Mandates.

The pandemic has also affected the mindsets of individuals owning EV's. Though there have been reports of people's interest to increase in the field of sustainable living conditions, and driving facilities, the situation of the pandemic has diverted the interest in owning EV's drastically. People's interest in relying upon a sustainable resource mobility system has caused great interest as it's the best alternative for the ever-increasing rates of fossil fuels.

While it might be a bad situation for Charging Station deployment across the places, it is expected that home charging facilities overnight would be more convenient for people to be relied upon. While most of the fact of Home-Charging-Facilities may seem the most convenient, over places, deployment of CS's may be risky as there might be a need for the same over different places in case of battery shortages and long driving days for the EV's owned by the customers.

A Covid-19 impacted sales of fuel-ICE-based vehicles would be a disaster for Global weather health and warming. While it isn't what is really happening, the chances for the same aren't even low. In 2021 even, the electric vehicle market is poised for growth.

All in all, the future of electric vehicles is going to be remarkable, and it's evident from all the papers discussed above.

12 Limitations and Future Scope of Work

India aspires to be a significant player in the worldwide electric car industry. The prevalence of BEVs has expanded dramatically in the previous five years, thanks to various automakers in the nation working on electric cars. Along with the traditional automotive manufacturers, a number of start-ups have risen in the market with their own goods and technology.

13 Conclusions

This review gives insight into an elaborated discussion about current types, trends and future scope of EV's. It is clear from this review that even though EV's are still an emerging technological issue under constant development, it has a bright future for the automobile industry. Considering all the other possible obstacles that need to be attained, it is a necessity for countries producing higher carbon emissions to replace traditional types of vehicles with modern types of electric vehicles as soon as possible to avoid further damage to the environment. EV's are great for this work as they provide transport with reduced carbon output. Garages could be used as a Home-CS facility for an individual who could afford the current type of EV. Even though it is hard to afford full-EV for citizens, many of the country's people could at least start adapting EV-driven types of vehicles as mentioned above to slowly compensate for the environmental effects.

The carbon footprint of an EV will depend upon the type and size of the battery as well. The demand for better quality depends mainly upon the type of battery being utilized. In recent years, the demand and price of raw materials used to manufacture lithium-ion batteries is getting higher and higher which is making the provision of EV's in the market to be reduced. The other materials such as Cobalt are also increasing, which is another material of significance in the battery evolution [24].

The main objective of this study is to provide a clear view of current types and trends of EV's in the market and classify the differences and benefits received for individual types for future customers. By the empirical results, it is confirmed that in the long run EV's show promising equilibrium for fighting against environmental issues.

With the ever-increasing demand for EV's in various developing countries, the expansion and development in the fields of renewable energy generating systems have to be increased and should be a topic of serious implementation and research subject. The demand for energy storage systems with effective storage capabilities is also yet to be achieved without using REE's. The demand for Batteries with such capabilities is increasing with a size range of mid-to-large in EV's. The policies for developing such technologies must be expanded. The extreme reliance on raw materials such as Lithium aand Cobalt will cause a huge halt in the further development of the technology [24]. New materials have to be examined and a superior element of replacement has to be discovered in order to accelerate the research and development. Newly emerging companies with great ambitions and innovative ideas should be given a chance to develop and for consideration in the market to accelerate the research. The limited and expensive supply of cobalt and lithium is causing companies to revert from the idea of utilization. In the use phase, LCA of EV's of every type has yet to be given importance as a research subject as very few papers have been seen in recent years [20, 21]. Cumulative efforts for increasing the work could promise a better future for humanity via automobile industries with the help of EV's.

References

1. Towoju OA, Ishola FA (2020) A case for the internal combustion engine powered vehicle. Energy Rep 6:315–321
2. Boston W (2019) Rise of electric cars threatens to drain German growth. WSJ. https://www.wsj.com/articles/rise-of-electric-cars-threatens-to-drain-german-growth-11565861401 (2019, Aug 16)
3. Xu X, Niu D, Li Y, Sun L (2020) Optimal pricing strategy of electric vehicle charging station for promoting green behavior based on time and space dimensions. J Adv Transp 1–16
4. Sneha Angeline P, Newlin Rajkumar M (2020) Evolution of electric vehicle and its future scope. Mater Today: Proc 33:3930–3936
5. Global greenhouse gas emissions data. US EPA. https://www.epa.gov/ghgemissions/global-greenhouse- gas- emissions-data (2021, March 25).
6. Nanaki EA (2021) Electric vehicles. Electric Veh Smart Cities 13–49
7. Larman C, Vodde B (2010) Practices for scaling lean and agile development: large, multisite, and offshore product development with large-scale scrum. Pearson Education, Boston
8. Zulkifli SA, Mohd S, Saad N, Aziz ARR (2015) Split-parallel through-the-road hybrid electric vehicle: operation, power flow and control modes. In: 2015 IEEE transportation electrification conference and expo (ITEC), pp 1–7

9. Doucette RT, McCulloch MD (2011) Modeling the prospects of plug-in hybrid electric vehicles to reduce CO2 emissions. Appl Energy 88(7):2315–2323
10. Chakraborty S, Vu HN, Hasan MM, Tran DD, Baghdadi ME, Hegazy O (2019) DC-DC converter topologies for electric vehicles, plug-in hybrid electric vehicles and fast charging stations: state of the art and future trends. Energies 12(8):1569
11. Gago RG, Pinto SF, Silva JF (2016) G2V and V2G electric vehicle charger for smart grids. In: 2016 IEEE international smart cities conference (ISC2)
12. Goel S, Sharma R, Rathore AK (2021) A review on barrier and challenges of electric vehicle in India and vehicle to grid optimisation. Transp Eng 4:100057
13. Kempton W, Tomić J (2005) Vehicle-to-grid power implementation: from stabilizing the grid to supporting large- scale renewable energy. J Power Sourc 144(1):280–294
14. NextEnergy. Vehicle-to-building (V2B). https://nextenergy.org/vehicle-building-v2b/. (2017, June 26)
15. Sami I, Ullah Z, Salman K, Hussain I, Ali SM, Khan B, Mehmood CA, Farid U (2019) A bidirectional interactive electric vehicles operation modes: vehicle-to-grid (V2G) and grid-to-vehicle (G2V) variations within smart grid. In: 2019 international conference on engineering and emerging technologies (ICEET)
16. Mahure P, Keshri RK, Abhyankar R, Buja G (2020) Bidirectional conductive charging of electric vehicles for V2V energy exchange. In: IECON 2020 The 46th annual conference of the IEEE industrial electronics society. Published
17. Attia PM, Grover A, Jin N, Severson KA, Markov TM, Liao YH, Chen MH, Cheong B, Perkins N, Yang Z, Herring PK, Aykol M, Harris SJ, Braatz RD, Ermon S, Chueh WC (2020) Closed-loop optimization of fast-charging protocols for batteries with machine learning. Nature 578(7795):397–402
18. Bonnema GM, Muller G, Schuddeboom L (2020) Electric mobility and charging: systems of systems and infrastructure systems. In: 2015 10th system of systems engineering conference (SoSE)
19. Runeson P, Höst M (2009) Guidelines for conducting and reporting case study research in software engineering. Empir Softw Eng 14:131–164
20. Helmers E (2020) Sensitivity analysis in the life-cycle assessment of electric vs. combustion engine cars under approximate real-world conditions. MDPI (2020, Feb 9).
21. Helmers E, Dietz J, Weiss M (2020) Sensitivity analysis in the life-cycle assessment of electric vs. combustion engine cars under approximate real-world conditions. Sustainability 12(3):1241
22. Nordelöf A, Messagie M, Tillman AM, Ljunggren Söderman M, Van Mierlo J (2014)Environmental impacts of hybrid, plug-in hybrid, and battery electric vehicles—what can we learn from life cycle assessment? Int J Life Cycle Assess 19(11):1866–1890
23. Jones B, Elliott RJ, Nguyen-Tien V (2020) The EV revolution: the road ahead for critical raw materials demand. Appl Energy 280:115072
24. Mo J, Jeon W (2018) The impact of electric vehicle demand and battery recycling on price dynamics of lithium- ion battery cathode materials: a vector error correction model (VECM) analysis. Sustainability 10(8):2870
25. U.S Department of Energy. (n.d.). All-electric vehicles. www.fueleconomy.gov - the official government source for fuel economy information. https://www.fueleconomy.gov/feg/evtech.shtml.
26. Triviño A, González-González JM, Aguado JA (2021) Wireless power transfer technologies applied to electric vehicles: a review. Energies 14(6):1547
27. Al Mamun MA, Istiak M, Al Mamun KA, Rukaia SA (2020) Design and implementation of a wireless charging system for electric vehicles. In: 2020 IEEE region 10 symposium (TENSYMP)
28. Mishra S, Verma S, Chowdhury S, Gaur A, Mohapatra S, Dwivedi G, Verma P (2021) A comprehensive review on developments in electric vehicle charging station infrastructure and present scenario of India. Sustainability 13(4):2396
29. Naik AR (2020) How electric vehicles will impact electricity demand, India's grid capacity. Inc42 Media. https://inc42.com/features/how-electric-vehicles-will-impact-electricity-demand-indias-grid-capacity/ (2020, April 3)

30. Electric vehicles. Deloitte insights. https://www2.deloitte.com/us/en/insights/focus/future-of-mobility/electric-vehicle-trends-2030.html (2020, July 28).
31. Sun S, Zhang J, Bi J, Wang Y (2019) A machine learning method for predicting driving range of battery electric vehicles. J Adv Transp 1–14
32. New machine learning method from Stanford, with Toyota researchers, could supercharge battery development for Electric vehicles. (2020, February 19). https://news.stanford.edu/press-releases/2020/02/19/machine-learninging- electric-car/
33. Vanitha V, Resmi R, Reddy KNSV (2020) Machine learning based charge scheduling of electric vehicles with minimum waiting time. Comput Intell 37(3):1047–1055
34. Jinil N, Reka S (2019) Deep learning method to predict electric vehicle power requirements and optimizing power distribution. In: 2019 fifth international conference on electrical energy systems (ICEES)
35. Andersson I, Börjesson P (2021) The greenhouse gas emissions of an electrified vehicle combined with renewable fuels: life cycle assessment and policy implications. Appl Energy 289:116621
36. Scott C, Ahsan M, Albarbar A (2021) Machine learning based vehicle to grid strategy for improving the energy performance of public buildings. Sustainability 13(7):4003
37. Zhang H, Jin X (2016) A method for new energy electric vehicle charging hole detection and location based on machine vision. In: Proceedings of the 2016 5th international conference on environment, materials, chemistry and power electronics
38. Lopez KL, Gagne C, Gardner MA (2019) Demand-side management using deep learning for smart charging of electric vehicles. IEEE Trans Smart Grid 10(3):2683–2691
39. Chandra S, Mazumdar S (2019) Road map for electric vehicle implementation in India". Int J Manage Comm 1(4):23–29
40. Zheng L, Li B, Zhang H, Shan Y, Zhou J (2018) A high-definition road-network model for self-driving vehicles. ISPRS Int J Geo-Inf 7(11):417
41. New early warning system for self-driving cars: AI recognizes potentially critical traffic situations seven seconds in advance. ScienceDaily. http://www.sciencedaily.com/releases/2021/03/210330121234.htm (2021, March 30)
42. Mattyus G, Luo W, Urtasun R (2017) DeepRoadMapper: extracting road topology from aerial images. In: 2017 IEEE international conference on computer vision (ICCV)
43. Mattyus G, Luo W, Urtasun R (2018) DeepRoadMapper: extracting road topology from aerial images. In: 2017 IEEE international conference on computer vision (ICCV)
44. Li Z, Wegner JD, Lucchi A (2019) Topological map extraction from overhead images. In: 2019 IEEE/CVF international conference on computer vision (ICCV)
45. Homayounfar N, Ma WC, Lakshmikanth SK, Urtasun R (2018)Hierarchical recurrent attention networks for structured online maps. In: 2018 IEEE/CVF conference on computer vision and pattern recognition
46. Liang J, Urtasun R (2018) End-to-end deep structured models for drawing crosswalks. In: Computer vision – ECCV 2018, pp 407–423
47. Liang J, Homayounfar N, Ma WC, Wang S, Urtasun R (2019) Convolutional recurrent network for roadboundary extraction. In: 2019 IEEE/CVF conference on computer vision and pattern recognition (CVPR)
48. Homayounfar N, Ma WC, Liang J, Wu X, Fan J, Urtasun R (2019) DAGMapper: learning to map bydiscovering lane topology. In: 2019 IEEE/CVF international conference on computer vision (ICCV) (2019).
49. Ma WC, Tartavull I, Bârsan IA, Wang S, Bai M, Mattyus G, Homayounfar N, Lakshmikanth SK, Pokrovsky A, Urtasun R (2019) Exploiting sparse semantic HD maps for self-driving vehicle localization. In: 2019 IEEE/RSJinternational conference on intelligent robots and systems (IROS)
50. Cordoba-Arenas A, Zhang J, Rizzoni G (2013) Diagnostics and prognostics needs and requirements for electrified vehicles powertrains. IFAC Proc Vol 46(21):524–529
51. Zhang M (2018) Battery charging and discharging research based on the interactive technology of smart grid and electric vehicle. In: Battery charging and discharging research based on the

interactive technology of smart grid and electric vehicle. Published. https://doi.org/10.1063/1.5041195
52. El-Bayeh CZ, Alzaareer K, Aldaoudeyeh AM, Brahmi B, Zellagui M (2021) Charging and dischargingstrategies of electric vehicles: a survey. World Electric Veh J 12(1):11
53. TVA electric vehicle survey. Stack path. https://www.tdworld.com/grid-novations/distribution/article/20963303/tva-electric-vehicle-survey-consumer-expectations-for-electric-vehicles
54. https://www.iea.org/commentaries/how-global-electric-car-sales-defied-covid-19-in-2020
55. McClone G, Kleissl J, Washom B, Silwal S (2021) Impact of the coronavirus pandemic on electric vehicle workplace charging. J Renew Sustain Energy 13(2):025701
56. Kothari V (2020) 4 reasons to prioritize electric vehicles after COVID-19. World Resources Institute (2020, October 14). https://www.wri.org/insights/4-reasons-prioritize-electric-vehicles-after-covid-19

Gold-ZnO Coated Surface Plasmon Resonance Refractive Index Sensor Based on Photonic Crystal Fiber with Tetra Core in Hexagonal Lattice of Elliptical Air Holes

Amit Kumar Shakya and Surinder Singh

1 Introduction

Photonic crystal fiber (*PCF*) is a compatible platform to design and develop a surface plasmon resonance (*SPR*)-based *RI* sensor [1]. The *PCF* is considered as an suitable candidate for sensor designing because it offers several advantages over conventional optical fibers. *PCF SPR* sensor offers advantages like design flexibility to obtain maximum sensing parameters, non-linearity, small analyte sample for detection, suitability to carry over to different places, and fit for remote sensing applications [2]. In *PCFSPR* sensors, plasmonic material deposition is an important task to perform. Gold (*Au*) [3], silver (*Ag*) [4], copper (*Cu*) [5], aluminum (*Al*) [6], titanium dioxide (TiO_2) [7], indium tin oxide (*ITO*) [8], etc. are some common plasmonic materials used in the sensor designing and fabrication. Recently in a quest for searching for new plasmonic materials, scientists and researchers have discovered materials like tantalum pentoxide (Ta_2O_5) [9], titanium nitrate (*TiN*) [10, 11], zinc oxide (*ZnO*) [12], palladium (*Pd*) [13], etc. These materials can be deposited over the *PCF* fiber using the chemical vapor deposition (*CVD*) technique [3]. The base material of the *PCFSPR* sensor design is mostly silica, because silica is easily and abundantly present in the environment. Besides silica, new background material like Topaz is also used in sensor designing these days [14].

The structural design of the *PCFSPR* sensor follows three different methodologies. The first one is *PCFSPR* sensor models in which plasmonic material coating is applied over the internal air holes of the *PCFSPR* design. This is a highly complicated methodology from a fabrication perspective. Because the size of the *PCFSPR* sensor itself is in the micrometer range, air holes have a more diminutive size. Therefore, applying a thin layer of plasmonic material in the nanometer range over the air holes

A. K. Shakya (✉) · S. Singh
ECE Department, Sant Longowal Institute of Engineering and Technology (SLIET), Longowal, Punjab, India
e-mail: pec1804_amit@sliet.ac.in

© The Author(s), under exclusive license to Springer Nature Singapore Pte Ltd. 2023

H. Muthusamy et al. (eds.), *Robotics, Control and Computer Vision*, Lecture Notes in Electrical Engineering 1009, https://doi.org/10.1007/978-981-99-0236-1_43

is complicated. The second methodology of the plasmonic material deposition is *D*—shaped fiber designing [12]. In *D*—shaped fiber, obtaining a polished flat surface to give fiber *D*—shape is again challenging [10]. Thus, most of the *D*—shaped fibers are again limited to theoretical designs. Finally, coating the *PCFSPR* sensor model with plasmonic material at the outer end is less challenging from a fabrication point of view and thus preferred over the other techniques of sensor designing. This technique is known as the external metal deposition (*EMD*) technique [3].

Ramola et al. [15] designed a *PCFSPR* biosensor for cancer detection. They used a merger of *Au* with TiO_2 as "plasmonic material". They detected six different cancer types from their designed sensor. They have obtained wavelength sensitivity of 12857.14 nm/RIU and 14285.71 nm/RIU for *TMmode* and *TEmode*, respectively. Amplitude sensitivity obtained from the designed sensor is 13240 RIU^{-1} and 15010 RIU^{-1} for *TMmode* and *TEmode*, respectively. They have obtained a "sensor resolution" of 7.77×10^{-6} RIU and 7.00×10^{-6} RIU for *TMmode* and *TEmode*, respectively. Popescu et al. [16] designed a honeycomb-based *PCFSPR* sensor having *Au* as the plasmonic material in their designed sensor. They have concluded that when plasmonic material thickness is increased to 38.75 nm, the wavelength sensitivity is increased from 1000 nm/RIU to 4500 nm/RIU. Their sensor design has obtained sensor resolution of 2.5×10^{-5} RIU when the detection is kept 0.1nm. Zhu et al. [17] designed a dual-core-based *PCFSPR* sensor having coating of *Au*. They have checked biochemicals having *RI* ranging from 1.33 to 1.44 RIU from their designed sensor. They have obtained a wavelength sensitivity of 29500 nm/RIU from their sensor design. They have obtained "sensor resolution" of 3.39×10^{-6} RIU. Yan et al. [18] designed a *PCFSPR* biosensor having elliptical shaped air holes. They have tested their sensor design with analytes having *RI* of 1.43 to 1.49 RIU. They have obtained a wavelength sensitivity of 12719.97 nm/RIU from their sensor design and obtained $R-square = 0.99927$ between resonant wavelength and RI. Falah et al. [19] designed a *D*—shaped *PCFSPR* biosensor having an eccentric core design. They have detected biochemicals having *RI* ranging from 1.33 to 1.42 RIU having *Au* layer as plasmonic material in their designed sensor. Their sensor has produced wavelength sensitivity of 21200 nm/RIU. They have obtained sensor resolution of 4.72×10^{-6} RIU. Full-wave half maximum ($FWHM$) of 29 nm and figure of merit (FOM) of 294 RIU^{-1} are also obtained from the proposed sensor design.

The proposed sensor consists of elliptical-shaped air holes having the combination of *Au* and *ZnO* as plasmonic materials. Since in many research articles related to *PCFSPR* sensors we have observed that *Au* is used with TiO_2 for plasmonic sensing applications, thus a quest for searching for alternate plasmonic material *ZnO* is used for sensing applications. Several sensor models reported to date consist of circular air holes, but the presented sensor design contains elliptical-shaped air holes. Elliptical air holes follow a complex design process compared to circular air holes. Thus, it will be interesting to observe the sensing behavior of the sensor for an elliptical-shaped air hole. Finally, in *PCFSPR* sensor analyte detection is the primary methodology along which sensor performance is examined. In the proposed sensor model, analytes vary from *RI* 1.40 to 1.48 *RIU*, which is *RI* range of the typical household oils, biochemicals, and analytes. Thus, the proposed sensor consists of several new features which will be interesting to observe during plasmonic sensing.

The paper is divided into four different sections. Sensor modeling and design parameters are explained in Sect. 2. Sensor simulation results and the future scope of the designed sensor are presented in Sect. 3. Finally, Sect. 4 offers a concluding remark on the research work.

2 Sensor Modeling and Description of Sensing Parameters

The sensor model is constituted of elliptical air holes arranged in a pattern to produce a tetra core within the *PCF* fiber. Silica material in a fused condition is used as the base material in the presented sensor. Elliptical air holes having 1.2 μm toward the semi-minor axis and 1.5 μm semi-major axes are created. The combination of *Au* and *ZnO* is examined as the plasmonic material in the presented sensor design. The thickness of plasmonic material *Au* is 35nm, and that of *ZnO* is taken as 75nm. A 1.25 μm thick analyte layer is placed over fused plasmonic material for analyte sensing. Finally, a 1.85 μm thick *PML* layer is placed over the fiber to prevent it from atmospheric disturbances [20]. The centers of two elliptical holes are separated by a distance called pitch which is selected as $\Lambda = 2.25$ μm. Figure 1a presents the 2*D* design of the presented *RI* sensor. The structural design of the presented *RI* sensor having a thin layer of plasmonic materials is zoomed to visually identify the thickness of the plasmonic material represented by Fig. 1b. Figure 1c presents the formation of the quad cores along *X*—polarization. Similarly, Fig. 1d shows the quad-core formation along with the *Y*—polarization modes.

The sensing methodology of the presented *RI* sensor is shown in Fig. 1e. Here, the light from the optical source passes in the proposed fiber through *IN* port along with the analytes. Those *RI* need to be investigated. The analyte is taken out of the *PCF* using the *OUT* port. An optical spectrum analyzer (*OSA*) is used to detect the variation developed in the light signal corresponding to different analytes which pass through the optical fiber. The output of *OSA* is connected with the computer to obtain the change produced in wavelength (*nm*). The shift in wavelength is different, corresponding to other analytes, oil samples, and chemicals. The capability of the setup can be enhanced by using a device known as a polarization controller. Thus, different analytes and chemicals can be analyzed from the proposed sensing setup. The *RI* range of 1.40 to 1.48 *RIU* belongs to household oils and analytes [8, 12]. The proposed system works efficiently in the presence of the computer system because the computer system is always required for reading the output obtained generated from the *OSA* device. No information about the chemical and oil behavior can be obtained if the computer read-out device is absent. Thus, this research work presents the sensing behavior of the proposed sensor with computer vision merged with optics.

The *Au* layer can be deposited over the *PCF* fiber using the "Drude-Lorentz Model" employing the *CVD* technique [3]. Sensing performance parameters for

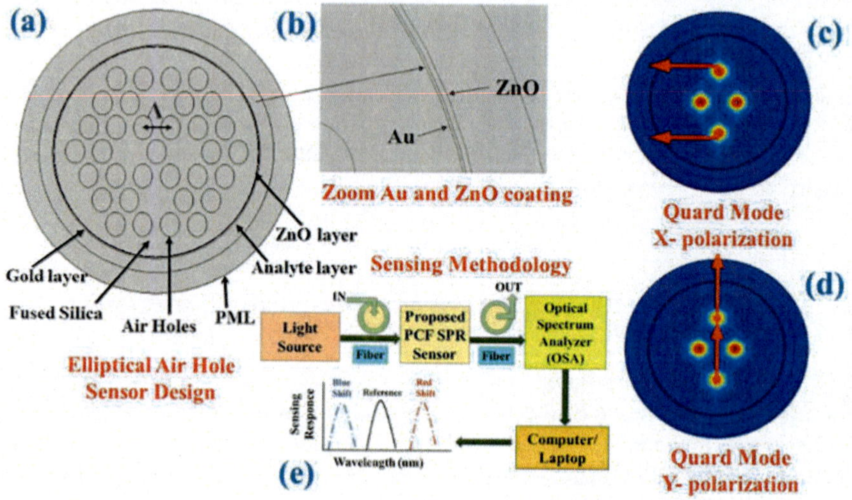

Fig. 1 Designed *PCFSPR* sensor model, **b** Zoom *Au* and *ZnO* layers, **c** Quad-core (*X*—polarization), **d** Quad-core (*Y*—polarization), and **e** Sensing setup for analyzing analyte using the proposed sensor

any designed sensor include confinement loss (*CL*), wavelength sensitivity (WS), amplitude sensitivity (*AS*), sensor resolution (*SR*), linear relationship between *RI*, and resonant wavelength [3]. They are expressed by Eqs. (1–4) [7].

1. Confinement loss (*CL*): It is defined as the amount of loss developed due to the non-perfect design of the sensor model. It can be understood as the power loss going outside the core of the designed *PCF*. It is expressed in terms of dB/cm and expressed by Eq. (1) [7]:

$$CL\left(\frac{dB}{cm}\right) = 8.686 \times k_0 \times \text{img}(n_{eff}) \times 10^4 \tag{1}$$

2. Wavelength sensitivity (WS): It is expressed as the ratio of peak difference between two consecutive *CL* peaks to the change in the *RI* of the biochemical. It is represented by Eq. (2) [7]:

$$WS = \frac{\Delta\lambda_{peak}}{\Delta RI\,(refractive\ Index)} \tag{2}$$

where $\Delta\lambda_{peak}$ is the difference in the *CL* between two consecutive analytes. It is assigned a unit nm/RIU [3].

3. Amplitude sensitivity (*AS*): It is expressed with the assistance of Eq. (3). It is assigned as a unit RIU^{-1} [3]:

$$AS(\mathrm{RIU})^{-1} = -\frac{1}{\alpha(\lambda, n_a)} \frac{\partial \alpha(\lambda, n_a)}{\partial n_a} \tag{3}$$

Here, ∂n_a represents the difference in the *RI* value of two consecutive analytes.

4. Sensor resolution (*SR*): It can be defined as the potential of the sensor to identify the slightest amount of drift in the "*RI* of the analyte." It is represented by Eq. (4) and assigned a unit *RIU* [3]:

$$SR(RIU) = \frac{\Delta n_a \times \Delta\lambda_{min}}{\Delta\lambda_{peak}} RIU \tag{4}$$

5. Linearity of resonance wavelength with *RI* can be assumed as a linear relationship between these two parameters. The goodness of the curve fitting between these parameters is expressed by the R^2.

3 Simulation Results

The *CL* (*dB*/*cm*) obtained from the proposed *RI* sensor corresponding to *X*—polarization is presented in Fig. 2a, respectively, for biochemical with *RI* varying from 1.40 to 1.48 *RIU*. The *CL* corresponding to different analytes is 48.64 dB/cm, 49.08 dB/cm, 49.20 dB/cm, 50.86 dB/cm, 50.89 dB/cm, 50.98 dB/cm, 51.15 dB/cm, 53.45 dB/cm, and 53.65dB/cm corresponding to analytes having *RI* 1.40, 1.41, 1.42, 1.43, 1.44, 1.45, 1.46, 1.47, and 1.48 respectively for *X*—polarization.

The *CL* corresponding to different analytes for *Y*—polarization is 42.18 dB/cm, 42.36 dB/cm, 42.36 dB/cm, 42.54 dB/cm, 42.68 dB/cm, 42.98 dB/cm,

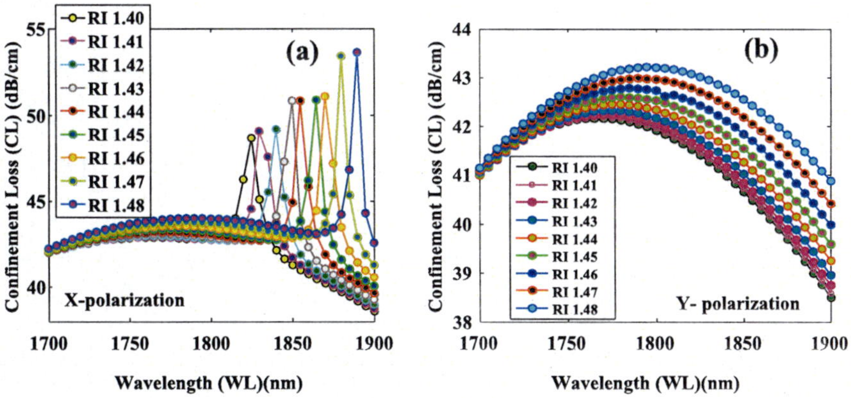

Fig. 2 Confinement loss versus wavelength **a** *X*—polarization and **b** *Y*—polarization

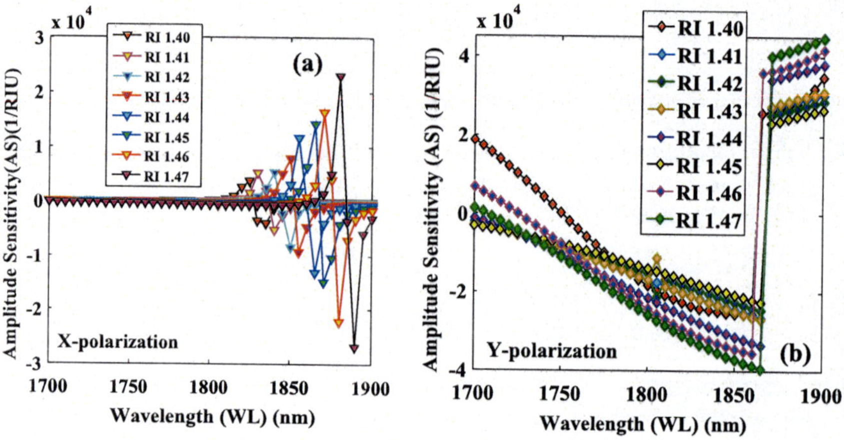

Fig. 3 Amplitude sensitivity versus wavelength **a** X—polarization and **b** Y—polarization

43.14 dB/cm, 43.18 dB/cm, 43.20 dB/cm, and 43.24 dB/cm, respectively, corresponding to biochemical having RI 1.40, 1.41, 1.42, 1.43, 1.44, 1.45, 1.46, 1.47, and 1.48 RIU respectively presented in Fig. 2b.

The amplitude sensitivity corresponding to wavelength for different analytes 3613 RIU^{-1}, 4107 RIU^{-1}, 5172 RIU^{-1}, 8380 RIU^{-1}, 9272 RIU^{-1}, 13074 RIU^{-1}, 14954 RIU^{-1}, 22150 RIU^{-1}, and 26834 RIU^{-1} corresponding to biochemical having RI 1.40, 1.41, 1.42, 1.43, 1.44, 1.45, 1.46, and 1.47 RIU, respectively, corresponding to X—polarization is presented in Fig. 3a.

The amplitude sensitivity corresponding to wavelength for different analytes is 21380 RIU^{-1}, 22630 RIU^{-1}, 24187 RIU^{-1}, 26178 RIU^{-1}, 26990 RIU^{-1}, 33580 RIU^{-1}, 35590 RIU^{-1}, and 39550 RIU^{-1} corresponding to biochemical having RI 1.40, 1.41, 1.42, 1.43, 1.44, 1.45, 1.46, and 1.47 RIU, respectively, for Y—polarization as illustrated in Fig. 3b.

The shift in resonance wavelength is 1805, 1810, 1815, 1820, 1830, 1840, 1850, 1860, and 1890nm for biochemical having RI 1.40, 1.41, 1.42, 1.43, 1.44, 1.45, 1.46, and 1.47 RIU respectively for X—polarization. The wavelength sensitivity for the proposed design is 500 nm/RIU, 500 nm/RIU, 500 nm/RIU, 1000 nm/RIU, 1000 nm/RIU, 1000 nm/RIU, 1000 nm/RIU, and 3000 nm/RIU for biochemical having RI 1.40, 1.41, 1.42, 1.43, 1.44, 1.45, 1.46, and 1.47 respectively for X—polarization.

The resonance wavelength for Y—polarization is shifted from 1760, 1765, 1770, 1775, 1780, 1785, 1795, 1810, and 1835nm for biochemical having RI 1.40, 1.41, 1.42, 1.43, 1.44, 1.45, 1.46, and 1.47 RIU respectively for Y—polarization. The wavelength sensitivity for the proposed design is 500 nm/RIU, 500 nm/RIU, 500 nm/RIU, 500 nm/RIU, 500 nm/RIU, 1000 nm/RIU, 1500 nm/RIU, and 2500 nm/RIU for biochemical having RI 1.40, 1.41, 1.42, 1.43, 1.44, 1.45, 1.46, and 1.47 respectively for Y—polarization.

The sensor resolution corresponding to X—polarization is $2 \times 10^{-4}\,RIU$, $2 \times 10^{-4}\,RIU$, $2 \times 10^{-4}\,RIU$, $1 \times 10^{-4}\,RIU$, $1 \times 10^{-4}\,RIU$, $1 \times 10^{-4}\,RIU$, $1 \times 10^{-4}\,RIU$, and $3.33 \times 10^{-5}\,RIU$ for biochemicals having RI1.40, 1.41, 1.42, 1.43, 1.44, 1.45, 1.46, and 1.47 RIU, respectively.

The sensor resolution corresponding to Y—polarization is $2 \times 10^{-4}\,RIU$, $2 \times 10^{-4}\,RIU$, $2 \times 10^{-4}\,RIU$, $2 \times 10^{-4}\,RIU$, $2 \times 10^{-4}\,RIU$, $1 \times 10^{-4}\,RIU$, $6.66 \times 10^{-5}\,RIU$, and $4.00 \times 10^{-5}\,RIU$ for biochemicals having RI 1.40, 1.41, 1.42, 1.43, 1.44, 1.45, 1.46, and 1.47 RIU, respectively.

The fitting between resonant wavelength and RI provides information about "sensor optimization." A value of R—square close to unity represents good fitting between resonance wavelength and RI. The fitting between RI and resonant wavelength produces $R^2 = 0.9839$ corresponding to X—polarization and $R^2 = 0.9758$ corresponding to Y—polarization illustrated by Fig. 4a and b, respectively. The value of R^2 is close to unity which represents great fitting to the sensor response.

The peak value of the sensing parameters is produced for the RI having a value of 1.47 RIU. Thus, the proposed sensor has justified various features based on which it can be considered an effective RI sensor.

Finally, Table 1 compares the parameters obtained for the proposed RI sensor with other reported sensors developed to date.

Besides the conventional sensor parameters, the figure of merit (FOM) can also be obtained for the designed sensor model. The FOM is dependent on the full-wave half maximum ($FWHM$). Today, $PCFSPR$ sensing field has been immensely revolutionized. Scientists and researchers have presented several applications related to $PCFSPR$ sensors like cancer detection, environmental monitoring, pregnancy detection, transformer oil monitoring, food pathogen detection, etc. These photonic sensors are working on the variation in the RI index values. Thus, there is the possibility that they can be used in several application areas where a change is determined on the basis of variation in the RI values. Household oils like coconut oil, gooseberry oil, and amla oil have RI varying in the range of 1.40–1.48 RIU, besides some biochemicals

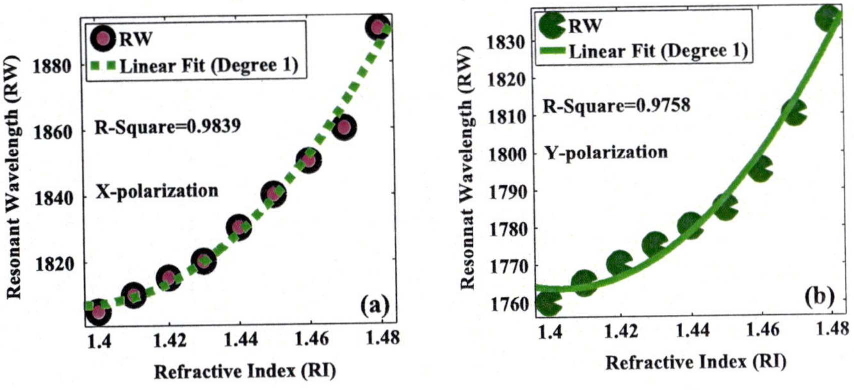

Fig. 4 Resonant wavelength versus RI **a** X—polarization and **b** Y—polarization

Table 1 Comparison of the sensing parameters with other sensors

References	Design	Wavelength sensitivity (nm/RIU)	Amplitude sensitivity (RIU^{-1})	Sensor resolution (RIU)
[22]	Birefringent PCF	2000	317	3.15×10^{-5}
[23]	Holly PCF	2000	370	2.70×10^{-5}
[24]	Surface core PCF	40	–	–
[25]	Birefringent PCF	–	860	4.00×10^{-5}
Proposed *PCFSPR* sensor		3000 2500	26834 39550	3.33×10^{-5} 4.00×10^{-5}

having the same operational range of *RI*. Thus, the proposed *RI* sensor is designed to cover *RI* range of various chemicals, household oils, and biochemicals. It is expected that with the evolution of *RI* sensing, *PCFSPRRI* sensors will be used in several new application areas.

4 Conclusion

The proposed *RI* sensor has presented reasonable sensing parameters, due to which it can be considered a suitable sensor for the detection of various analytes, oils, and biochemicals. It has produced wavelength sensitivity of 3000 nm/RIU and 2500 nm/RIU, corresponding to *X*—polarization and *Y*—polarization, respectively. An extreme peak amplitude sensitivity of 26834 RIU^{-1} and 39550 RIU^{-1} is presented corresponding to *X*—polarization and *Y*—polarization, respectively. The proposed sensor has delivered a sensor resolution in the range of 10^{-5} RIU, more specifically, the sensor resolution of 3.33×10^{-5} RIU and 4.00×10^{-5} RIU is obtained corresponding to *X*—polarization and *Y*—polarization, respectively. The value of R^2 is 0.9839 for *X*—polarization and 0.9758 corresponding to *Y*—polarization which is close to unity, resulting in the good fitting of the sensor parameters. Besides, the combination of plasmonic material *Au* with *ZnO* is also reported in this research work. Thus, the proposed sensor is an effective sensor for the *RI* range of 1.40 RIU to 1.48 RIU.

Acknowledgements "This work is performed under the All India Council of Technical Education (AICTE), National Doctoral Fellowship (NDF). Authors are further thankful to AICTE for the AICTE NDF RPS project, sanction order no: File No.8-2/RIFD/RPS-NDF/Policy-1/2018-19 dated March 13, 2019".

References

1. Liu W, Wang F, Liu C, Yang L, Liu Q, Su W, Lv J (2020) A hollow dual-core PCF-SPR sensor with gold layers on the inner and outer surfaces of the thin cladding. Results Opt 1:100004. https://doi.org/10.1016/j.rio.2020.100004
2. Khanikar T, De M, Singh VK (2021) A review on infiltrated or liquid core fiber optic SPR sensors. In: Photonics and nanostructures —fundamentals and applications, vol 46, p 100945. https://doi.org/10.1016/j.photonics.2021.100945
3. Shakya AK, Singh S (2021) Design of dual-polarized tetra core PCF based plasmonic RI sensor for visible-IR spectrum. Opt Commun 478:126372. https://doi.org/10.1016/j.optcom.2020.126372
4. Yang H, Wang G, Lu Y, Yao J (2021) Highly sensitive refractive index sensor based on SPR with silver and titanium dioxide coating. Opt Quantum Electron 53:341. https://doi.org/10.1007/s11082-021-02981-1
5. Butt M, Khonina S, Kazanskiy N (2021) Plasmonics: a necessity in the field of sensing-a review (invited). Fiber Integrat Opt 40:14–47. https://doi.org/10.1080/01468030.2021.1902590
6. Liu Q, Ma Z, Wu Q (2020) The biochemical sensor based on liquid-core photonic crystal fiber filled with gold, silver, and aluminum. Opt Laser Technol 130:106363. https://doi.org/10.1016/j.optlastec.2020.106363
7. Shakya AK, Singh S (2021) Design and analysis of dual-polarized Au and TiO2-coated photonic crystal fiber surface plasmon resonance refractive index sensor: an extraneous sensing approach. J Nanophotonics 15(1):016009
8. Liu A, Wang J, Wang F, Su W, Yang L, Lv J, Fu G (2020) Surface plasmon resonance (SPR) infrared sensor based on D-shape photonic crystal fibers with ITO coatings. Opt Commun 464:125496. https://doi.org/10.1016/j.optcom.2020.125496
9. Danlard, Akowuah EK (2021) Design and theoretical analysis of a dual-polarized quasi D-shaped plasmonic PCF microsensor for back-to-back measurement of refractive index and temperature. IEEE Sens J 21(8):9860 —9868
10. Shakya K, Singh S (2022) Design of novel Penta core PCF SPR RI sensor based on the fusion of IMD and EMD techniques for analysis of water and transformer oil. Measurement 188:110513. https://doi.org/10.1016/j.measurement.2021.110513
11. Monfared YE (2020) Refractive index sensor based on surface plasmon resonance excitation in a D-shaped photonic crystal fiber coated by titanium nitride. Plasmonics 15:535–542. https://doi.org/10.1007/s11468-019-01072-y
12. Liang H, Shen T, Feng Y, Liu H, Han W (2021) A D-shaped photonic crystal fiber refractive index sensor coated with graphene and zinc oxide. Sensors 21(1):71
13. Chen DY, Zhao Y (2021) Review of optical hydrogen sensors based on metal hydrides: Recent developments and challenges. Opt Laser Technol 137:106808. https://doi.org/10.1016/j.optlastec.2020.106808
14. Hasan MM, Barid M, Hossain MS, Sen S, Azad MM (2021) Large effective area with high power fraction in the core region and extremely low effective material loss-based photonic crystal fiber (PCF) in the terahertz (THz) wave pulse for different types of communication sectors. J Opt 50:681–688. https://doi.org/10.1007/s12596-021-00740-9
15. Ramola A, Marwaha A, Singh S (2021) Design and investigation of a dedicated PCF SPR biosensor for CANCER exposure employing external sensing. Appl Phys A 127:643. https://doi.org/10.1007/s00339-021-04785-2
16. Popescu V, Sharma AK, Marques C (2021) Resonant interaction between a core mode and two complementary supermodes in a honeycomb PCF reflector-based SPR sensor. Optik 227:166121. https://doi.org/10.1016/j.ijleo.2020.166121
17. Zhu M, Yang L, Lv J, Liu C, Li Q, Peng C, Li X, Chu PK (2021) Highly sensitive dual-core photonic crystal fiber based on a surface. Plasmonics 1:1–8. https://doi.org/10.1007/s11468-021-01543-1
18. Yan X, Wang Y, Cheng T, Li S (2021) Photonic crystal fiber SPR liquid sensor based on elliptical detective channel. Micromachines 12(4):408

19. Falah AS, Wong WR, Adikan FRM (2022) Single-mode eccentric-core D-shaped photonic crystal fiber surface plasmon resonance sensor. Opt Laser Technol 145:107474. https://doi.org/10.1016/j.optlastec.2021.107474
20. Shakya AK, Singh S (2022) Design of biochemical biosensor based on transmission, absorbance, and refractive index. Biosens Bioelectron X 10:100089. https://doi.org/10.1016/j.biosx.2021.100089
21. Society G (2021) Refractive index list of common household liquids, IGS, 01 January 2021. https://www.gemsociety.org/article/refractive-index-list-of-common-household-liquids/. [Accessed 01 Nov 2021].
22. Otupiri R, Akowuah EK, Haxha S, Ademgil H, AbdelMalek F, Aggoun A (2014) A novel birefringent photonic crystal fiber surface plasmon resonance biosensor. IEEE Photonics J 6(4):6801711
23. Gao D, Guan C, Wen Y, Zhong X, Yuan L (2014) Multi-hole fiber-based surface plasmon resonance sensor operated at near-infrared wavelengths. Opt Commun 313:94–98. https://doi.org/10.1016/j.optcom.2013.10.015
24. Osório H, Oliveira R, Aristilde S, Chesini G, Franco MAR (2017) Bragg gratings in surface-core fibers: refractive index and directional curvature sensing. Opt Fiber Technol 34:86–90. https://doi.org/10.1016/j.yofte.2017.01.007
25. Dash N, Jha R (2014) Graphene-based birefringent photonic crystal fiber sensor using surface plasmon resonance. IEEE Photon Technol Lett 26(11):1092–1095

Fault Detection and Diagnostics in a Cascaded Multilevel Inverter Using Artificial Neural Network

Stonier Albert Alexander, M. Srinivasan, D. Sarathkumar, and R. Harish

1 Introduction

In industrial applications, the inverters play a major role in adjustable speed control of AC drives, induction heating, air-craft stand-by power supplies, UPS for computers, etc. The phase-controlled converter operated in the inverter mode is called a line commutated inverter that requires the existing AC supply for the commutation purpose. This implies that the line commutated inverter cannot be operated as an isolated AC voltage source or a variable frequency generator with the input as DC power. Thus, the AC side voltage of the line commutated inverter cannot be changed by its voltage and frequency. Hence, the forced commutated inverters are used to provide adjustable voltage and frequency for independent AC output that are used in wider applications. The DC power input to the inverter is fed from different kinds of sources like battery, photovoltaic array and fuel cell. This can be done by using the DC link which comprises an AC to DC converter and a DC to AC inverter connected to the DC link. Most of the rectification process is performed using diodes or thyristor converter circuits.

Basically, the inverters are classified into two different types such as voltage source inverters (VSI) and current source inverters (CSI). For the reduction of harmonics, multilevel inverters are highly preferred whose types are (i) Flying capacitor inverter, (ii) Diode-clamped system inverter and (iii) Cascade H-type level inverter [1–5]. Among the various types, owing to the advantages of cascaded multilevel inverter is taken into consideration in this paper. A cascaded H-bridge multilevel inverter can

S. Albert Alexander (✉) · M. Srinivasan · D. Sarathkumar · R. Harish
Electrical and Electronics Engineering, Kongu Engineering College, Perundurai 638060, India
e-mail: albert.alexander@vit.ac.in

S. Albert Alexander
School of Electrical Engineering (SELECT), Vellore Institute of Technology, Vellore 632014, India

© The Author(s), under exclusive license to Springer Nature Singapore Pte Ltd. 2023

H. Muthusamy et al. (eds.), *Robotics, Control and Computer Vision*, Lecture Notes in Electrical Engineering 1009, https://doi.org/10.1007/978-981-99-0236-1_44

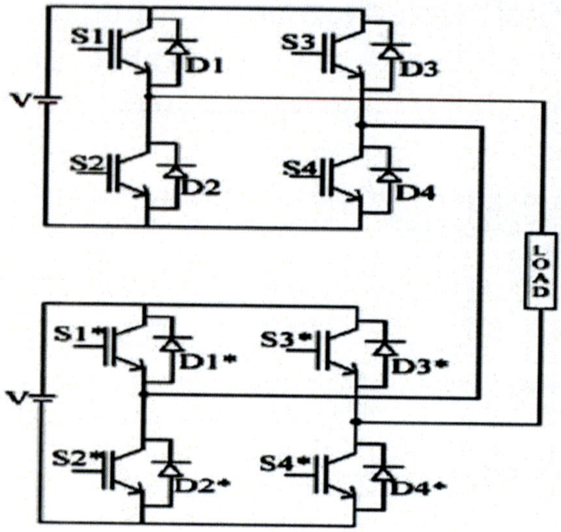

Fig. 1 Cascaded five-level multilevel inverter

be used for both single-phase and three-phase systems. Each H-bridge cell consists of four switches and fly-wheeling diodes.

The proposed method deals with the implementation of a five-level cascaded multilevel inverter employed with multilayer perceptron networks to identify the fault location from inverter output voltage measurement and the corresponding diagnosis for the same. Figure 1 shows the five-level cascaded multilevel inverter comprising 8 semiconductor switches. The objective of the work is to appropriately detect the various faults existing in the system. In addition, the system should locate the fault and diagnose it by stimulating the auxiliary circuit for providing continuous power even under fault conditions. Most of the literature dealt with the faults by considering only the common short-circuit and open-circuit faults [6–15]. In this paper, an intelligence-based ANN is proposed to detect and diagnose the various faults in an inverter configuration.

2 Proposed Methodology

The structure of a fault diagnostic system is illustrated in Fig. 2. The structure has four main blocks such as feature extraction, network configuration, fault diagnosis and switching pattern calculation. The feature extraction block extracts the output voltage of a five-level inverter and transfers the same to the ANN. The ANN is trained with normal and fault data and provides the corresponding binary code that if "1" arrives it is a normal condition, and if "0" it is a fault condition. Hence, the output of the network configuration is merely the binary code of either 0 or 1.

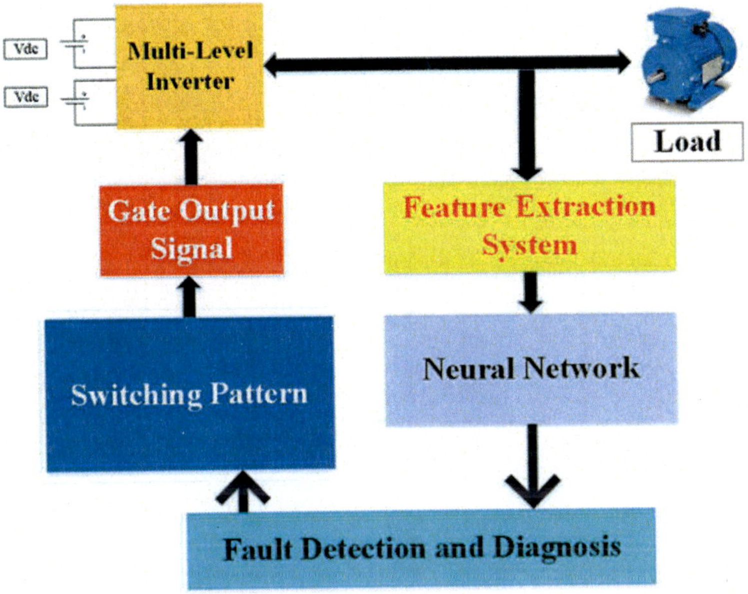

Fig. 2 Functional block diagram

The location corresponding to the code is then sent to the fault diagnosis to interpret the condition. Based on this, the switching pattern is calculated which is then provided to the inverter switches. A single-phase cascaded multilevel inverter with 10 V DC and MOSFET as the switching device is used. The level of an inverter is given by m = 2Ns + 1. Here, m denotes the level of an inverter and Ns denotes the number of stages included. In the proposed configuration, m = 5 and Ns = 2.

The types of faults considered and their conditions are as follows:

- Open-circuit fault (V = 10 V; I = 0.09693A)
- Short-circuit fault (V = 0 V; I = 10.32A)
- Over-voltage fault (V = 99.96 V; I = 9.63A)
- Losing drive pulse fault (V = 19.99 V; I = 1.907A).

The losing drive pulse fault occurs when the pulse given to the circuit is lost or if the pulse is not given properly. If the given pulse is wrong, the normal output will not be displayed. The output may vary based on the pulse provided.

MATLAB/Simulink simulation tool is used to simulate the proposed system. The selection of an approximate signal is much essential for feature extraction and will have a significant insight to make a decision, and the highest degree of accuracy is obtained by a neural network. Features concentrate on voltage, current and error signals at various normal and abnormal conditions. The dataset is the first pre-requisite for the process of ANN. Once the dataset is obtained, the next stage is training which is done with the aid of a backpropagation algorithm. Once the training is completed, the testing process is followed to check the accuracy of the system. The network is examined by the test data values given to the network and is trained to achieve the desired goal. Testing of the system network is based on the way by which the system responds to normal and fault conditions. The trained system covers the entire fault detection and diagnosing of the network to the required level of the output requirement. Figure 3 shows the simulation of a five-level inverter without an ANN-based controller. Figure 4 shows the simulation of the inverter with ANN.

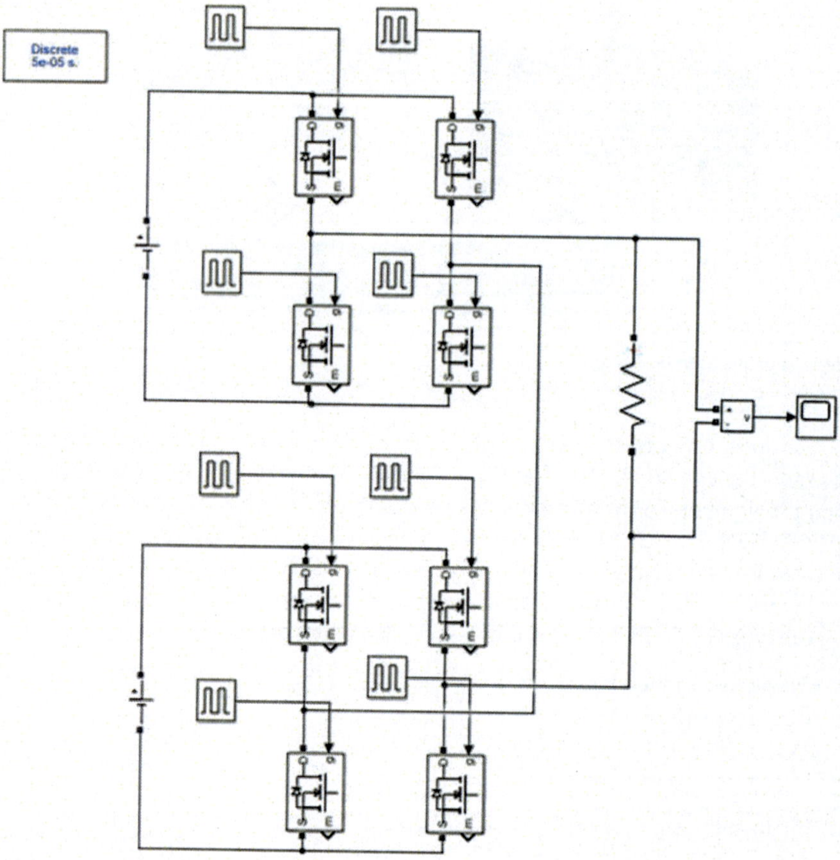

Fig. 3 Simulation of cascaded five-level inverter

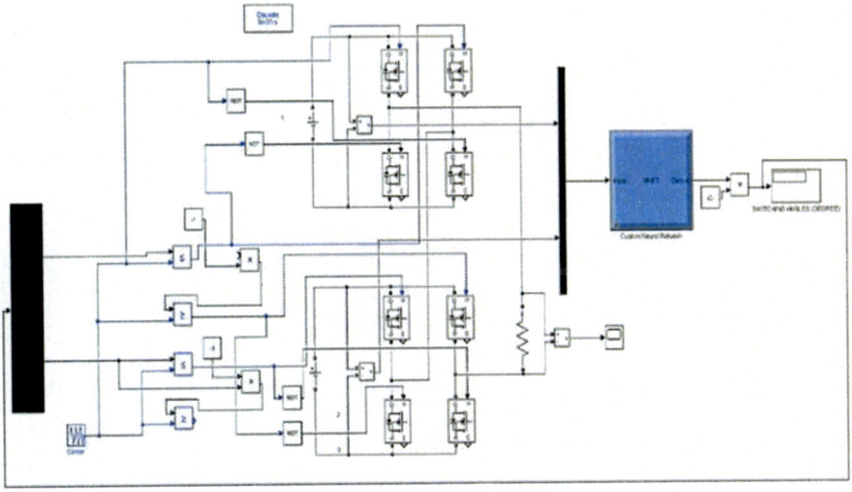

Fig. 4 Simulation of cascaded five-level inverter with ANN

3 Artificial Neural Network

Neural networks comprise different layers such as input, hidden and output. Figure 5 shows the network architecture. The layers are interconnected with the aid of activation functions to perform mathematical calculations and corresponding scaling processes. Input layers are linked with each other in the form of a hidden layer and an output layer. The function used is the sign activation function for input layer nodes, tangent for hidden nodes, and log segment for the output node. Among the various algorithms used for the implementation of ANN, the BPN algorithm is predominately used for complex applications. The functions performed in the BPN algorithm are feed-forward of data, error backpropagation and weight (connection links between the layers) updating [16–20]. The algorithmic involved for the implementation of fault detection and diagnosis is given as follows:

- Two-stage five-level inverter is simulated using MATLAB/Simulink.
- Voltage and current values were collected by varying the load conditions.
- With the aid of a dataset, the neural network was trained to get the best training performance curve.
- The network is trained to detect and diagnose the various faults.
- The trained system is tested to check its accuracy.
- Five-level inverter is now implemented with ANN.

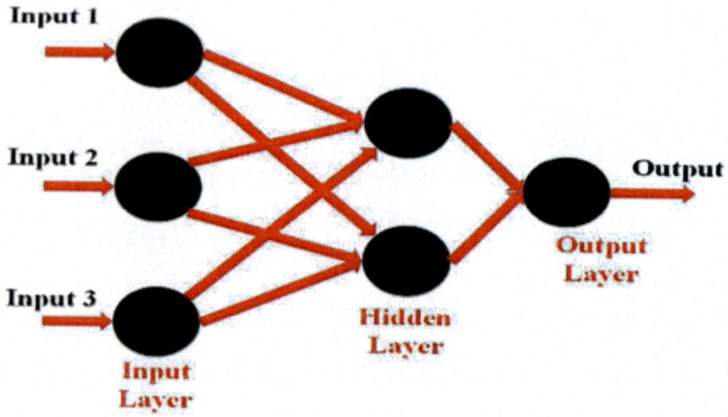

Fig. 5 ANN architecture

4 Results and Discussion

The simulation results using MATLAB for various fault conditions using the ANN controller are shown in the following figures. Without introducing any fault, the waveform obtained under normal conditions is obtained as shown in Fig. 6. It clearly depicts the five-level output voltage waveform. By introducing various faults like an open circuit, short circuit, losing drive pulse and overvoltage faults, the waveforms are obtained as shown in Figs. 7, 8, 9 and 10, respectively. Figure 11 shows the training performance curve trained with the ANN-based controller. For the different time intervals, the faults are introduced, tested and analyzed. Figure 12 shows the waveform obtained in the five-level inverter with ANN after introducing a fault in the system.

The various types of faults are detected by the corresponding binary values of ANN (as per its training) as displayed in Table 1. Various fault detections are observed during the simulation process using the Artificial Neural Network with the reference output voltage waveform compared with the actual waveform obtained during the different fault conditions. The different types of faults have been detected by comparing the output waveforms of actual and desired ones. According to the result, the values assigned to each fault are 00, 01, 10 and 11 by the ANN controller, and the fault detection processed can be easily assessed.

5 Conclusion

In this article, the fault detection and diagnosis of the cascaded five-level inverter using a backpropagation algorithm-enabled artificial neural network is performed. Different types of faults are induced in the cascaded multilevel inverter, and fault detection and diagnosis are undertaken with reduced computation complexity. The

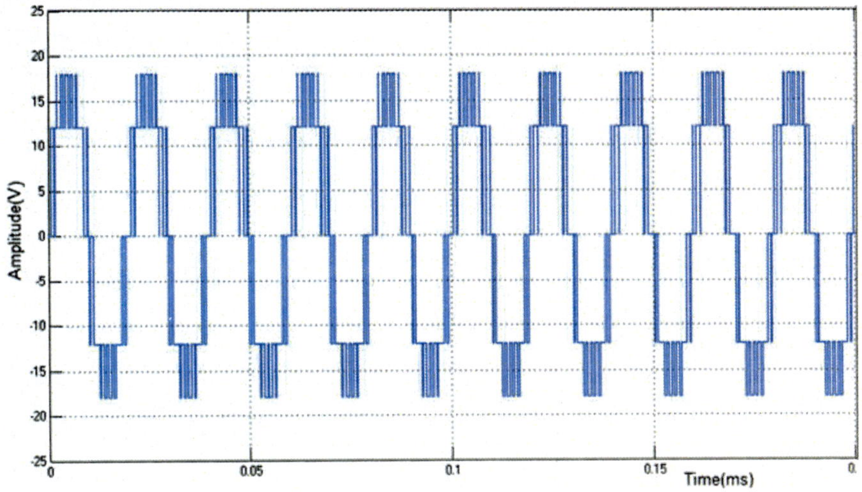

Fig. 6 Normal five-level waveform

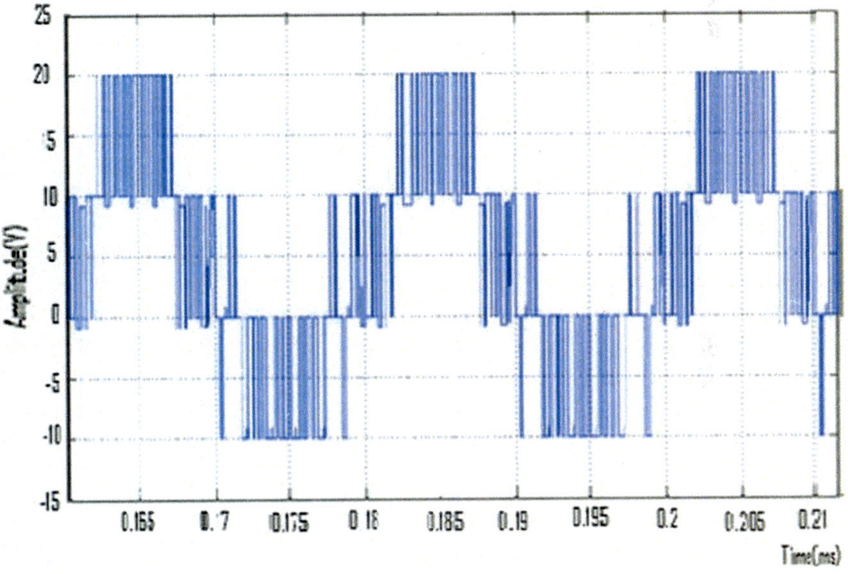

Fig. 7 Open-circuit fault

fault conditions considered in the paper are short-circuit fault, open-circuit fault and overvoltage fault along with other common faults.

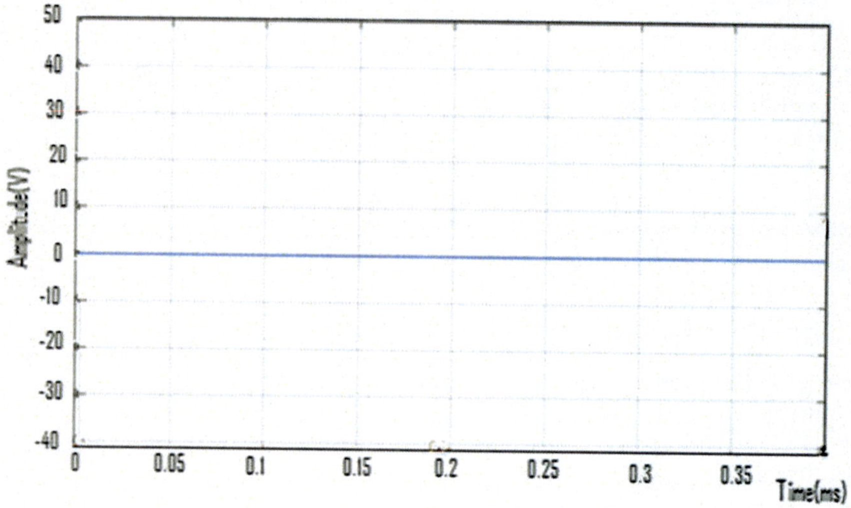

Fig. 8 Short-circuit fault

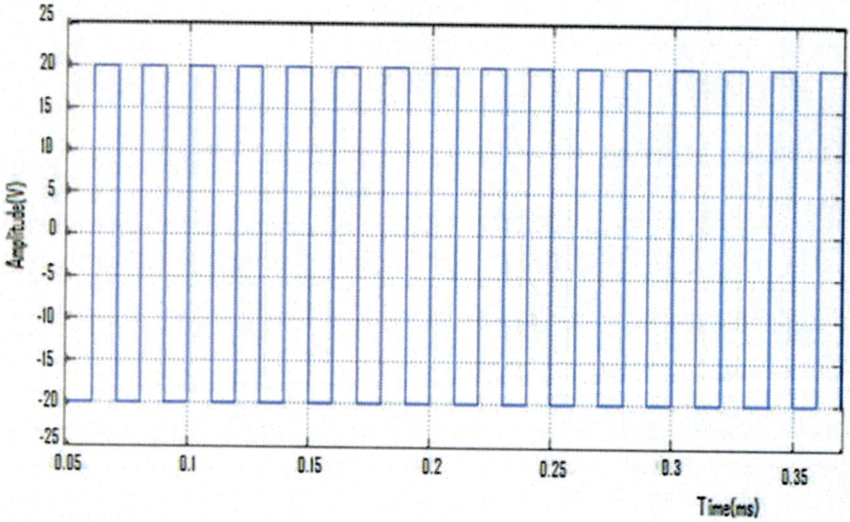

Fig. 9 Losing gate drive pulse fault

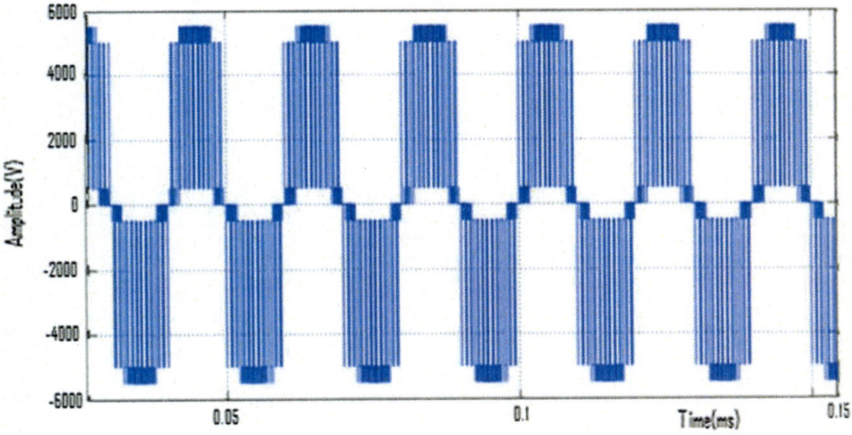

Fig. 10 Overvoltage fault

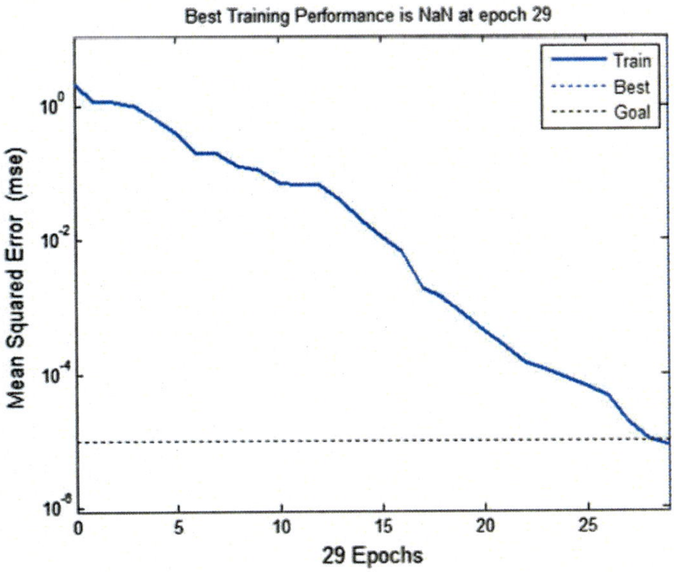

Fig. 11 Neural network training curve

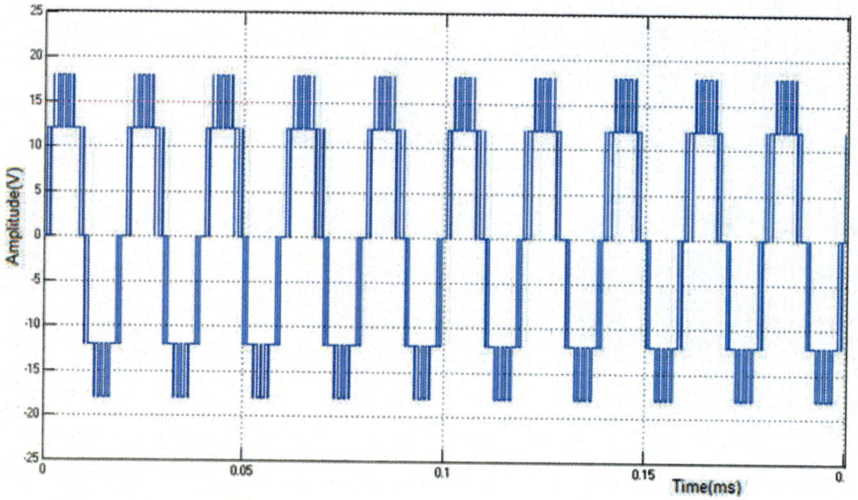

Fig. 12 Output voltage waveform after ANN training

Table 1 Various fault and detection values using ANN

S. no	Type of the fault	Values displayed by ANN
1	Open-circuit fault	00
2	Short-circuit fault	01
3	Losing gate drive pulse fault	10
4	Overvoltage fault	11

Funding The authors acknowledge and thank the Department of Science and Technology (Government of India) for sanctioning the research grant for the project titled, "Design and Development of Solar Photovoltaic Assisted Micro-Grid Architecture with Improved Performance Parameters Intended for Rural Areas" (Ref. No. DST/TMD/CERI/RES/2020/32 (G) dated 03.06.2021) under TMD-W&CE Scheme for completing this work.

References

1. Vanaja DS, Stonier AA, Mani G, Murugesan S (2021) Investigation and validation of solar photovoltaic fed modular multilevel inverter for marine water pumping applications. Electr Eng. https://doi.org/10.1007/s00202-021-01370-x
2. Jalhotra M, Sahu LK, Gupta S, Gautam SP (2021) Highly resilient fault-tolerant topology of single-phase multilevel inverter. IEEE J Emerg Select Topics Power Electron 9(2)
3. Kumar M (2021) Open circuit fault detection and switch identification for LS-PWM H- bridge inverter. IEEE Trans Circuits Syst—Ii: Express Briefs 68(4)
4. Majumder MG, Rakesh R, Gopakumar K, Umanand L, Al-Haddad K, Jarzyna W (2021) A fault-tolerant five-level inverter topology with reduced component count for OEIM drives.

IEEE J Emerg Select Top Power Electron 9(1)
5. Huang Z, Wang Z, Song C (2021) Complementary virtual mirror fault diagnosis method for microgrid inverter. IEEE Trans Indust Inform 17(11)
6. Mhiesan H, Wei Y, Siwakoti YP, Mantooth HA (2020) A fault-tolerant hybrid cascaded h-bridge multilevel inverter. IEEE Trans Power Electron 35(12)
7. Fard MT, Khan WA, He J, Weise N, Abarzadeh M (2020) Fast online diagnosis of open-circuit switching faults in flying capacitor multilevel inverters. Chinese J Electr Eng 6(4)
8. Shi X, Zhang H, Wei C, Li Z, Chen S (2020) Fault modeling of IIDG considering inverter's detailed characteristics. In: IEEE power and energy society section received, September 14, 2020, accepted September 23, 2020, date of publication September 28, 2020, date of current version, (October 19, 2020)
9. Guo X, Sui S, Wang B, Zhang W (2020) A current-based approach for short-circuit fault diagnosis in closed-loop current source inverter. IEEE Trans Indust Electron 67(9)
10. Zhang Z, Luo G, Zhang Z, Tao X (2020) A hybrid diagnosis method for inverter open-circuit faults in PMSM drives. CES Trans Electr Mach Syst 4(3)
11. Chao KH, Chang LY, Xu FQ (2020) Three-level T-type inverter fault diagnosis and tolerant control using single-phase line voltage. In: IEEE access received February 11, 2020, accepted February 24, 2020, date of publication March 3, 2020, (March 13, 2020)
12. Cheng Y, Dong W, Gao F, Xin G (2020) Open-circuit fault diagnosis of traction inverter based on compressed sensing theory. Chinese J Electr Eng 6(1)
13. Praveen Kumar N, Isha TB (2019) FEM based electromagnetic signature analysis of winding inter-turn short-circuit fault in inverter fed induction motor. CES Trans Electr Mach Syst 3(3)
14. de Mello Oliveira AB, Moreno RL, Ribeiro ER (2019) Short- circuit fault diagnosis based on rough sets theory for a single-phase inverter. IEEE Trans Power Electron 34(5)
15. Wu X, Chen TF, Cheng S, Yu T, Xiang C, Li K (2019) A non- invasive and robust diagnostic method for open-circuit faults of three-level inverters. In: IEEE access received November 8, 2018, accepted December 10, 2018, date of publication December 17, 2018, date of current version January 7, 2019
16. Stonier AA, Lehman B (2017) An intelligent-based fault-tolerant system for solar- fed cascaded multilevel inverters. IEEE Trans Energy Convers 33(3):1047–1057
17. Alexander A, Thathan M (2013) Modelling and simulation of artificial neural network based harmonic elimination technique for solar-fed cascaded multilevel inverter. International review of modelling and simulations (IREMOS) 6(4):1048–1055
18. Alexander SA, Manigandan T (2014) Power quality improvement in solar photovol- taic system to reduce harmonic distortions using intelligent techniques. J Renew Sustain Energy 6(4):043127
19. Alexander A, Thathan M (2014) Design and development of digital control strategy for solar photovoltaic inverter to improve power quality. J Control Eng Appl Inf 16(4):20–29
20. Kumar AL, Alexander SA, Rajendran M (2020) Power electronic converters for solar photovoltaic systems. Academic

Identification of Multiple Solutions Using Two-Step Optimization Technique for Two-Level Voltage Source Inverter

M. Chaitanya Krishna Prasad, Vinesh Agarwal, and Ashish Maheshwari

1 Introduction

VSIs are typically used for generating alternating three-phase voltages of variable magnitude and frequency voltages from a fixed DC-source for the different applications such as variable speed or torque drives [1] traction drives or electrical vehicles [2] STATCOM [3] power system distributed generation [4] solar photo voltaic cells [5]. VSIs in electrical industrial markets have shown to be more efficient, dependable, and quicker in dynamic reaction, as well as capable of operating motors that have been de-rated [6] for low power applications, To increase the quality of the voltage source inverter output line voltage, the number of pulses is increased [7], i.e., P = 2N + 1, where 'N' represents the number of triggering instants for a quarter cycle of the fundamental voltage. However, due to higher switching losses in power semiconductor devices, low-frequency device switching is favored at higher levels [8]. At low switching frequency, Odd harmonics surround the fundamental component in pole voltage of a voltage source inverter (VSI) [9]. Various PWM techniques, such as the traditional SPWM, SVPWM, and SHE PWM, have been proposed for enhancing the inverter performance [10]. This paper presents the SHE technique, and several solutions for bipolar PWM waveform are examined. The primary distinction among the discussed modulation systems is the generation of Pulse Width Modulation(PWM) signals to switch ON and OFF the corresponding power electronic devices [11]. In the early 1970s the SHE PWM method was established with inverter angles switching based on off-line calculations [12]. This strategy is based

Vinesh Agarwal, and Ashish Maheshwari These authors contributed equally to this work.

M. Chaitanya Krishna Prasad (✉) · V. Agarwal
Electrical Engineering, Sangam University, NH-79, Chittor road, Bhilwara 311001, RJ, India
e-mail: chaitanyaprasad92@gmail.com

A. Maheshwari
Electrical Engineering, Government Polytechnic, UT of Daman, Diu and DNH, Varkund, Daman 396210, India

© The Author(s), under exclusive license to Springer Nature Singapore Pte Ltd. 2023

H. Muthusamy et al. (eds.), *Robotics, Control and Computer Vision*, Lecture Notes in Electrical Engineering 1009, https://doi.org/10.1007/978-981-99-0236-1_45

on the construction of a transcendental set of non-linear mathematical equations and the application of relevant SHE techniques, to get optimal switching angles [13]. Numerous algebraic, numerical, and optimization strategies have been presented in the literature to solve the SHE [14]. Newton Raphson (NR) iterative method is frequently used because of its quick convergence rate and precise results [15]. However, appropriate initial switching angle assumptions are required to attain global optimal solutions. And the resultant theory [16] transforms the non-linear equations of SHE to algebraic equations in order to get the real values of optimum switching angles. However, the method's complexity increases with an increase in the number of levels of the inverter. Several optimization-based approaches, such as the Genetic Algorithm(GA) [17] Bee Algorithm (BA) [18], Particle Swarm Optimization (PSO) [19], and Artificial Neural Network (ANN) [20], have been created and proposed to identify optimum solutions. Optimization-based solutions do not require sophisticated derivation and can potentially be used for multi-level inverter voltage. Furthermore, for each modulation index value, these algorithms identify several solutions (M). To determine the unique solutions of the bipolar switching waveform, a sequential-homotopy technique is described in [21]. The process of getting progressive angles for N intervals is find out using solutions of (N–1) intervals. Despite the fact that a comparison of the performance of distinct solution sets is not presented in experimental studies, and the process is lengthy and complex. Later, authors introduced a mathematical approach resultant theory [22], in which the non-linear equations for harmonic elimination are expressed as polynomial equations in order to find out all feasible sets of switching angles. However, computation complexity rises owing to higher order polynomials throughout the procedure of calculating optimal angles to larger switching angles and harmonics to be minimized or eliminated. The minimization technique is presented in [23] taking into consideration of selective harmonic elimination for single-phase and polyphase systems. However, this technique discovers unique required solutions by taking into account two fundamental waveforms of 180°C phase shifted with one another. Instead of completely eliminating non-tripled harmonics, a minimization approach is used, and the modulation index limits for total harmonic removal are not shown. The quarter-wave limits were removed in order to broaden the solution range and present several solutions sets within a single report [24]. However, these waveform constructions considerably increase the issue of complexity, resulting in a longer convergence time. The present research provides a comprehensive investigation and comparison of several sets of solutions linked to the SHE PWM approach of a two-level voltage source inverter across a linear range of modulation index for the N = 2 switching state instants. And Sect. 2 provides the outline and working principle of the two-level voltage source inverter. Section 3 has a full explanation for determining switching angles of 5th harmonic removal utilizing the SHE PWM method, as well as specifics on the combination of NR and GA methods. In Sect. 4 the analyses on the working of two sets of solutions at various modulation levels. with the help of MATLAB/Simulink, the simulation results based on the line voltage the THD, and the per unit value of fundamental voltage is produced. Section 5 contains the last concluding remarks.

The performance evaluation for reducing V_{WTHD} in the instance of a simple 2-level VSI with the switching angles N = 2, 3 and two distinct PWM waveforms type A and type B are provided. The transition of the voltage waveform value at the instance where the fundamental value of voltage is having the greatest positive slope may be changed to generate different sorts of waveforms. Obtained results revealed two distinct angle solutions for differing voltage distortion V_{WTHD} values for various modulation ranges associated with every waveform and type. However, only any one of the angle solution sets associated with every waveform type was considered to evaluate THD, limiting total inverter performance. The optimal switching angles for type A PWM current and voltage waveforms for I_{THD} and V_{WTHD} reduction with pulse-number P = 5 are evaluated in this study. The best optimum solution from the two available solution sets is chosen for the SHE PWM technique. Furthermore, with the same number of pulse, i.e., every quarter have N = 2 switching states in the fundamental waveform, a comparative study for maximum range with 5th harmonic elimination is performed. Furthermore, the theoretical results are validated using MATLAB simulations on the three-phase RL load linked to a two-level voltage source inverter.

2 Two-Level Voltage Source Inverter

Figure 1 depicts the setup of a two-level voltage source inverter. Every leg of the inverter contains two number of power electronic switches, pole voltage of any one phase is calculated using the DC bus's midpoint 'O,' i.e., V_{RO}, V_{YO} and V_{BO}. It is observed that the upper and lower switches are to be operated in complimentary ways to minimize the short circuit condition during DC supply transients. It has been recommended for keeping a minimal delay period for both the switches of the similar leg of the inverter must be turned OFF. While S_{R1} is turned ON and S_{R2} is turned OFF, the Pole voltage $V_{RO} = V_{dc}/2$; when S_{R2} is turned on and S_{R1} is turned off, the pole voltage $V_{RO} = -V_{dc}/2$. The voltage waveform of phase R is shown in Fig. 2

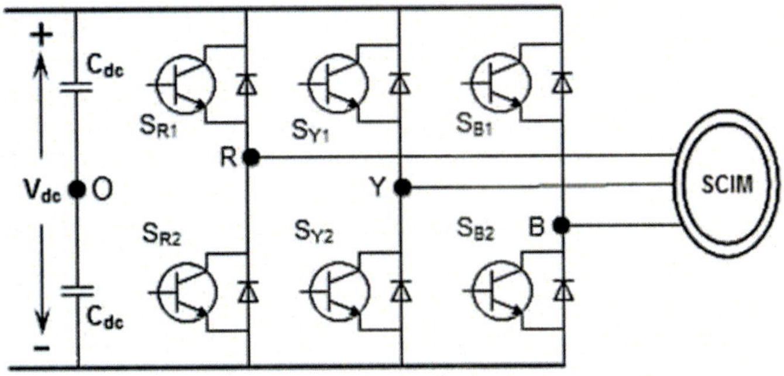

Fig. 1 The 3 phase two-level Voltage Source Inverter fed with a squirrel cage induction motor

Fig. 2 Bipolar waveforms for the pole voltage V_{RO} with N = 2

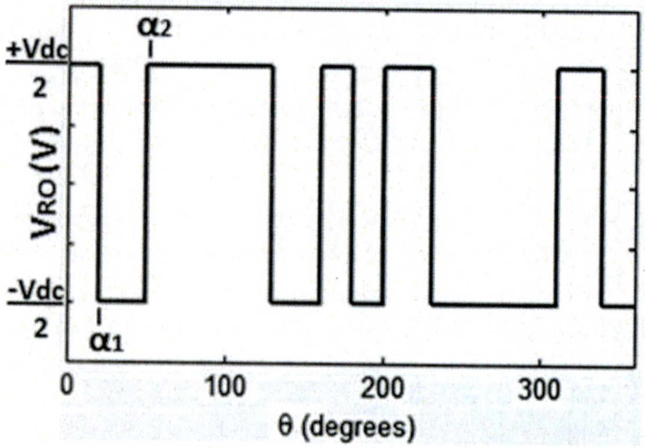

where two switching instants (α_1 and α_2) each quarter waveform, i.e., N is equal to 2. The number of Pulses (P) for the two switching angles might be calculated using the expression P = 2N + 1, where, P = 5 indicates the switching value of frequency is 5 times that of the basic inverter value of frequency.

It should be noticed that the symmetric characteristics of two-level PWM waveforms are retained as shown in Fig. 2. For both the quarter cycle(QWS) and half cycle (HWS) periods in each cycle of the waveform. Equations (1) and (2) show the mathematical expressions that represent QWS and HWS requirements.

$$V_{RO}(\theta) = V_{RO}(\theta + 180^{\circ}) \tag{1}$$

$$V_{RO}(\theta_m - \theta) = V_{RO}(\theta_m + \theta) \tag{2}$$

where θ_m indicates either the positive or the negative maximum angle with respect to fundamental R-phase voltage.

3 Optimum Solutions for SHE PWM

The SHE PWM approach may totally minimize odd (N–1) non-triplen unwanted harmonics from the output line voltage, Here N denotes the total number of the switching angles in a quarter-wave cycle. In the present research paper, two switching angles are employed to eliminate the 5th odd harmonic while retaining the correct fundamental required voltage value. SHE PWM approach is based on the Fourier-series formula for pole voltage V_{RO} shown in Fig. 2, It is written as:

$$V_{out} = \sum_{n=1}^{\infty} (a_n cos(n\theta) + b_n sin(n\theta)) \tag{3}$$

where Fourier coefficients denoted by a_n and b_n. Due to the half-wave symmetry and the odd symmetry, coefficients in the cosine series, as well as the even number harmonics of the sine values, are lacking in the Fourier-series formulation of the pole voltage. As a result, for a two-level inverter, the sine series coefficient with switching angles α_1 and α_2 can be represented as:

$$b_n = \frac{2V_{dc}}{n\pi}\left[1 - 2cos(n\alpha_1) + 2cos(n\alpha_2)\right] \tag{4}$$

here, V_{dc} is the DC-source voltage. The values of switching angles are found by solving the following non-linear equation sets, stated in Eqs. (5) and (6) for the elimination of the 5th harmonic component while keeping a specified fundamental component.

$$V_1 = \frac{2V_{dc}}{\pi}\left[1 - 2cos(\alpha_1) + 2cos(\alpha_2)\right] \tag{5}$$

$$V_5 = \frac{2V_{dc}}{5\pi}\left[1 - 2cos(5\alpha_1) + 2cos(5\alpha_2)\right] = 0 \tag{6}$$

The optimal switching angles determined by Eqs. (5), (6) are constrained by inequality constraint mentioned in Eq. (7). Switching angles can be symmetrically summarized in the following table enabling continuous inverter operation across the whole modulation range.

$$0 \leq \alpha_1 \leq \alpha_2 \leq \pi/2 \tag{7}$$

The precision of switching angles, as well as the amount of iterations necessary for the global optimal solutions, is determined by starting switching angle values. The findings of the GA technique are used as starting values for the NR iterative algorithm. Two non-linear equations may be developed for the elimination of 5th harmonics, while preserving fundamental voltage value.

$$1 + 2cos(\alpha_2) - 2cos(\alpha_1) = M^* \tag{8}$$

$$1 - 2cos(5\alpha_1) + 2cos(5\alpha_2) = 0 \tag{9}$$

Here, M^* indicates the preferred random value of the modulation changing in between 0 and 1

$$F(\alpha) = H \tag{10}$$

Fig. 3 Optimum switching angles for type A PWM solutions set 1

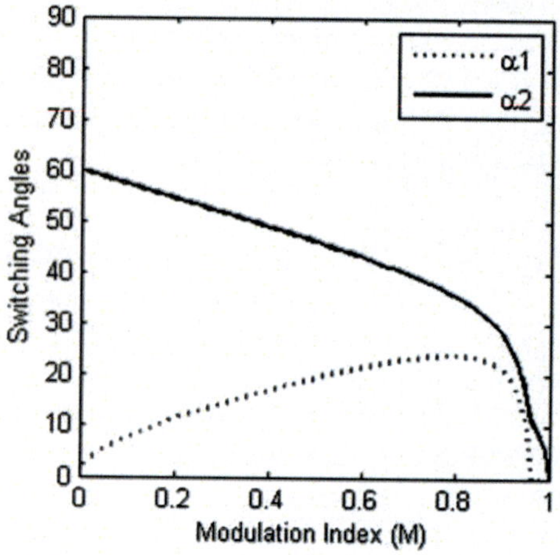

where,

$$F(\alpha) = \begin{bmatrix} 1 + 2cos(\alpha_2) - 2cos(\alpha_1) \\ 1 + 2cos(5\alpha_2) - 2cos(5\alpha_1) \end{bmatrix},$$

$$H = \left[M^* \; 0\right]^T, and$$

$$\alpha = \left[\alpha_1 \; \alpha_2\right]$$

Next, Jacobian matrix for non-linear equation set is solved by using Eq. (11).

$$J^i(\alpha) = \begin{bmatrix} \frac{\partial F_1^i(\alpha)}{\partial \alpha_1} & \frac{\partial F_1^i(\alpha)}{\partial \alpha_2} \\ \frac{\partial F_2^i(\alpha)}{\partial \alpha_1} & \frac{\partial F_2^i(\alpha)}{\partial \alpha_2} \end{bmatrix} = \begin{bmatrix} 2sin(\alpha_1) & -2sin(\alpha_2) \\ 10sin(5\alpha_1) & -10sin(5\alpha_2) \end{bmatrix} \quad (11)$$

Initial values are used for the switching angles of $M^* = 0.01$, displacement vector $\Delta\alpha$ might be obtained as follows:

$$\Delta\alpha^i = (J^i(\alpha))^{-1}[H - F^i(\alpha)] \quad (12)$$

Finally, the switching angles are modified by using Eq. (13).

$$\alpha^{(i+1)} = \alpha^i + \Delta\alpha^i \quad (13)$$

Increase the value of M^* in 0.01 in increments to obtain the ideal switching angles for the whole modulation index. Figures 3 and 4 show two distinct solution sets. The first set of the switching angle solutions was discovered within 60°, whereas the second set

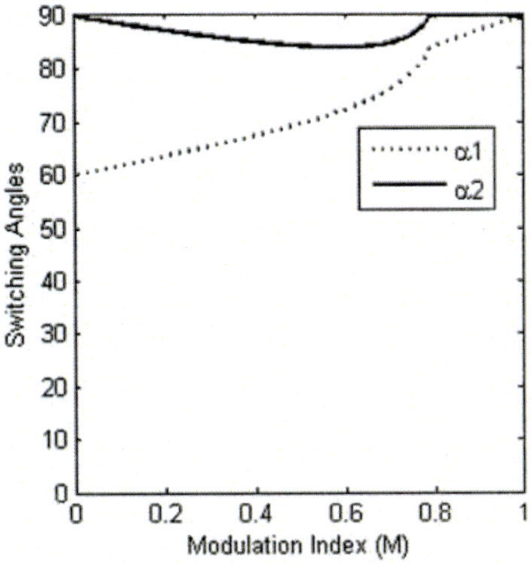

Fig. 4 Optimum switching angles for type A PWM solutions set 2

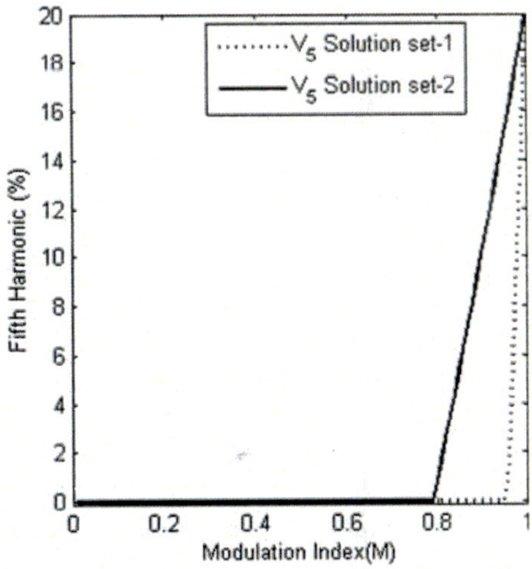

Fig. 5 5th harmonic voltage w.r.t solutions set 1, solutions set 2

was discovered within 90°. Which is shown in Fig. 5, the removal of first significant 5th harmonics are achieved for solutions set1 for the modulation index range M = 0 to 0.95. When compared the solutions set 1 with solutions set 2 eliminates the 5th harmonic across a relatively limited range of the M, i.e., $M <= 0.80$. Figure 6 depicts the accomplishment of the two solutions sets with reference to weighted THD, In terms of V_{WTHD}, and it can be demonstrated that the solutions set 2 distinctly exceeds solutions set 1 for the range $0 < M < 0.8$. However, within a small range, solution-set1 provides a marginal gain in higher M values above 0.8. Figure 7 depicts two dimension of $\alpha 2$ and $\alpha 1$ plane denoted with restrictions. $0 < \alpha 1 < \alpha 2 < \pi/2$. Furthermore, solutions constrained to the typical triangular area denoted with $0 < \alpha 1$ and $\alpha 1 < \alpha 2$, $\alpha 2 < \pi/2$ in Fig. 7. Solid and dotted lines indicate the curves that

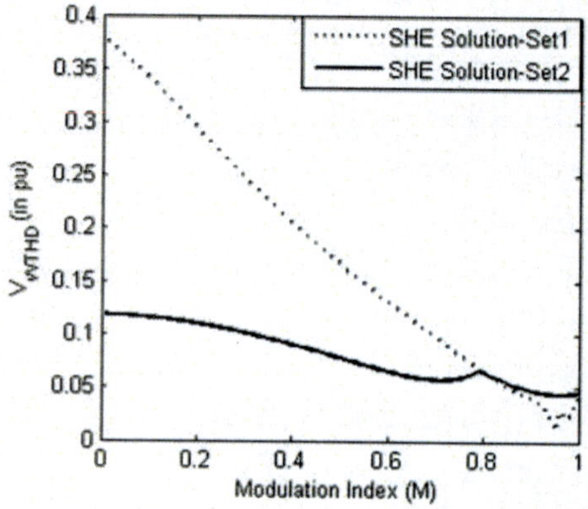

Fig. 6 V_{WTHD} values w.r.t solutions set 1 and solutions set 2 for P = 5

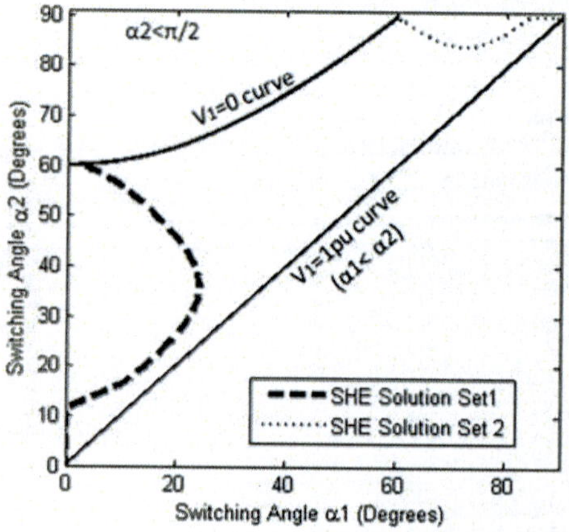

Fig. 7 2D representation of solutions for $\alpha 2$ and $\alpha 1$

reflect two sets of solutions for SHE where N = 2. To assess quality in line voltage, current, and total harmonic distortion V_{THD}, I_{THD} are to be determined by Eqs. (14), (15). Furthermore, uncontrolled voltage harmonics introduce harmonic distortion into the inverter's line voltages.

$$I_{THD} = \sqrt{\frac{\sum_{n=6k\pm1} I_n^2}{I_1^2}} \tag{14}$$

$$V_{WTHD} = \sqrt{\frac{\sum_{n=6k\pm1} F_n^2/n^2}{F_1^2}} \tag{15}$$

k = 1, 2, 3, 4, 5....

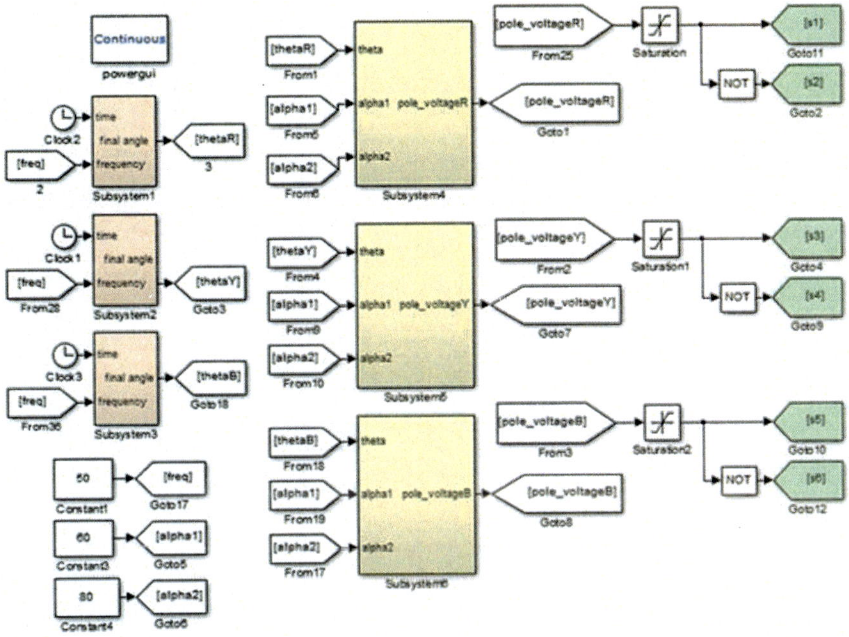

Fig. 8 Simulink model for gate signal generation

Here, I_1, I_n denotes fundamental, RMS nth harmonic current. And, F_1, F_n represents fundamental nth harmonic voltage.

4 Discussion on Simulation Results

The simulation of two-level and three-phase voltage source inverters is done by using the MATLAB and simulation package to validate the accuracy of the numerous solutions using SHE. To perform simulation work with the DC fixed voltage source of the 30V and the required fundamental frequency 50 Hz, two unique operating point values were chosen. Three-phase gate signals to power semiconductor devices of two-level inverter are generated by the Simulink model given in Fig. 8.

Case 1: for M = 0.65, Fig. 9a, c show the line voltage waveform for solution sets 1 and 2 for the two switching angles in the fundamental quarter cycle. Figure 9b, c show the associated frequency spectra, which clearly show that the 5th voltage harmonic is fully removed, validating the theory. When examining FFT findings, it is observed that solution set 2 delivers a superior V_{WTHD} value of 0.06 as opposed to the 0.114 for the solution set 1.

Case 2: for M = 0.9, As previously stated, none of the solutions sets outperforms the other throughout the whole modulation index range. To perceive the harmonic

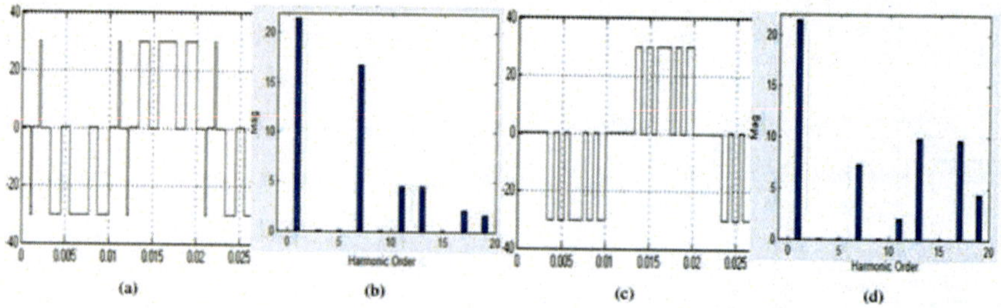

Fig. 9 V_{RY} voltage waveform of the **a** solutions set 1, **b** solutions set 1 harmonics spectra and **c** solutions set 2, **d** solutions set 2 harmonics spectra for the M = 0.65 and P = 5

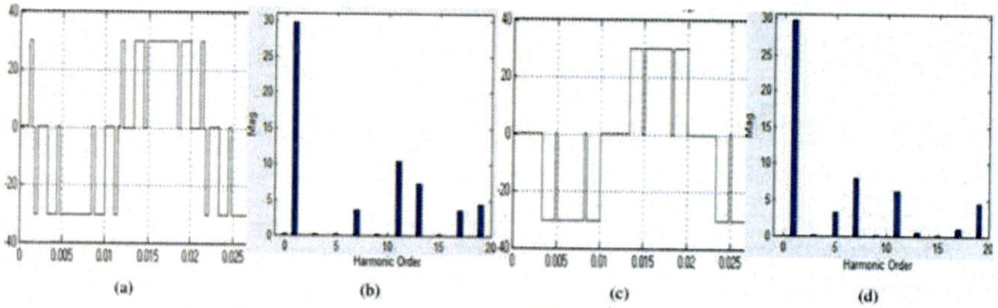

Fig. 10 V_{RY} voltage waveform of the **a** solutions set 1, **b** solutions set 1 harmonics spectra and **c** solutions set 2, **d** solutions set 2 harmonics spectra for the M=0.9 and P = 5

Table 1 Weighted harmonic distortion values and Optimum switching angle values for the two solution sets

Optimum Solution	Switching Angles $\alpha 1$	Switching Angles $\alpha 2$	Simulated voltage value V_{WTHD}
M = 0.65			
Solution Set 1	22.35	41.53	0.1140
Solution Set 2	73.951	84.20	0.0591
M = 0.9			
Solution Set 1	21.08	27.96	0.0409
Solution Set 2	87.15	89	0.0481

index of the two solutions sets, the modulation index of M = 0.9 is used in this example.

Figure 10a depicts the line voltage waveform for solution sets 1 and 10(c) for solution sets 2. In particular, no solution for claimed 5th harmonic removal possible at the M = 0.9 for solutions set 2, resulting in the greater amplitude for the 7th harmonics are compared with a solution set 1, as represented the FFT ranges of Fig. 10b, d, clearly shows that the solution set 1 results have full 5th harmonic removal, showing higher harmonics minimization or elimination at the M = 0.9. Table 1 summarizes switching angle instances and overall distortions value of two solutions sets. Figures 5 and 6 indicate the modulation index range is extended to 0.95 from 0.8 with reduced total harmonic distortion. These findings can be further implemented in induction motor drives, renewable energy integration, real-time microgrid optimization for home appliances, etc.,

5 Conclusion

This work reports on an expansion of the solutions linked to SHE approaches for two-level inverters. In a quarter cycle, a hybrid GA-NR approach is used to identify various sets of solutions for the two switching angles. The comparison and analysis of several solution sets were performed. As compared to solution set 1, solution set 2 significantly decreases the 5th harmonic minimization solutions ranging from (0–0.95) to (0–0.79). With regard to the voltage THD performance, solutions set 2 distinctly outperforms the solution set 1 up to the $M = 0.8$, and solution set 1 assures ideal performance behind $M = 0.8$. The simulation findings for the bipolar-type waveform are in order and good enough for confirming the robustness of findings and this recommended solution augmentation.

References

1. Iqbal A, Khan MA (2008) A simple approach to space vector PWM signal generation for a five-phase voltage source inverter. Ann IEEE India Conf 2008:418–424. https://doi.org/10.1109/INDCON.2008.4768760
2. Su G, Tang L (2011) Current source inverter based traction drive for EV battery charging applications. IEEE Veh Power Propuls Conf 2011:1–6. https://doi.org/10.1109/VPPC.2011.6043143
3. Kantaria RA, Joshi SK, Siddhapura KR (2011) A novel hysteresis control technique of VSI based STATCOM. India Int Conf Power Electron 2010(IICPE2010):1–5. https://doi.org/10.1109/IICPE.2011.5728110
4. Kantaria RA, Joshi SK, Siddhapura KR (2011) A novel hysteresis control technique of VSI based STATCOM. India Int Conf Power Electron 2010(IICPE2010):1–5. https://doi.org/10.1109/IICPE.2011.5728110
5. Meshram S, Agnihotri G, Gupta S (2012) The steady state analysis of Z-source inverter based solar power generation system. In: 2012 IEEE 5th India international conference on power electronics (IICPE), pp 1–6. https://doi.org/10.1109/IICPE.2012.6450366
6. Kharjule S (2015) Voltage source inverter. Int Conf Energy Syst App 2015:537–542. https://doi.org/10.1109/ICESA.2015.7503407
7. Holmes D, Lipo T (2003) Pulse width modulation for power converters: principles and practice. Wiley
8. Tripathi A, In: Student member, IEEE, G Narayanany; investigations on optimal pulse-width modulation to minimize total harmonic distortion in the line current
9. Abdul Azeez N, Mathew J, Gopakumar K, Cecati C (2013)A 5th and 7th order harmonic suppression scheme for open-end winding asymmetrical six-phase IM drive using capacitor-fed inverter. In: IECON 2013—39th annual conference of the IEEE industrial electronics society, pp 5118–5123. https://doi.org/10.1109/IECON.2013.6699966

10. Sinha A, Jana KC, Das MK, An inclusive review on different multi-level inverter topologies, their modulation and control strategies for a grid connected photo-voltaic system
11. Corzine KA, Wielebski MW, Peng F, Wang J (2003) Control of cascaded multi level inverters in electrical machines and drives conference, IEMDC'03. IEEE Int 149–1555
12. Omara AM, Moschopoulos G (2018) Implementation of SHE-PWM technique for parallel voltage source inverters employed in uninterruptible power supplies. IEEE Int Telecommun Energy Conf (INTELEC) 2018:1–6. https://doi.org/10.1109/INTLEC.2018.8612396
13. Yang K, Fu S, Hu H, Yuan R, Yu W (2010) Real solution number of the nonlinear equations in the SHEPWM technology. Int Conf Intell Control Inf Process 2010:446–450. https://doi.org/10.1109/ICICIP.2010.5565322
14. Omara AM, Sleptsov M, El-Nemr MK (2018) Genetic algorithm optimization of SHE-PWM technique for paralleled two-module VSIs employed in electric drive systems. In: 2018 25th international workshop on electric drives: optimization in control of electric drives (IWED), pp 1–6. https://doi.org/10.1109/IWED.2018.8321380
15. Ahmad S, Ashraf I, Iqbal A, Fatimi MAA (2018) SHE PWM for multilevel inverter using modified NR and pattern generation for wide range of solutions. In: 2018 IEEE 12th international conference on compatibility, power electronics and power engineering (CPE-POWERENG 2018), pp 1–6. https://doi.org/10.1109/CPE.2018.8372498
16. Artificial Neural Network and Newton Raphson (ANN-NR) Algorithm Based Selective Harmonic Elimination in Cascaded Multilevel Inverter for PV Applications SANJEEVIKUMAR PADMANABAN, (Senior Member, IEEE), C. DHANAMJAYULU 2 , (Member, IEEE), AND BASEEM KHAN 3 , (Member, IEEE)
17. Patil SD, Kadwane SG (2017) Application of optimization technique in SHE controlled multilevel inverter. In: 2017 international conference on energy, communication, data analytics and soft computing (ICECDS), pp 26–30. https://doi.org/10.1109/ICECDS.2017.8390050
18. Kavousi A, Vahidi B, Salehi R, Bakhshizadeh MK, Application of the Bee Algorithm for selective harmonic elimination strategy in multilevel inverters
19. Jiang Y, Li X, Qin C, Xing X, Chen Z (2022) Improved particle swarm optimization based selective harmonic elimination and neutral point balance control for three-level inverter in low-voltage ride-through operation. IEEE Trans Ind Inf 18(1):642–652. https://doi.org/10.1109/TII.2021.3062625
20. Deniz E, Aydogmus O, Implementation of ANN-based selective harmonic elimination PWM using hybrid genetic algorithm-based optimization
21. Kato T (1999) Sequential homotopy-based computation of multiple solutions for selected harmonic elimination in PWM inverters. IEEE Trans Circ Syst I: Fund Theory Appl 46(5):586–593. https://doi.org/10.1109/81.762924
22. Guan Eryong, Song Pinggang, Ye Manyuan, Bin Wu (2005) Selective harmonic elimination techniques for multilevel cascaded H-bridge inverters. Int Conf Power Electron Drives Syst 2005:1441–1446. https://doi.org/10.1109/PEDS.2005.1619915
23. Mythili M, Kayalvizhi N (2013) Harmonic minimization in multilevel inverters using selective harmonic elimination PWM technique. Int Conf Renew Energy Sustain Energy (ICRESE) 2013:70–74. https://doi.org/10.1109/ICRESE.2013.6927790
24. Dahidah MSA, Agelidis VG (2007) Non-Symmetrical selective harmonic elimination PWM techniques: the unipolar waveform. IEEE Power Electron Spec Conf 2007:1885–1891. https://doi.org/10.1109/PESC.2007.4342290

A Review on Recent Trends in Charging Stations for Electric Vehicles

Vinaya Chavan Thombare, Kshitij Nerlekar, and Juhi Mankumbare

1 Introduction

The international introduction of electric vehicles (EVs) will see a change in pri- vate passenger car usage, operation, and management [1]. To construct a large amount of electric vehicle charging stations with appropriate locations, a multi- level layout planning model, which minimizes the initial construction investment and the users charging cost at the same time, is necessary [2]. With the increased popularity of electric vehicles and increasing awareness of renewable energy sys- tems, charging of EVs should be crucial to enable EVowners to align with available RES generation and available charging time, and at the same time, maximize profits by taking into consideration the variance in grid prices [3]. The renewable energy-supported system has the potential to meet the increasing charging demand to reduce the effect of EV charging on the grid [4]. The rise of renewable energies on the one hand, and the rise of electric cars on the other hand, requires expensive expansion projects of the low voltage network in many cases. With the help of alternative approaches, such as Vehicle to Grid (V2G) applications, the expansion actions can be circumvented [5].

V. C. Thombare · K. Nerlekar (✉) · J. Mankumbare
Department of Electrical Engineering, Ashokrao Mane Group of Institution Vathar, Vathar Tarf Vadgaon, Maharashtra 416112, India
e-mail: kshitijnerlekar9@gmail.com

© The Author(s), under exclusive license to Springer Nature Singapore Pte Ltd. 2023

H. Muthusamy et al. (eds.), *Robotics, Control and Computer Vision*, Lecture Notes in Electrical Engineering 1009, https://doi.org/10.1007/978-981-99-0236-1_46

2 EV Charging Technology

2.1 *Electrical Vehicle Charging Standards and Requirements*

For EV charging stations, the following standards have been mentioned in the literature.

IES Standards

AC charging—3 levels of AC power charging.
Level 1—The voltage is 120 v, 1 phase supply, and the current used to charge is 12A to 16 A.
Level 2—The 1 phase charging voltage 240 v, current up to 60 A.
Level 3—3 phase supply and charging voltages is 400 v ac, current from 32 to 63 A. Level 3 is used when charging time is fast around 30 min.

DC fast charging—The electric current is from 400 A, and the power rating is from 100 to 200 kW. The charging time of the DC charger is faster than the AC charger. The AC charging system of the SAE standards is the same as the IEC standards. But, the DC charging system of the SAE standards is different from the IEC one. It is divided into 3 levels, and the DC output voltage can change to suit various EVs and batteries.

SAE Standards

Level 1—80 A of electric current and the rated power capacity is 40 KW.
Level 2—electric current is 200 A and the rated power capacity is 90 KW.
Level 3—electric current is 400 A and the rated power is 240 KW.

CHAdeMO Standards

This standard is made by the Tokyo Electric Power Company (TEPCO), The electric current is 400 A and the rated power capacity is 240 kW [6].

2.2 *Energy Storage System for EV*

One important reason is energy storage because nothing can be compared with the specific energy found in gasoline, which is around 10,000 Wh/kg against 150 Wh/kg found in the best Li-ion battery. Other alternatives instead of batteries are flywheels and ultracapacitors, which present the same energy limitation as electrochemical batteries [7] (Figs. 1 and 2).

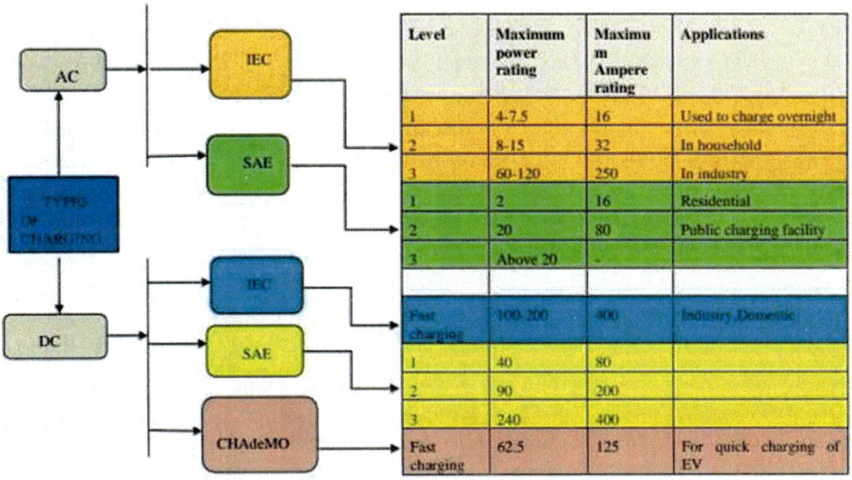

Fig. 1 Charging standards for EV

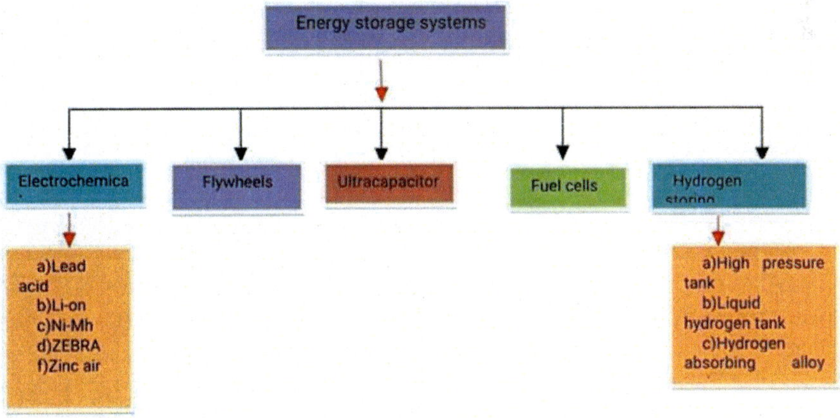

Fig. 2 Classification of energy storage systems

3 Design of EV Charging Station

Standalone PV system term means the system has no connection to the grid. Off-grid systems are suitable for EV charging stations on roads. Backup batteries are used most commonly for the off-grid system for energy storage. The PV-Battery Energy Storage Charging stations either work on grid or solar power, and both of them have advantages and limitations. Grid power increases the reliability of the system, but at the same time, increases the cost of the energy; and solar power decreases the cost of energy, but at the same time, it will decrease the reliability of the system [8]. The PV array delivers power at the DC link through boost converters, whereas the DG set and grid exchange power with the PCC. The local loads also take power from the PCC. On both sides of the switch, a filter is placed, which filters the switching noise in islanded and DG set or grid-connected modes [9]. The process of Maximum PowerPoint Tracking is performed by a boost converter which facilitates the wind turbine to operate in maximum PowerPoint. The energy storage unit is connected to the DC bus through the bidirectional buck-boost converter. The excess power from renewable energy sources is used to charge the ESU [10]. This paper proposes a hybrid charging station for an electric vehicle, it uses both solar energy and conventional energy. The charging station will charge the EV when solar power is available. When solar is not available, it will use grid power for charging the EV. When solar is available, but EV is not present for the charging, then grid-tie inverter technology is used to feed power back to the grid. Generally, a grid-tie inverter uses a transformer to step the voltage, the use of a transformer makes the system costly. Solar power is fed to the grid by using a voltage source PWM inverter. When the dc bus charging station is fed from the PV array and operates in islanded mode, the power supply is limited by the PV system. Hence, the MPPT algorithm is applied to the PV array. A dc bus voltage can be affected by the increase in the number of PEVs or due to weather conditions (Table 1).

Charging stations for EV, supplied by small-scale wind energy systems, is reasonable because of the following reasons:

(1) Immense advancement in the innovation of power converter topology for small-scale wind energy system
(2) Excess electricity productivity of the system from slow winds which are frequent Economics of the wind energy system can be increased by absorbing excess wind production. The EV can absorb that wind production [11].

The process of MPPT is performed by a boost converter that fascinates the wind turbine to operate at maximum power. The energy storage system is connected to the DC bus through a bidirectional buck-boost converter. The energy storage system can charge the EV when power from the wind turbine is not available. This charging

Table 1 Various batteries used for energy storage for EV

Sr. no	Battery	Advantages	Disadvantages
(1)	Lead-acid	The oldest and cheapest method	Frequent replacement is required Lifecycle is low
(2)	Ni-MH	Environment-friendly Maintenance-free	High cost
(3)	Li-on	Lifecycle is more Fast charging can be done	High cost
(4)	ZEBRA	Works at high temperature	Cost is high, requires a special container

station can be installed at shopping malls, universities, etc. To ensure the reliable operation of the charging station, there is a need to consider different operating modes (Figs. 3, 4 and Table 2):
Mode 1—WPCS with grid connection.
Mode 2—Inversion operation.
Mode 3—Rectification operation.

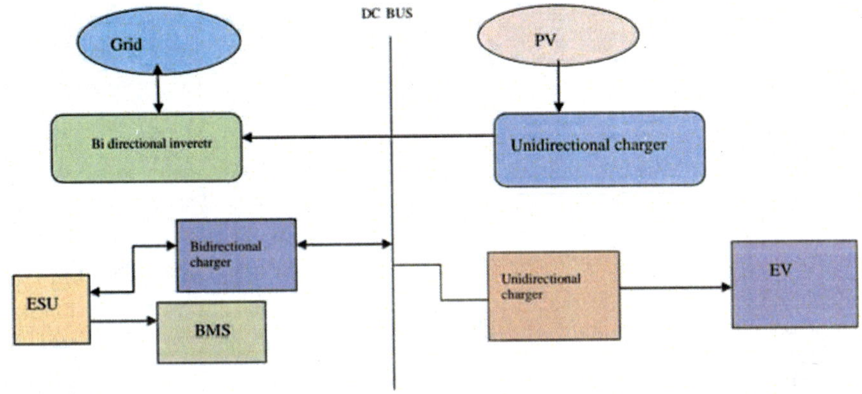

Fig. 3 Charging system for EV

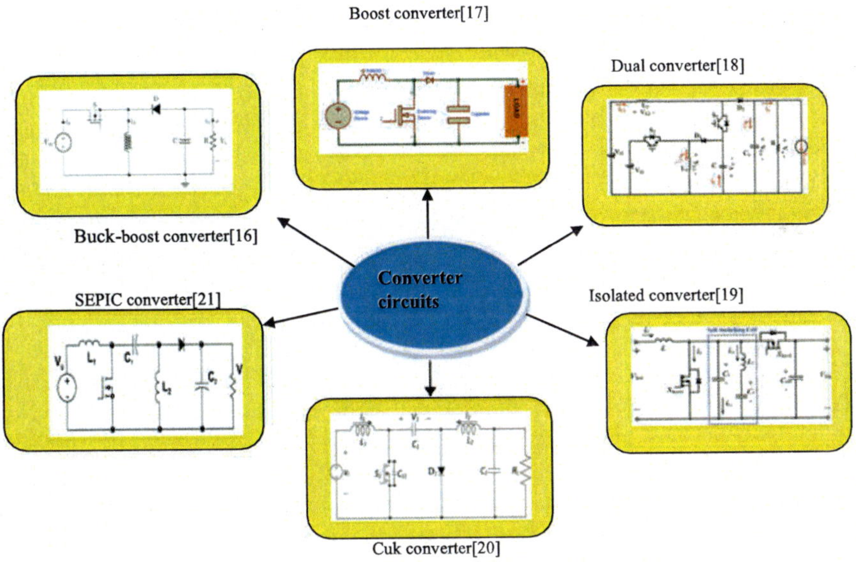

Fig. 4 Circuit diagrams of converter

Table 2 ReviewCharging stations for EV

Ref. No	Year	Source of power generation	Grid connection	Energy storage	Converter
[12]	2018	PV	Yes	–	T-source converter
[12]	2019	PV	–	No	Switched capacitor converter AC/DC converter
[13]	2020	PV	Yes	Yes	AC–DC Converter
[14]	2020	PV	Yes	Yes	Bi-directional DC/DC converter
[15]	2020	PV	–	Yes	Buck-boost converter
[16]	2020	PV	Yes	Yes	DC-DC Boost converter
[17]	2020	PV	No	Yes	Dual converter
[18]	2020	PV	Yes	Yes	Isolated DC-DC converter

4 Converters Used In EV Charging Stations

In Electrical Vehicles, Hybrid Energy Storage System presents a Deadbeat-based method using a Three-Level Cascaded Converter and also consists of PV panels. This method is used because of achieves a longer driving range. In Electrical Vehicles, the Bidirectional Three-Level Cascaded Converter can integrate the battery, and the supercapacitor is used to mismatch the power between the PV Power generation and load consumption. The Power Electronic Converters can either be isolated or non-isolated from the grid. The isolated arrangement uses a transformer at the inverter output to step up the voltage [18]. Cuk converter does not need a battery because it is di- rectly used for DC loads so that it can save investment. Cuk converter has the same character as a buck-boost converter [19]. Multiple-input converters can provide a single-unit solution interfacing multiple energy sources and common loads. They perform better than traditional two-port converters due to their lower part count and smaller converter size with low cost (Table 3).

The reference signal is used to compare the SEPICs output, and to achieve the maximum power. The SEPICs output signal is thus compared with the adaptive reference signal to feed the inverter with the most suitable power. The inverter's input signal should be as smooth as possible, but the SEPIC MPPT generates a non-smooth signal, owing to its tracking of maximum power. This problem is not big, as the non-smooth signal can be enhanced by the inverter fuzzy controller and the low-pass filter connected to the inverter [20].

Table 3 Review on converters

Ref. No	Type of converter	Components used	Advantages	Disadvantages
[15]	Buck-Boost converter	Inductor capacitor Diode Switch	1. It steps up and steps down voltage 2. low operating duty cycle 3. Less expensive as compared to others	1. Discontinuous Input current 2. Power loss is high
[16]	Boost converter	Inductor(1) capacitor(1) Diode Switch	1. High output voltage 2. Low operating duty cycle	1. large inrush current 2. Less reliability
[17]	Dual converter	Inductor(2) capacitor(2) Diode Switch(2)	1. It steps up and steps down voltage 2. Output current is continuous and and free from ripples	1. High cost 2. High complexity
[18]	Isolated converter	Inductor capacitor Diode Switch	1. It steps up and steps down voltage 2. Output current is continuous	1. High cost 2. High complexity
[19]	Cuk converter	Inductor(2) Capacitor(2) Diode	1. Continuous input and output current	1. It has high ripples 2. Current limit is difficult
		Switch	2. Zero ripple output current 3. It uses a capacitor for power transfer	
[20]	SEPIC converter	Inductor(2) Capacitor(2) Diode Switch	1. Continuous output current 2. High efficiency as compared to buck-boost	1. large inrush current 2. Less reliability

5 Conclusion

This review has covered some of the designs and converter topologies required for the effective operation of charging stations of electrical vehicles, which focus on the design of MPPT controllers, hybridization of charging stations, the software required for the greater operation of the charging station, and grid synchronization. The demands stated by PV modules have also been reviewed; in particular, the effective design for the charging station has been investigated. According to the above discussion, the converter should boost the voltage as per the demand for the electrical vehicle. The specifications of the DC-DC converters for the application of this paper were investigated. Compared with the standalone charging station, a hybrid charging station promises an uninterrupted power supply to EV. The presented concept provides a method to integrate various energy sources for the charging station.

References

1. Foley AM, Winning IJ, Ó Gallachóir BPÓ (2010) State-of-the-art in electric vehicle charging infrastructure. In: 2010 IEEE vehicle power and propulsion conference, pp 1– 6
2. Jin M, Shi R, Zhang N, Li Y (2012) Study on multi-level layout planning of electric vehicle charging stations based on an improved genetic algorithm 5–10
3. Li H, Liu H, Ji A, et al (2013) Design of a hybrid solar-wind powered charging station for electric vehicles. In: 2013 international conference on materials for renewable energy and environment, pp 977–981
4. Wang R, Wang P, Xiao G (2014) Two-stage mechanism design for electric vehicle charging involving renewable energy. In: 2014 international conference on connected vehicles and expo (ICCVE), pp 421–426
5. Aldejohann C, Maasmann J, Horenkamp W, et al (2014) Testing environment for vehicle to grid (V2G) applications for investigating a voltage stability support method. In: 2014 IEEE transportation electrification conference and expo (ITEC), pp 1–6
6. Dost P, Bouabana A, Sourkounis C (2014) On analysis of electric vehicles DC-quick- chargers based on the CHAdeMO protocol regarding the connected systems and security behaviour. In: IECON 2014 - 40th annual conference of the IEEE industrial electronics society, pp 4492–4497
7. Takeda K, Takahashi C, Arita H, et al (2014) Design of hybrid energy storage system using dual batteries for renewable applications. In: 2014 IEEE PES general meeting | conference exposition, pp 1–5
8. Nizam M, Wicaksono FXR (2018) Design and optimization of solar, wind, and dis- tributed energy resource (DER) hybrid power plant for electric vehicle (EV) charging station in rural area. In: 2018 5th international conference on electric vehicular technology (ICEVT), pp 41–45
9. Verma A, Singh B (2018) A Solar PV, BES, Grid and DG set based hybrid charging station for uninterruptible charging at minimized charging cost. In: 2018 IEEE industry applications society annual meeting (IAS), pp 1–8
10. Vijayakumar R (2018) Design of public plug-in electric vehicle charging station for improving LVRT capability of grid connected wind power generation. International conference on soft-computing and network security (ICSNS) 2018:1–6
11. Koochaki A, Divandari M, Amiri E, Dobzhanskyi O (2018) Optimal design of solar- wind hybrid system using teaching-learning based optimization applied in charging station for electric vehicles. In: 2018 IEEE transportation electrification confer- ence and expo (ITEC), pp 1–6
12. Narula A, Verma V (2018) PV fed cascaded modified T source converter for DC support to grid coupled inverters. In: 2018 IEEE international conference on power electronics, drives and energy systems (PEDES), pp 1–6
13. Uno M, Sugiyama K (2019) Switched capacitor converter based multiport converter integrating bidirectional PWM and series-resonant converters for standalone photovoltaic systems. IEEE Trans Power Electron 34:1394–1406. https://doi.org/10.1109/TPEL.2018.2828984
14. Jensanyayut T, Phongtrakul T, Yenchamchalit K, Kongjeen Y (2020) Design of solar- powered charging station for electric vehicles in power distribution system 7–10. https://doi.org/10.1109/iEECON48109.2020.229545
15. Fareed N, Kumar MVM (2020) Single stage grid tied solar PV system with a high gain bi-directional converter for battery management. In: 2020 international conference on power electronics and renewable energy applications (PEREA), pp 1–6
16. Nesrin AKN, Sukanya M, Joseph KD (2020) Switched dual input buckboost inverter for continuous power operation with single stage conversion. In: 2020 international conference on power electronics and renewable energy applications (PEREA), pp 1–6

17. Singh S, Manna S, Hasan Mansoori MI, Akella AK (2020) Implementation of perturb amp; observe MPPT technique using boost converter in PV system. In: 2020 international conference on computational intelligence for smart power system and sustainable energy (CISPSSE), pp 1–4
18. Tayebi SM, Chen X, Batarseh I (2020) Control design of a dual-input LLC converter for PV-battery applications. In: 2020 IEEE applied power electronics conference and exposition (APEC), pp 917–921
19. Wei Y, Luo Q, Mantooth A (2020) A function decoupling partially isolated high voltage gain DC/DC Converter for PV application. In: 2020 IEEE transportation electrifi- cation conference expo (ITEC), pp 1–5
20. Sudiharto I, Murdianto FD, Budikarso A, Wibisana A (2020) CUK converter using FLC to manage power consumption from PV directly. In: 2020 international conference on applied science and technology (iCAST), pp 575–579
21. Manikandan K, Sivabalan A, Sundar R, Surya P (2020) A study of landsman, sepic and zeta converter by particle swarm optimization technique. In: 2020 6th international conference on advanced computing and communication systems (ICACCS), pp 1035–1038

IoT-Based Vehicle Charging Eco System for Smart Cities

N. Dinesh Kumar and F. B. Shiddanagouda

1 Introduction

The retained type of electrically operated driven quality storage in electric vehicles ended up being batteries (EVs). Over the past couple of years, the development around transportation has essentially changed, culminating in the relatively critical enhancement of social demands amid the local situation [1]. Since the battery is a regularly finished gadget for the unity garage locale, finding Charge's worth now plays a fundamental significance. Satisfactory behaviour is being established by strategies for the use of electric cars to eliminate the devouring engines as a result of the upward drive in CO_2 accomplished through business parts and transportation.

This course of action has been professed to lower the amount of CO_2 and the disclosures for new cleaner efficiency technologies have been delayed for the sake of the fact. Electric vehicles (EVs) tended to be a decision to force CO_2 discharges as a finding. As the proportion of electric cars is improving, electric vehicles are redesigning bit by bit around the globe, there might be excitement to bring in power electric vehicle charging machines in the car leaving systems or networks. With Engel, a worldwide force association, in the UK, Nissan distributed a vehicle-to-grid (V2G) procedure to full-size cars. Nissan has genuinely been researching and despite this endeavour, in addition to scrutinizing the topic of V2G networks, they are undeniably the first of their kind inside the UK and a portion of the most outrageous stunning partnerships to this stage. Smart Grid gear's compromise in power frameworks will affect unthinkable trade in ownership of structures [2]. In addition to wandering forward dissipating non-conventional power tools, the conscious trade load fundamental belief implies an optimal propensity to concentrate on conditions in planning rate and in addition to output.

N. Dinesh Kumar (✉) · F. B. Shiddanagouda
Department of ECE, Vignan Institute of Technology and Science, Telangana 508284 Hyderabad, India
e-mail: nalidinesh@gmail.com

© The Author(s), under exclusive license to Springer Nature Singapore Pte Ltd. 2023
H. Muthusamy et al. (eds.), *Robotics, Control and Computer Vision*, Lecture Notes in Electrical Engineering 1009, https://doi.org/10.1007/978-981-99-0236-1_47

Instead of vocations on–off procedure to set up the EV charging, the triggered persuasive artwork strives to monitor the EV charging output inconsistent essential. In the same way, the day-by-day schedule for regular charging capacity is masterminded to be abundant extra consoling for getting the option to use it absolutely without mishandling the goals of the system. This paper considers the impacts of the person on the electric power rate paid, the time saved by infrequent checking of the battery, besides power charges in understanding the EV charging improvements. If these facilities are provided to the users, automatically the client's comfort will increase along with the battery performance.

2 Literature Survey

This work focuses on creating an automated and effective charging management system (CMS) for Electric vehicles (EVs) by leveraging the advantages of IoT technologies to provide multiple sensors and radio systems with omnipresent perception skills and a real-time dynamic vision of the physical world. It is also proposed to organize the charging of large-scale EVs in multiple residential buildings by exploiting the distributed optimization by exploiting the distributed potential of an enhanced decartelized charging scheme alternating direction method of multiplier (ADMM).

Extensive tests have shown that by generating higher final mean state-of-charge (SOC) amounts for the linked EVs and incurring less power bills charged to the utility, the proposed work outperforms the two most comparable decentralized approaches. In Comparison, because of its potential to absorb less overhead communication and deliver greater robustness [2–4].

As we are aware of the mass adoption of EVs on the horizon, smart electric vehicle charging will become essential for both the charging point network operators and the national electricity grid. Across the planet, EVs are becoming famous. EV charging facilities would also be a basic requirement as the number of EVs grows. The task is to build a smart application by connecting to the grid to consider the multiple tariff rates of the grid. Both the electricity consumption rate and the outgoing power rate will be used in the tariff prices. The program can also show the battery SOC when the consumer comes to the grid. The main agenda is to optimize low-carbon technologies through one connected platform using rule-based algorithms, helping to decarbonize both the production and consumption of energy [5].

This paper proposes a real-time Battery Monitoring System (BMS) using the coulomb counting technique as the coordination protocol for SoC estimation and messaging-based MQTT. The proposed BMS is applied using sufficient sensing technologies, central processors, interfacing systems, and Node-RED environments on the hardware platform. An optimization model aimed at optimizing the commercial revenue of the aggregator for EVs is introduced to allow smart charging [6].

As more countries move towards pollution-free traffic, EVs are gaining more momentum around the world. EV charging facilities would also be a basic requirement as the number of EVs grows. An IoT device will certainly streamline EV charging efficiency and look at the consequences. For transportation systems and V2G services, this approach is useful. This new system would boost public planning and make it easier for the city. We can effectively handle the whole V2G infrastructure from IoT, which would certainly save time as well as resources. The job is to design a smartly programmed application for grid communication and understand the different tariff rates of the grid. Both the grid energy delivery rate and the grid power take-off tariff rate will be used in the tariff rates. If the customer has charged the car battery full, he will supply some power to the grid and collect some cash. Using the ARM Mbed controller, the SoC is measured and sent to the cloud. When it comes to the grid, the program can also show the user's battery status (SoC) [7].

3 Methodology

When more nations are heading towards free traffic defilement, EVs are raising more prestige across the globe. In addition, as the number of EVs increases, the EV charging structure would be an integral necessity. For transportation networks and V2G structures, this technique is useful. This suggested system would strengthen the masterminding of the city and make life easier for the city. Without a very striking stretch deal for the whole V2G scheme that sets time and resources back. This job is to develop a wonderful application to communicate with the system and to consider the system's real cost rates. The application program will display the client's battery status (Sock) as he goes to the structure in the same way.

A. *Existing System*

In the existing system, the Electric Car battery-charging system and the Wired Charging and Control Infrastructure are provided. The rise in oil prices and environmental problems has led to growing demand for technology for clean vehicles such as EVs and EV fuel cells. Electric vehicles (EVs) are now becoming a more attractive choice than conventional automobiles (CV). EVs are powered by electric batteries which have to be recharged with energy from the grid. It is evident that EVs are a direct link between the sectors of transport and electricity. Moreover, since they have low energy usage and zero emissions, EVs would be better positioned to reduce the environmental impacts of transport and energy dependency. Off-board and on-board, unidirectional, and bi-directional power flow battery chargers are widely used for two types of battery chargers. During the daytime, on-board chargers can be used to charge from the electrical socket at the office or home socket, or shopping centre. Off-board charging is similar to those used by conventional vehicles at a gas station, but the purpose is to charge easily. In contrast to off-board charging, on-board charging infrastructure is less suitable. The existing system is shown in Fig. 1.

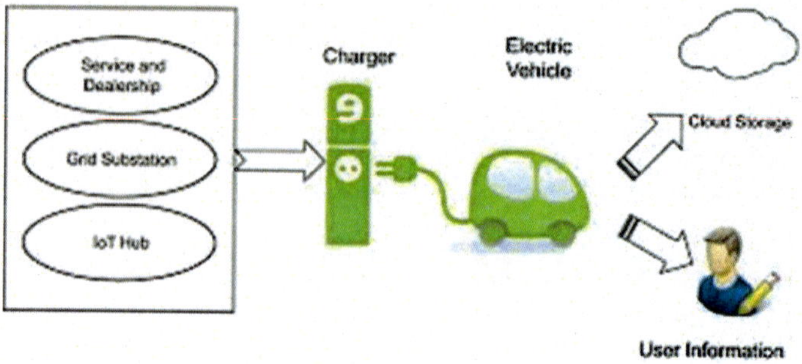

Fig. 1 Electric vehicle charging system

2 *Proposed System*

We use the Raspberry pi controller board in the proposed framework, which functions as a small-sized computer. It can be used for much of the same things as a PC after you add a keyboard, mouse, or trackpad. To create a vehicle charging device, the Raspberry pi controller interface with external modules is used. Three separate types of RFID passive tags are been used, two of which are permitted and one is unauthorized. The RFID tags enable customer details and automated billing to be identified. In the Thing speak IoT server, consumer data is stored. In the Thing speak server, the car charging battery state can be checked, so data can be downloaded and analyzed. Every cloud has its own special API Key and IP Address. Your cloud channel can be made private or public.

3 *Working Principle*

Electric Vehicles today have an extraordinarily significant charge; together it has been an important verbal trade media for public use in addition to business use. At the same time, as verbal exchange, the battery can often become vain, particularly in emergency cases where access to an essential charger is not feasible. To cope with this problem, coin-based fully mobile battery chargers are made. This computer is identical to coin-based mobile phones, which were popular at the beginning of the twentieth century. Initially, while we set the coin properly into the coin insertion slot, then the mobile cell phone will really be paid; it will compare the coin picture stored within the source of information. If the current picture of the logo and the saved photograph are matched, it will screen on the broadcast. After that, in addition to spending, we will really start joining our device to the billing plug. With the aid of fixed charges, billing relies on the coin. Similar to the coin-based technology, RFID tag—plastic card secured and easy-to-use technology is chosen in the proposed method (Fig. 2). The heart of the block is raspberry pi which interfaces with different modules such as voltage sensor, RFID reader, LCD, chargeable battery, and Wi-Fi. The Raspberry pi controller and the external module are attached to a 230 V Step-down transformer that is transformed to a 12VAC supply at the bridge rectifier with near-ideal filtering to 12VDC (EM-18 Reader Module TTL Pin). Three separate Passive RFID (Radio

Frequency Identification) tag forms are used with protected authorization. The LCD monitor displays a card that has been swiped. Using the EM-18, Reader Module first swipe the RFID tag and input the number to be deducted for charging the car. It is assumed that the tags are all recharged with RS 1000. The complete flow of charging a battery is shown in Fig. 3.

4 *Simulation Tools*

5 Operating system—Raspberries OS

It's a single-board computer, based on an ARM processor with built-in graphics and sound. If a keyboard and mouse or trackpad is attached, it can be used for much of

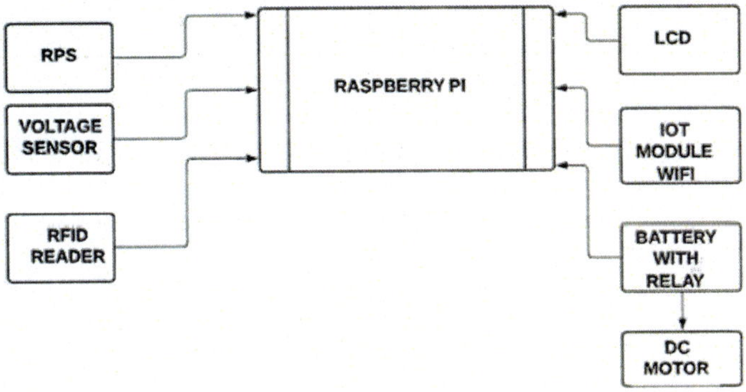

Fig. 2 Proposed block diagram

Fig. 3 Flow chart

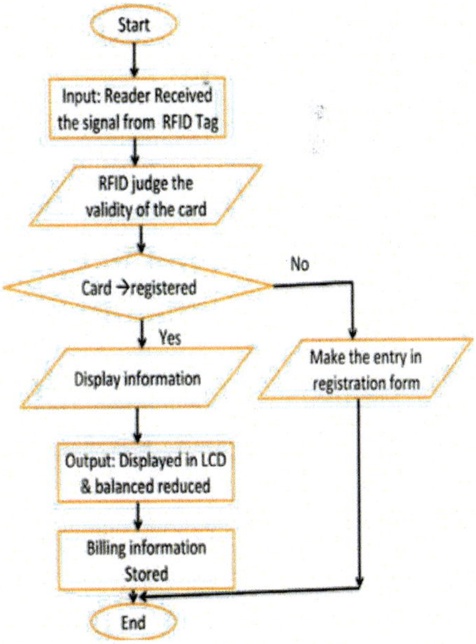

the same tasks as a PC. It is possible to use storage in the form of an SD card, or hard drive using USB. We can import images of operating systems from Raspberry Pi downloads or the Noobs system, which enables several operating systems, and OSMC which is like a Kodi media player. It can also be used in several languages to practice programming, some of which are ideal for a ten-year-old, up to advanced Python and Java. External Electronics like musical instruments, lights, motors, and robots can be operated using R-Pi.

2. Editor and Compiler—Python

Python is a high-level, disrupted, programming language that is open source and very easy to use. It is often known to be a very powerful language, too. Python is a perfect language for programming. With the help of which consumer applications and gaming applications can be made very fast.

3. Thing speak

Thing speak is a cloud platform that is open source. This is the location where information about a real-time sensor is uploaded. Download and review the information and it can be used for our own purposes. Per cloud have its own special API Key and IP Address. Cloud channels may be made public or private.

4 Results and Discussions

In the application, the user can see the realities. The user may also use the product application to identify the places of the charging station. As easily as the consumer learns about the circumstances of his vehicle battery because of the fact, he will determine without problems whether or not to continue giving power to the matrix or to take power from the system depending primarily on the levy costs. To get the desired results, IoT architecture uses sensors. It uses sensors or controls so the essential operating system is used. Gadgets like a cell phone or tablet PC are done to examine the ultimate results or a final product which lessens the endeavour to get the measurements. Figure 4 shows the complete hardware circuit.

LCD display in Figs. 5 and 6 shows a particular RFID tag C1 debited with 10 Rupees for charging purposes and the remaining balance of rupees 990 left in the card. The voltage level is 13 v in the battery and T indicates the Thing speak web server where the data is stored in the cloud. Hence, the propositional amount of charging will be made to the battery of the C1 user. And if any unauthorized tag is used, no user data will be available and hence the battery charging will not be successful.

Figure 7 shows the user C1 data graphical representation in the IoT server "Thing Speak" web page. It displays the balance amount on the card along with the date and time. Similarly, Fig. 8, shows the user C2 data graphical representation in the IoT server "thing speak" web page. Figure 9 shows unauthorized user RFID Tag C3 output data graphical representation in the IoT server "thing speaks" web page.

Fig. 4 Hardware kit

Fig. 5 LCD display

Fig. 6 User C1 RFID TAG Output

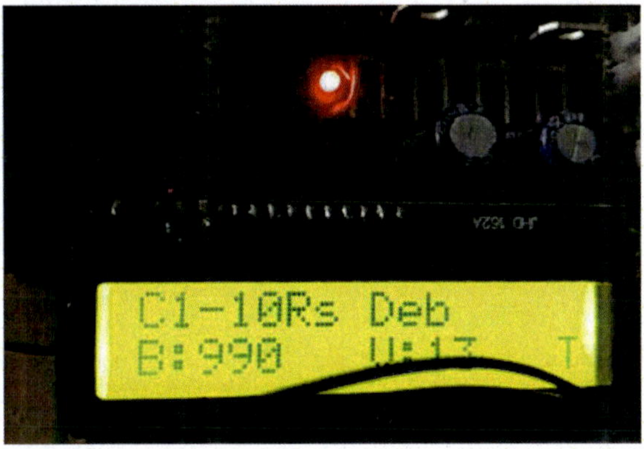

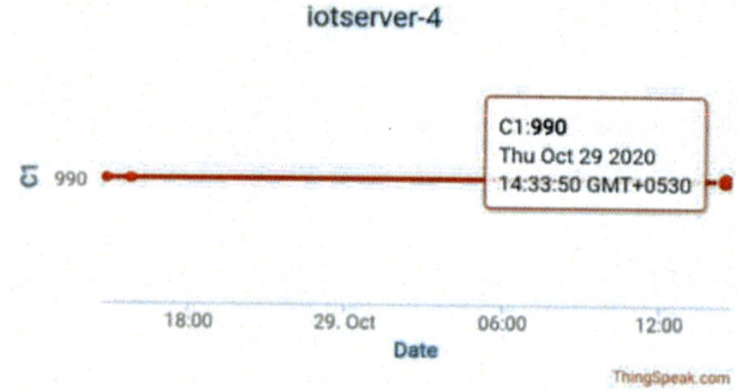

Fig. 7 User C1 RFID TAG output in server with respect to time

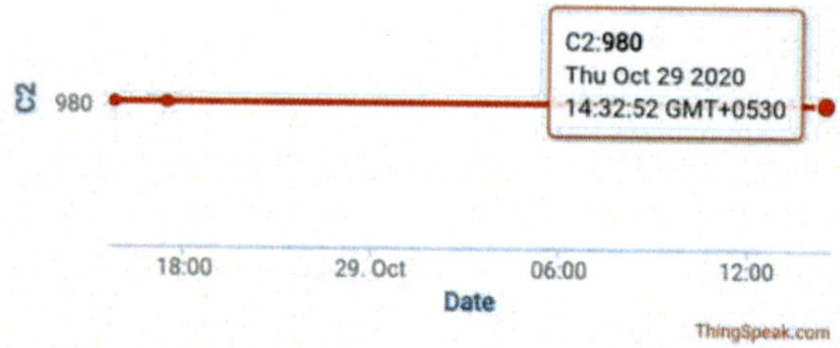

Fig. 8 User C2 RFID TAG Output in server with respect to time

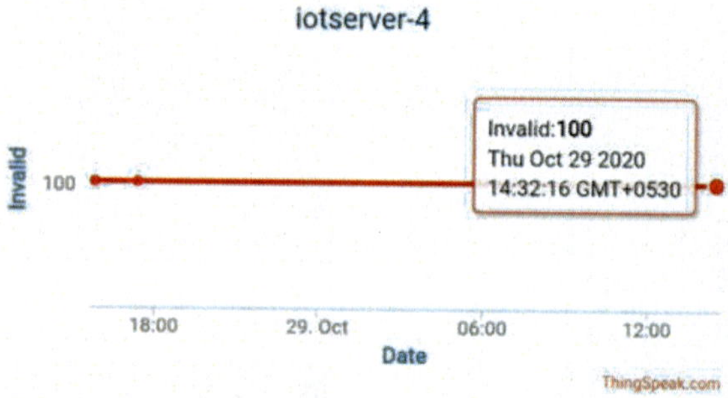

Fig. 9 User C3 RFID TAG Output in server with respect to time

Voltage levels of the battery before and after charging are shown with graphical representation in Thing Speak as shown in the Figs. 10 and 11. In Thing Speak graphical representation, it is clearly visible the time needed to charge the battery for deducted amount.

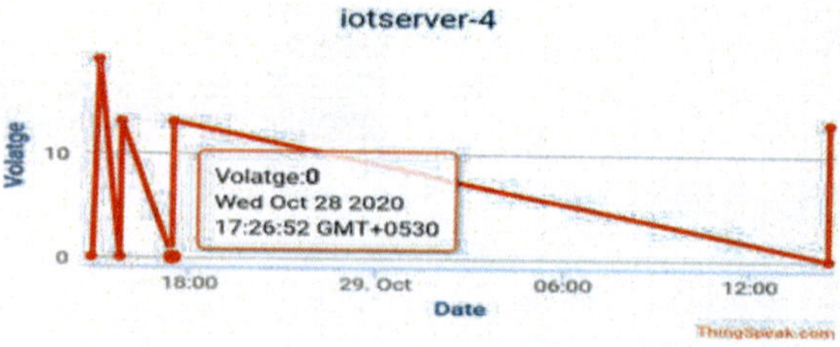

Fig. 10 Voltage level of battery before and after recharging

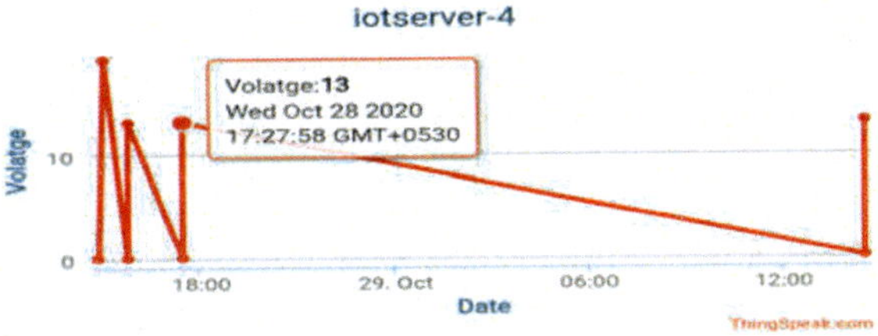

Fig. 11 Voltage level of battery before and after recharging

5 Conclusion

The battery-charging infrastructure for an electric vehicle was reviewed in this article. Transportation is the greatest cause of environmental emissions in any region. To address the climate crisis, we need to make the cars on our highways as safe as possible. Vehicle pollutants are not only bad for our atmosphere; they're bad for our wellbeing. Gasoline and diesel-powered automotive air emissions cause asthma, bronchitis, cancer, and premature death. The Electronic Vehicle charging system shows promising results. In this paper, passive RFID tags permit to detect the customer information and automatic billing. The customer data is stored in the Thing speak IoT cloud server so that the server will always be updated and will be aware of the status of every customer. Therefore, the implementation of an EV charging management system (CMS) is important to automatically and effectively organize these large charging demands by leveraging the benefits of IoT technology.

Acknowledgements We would like to be thankful to Vignan Institute of Technology and Science, Hyderabad, for providing laboratory facility to complete the entire work.

References

1. Benedetto M, Ortenzi F, Lidozzi A, Solero L (2021) Design and implementation of reduced grid impact charging station for public transportation applications. World Electr. Veh. J. 12:28
2. Vermesan O, Friess P (2013) Internet of things: converging technologies for smart environments and integrated ecosystems. River Publishers
3. Yao L, Chen YQ, Lim WH (2015) Internet of things for electric vehicle: an improved decentralized charging scheme. In: Proceedings of the 2015 IEEE international conference on data science and data intensive systems, Sydney, Australia, 11–13 December 2015; pp 651–658
4. Sousa RA, Monteiro V, Ferreira JC, Melendez AA, Afonso JL, Afonso JA (2018) Development of an IoT system with smart charging current control for electric vehicles. In: Proceedings of the IECON 2018–44th annual conference of the IEEE industrial electronics society, Washington, DC, USA, 21–23 October 2018; pp. 4662–4667.
5. Yao L et al. (2015) Internet of things for electric vehicle: an improved decentralized charging scheme. In: 2015 IEEE international conference on data science and data intensive systems, pp 651–658
6. Sharma E, Bharath S, Devaramani A, Deepti SR, Kumar S (2019) IOT enabled smart charging stations for electric vehicles. J Telecommun Study 2:34–39

7. Huang S, He L, Gu Y et al (2015) Design of a mobile charging service for electric vehicles in an urban environment. IEEE Trans Intell Transp Syst 16:787–798
8. Arun kumar P, Vijith K (2018) IOT enabled smart charging stations for electric vehicle. Int J Pure Appl Math 119:247–252
9. Savari GF, Krishnasamy V, Sathik J, Ali ZM, Aleem SH (2020) Internet of things based real-time electric vehicle load forecasting and charging station recommendation. ISA Trans 97:431–447